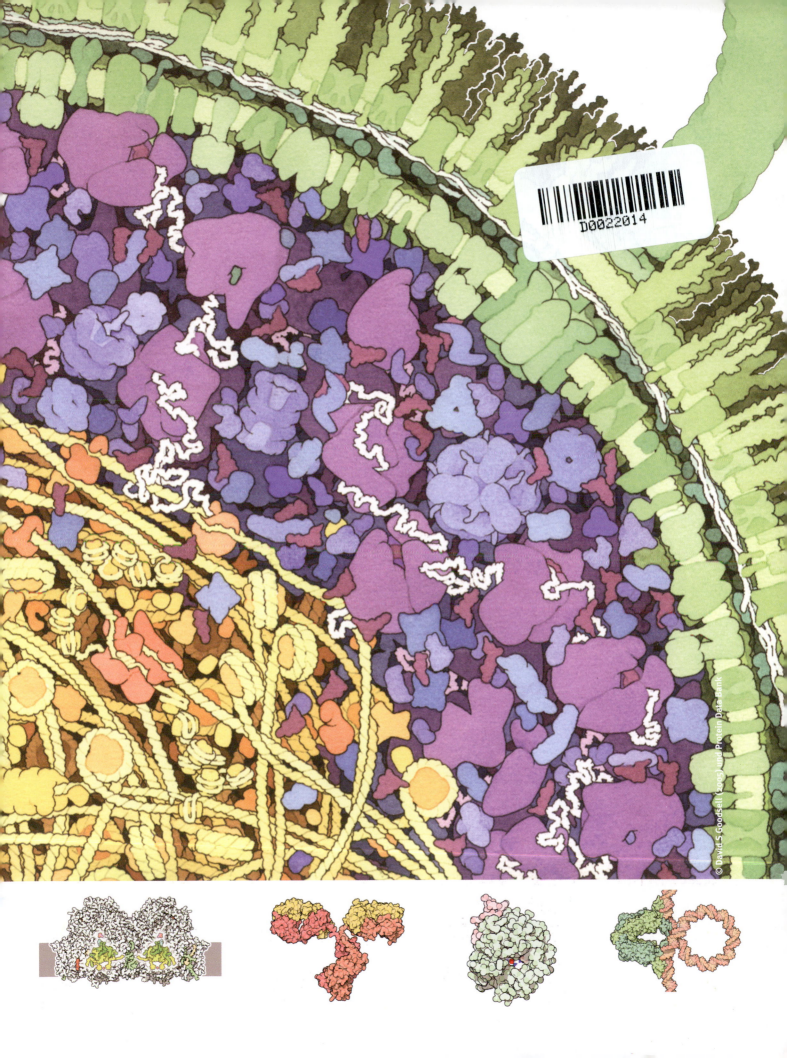

Reinhard Renneberg
The Hong Kong University of Science and Technology

Darja Süßbier (Illustrations)

BIOTECHNOLOGY FOR BEGINNERS

Edited by Arnold L. Demain
Professor emeritus of Massachusetts Institute of Technology (MIT),
at present Research Fellow, Research Institutes for Scientists Emeriti
(RISE), Drew University

Foreword by Tom A. Rapoport
Harvard Medical School

Translated from German by
Renate FitzRoy
and
Jackie Jones

ELSEVIER

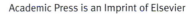
AMSTERDAM • BOSTON • HEIDELBERG • LONDON
NEW YORK • OXFORD • PARIS • SAN DIEGO
SAN FRANCISCO • SINGAPORE • SYDNEY • TOKYO

Academic Press is an Imprint of Elsevier

Acquisition Editor: Merlet Behncke-Braunbeck, Heidelberg, Germany
Production Manager: Ute Kreutzer, Detlef Mädje, Heidelberg, Germany
Copy editor: Theodor H. C. Cole, IPMB University of Heidelberg
Composed by: Darja Süßbier, Berlin, Germany
Illustrations: Darja Süßbier, Berlin, Germany (if not stated otherwise in the Credit List)
Printed and bound by: Stürtz GmbH, Würzburg, Germany
Cover Illustration: Darja Süßbier, Berlin, Germany
Cover Design: SpieszDesign, Neu-Ulm, Germany

Academic Press is an Imprint of Elsevier
30 Corporate Drive, Suite 400, Burlington, MA 01803, USA
525 B Street, Suite 1900, San Diego, California 92101-4495, USA
84 Theobald's Road, London WC1X 8RR, UK

This book is printed on acid-free paper.

Springer is a part of Springer Science+BusinessMedia
springer.de

Copyright © Springer-Verlag Berlin Heidelberg 2008
Spektrum Akademischer Verlag is an Imprint of Springer SBM

This edition of **Biotechnology for Beginners** by **Reinhard Renneberg** is published by arrangement with Elsevier Inc.
and commissioned and translated by Spektrum Akademischer Verlag from the original German edition:
Biotechnologie für Anfänger, 2nd ed., by Reinhard Renneberg with illustrations by Darja Süßbier.
Copyright © 2006 by Elsevier GmbH, München, Germany, ISBN 978-3-8274-1847-0

Library of Congress Cataloging-in-Publication Data
CIP applied for and in process.

British Library Cataloguing in Publication Data
A catalogue record for this book is available from the British Library

Bibliographic information published by Die Deutsche Bibliothek
Die Deutsche Bibliothek lists this publication in the Deutsche Nationalbibliografie;
Detailed bibliographic data is available in the Internet at http://dnd.ddb.de.

ISBN: 978-0-12-373581-2

**For information on all Academic Press publications
visit our Web site at www.books.elsevier.com**

08 09 10 11 12   5 4 3 2 1

THERE IS NOTHING SO POWERFUL
AS AN IDEA WHOSE TIME HAS COME.
Victor Hugo

IN THE FIELD OF OBSERVATION,
CHANCE FAVORS ONLY THE PREPARED MIND.
Louis Pasteur

I COULDN'T DO IT.
I COULDN'T REDUCE IT TO THE FRESHMAN LEVEL.
THAT MEANS WE DON'T REALLY UNDERSTAND IT.
Richard Feynman
(on why spin one-half particles obey Fermi-Dirac statistics)

EVERYTHING SHOULD BE MADE
AS SIMPLE AS POSSIBLE, BUT NOT SIMPLER.
Albert Einstein

I PREDICT THAT THE DOMESTICATION OF BIOTECHNOLOGY
WILL DOMINATE OUR LIVES DURING THE NEXT 50 YEARS
AT LEAST AS MUCH
AS THE DOMESTICATION OF COMPUTERS
HAS DOMINATED OUR LIVES
DURING THE PAST 50 YEARS.
Freeman Dyson (2007)

FOR MY WONDERFUL
PARENTS,
ILSE AND HERBERT RENNEBERG

CONTRIBUTORS

Contributions to the Whole Book

Francesco Bennardo, Liceo Scientifico S. Valentini, Castrolibero, Cosenza

David S. Goodsell, The Scripps Research Institute, La Jolla

Oliver Kayser, Rijksuniversiteit Groningen

Oliver Ullrich, Hochschule für Angewandte Wissenschaften Hamburg

Contributions to Single Chapters

Rita Bernhardt, Universität des Saarlandes, Saarbrücken

Alan Blake, GloFish, Yorktown Technologies

Uwe Bornscheuer, Ernst-Moritz-Arndt-Universität, Greifswald

George Cautherley, R&C Biogenius Hong Kong

Ananda Chakrabarty, University of Illinois

Maia Cherney, University Alberta

Theodor Dingermann, Johann Wolfgang Goethe-Universität, Frankfurt am Main

Stefan Dübel, Technische Universität Braunschweig

Roland Friedrich, Justus-Liebig-Universität Gießen

Peter Fromherz, Max-Planck-Institut für Biochemie, Martinsried/München

Dietmar Fuchs, Universität Innsbruck

Saburo Fukui (†), Kyoto University

Oreste Ghisalba, Novartis AG, Basel

Horst Grunz, Universität Duisburg Essen

Georges Halpern, University of California at Davis

Albrecht Hempel, Zentrum für Energie- & Umweltmedizin, Schloss Heynitz, Nossen

Choy-L. Hew, National University of Singapore

Franz Hillenkamp, Universität Münster

Bertold Hock, Technische Universität München

Martin Holtzhauer, IMTEC, Berlin-Buch

Frank Kempken, Christian-Albrechts-Universität Kiel

Albrecht F. Kiderlen, Robert-Koch-Institut, Berlin

Uwe Klenz, Institut für Physikalische Hochtechnologie e.V. Jena

Louiza Law, Hong Kong

Matthias Lehmann, 8sens.biognostic GmbH, Berlin-Buch

Inca Lewen-Dörr, GreenTec., Köln

Hwa A. Lim, D'Trends Inc., Silicon Valley

Jutta Ludwig-Müller, Technische Universität Dresden

Stephan Martin, Deutsches Diabetes-Zentrum an der Heinrich-Heine-Universität Düsseldorf

Wolfgang Meyer, Berlin

Marc van Montagu, Max-Planck-Institut für Pflanzenzüchtung, Köln

Reinhard Niessner, Technische Universität München

Susanne Pauly, Hochschule Biberach

Jürgen Polle, Brooklyn College of the City University of New York

Tom A. Rapoport, Harvard Medical School, Boston

Matthias Reuss, Universität Stuttgart

Hermann Sahm, Universität Düsseldorf and Institut for Biotechnologie 1 des Forschungszentrums Jülich

Frieder W. Scheller, Universität Potsdam

Steffen Schmidt, Neues Deutschland, Berlin

Andreas Sentker, Die Zeit, Hamburg

Matthias Seydack, 8 sens.biognostic GmbH, Berlin

Georg Sprenger, Universität Stuttgart

Eric Stewart, INSERM – University Paris 5

Gary Strobel, Montana State University, Bozeman

Kurt Stüber, Max Planck Institute for Plant Breeding Research, Köln

Atsuo Tanaka, Kyoto University

Dieter Trau, National University of Singapore

Thomas Tuschl, Rockefeller University, New York

Virginia Unkefer, Hong Kong

Larry Wadsworth, Texas A&M University

Terence S. M. Wan, Head of Racing Laboratory, The Hong Kong Jockey Club

Zeng-yu Wang, The Noble Foundation, Ardmore, Oklahoma

Eckhard Wellmann, Universität Freiburg

Michael Wink, Ruprecht-Karls-Universität Heidelberg

Dieter Wolf, Boehringer-Ingelheim, Biberach

Mengsu Yang, City University of Hong Kong

Leonhard Zastrow, Coty International Inc., Monaco

Experts' Boxes and Biotech History Contributions

Wolfgang Aehle, Genencor, Leiden, The Netherlands

Gro V. Amdam, Arizona State University, Phoenix, and Norwegian University of Life Sciences, Aas, Norway

Susan R. Barnum, Miami University, Oxford

Ananda M. Chakrabarty, University of Illinois College of Medicine, Chicago

Cangel Pui Yee Chan, R&C Biogenius Ltd, Hong Kong

Charles Coutelle, Imperial College, London

Jared M. Diamond, UCLA

Carl Djerassi, Professor em. Stanford University

Michael Gänzle, University Alberta

Oreste Ghisalba, Universität Basel, and Novartis Basel

David S. Goodsell, Scripps Research Institute, La Jolla

Susan A. Greenfield, Oxford University, and Director Royal Institution, London

Alan E. Guttmacher, National Human Genome Research Institute (NHGRI), Washington, DC

Frank Hatzak, Novozymes, Denmark

Sir Alec Jeffreys, University of Leicester, U.K.

Shukuo Kinoshita, Kyowa Hakko, Tokyo

Jörg Knäblein, Head Microbial Chemistry, Bayer Schering AG, Berlin

Stephen Korsman, National Health Laboratory Service, and Walter Sisulu University, South Africa

Karl K. Kruszelnicki, University of Sydney

James W. Larrick, Panorama Research Institute, Silicon Valley

Frances S. Ligler, US Naval Research Lab, Washington, DC

Alan MacDiarmid, University of Pennsylvania, Philadelphia

Uwe Perlitz, Deutsche Bank Research, Frankfurt am Main

Ingo Potrykus, Prof em. ETH Zurich

Wolfgang Preiser, Stellenbosch University, South Africa

Timothy H. Rainer, Chinese University of Hong Kong, Prince of Wales Hospital, Hong Kong

Jens Reich, Max-Delbruck-Center for Molecular Medicine, Humboldt Universität, Berlin

Michael K. Richardson, University Leiden, The Netherlands

Sujatha Sankula, National Center for Food and Agricultural Policy, Washington, DC

Gary A. Strobel, Montana State University, Bozman

Christian Wandrey, Institut für Biotechnologie, Forschungszentrum Jülich

James D. Watson, Cold Spring Harbour Laboratory, Watson School of Biological Sciences

Ian Wilmut, Edinburgh University

Eckhard Wolf, Universität München

Harold Boyd Woodruff, Merck, Inc., and Soil Microbiology Associates

Rainer Zocher, Technische Universität Berlin

Table of Contents

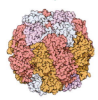

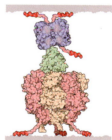

Boxes

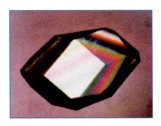

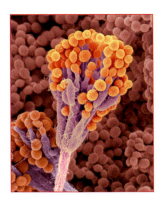
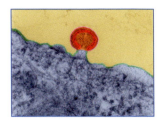

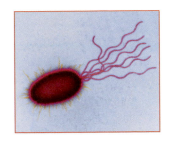
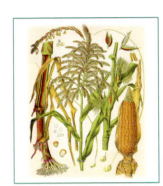

FOREWORD BY THE US-EDITOR ARNOLD DEMAIN

Arnold L. Demain in his office at Drew University

There are no two sciences,
there is only one science
and its application
and these two activities
are linked
as the fruit is to the tree.

Louis Pasteur

We have the unity
of biochemistry
on one hand,
and the diversity
of microbial life
on the other;
we have to understand
and appreciate both.

Arnold L.Demain

It is hard to decide when biotechnology began. Some would consider dates such as 3500-7000 BC when microbes were first used to preserve milk, fruits and vegetables and to make cheese, bread, beer, wine pickled foods and vinegar.

If that was the proper decision, then I have been in biotechnology for my entire adult life, i.e., from 1949. At that time, I was 22 years of age and starting my Master's degree research at Michigan State College (now University) on the spoilage of pickles. Also, I would have to consider my grandfather and father as biotechnologists since they were both "pickle men".

On the other hand, many consider the above as industrial microbiology and assume that biotechnology was born in 1972-1973, when Paul Berg, Stanley Cohen of Stanford University and Herbert Boyer of the University of California at San Francisco discovered recombinant DNA. If this latter view is correct, then I started my biotechnology career in 1972 when I was asked to be a consultant for the first biotechnology company, i.e., Cetus Corporation in Berkeley, California. Indeed, it was an exciting time during which efforts were put forth to commercialize the discovery of recombinant DNA technology. Within ten years, companies such as Genentech, Biogen, Amgen, Genetics Institute, Chiron and Genzyme were born and these exciting developments resulted in the production of recombinant proteins which solved medical problems of huge importance to patients throughout the world. They also led to the development of a huge and important biotechnology industry which today generates annual product revenue of over 60 billion dollars.

I have enjoyed over 50 years of participation in both industrial microbiology and biotechnology in industry (at Merck & Co), in academia (at M.I.T. and recently at Drew University). It always excited me to design microbes to do important things and then discover the ways, both genetic, biochemical and nutritional, to force these fantastic microbes to make industrial quantities of their valuable products, whether they be antibiotics, amino acids, purine nucleotides, immunosuppressants, cholesterol-lowering agents, toxins or recombinant proteins.

I have been fortunate to have mentors, colleagues and students to help me in these efforts. I have always felt that the genetic aspects that propelled industrial microbiology into the new world of biotechnology were both fascinating and fantastic.

Not being a geneticist myself and never even having taken a course in genetics, I have been in awe of the developments of biotechnology for many years. Over this time, I have struggled to understand genetic concepts and the new aspects introduced by biotechnologists.

What I needed and never had was a basic introductory book on biotechnology.

Then, it happened: the book "Biotechnologie fur Einsteiger" by Reinhard Renneberg was born! Unfortunately, my German language skills were virtually non-existent. However, even without the ability to read the book, I was fascinated by the many photos of famous biotechnologists and scientists and the beautiful color illustrations of Darja Süßbier.

How lucky I am that Professor Renneberg asked me to be the Editor of the English edition. After translation of the German text into the English by translators Renate FitzRoy and Jackie Jones, I have thus read, word by word, the entire ten chapters of this fantastic effort by Reinhard Renneberg.

I enjoyed every minute of this activity and editing the formal English translation into a version suitable to both young English and American students has truly been a labor of love.

Arnold L. Demain
May 26, 2007

Dear Master Arny,

the whole team of Biotechnology for Beginners wishes to dedicate this book to your 80th BIRTHDAY celebrated in 2007. We are honored to know you personally and admire your contribution to microbiology and biotechnology, as well as your kindness and wisdom!

Reinhard and Merlet,
Dascha, Renate, Jackie, Christoph, Ute, and Ted

A PERSONAL FOREWORD BY TOM RAPOPORT

I admit that I don't like reading textbooks, and I know that many students share my reluctance: it requires frustrating diligence to plow through several hundred pages of accumulated dry knowledge. Modern textbooks try their best with their ever-increasing number of colorful illustrations, but that still won't turn them into thrillers.

Thus, when Reinhard Renneberg told me that he had written a textbook about biotechnology, I was sceptical. And when he even asked me to write a foreword for it, my first reaction was to turn him down: it's going to be boring, about a subject I have no clue about, and, most important of all, I am busy.

But then I did it anyway, and the bottom line is: this book is different! Sure, it's still a textbook and conveys knowledge, but it's real fun to read.

Reinhard Renneberg and his illustrator Darja Süßbier have packed the facts into excellent illustrations, interesting historical discourses, funny cartoons, and concise texts. An entertaining textbook – this must be the first of its kind!

I would call it "biotechnology in a nutshell", addressed to everyone who is curious about the latest developments as well as historic foundations. You really don't need any particular background for the book – something like a high school diploma is sufficient. I am sure you will have the same experience – being amazed about the wonders of biology and biotechnology. It's clear that Reinhard is addicted to the subject, and his enthusiasm is contagious.

Let me tell you about the author. Reinhard grew up in the German Democratic Republic, known to people in the West as "East Germany". His young parents became school teachers after World War II, thrown into the job without much training, simply because there was a teacher shortage, due to the ongoing denazification procedures. Like many others, his parents embraced the fresh start with enthusiasm and determination. This upbringing explains Reinhard's dedication to teaching.

Reinhard had an early love for science books. He told me that his first hero was **Ernst Haeckel**, probably because they both attended the same school in his hometown Merseburg. He imitated Haeckel by establishing a herbarium at home and by becoming a keen bird and insect spotter. Later, his interest switched to DNA when a friend gave

him the "Double helix" by **Jim Watson**, a book that was not available in the country. He only had it for one night, then he had to pass it on to the next person in line. The next day, he made his own DNA model out of colorful plastic balls used in baby strollers.

After finishing high school, Reinhard wanted to leave the country for the big wide world. Well, only half the world was accessible to East Germans, and his first choice was China. However, in 1975, relations with China were at a low point, and China was undergoing its "Great Cultural Revolution" anyway. So, he went to the Soviet Union instead. What attracted him was the big heart of the Russians, their hunger for books and art, and their talent for improvisation. He studied at the Institute for Bioorganic Chemistry in Moscow, a place that was then a top address.

After his return to East Germany, he joined the Central Institute for Molecular Biology in Berlin-Buch, and this is where I met him. He was in **Frieder Scheller**'s group, working on the development of a biosensor for glucose, a device needed by diabetes patients for the control of their blood sugar, and although the USA, Great Britain, and Japan got there first, their success was remarkable, given the lack of biochemicals, instruments, copying machines, and many other things that we now take for granted. Improvisation and a lot of enthusiasm made up for it!

Soon after the unification of Germany, Reinhard took a position at the Hong Kong University of Science and Technology. His current boss, Professor **Nai-Teng Yu**, met him 1994 in East-Berlin and enthusiastically told his Department: "*Hire Renneberg immediately; the combination of the high motivation of an East German with the modern technology in Hong Kong is unbeatable!*" So finally, Reinhard's dream of going to China became true.

Prof. Renneberg directs a pretty large research group that works on the development of biotests. He also runs a biotech company, owns two cats and a rabbit, and has a subtropical garden that overlooks a beautiful bay. Apparently, he has a lot of spare time because he writes a regular column, the "Biolumne", for a major newspaper, draws cartoons, and always has a new book project.

The current textbook does not come as a surprise to me. It's actually Reinhard's fourth book. He

started out in East Germany with two biotechnology books for general readers and children, and this is where he developed his gift for clear and entertaining writing.

For me, the most important task of a teacher is to convey to the students our passion for science. Nature has so many wonders and puzzles, and it's a privilege that we can contribute to solving them.

Reinhard's book certainly conveys the enthusiasm for science and the belief that it can do a little bit to improve the world. I wish the book all the success it deserves.

Tom Rapoport
Boston, June 1, 2007

Tom Rapoport was born 1947 in Cincinnati (USA) but grew up in East-Berlin. He studied chemistry and biochemistry at the Humboldt-University. After obtaining his Ph.D, he joined the Central Institute for Molecular Biology of the Academy of Sciences of the GDR in Berlin-Buch. In 1985, he became Professor for Cell Biology, and after the unification of Germany, group leader at the Max Delbrueck Center for Molecular Medicine. In 1995, he accepted an offer from the Harvard Medical School in Boston. Since 1997, he is also Investigator of the Howard Hughes Medical Institute.

Tom Rapoport is a member of the National Academy of Sciences of the USA, of the American Academy of Arts and Sciences, and of the Leopoldina Academy, and was awarded several prestigious awards and prizes.

PREFACE

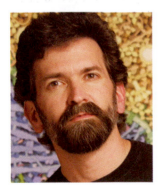

The author after cloning experiments
of his tomcat Fortune
(with Fortune Zero, Fortune I,
Fortune II, and Fortune III).

Darja Süßbier with tomcat
Asmar Khan

David Goodsell, Molecular graphics
wizard, yoga practitioner and
bionanotech visionary.

Who wants to read a long preface anyway - so let us get straight to the point: What made me write this book?

Curiosity and Enthusiasm. Even as a young boy, I loved reading everything that explained the world to me. Today, as a scientist, I cannot think of a more fascinating subject than biotechnology - our future is at stake here! What could be more exciting?

Wanting to know it all. During my forays into scientific literature, I realized that "I know nothing except the fact of my ignorance" – as Socrates put it. Becoming a well-versed Renaissance scholar would have been my ideal, but, alas, this is not an option these days. It is just about possible to get an overview of one field of knowledge. Beyond that, one has to rely on cooperation with scientists who are specialists in neighboring fields. In that respect, I was lucky enough to find the two Olivers – Oliver Kayser from Berlin, now working in Groningen, the Netherlands, and Oliver Ullrich from Hamburg, Germany. Both of them cover a vast area of knowledge. They agreed to read through the whole book and largely contributed to its present shape. Thank you! Where the subject matter became more complex, I turned to experts in the field and put their – often very abridged – views in a nutshell, i.e. in a box. You will find the names of these experts on p. X. I am very grateful to them and just hope that I did not leave out anybody!

Laziness. I have been teaching analytical biotechnology and chemistry in Hong Kong for ten years. My Chinese students know next to nothing about beer-brewing, enzyme-containing detergents, DNA, oil-eating bacteria, "golden rice",

GloFish®, heart attacks, or the human genome project. As a result, my seminars tend to include long, time-consuming deviations into biotechnology. Pointing the students to an 88-title bibliography is of no avail – it is just one book they are prepared to read. Now I will be able to say: "Just buy my book and read it – it covers all you need to know."

Enjoyment. "Everything you can imagine is real" said Picasso, and turning this new type of textbook into reality with the help of Darja Süßbier, to me the best bio-graphics artist in Germany – was pure joy. She found ingenious and sympathetic ways of transforming my ad hoc improvisations into brilliant graphics. Any other graphics artist would have been driven to despair by my chaotic workstyle. Many thanks, Dascha!

Being able to use David Goodsell's amazing graphics of molecules was a dream come true, and when I was just getting fed up counting the carbon atoms in taxol, Francesco Bernardo from Italy stepped in, creating 3D models of essential molecules. Oh what fun!

A **passion for images**. Asia has a long tradition of pictorial representation. Searching Google Images for biotech illustrations, I worked myself into a frenzy. At first, the publisher was a bit shocked to see the nice white textbook with two-color illustrations gradually turn into a riot of color. In the end, there was hardly any white left!

With the images came the problem of sorting out copyrights, but most copyright owners were very helpful. Ringier publishers, Switzerland, transferred all the rights of an earlier book of mine, Bio-Horizonte, formerly owned by Urania Verlag, Leipzig.

Others, such as GBF Braunschweig, Roche Penzberg, Degussa, Transgen, and Biosafety Networks agreed to let me use dozens of their images while testing the capacity of my server by sending 10 MB mails. Larry Wadsworth from Texas provided me with a host of photos of cloned animals. Anybody whom I have not mentioned as the author of an illustration or whom I have not been able to contact, please come forward. Any oversight has been unintentional.

Readers will also notice that I have been using my own photos of cats, birds, frogs, dolphins, food, China, and Japan – everything was photographed with the biotech book in mind. I hope you don't mind seeing me in some self-experiment rather than a professional model.

Communication mania. What could be more wonderful than sitting down with a cup of coffee, overlooking the South Chinese Sea, and opening my laptop to see what mail has arrived overnight. Perhaps news from bigwigs and smallfry in the biotech world or some new layouts from Dascha in Berlin. Some ten thousand e-mails have been going back and forth, and it was like magic. This book is a child of the internet. There I was sitting on a subtropical island, pressing a few keys - and lo and behold - a beautiful book emerged at the other end of the world. Jules Verne would have been impressed.

Whose **idea** was it? Merlet Behncke-Braunbeck from Spektrum Akademischer Verlag, Germany, made me promise to turn my ideas into a textbook. Imme Techentin-Bauer, Bärbel Häcker, and Ute Kreutzer were a highly motivated, efficient, and charming editing team, perhaps sometimes cursing under their breath when I had done it again and completely changed or "enhanced" an almost finished chapter. However, they managed to give Dascha and me wonderful support. Thank you, ladies!

How is this book to be used?

As an **introduction** for students starting at college or university, for teachers, journalist, or any interested layperson.

As a **textbook for students**. You can work through the chapters systematically and check whether you are able to answer the eight questions at the end of each chapter.

As a **eureka experience**. Just flick through the book, and I hope you will become intrigued and feel inspired to seek further information in specialist books or on the Internet.

As a **reference book**. This may be a first port of call when looking for an answer to some biotech question that is bugging you. You can then follow it up on the Internet or in specialist textbooks.

Will it work? Some of my colleagues may turn up their noses at this book, which is, admittedly, an experiment, but I have no patience with boring books for which trees have died in vain.

Comments by users/readers will be very welcome. Please send your mail to
chrenneb@ust.hk

Reinhard Renneberg, August 2005

About this English edition

I do not wish to burden you with more words to read, dear reader: three forewords, two by eminent scientists, are more than enough. I simply wish to use this short space to thank my amazing team: without the incredible, cheerful Merlet Behncke-Braunbeck as our Good Spirit (guter Geist in German), this project would not exist or it would have failed at least six times.

While sitting in Hong Kong in my subtropical garden with two cats blocking the computer and eight love birds making background music, I had the easiest task in the world, the FUN PART, to provide 24-hour a day work to

- Heidelberg: perhaps the most romantic German city (no wonder that the headquarters of the US Army moved there after the war) and now (no wonder!) the German headquarters of SAV and its team of enthusiasts (see my Foreword to the German edition!);

- the "Queen of Bio-graphics," Darja Süssbier, in Berlin;

- the wise Great Old Man of Biotech and my extremely precise editor, Arny Demain, at Drew University in New Jersey;

- the ingenious translators, Renate FitzRoy (almost a coauthor) and Jackie Jones in the U.K;

- the Masters of Molecules, David Goodsell in La Jolla (California) and Francesco Bennardo In Cosenza (Italy);

- the 25 new contributors to this book, mainly from the US, UK, Japan, Switzerland and India….

- Ted Cole, our native American copy editor, dug out another 887 (I lost count!) comma and spelling mistakes and taught me that American English is indeed different from what I learned at school...

It would have been 100% less trouble if I had simply stuck to the German original. Because biotech is changing by the day and because the needs of students in the English-speaking world differ from those in Germany, I decided to rewrite whole parts of the book. I "de-Germanized" it at the same time as I globalized the text and extended the book by adding more than a dozen new Expert Boxes (for which poor Christoph Iven in Heidelberg had to fight for copyrights of books, pictures, and stamps).

The result: This is a completely NEW BOOK! "My hope is that YOU, THE READER, will not only find this book useful, but will get hooked …

Reinhard Renneberg Hong Kong, July 20th 2007 (a birthday gift to myself)

Oliver Ullrich with non-engineered mouse named Ollie...

Oliver Kayser with his kids.

Biotech History: Francesco Bennardo enjoying a fermentation product (red arrow) in the organic chemistry lab (highly flammable!). After this, Francesco was suspended from lab, and smokes now in front of his computer modelling molecules.

Some personal comments about our US-Editor Professor Arnold Demain

Soon after the German edition was launched and a few enthusiastic readers had replied, Spektrum Akademischer Verlagstarted to consider to go West...

The problem was: the "Germanized" version should be edited to be "digestable" for the US-reader and also globally.

When thinking about a qualified person in US-biotechnology AND a person who would appreciate my style of design and humor, I came up with one name immediately: Arnold Demain.

I first started to read his BIOs. Here is an extract of the hundreds of pages I found:

"Dr. Arnold Lester Demain has done a tremendous amount of work on the application of microbiology, which includes microbial toxins, enzyme fermentations, biosynthesis of antibiotics, vitamins, amino acids and nucleotides, microbial nutrition, industrial fermentations, regulatory mechanisms, and genetic engineering of fermentation microorganisms."

To call him a giant of industrial microbiology is very appropriate. His life is a witness to the progress of biotechnology!"

Could I approach such a GIANT? I never met him in person!

"Arnold ("Arny") was born on April 26, 1927 in Brooklyn, New York. All of his grandparents were immigrants from the Austrian-Hungarian Empire."

Arny's European origin: Lemberg (Lvov)

As my grandparents came also from the same Empire, I was thinking (self-flattering): these people are well-known for their witty charm (think about Vienna!), their love of arts AND sharp minds in science and technology plus hard work...

Brooklyn Bridge, graphics art by the Czech painter Tavik Frantisek Simon (1877-1942) in 1927, the year of Arny's birth.

"Arny was raised in Brooklyn and the Bronx in years dominated by the depression. The family moved often and he attended some five elementary schools and three high schools before graduating in February 1944.During those years, Arny worked very hard to help the family. He had many jobs such as a grocery delivery boy (2 cents per delivery plus an occasional 5 cent tip), and as a stock boy for Lord & Taylor's Fifth Avenue department store (at 40 cents per hour during the school months and $17 per week during the summer). It turns out that his exposure to and experience working with pickles had a great influence on his later career development."

I must say, I liked this very much! I was born in the poorer part of Germany and learned to appreciate practical work (as a cowhand in agriculture) in my young years, too.

Arny as undergraduate at Michigan State College (now University) in 1948.

"At the age of 16 (1944), Arny's mother and father took him by train to Michigan State College (MSC) (now Michigan State University) in East Lansing. This was because Henry, his father and a pickle manufacturer, knew that the leading investigator of cucumber fermentations in the US was Professor **Frederick W. Fabian** (1888–1963) at MSC.

Arny was virtually "left behind" on the steps of Professor Fabian's food fermentation laboratory at MSC.

Charles Lindbergh' s flight in 1927, a month after Arny was born.

Arny enlisted in the US Navy for service in February 1945, several months before his 18th birthday, because World War II was still going on.

He returned to MSC's Department of Microbiology and Public Health in 1947 and obtained his BS degree in 1949 and his MS degree in 1950. His Master's thesis dealt with the microbial spoilage of cucumbers by softening during fermentation. The reason for Arny's choosing this topic was definitely influenced by his father's profession. **Henry Demain** had established a canning and pickling plant for the Vita Foods Corp. in Chestertown, Maryland, and Arny's uncles Ben and Seymour had opened **Demain Foods Co.**, another pickling operation, in Ayden, North Carolina. Henry was a leader in the pickle business, working for Fields and then Bloch and Guggenheimer in New York before setting up the pickle plant in Chestertown, Maryland. "My grandfather, **Joseph Demain**, had sold pickles for years in one of New York's major market areas. Thus, I was well primed to become a pickle man myself."

Arny himself remembers: "I have no trouble remembering the year of my birth because in that year two important records were set. **Babe Ruth** hit 60 home runs of the New York Yankees and **Charles Lindbergh** flew the first solo flight across the Atlantic Ocean to Paris.

My father convinced me that I should become a food fermentation expert. He knew only one professor, **Frederick W. Fabian**, who conducted annual one-week summer courses at Michigan State titled The Pickle and Kraut Packers' School.

After the war, I returned to East Lansing in early 1947 and resumed my studies. During my stay at Michigan State, I worked in Fred Fabian's lab on spoilage (softening) of pickles. In the summers, I worked for my father and my uncles. My summer responsibilities from 1947 to 1950 were to start in South Carolina, buying cucumbers from farmers and shipping them by truck to my father's pickle factory. I followed the cucumber crop virtually state by state from South Carolina to Wisconsin. During this trip north, I also worked at my uncle's pickle plant in North Carolina, my father's plant in Maryland, and ran a pickling plant in Brodhead, Wisconsin. During my periods of work at my uncle's plant, I met Professor **John** ("**Jack**") **Lincoln Etchells** of the US Department of Agriculture and North Carolina State University. When my uncle's fermentations went awry, they often called on Jack to come to Ayden and recommend measures to correct the problems. Jack became my first mentor.

Demain's pickles are the best, according to wife Jody during graduate student days in California.

And: Except the microbes, the most important event?

Arny had met incoming freshman **Joanna** ("**Jody**") **Kaye** from Youngstown, Ohio, and fell in love with her! Together, they headed for continued studies as a Ph.D. candidate at the University of California (Berkeley) in the fall of 1950.

His main activity in Berkeley, other than studying, was to transfer and maintain the viability of the cultures of the famous UC yeast collection.

Arny then spent four years under the tutelage of the prominent yeast scholar, **Herman J. Phaff** (born 1913), working on yeast polygalacturonase. He obtained his PhD from UC Davis in 1954. His dissertation was on the nature of pectic enzymes, which were responsible for pickle

A giant of Industrial Biotech in the lab.

softening. Professor Phaff became Arny's second mentor.

Arny: "We worked from early morning until evening, went home for dinner with our spouses, and returned to the lab for research and discussions that lasted until the wee hours. With Herman's help, I elucidated the mechanism of pectic acid degradation by the extracellular polygalacturonase (YPG) of the yeast *Klyveromyces fragilis*.

We apparently were the first in the world to carry out affinity chromatography, using a pectic acid gel to selectively adsorb YPG from culture filtrates. We proved that the entire hydrolysis of polymer to dimer was accomplished by a single enzyme in contrast to current thought that multiple enzymes were necessary.

The work was published in four publications, one appearing in *Nature*. I didn't realize how significant that was, but I learned later in life (after receiving many rejection notices form *Nature*) that for a graduate student to publish one of his/her first papers in *Nature* was an unusual feat."

In March 1954, Arny accepted a position as a research microbiologist at the penicillin factory of **Merck Sharp & Dohme** in Danville, Pennsylvania.

This was a major shift for his research, because now he had to focus on penicillin biosynthesis. He educated himself about penicillin fermentation by reading extensively. At the end of 1955, he moved to the main Merck Sharp & Dohme Research Lab-

oratories (MSDRL) in Rahway, New Jersey, where he spent the next 13 years.

During his employment at Merck, he innovated methods to enhance the production of secondary metabolites using starved resting cells. He was the first to detect feedback inhibition of penicillin production by the amino acid lysine (see Chapter 4 of this book!), and originated the study on the effects of primary metabolites on the secondary metabolism of microorganisms.

Arny: "My research in the penicillin factory showed that over 99% of the penicillin formed appeared in the liquid portion of the broth and that penicillin was partially degraded during its production. Thus, I showed that the apparent rate of penicillin production was the net result of synthesis and inactivation during fermentation. I also confirmed a controversial claim by **Koichi Kato** of Japan in 1953 that he had isolated the "penicillin nucleus" from fermentations conducted without the addition of the side chain precursor, phenylacetate". His discovery was of great economic importance, since the nucleus was later used and is still used to produce all semisynthetic penicillins of commerce. Workers at the Beecham Laboratories in the United Kingdom identified the compound as 6- aminopenicillanic acid (6APA) in 1959.

During the first 10 years in Rahway, he and his technical assistant **Joanne Newkirk** made great progress on microbial growth factors, cephalosporin biosynthesis, protein synthesis, and nucleotide fermentation. His group was the first to discover that one of the growth-stimulating effects of protein digests was due to lipids associated with proteins such as casein, and not always due to their peptides or amino acid components. He demonstrated that *Lactobacillus homohiochi* and *L. heterohiochi*, contaminants of the wine making process, not only tolerated over 20% ethanol, but required ethanol for optimal growth. His superiors in Merck's Microbiology Department were **David Hendlin** and **Boyd Woodruff**. These well-known microbiologists became Arny's third and fourth mentors.

Arny: "My studies on the biosynthesis of the very important β-lactam antibiotic cephalosporin C began around 1961, some six years after its discovery in Oxford by **Abraham** and **Newton**.

My participation in Merck's Microbiology Department was brought to a close in 1964 when I was asked to form a new department at Merck that would involve the improvement of product biosynthesis in microbial strains."

Boyd Woodruff (on the left), Arny's supervisor at Merck and later his friend, as a student with the great teacher of American microbiologists, Selman A. Waksman (1888-1973).

Arny was asked by Merck Vice President **Karl Pfister** (1919-) to found a new department which he named Fermentation Microbiology. This gave him an opportunity to set up a new group of over 30 people dealing with improvement of current Merck fermentations and development of fermentations for new products. He elucidated the mechanism by which the biosynthesis of cephalosporin in *Cephalsporium acremonium* was stimulated by the presence of methionine -a new mechanism which had not been reported before.

One of Merck's fermentation factories

Arny studied with his group the overproduction of L-glutamic acid (GA) by *Corynebacterium glutamicum*, since Merck was selling monosodium glutamate (MSG) at the time. His colleagues at the penicillin plant had discovered that one way to trigger this process was to add penicillin to the culture after growth.

One of the other major products of Merck at the time was vitamin B_{12}. The organism used for production at Merck was *Pseudomonas denitrificans*. Strain improvement of production was a very successful endeavor. One of the new strains, a revertant of a pantothenate auxotroph, was found to produce twice as much B_{12} as the commercial strain from which it was derived by mutagenesis. Such a major increase was virtually unknown in strain improvement programs on mature fermentation processes.

Arny: "Around 1966, **Max Tishler**, the President of MSDRL, gave me the responsibility to develop a microbiological riboflavin (vitamin B_2) process for the company. We chose Ashbya gossypii as our organism because it was already known to make 5g/liter. By the time I left Merck in 1969, we had developed a process capable of producing 12g/liter. To be economical, the process had to reach about 25g/liter and I assume this was accomplished a few years later by my former group since biosynthetic riboflavin became a commercial product of Merck for both human and animal use."

In 1968, Arny was invited by **Nevin Scrimshaw** to become a Full Professor at MIT. In 1969, he set up a fermentation microbiology laboratory and worked at MIT for 32 years. The research of his group resulted in a breakthrough discovery of a key enzyme in cephalosporin biosynthesis – deacetoxycephalosporin C synthetase ("expandase"). The discovery of this enzyme established the role of penicillin as an intermediate in cephalosporin C biosynthesis and disproved the previous hypothesis that these two separate end products of *C. acremonium* were formed by a branched secondary metabolic pathway.

"At the time I arrived at MIT, there was no work going on with antibiotics in the Nutrition and Food Science Department. At that point, I was anxious to introduce such studies into the department. My student picked up on my earlier observation of lysine interference in penicillin production and found the effect of lysine addition to be mainly one of feedback inhibition, not enzyme repression. I had an interest not only in the "good" products made by microorganisms, but also in the "bad" ones. In collaboration with **George Buchi**, MIT's great natural product chemist, and **Gerald Wogan**, a toxicologist, we

found that a mold isolated from a Thai household, which had killed a child there, produced three toxins: cytochalasin E and the tremorgens, tryptoquivalin and tryptoquivalone. Our cooperation led to the discovery and structure elucidation of ditryptophenaline from *Aspergillus flavus* and aspersitin from *Aspergillus parasiticus*, and the isolation of cyclopiazonic acid.

Arny's group at MIT in the early 1970s.

"I was drawn into the cellulase area by some comments I was asked to present at a symposium dealing with cellulase of the fungus *Trichoderma. viride* (later *Trichoderma reesei*) (see Chapter 6). We found that the extracellular cellulase of *Clostridium thermocellum* ATTC 27405 could attack crystalline cellulose. This was the first demonstration of a true cellulase produced by a bacterium.

Using advice from **Elwyn T. Reese** (of *Trichoderma* cellulase fame; from the U.S. Army Research & Development Laboratories in Natick, Massachusetts), we found the enzyme to be inhibited by cellobiose when attacking crystalline cellulose but much less so when attacking amorphous cellulose. Later, we found that the enzyme was very useful in the conversion of cellulose to glucose, since it degraded cellobiose, the inhibitor of cellulase action."

Arny also postulated that antibiotics and other secondary metabolites are produced in nature, which serves the survival of the producing organisms. This theory overthrew the previous concept, which considered that the production of antibiotics was an artifact of microbial growth in the laboratory. It turns out that the antibiotic-producing microorganisms also develop means to protect themselves from their own antibiotics, i.e., to prevent suicide. This is a very fundamental principle of antibiosis in natural environments.

Receiving the Honorary Doctorate from Ghent University, Belgium.

Arny also worked on the production of gramicidin S by *Bacillus brevis* and found that the antibiotic acted as an inhibitor of germination outgrowth of the spore in the producing organism. Arny and his group actively studied the effect of nutrition on the production of red pigments by *Monascus sp.* and of rapamycin by *Streptomyces hygroscopicus*. They also developed important bioconversions of compactin to pravastatin, and penicillin G to deacetoxy-cephalosporin G.

"My earlier work on vitamin B_{12} at Merck stimulated my interest in continuing such studies at MIT and an active program on this vitamin was set up.

In the early 1990s, a study of biosynthesis of the immunosuppressant rapamycin (sirolimus) began in my laboratory. In 1995, we began a program on the bioconversion of the hypocholesterolemic agent compactin to pravastatin, a potent inhibitor of -hydroxy-methylglutaryl CoA-reductase, and discovered a new method by which such a reaction was accomplished. It currently is used in industry."

Finally, Arny went to Space!

"In the mid-1990s, we started a series of NASA-sponsored experiments to determine the effect of simulated microgravity (SMG) on secondary metabolism. We used rotating wall bioreactors (RWBs) designed at NASA's Johnson Space Center in Houston, Texas. We found that regulation of microbial processes under SMG was quite different from that at normal gravity.

My last projects at MIT dealt with nutritional studies on *Clostridium tetani* and *C. difficile* with the aim of devising media for culture preservation, inoculum development, and fermentation that lacked animal-and dairy-derived products but supported good growth and toxin produc-

tion. The purpose was to facilitate the production of vaccines against tetanus and antibiotic-associated diarrhea free of potential problems with prions. We were successful in the case of both toxin production processes."

After retirement

Over the years, Arny has been successful in combining knowledge of basic principles and potential applications. He has published more than 500 research and review papers, and more are on their way, since he is still actively directing research with undergraduate students at Drew University in Madison, New Jersey.

"After retirement from MIT in 2001, I was fortunate to join a unique institute at **Drew University**. Drew is a small, picturesque, and academically strong university mainly at the undergraduate level. It houses the Charles A. Dana Research Institute for Scientists Emeriti, also known as R.I. S. E. The Institute is made up of ten Research Fellows who have had experience in industry. Our main purpose is to train undergraduate science students in modern day research.

I owe a lot to my 15 years in industry, because working at Merck gave me an appreciation of Pasteur's statement that 'There are not two sciences, there is only one science and its applica tion, and these two activities are linked as the fruit is to the tree'. As I stated in an interview published in 1991, 'It's clear that those who understand basic biology are the ones most able to apply it'.

At Merck, I learned early in my career that there was a lot more to applied microbiology than pickles and sauerkraut, and that microbes could be used to make antibiotics, produce commercial enzymes, convert steroids to be used for arthritis and rheumatism, make amino acids and vitamins for human and animal nutrition, and make many medicinals for human and animal health.

I was very lucky at MIT to have had a fantastic group of bright and hard-working visiting scientists, postdoctoral associates, graduate students, undergraduate students and high school students. I owe all my success to them and my two amazing lab supervisors, Nadine A. Solomon and Aiqi Fang. These two scientists not only did research but also supervised my group and

helped me in the office. Success at MIT would not have been possible without them.

In the early seventies, the explosion of the biotechnology field really turned me on to the applications of sciences. In 1972-1973, the development of recombinant DNA by **Paul Berg**, **Stan Cohen**, and **Herb Boyer** at Stanford University and the University of California, at San Francisco, triggered the birth of modern biotechnology. A physician (**Peter Farley**), a biochemist (**Ronald Cape**), and a Nobel Laureate physicist (**Donald Glaser**), with several others, conceived of the commercialization of recombinant DNA technology and established the first biotechnology company, the Cetus Corporation, in Berkeley in 1971.

I was lucky enough to become the second consultant in the biotechnology industry, joining Nobel Laureate **Joshua Lederberg,** then at Stanford University, as Advisor to Cetus. Thus begun one of the most exciting adventures that I have ever experienced. The vision of these Cetus founders resulted in a major industry, serving the needs of patients throughout the world and revolutionizing the practice of industrial microbiology and agricultural technology.

Although Cetus is no longer around (it was incorporated into Chiron Corporation, now Novartis, in the mid-1990s), it should long be remembered as the founder of modern biotechnology and the developer of the polymerase chain reaction (PCR), a technique of enormous importance today. Indeed, the PCR principal investigator, **Kary Mullis**, holds the only Nobel Prize ever awarded to a scientist in the biotechnology industry."

The first genetic engineering firm was not Genentech, as is often assumed, but the Cetus Corporation, founded in Berkeley, California in 1971. Professor Arnold Demain (left) in 1974 with Peter Farley (first Cetus President) and Robert Fildes (second Cetus President) at a Genetics of Industrial Microorganism Symposium in Sheffield, UK..

BEER, BREAD AND CHEESE –
The Tasty Side of Biotechnology

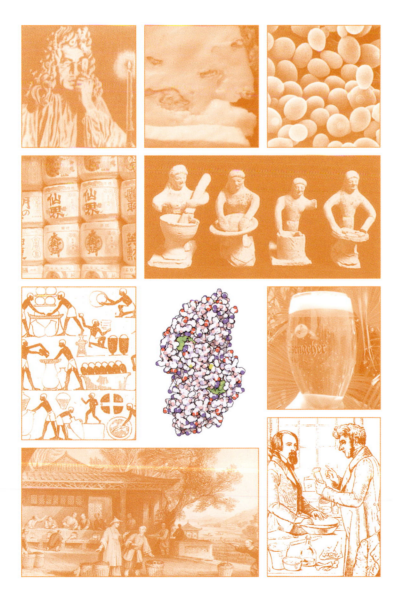

CHAPTER **1**

Fig. 1.1 Brewing beer in Egypt 4,500 years ago.

Fig. 1.2 Boeotian women from ancient Greece baking bread 6,000 years ago.

Fig. 1.3 Yeast, as drawn by Leeuwenhoek (above) and seen under a modern-day scanning electron microscope, clearly showing the budding daughter cells.

Fig. 1.4 Medieval beer-brewing.

1.1 In the Beginning, There Was Beer and Wine – Nurturing Civilization

The first beer we know of was brewed by the Sumerians who lived in Mesopotamia between the rivers Euphrates and Tigris (modern-day Iraq), in 8000 to 6000 B.C. They produced a nutritious, non-perishable and intoxicating beverage by soaking barley or Emmer wheat (an ancient wheat cultivar grown in the region) in water and letting it germinate. The process of husking wheat around 3000 B.C is shown on a Sumerian clay tablet, known as *Monument bleu*, in the Louvre museum in Paris. The germinated grain was then kneaded into beer bread which was only lightly baked and then crumbled and stirred into water. The mixture was later poured through a wicker sieve and kept in sealed clay vessels. Soon, gas bubbles began to rise, as fermentation set in.

Fermentation is an anaerobic process which transforms the sweet juices into an alcoholic drink – in this case, beer.

Part of the germinated grain was dried in the sun – a process equivalent to modern kiln-drying – to preserve it for periods when fresh grain was in short supply. The beer brewed later by the Babylonians in the same region had a slightly sour taste, due to **lactic acid fermentation** taking place simultaneously. Lactic acid fermentation helped greatly to prolong the storage life of beer, as many microbes cannot survive in an acidic environment. This was immensely important in the hot climate of the Middle East where hygienically safe beverages were at a premium.

Alcohol is fermented sugar, a final metabolite of yeast. Even an alcohol content of 2-3% affects the permeability of the cytoplasmic membrane in bacteria and inhibits their growth. In the hot climate of the Middle East, **slowing down the growth of microbes** through fermentation is a definite – perhaps even decisive – advantage. The development of agriculture brought about a dramatic rise in population, and with it a dearth of clean drinking water. It is a problem that the Western world had also been struggling with right to the end of the 19th century, and which many other parts of the world are still facing. Think of the familiar images of the river Ganges where animal and human carcasses and feces pollute the drinking water. Contaminated

water can be highly dangerous, whereas fermented produce such as **beer, wine, or vinegar were virtually germ-free**. They could even be used to make slightly contaminated water safe, as not only alcohol, but also organic acids in the fermented product inhibit the growth of potential pathogens.

The thirst of our ancestors was quenched by beer, wine, or vinegar rather than water. The oldest biotechnology in the world provided safe and stimulating drinks that nurtured civilization. Such revolutionary technology was bound to succeed.

Not only the people of Mesopotamia, but also the ancient Egyptians were beer brewers. A mural dating from around 2400 B.C shows the production process (Fig. 1.1).

The Egyptians were already aware that using the sediment from a successful batch would speed up the new fermentation process. Their beers tended to be dark, as they used roasted beer bread, and they reached an alcohol content of 12 to 15%. Bottled beer was also an Egyptian invention. When the pyramids were built, beer in clay bottles was delivered to the building site.

While the Celtic and Germanic tribes were happy with mead, a sour-tasting beer stored in vessels in the ground holding up to 500L (132 US gallons) at a temperature of around 10°C (50°F) – the art of beer-brewing reached another heyday when it was developed by monks in the 6th century A.D. Their motivation lay in their fasting rules – *liquida non fragunt ieunum* (liquids do not break the fast) was the motto that led them to brew strong beers with a high alcohol content.

The word beer itself is supposed to be derived from the ancient Saxon word bere, meaning barley. Being able to brew good beer is one thing – knowing how it happens, is quite another.

The first person to spot the yellow yeast blobs in a beer sample (Fig. 1.3) through his single-lens microscope was **Antonie van Leeuwenhoek** (1632-1723), who had also been the first person to see bacteria. In those days, yeast was available in concentrated and purified form and used for baking, brewing, and wine-making.

1.2 Yeasts – The Secret Behind Alcoholic Fermentation

Yeasts are members of the largest and most varied phylum of fungi – *Ascomycota*.

Box 1.1 Biotech History: Leeuwenhoek

Antonie van Leeuwenhoek working with his single-lens microscope, capable of 200x magnification.

"What we saw was vigorous and rapid movement, similar to that of a pike through water. The creatures were low in number. A second species, looking like Fig.B, frequently spun round themselves like spinning tops and finally began to move as shown in Figs. C and D. These were more common. A third species did not seem to have a defined shape – they seemed to be oval-shaped, but then again, they looked like circles. They were so small – no larger than what you see in Fig. E – while moving amongst themselves like a swarm of flies or midges.

They were so numerous that I thought there were several thousands of them in the water (or in water mixed with spittle). The sample – which I had scraped out between my canines and molar – was no larger than a grain of sand. It mainly consisted of slimy masses of different lengths, but the same thickness. Some were curved, others straight, as in Fig. F, in no particular order."
(Source: Paul de Kruif, Microbe Hunters, 1930).

This description of a view of the deposit on his teeth under the microscope opened a whole new world to humankind. **Leeuwenhoek** (1632-1723) was merchant and a self-taught scientist. He was the first person to see bacteria and draw them. It all began when he watched a spectacle maker grind lenses at a fair and learnt how to do it himself. It became an obsession, and he began to make ever stronger glass lenses, building microscopes with up to 200x magnification. He could spend hours watching a hair from a sheep that looked like a strong rope under his single-lens microscope.

One day, Leeuwenhoek hit on the idea of looking at a drop of water from his rain barrel and had the most terrible shock when he saw it under the microscope. It was teaming with tiny creatures, swimming about as if playing with each other. Leeuwenhoek estimated their size was a thousandth of the eye of a louse.

Leeuwenhoek's first drawings of bacteria. The British science writer Brian F. Ford tried out one of Leeuwenhoek's microscopes and discovered that he was able to see spirilli.

Encouraged by a friend, Leeuwenhoek wrote an enthusiastic letter in Dutch to what was then the most prestigious organization of scientists, the Royal Society of London, in 1673. The learned gentlemen were most surprised to read a description of 'miserable wee beasties', as Leeuwenhoek chose to call them in his letter.

Leeuwenhoek's microscope. He built around 500 single-lens microscopes.

Robert Hooke (1635-1703) was a member of the Royal Society at the time and in charge of organizing new experiments for every meeting of the society. When looking at slices of cork through his own multi-lens microscope, he discovered a regular pattern of small holes which he named 'cells'. Hooke also built a microscope to Leeuwenhoek's specifications and was able to confirm the Dutchman's observations. Little did he know that the 'wee beasties' also consisted of cells, in most cases just one single cell.

The scientists of the Royal Society saw with their own eyes that microscopic creatures existed and became very excited about them. Leeuwenhoek who had never attended a university was unanimously voted in as a member of the Royal Society in 1680.

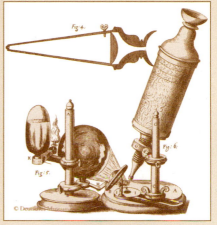

Robert Hooke's multi-lens microscope under which he examined thin sections of cork which he described in terms of 'cells'.

Through his dexterity, curiosity, and persistence, he had achieved more than many of his academic contemporaries who, when asked about the number of teeth a donkey had in its mouth, would much rather look it up in Aristotle's writings than look into a donkey's mouth.

Kings, princes, and scientists of all countries were interested in Leeuwenhoek's discoveries. The Queen of England, King Frederic I of Prussia as well as the Russian Czar Peter the Great (who had come to the Netherlands incognito to study the art of shipbuilding) all came and paid him a visit.

The "beasties" were a fascinating novelty for a while, but then were forgotten again.

Box 1.2 Modern Beer-Brewing

Beer brewing is most straightforward in Germany because it must comply with the **purity requirements** established in Bavaria in 1516 (see p. 26), i.e., it must not contain anything other than barley malt, hops, and water and must be prepared with yeast.

The starch contained in grain cannot be fermented straightaway, but must first be **broken down by enzymes** (**amylases**). These develop while the grain is left to germinate. In a process known as saccharification, they break down starch into maltose and glucose. Malting is the first step in beer-brewing, followed by the preparation of wort and fermentation.

In the **malting house**, barley is sorted and cleaned and left to soak in water for up to two days. The grain is then transferred to large germinating trays where it remains for seven days at a temperature of 59 to 64°F (15 to 18°C). The grain is turned automatically during this period. The germination process is then brought to a halt.

In the resulting **green malt**, the starch has only partially been broken down into maltose. The malt is now kiln-dried at slowly rising temperatures (initially 113°F/45°C), then rising to 140-176°F (60-80°C) and – for dark beer – up to 221°F (105°C). The result is called **brown malt**.

In the brewhouse, wort is prepared by mashing the shredded malt. Large quantities of water are stirred in, and the mixture is heated. The mashing process is interrupted at scheduled intervals or steps for **enzymatic breakdown** to take its course. Below 50°C, β-glucanases degrade gumlike substances that obstruct filtration.

The **protein breakdown** step is scheduled at 122-140°F (50 and 60°C), and the **saccharification step** is reached at 140-165°F (60-74°C) where starch-cleaving enzymes (α and β-amylases) break down the remaining starch into glucose, maltose, and larger starch molecule fragments (dextrins).

Once the mash has settled at the bottom of a **lauter tun** or the clear liquid has been strained through a mash filter, it is boiled, and hops are added to produce concentrated and germ-free **wort**. Hops contain bitter substances as well as resins and essential oils, which all contribute to the fresh bitter taste and better storability.

When the wort has cooled and taken up oxygen in a whirlpool, **pitching yeast** is added. These yeasts are pure cultures of *Saccharomyces cerevisiae* which are first encouraged to grow by the increased oxygen supply and then transferred to a fermenter.

Lager beers are the result of slow (8-10 days) **bottom fermentation**. The German word "Lager" means "storage", and indeed, these beers store very well and are very stable when shipped. The faster top fermentation (4-6 days) produces weissbier, ale, porter, and stout.

The beer is then left to **mature** in storage tanks over several weeks at a temperature of 32 to 35.6°F (0 to 2°C) before it is transferred into small kegs or bottles.

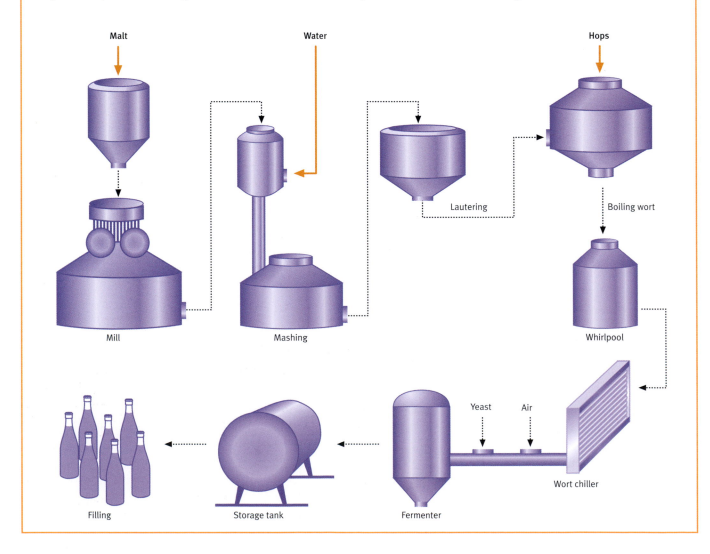

Malt

Water

Hops

Mill

Mashing

Lautering

Boiling wort

Whirlpool

Yeast

Air

Wort chiller

Filling

Storage tank

Fermenter

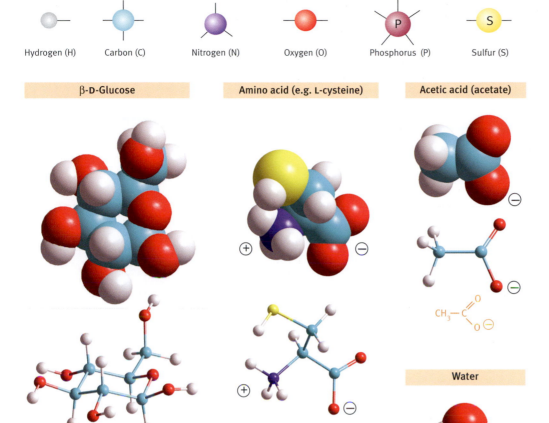

Space-Filling Models

Hydrogen (H) Carbon (C) Nitrogen (N) Oxygen (O) Phosphorus (P) Sulfur (S)

β-D-Glucose

Amino acid (e.g. L-cysteine)

Acetic acid (acetate)

Water

Fig. 1.5 Molecular representations of the various atoms and molecules shown in this book: structural formulas, ball-and-stick models, and space-filling representations are shown.

Unlike bacteria which are prokaryotes, yeasts have a complex eukaryotic cell structure, including a **genuine nucleus** and compartments such as mitochondria. They have also been called budding fungi in reference to their asexual propagation. Yeasts, however, can also multiply through sexual reproduction, i.e., through the copulation of two haploid budding cells, each of which contains a complete set of chromosomes. Yeasts are classified on the basis of their mode of reproduction.

Yeasts are single-celled organisms, also called mother cells. Each of them sprouts several buds which grow into viable daughter cells. These are then pinched off and can, in turn, form new cells (Fig. 1.3). They are heterotrophic, i.e., they depend on organic nutrients and cannot perform photosynthesis, preferring an acidic environment in their hosts. Their cell walls contain a substance other-wise found in insect skeletons – chitin – as well as hemicellulose. The alcoholic fermentation process in beer production depends on the carbohydrates in the grain. These are mainly polysaccharides and can only be digested by the glycolytic enzymes in the yeast cells (Fig. 1.15) if first broken up into disaccharides and monosaccharides.

■ 1.3 Now as Ever, Beer is Brewed From Yeast, Water, Malt And Hops

As in Sumerian times, germinating barley stands at the beginning of the brewing process since the germination process turns barley into **malt** by enhancing the production of certain enzymes.

The malt is then crushed, mixed with warm water, and filled into **mash tuns**. Within a few hours, the starch stored in the grain is broken down into malt-

Fig. 1.6 Beer-brewing in ancient Egypt.

Fig. 1.7 Beer – the carbon dioxide bubbles are clearly visible.

5

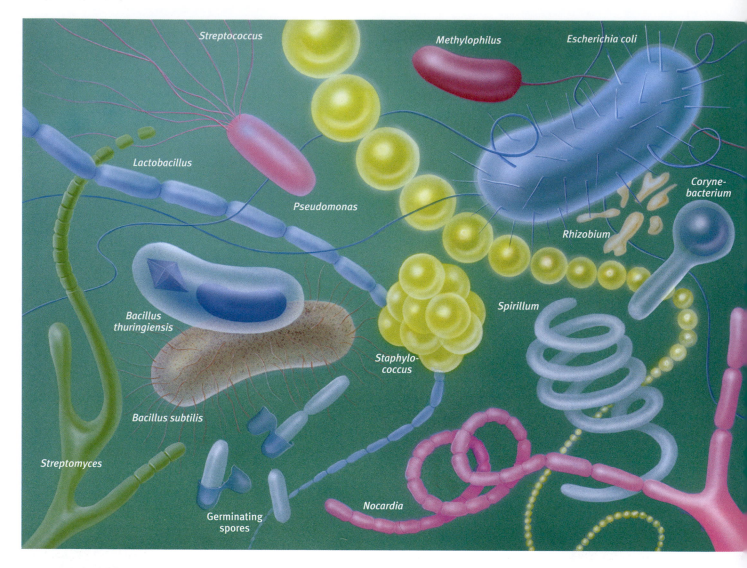

Streptococcus

Methylophilus

Escherichia coli

Lactobacillus

Coryne-
bacterium

Pseudomonas

Rhizobium

Bacillus
thuringiensis

Spirillum

Staphylo-
coccus

Bacillus subtilis

Streptomyces

Germinating
spores

Nocardia

Bacteria colony pattern formation.

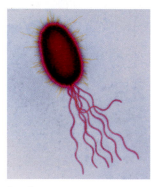

Pseudomonas

Fig. 1.8 Bacteria

Bacteria are prokaryotes, i.e., their genetic information is not contained in a nucleus, but is found in the cytoplasm – mostly as double-stranded DNA loops.

They are lacking organelles typically found in eukaryotic (yeasts, molds, higher animal and plant) cells, such as mitochondria (for respiration) or chloroplasts (for photosynthesis), and an endoplasmic reticulum.

Most bacteria are heterotrophic, i.e., they draw their energy from organic matter, whereas other species get their energy through photosynthesis or from inorganic (e.g., sulfur) compounds. Bacteria comprise mobile as well as immobile single-celled organisms (e.g., rods and cocci), but also multicellular filaments on the substrate as in *Nocardia* and the fungus-like structures (mycelia) of *Streptomyces* spp. in the air.

The color tables show biotechnologically relevant bacteria species. The structure of bacterial cell walls may vary (Ch. 4). According to their ability to take up stain that makes them visible under the microscope, bacteria are divided into Gram-positive and Gram-negative species.

Gram-negative aerobic rods and **cocci** include *Pseudomonas* spp. (utilization of hydrocarbons, steroid oxidation) and *Acetobacter* (production of acetic acid), *Rhizobium* (nitrogen fixation) and *Methylophilus* (Single Cell Protein, methanol oxidation).

By contrast, the intestinal bacterium *Escherichia coli*, the microbial equivalent of guinea pigs in research, are also Gram-negative rods, but **facultatively anaerobic**.

Bacillus (enzyme production) and *Clostridium* (acetone and butanol production) spp. are **spore-forming rods** and cocci. Club-shaped Gram-positive *Corynebacterium* spp. produce amino acids.

Lactic acid-producing *Streptococcus* spp., *Staphylococcus* (food poisoning),

Propionibacterium (vitamin B$_{12}$), cheese production), *Nocardia* spp. (hydrocarbon oxidation) and *Streptomyces* (antibiotics and enzymes) are all Gram-positive.

Lactobacillus spp. (lactic acid-producing) are Gram-positive, non-spore-forming bacteria.

While some bacteria are able to cause serious damage through disease and food spoilage, others have large economic potential in biotechnological processes.

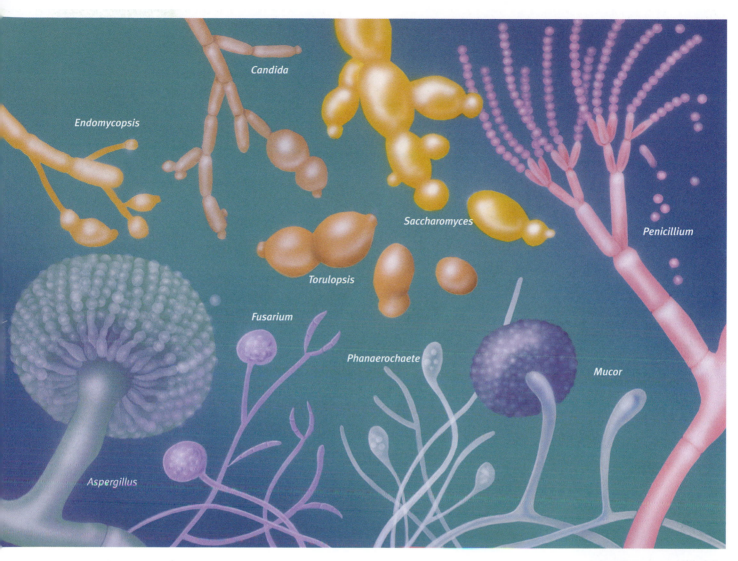

Fig. 1.9 Fungi

Fungi have an eminent role to play in the natural cycle, particularly in degradation processes. To date, around 70,000 fungal species have been classified. **Yeasts**, like all fungi, are eukaryotes with a nucleus that contains all genetic information.

They are **sprouting fungi** (*Endomycota*). We distinguish wild yeasts from cultured **yeasts** grown at an industrial scale, such as brewers' yeast (e.g., *Saccharomyces carlsbergensis*), wine and bakers' yeast (*S. cerevisiae*) or fodder yeast (*Candida*). *Candida utilis* is grown on the sulfite-containing effluents of paper mills. *Candida maltosa* feeds on alkanes (paraffins) of crude oil, producing fodder protein. *Trichosporon cutaneum* degrades phenol in effluents which would be toxic to other fungi. *Trichosporon* and the yeast *Arxula adeninivorans* are used in microbial effluent sensors (Ch.s 6 and 10).

Mold fungi are members of the largest group of fungi, **Ascomycota**, comprising 20,000 different fungi. Unlike the spherical yeasts, their cells are long and thin, and most of them are strictly aerobic. Mold spores are formed asexually by division of the cell nuclei in the mycelia. These mycelia extend into the air and are called sporangia. The mature spores travel easily in the air, and when they land on a suitable substrate, they germinate to form new mycelia. Ascomycota are further subdivided according to morphology and color of asci.

Their mycelia all look pretty much the same – a tangle of threadlike hyphae. Many of the industrially cultured fungi are grown in submerged cultures in tanks where they produce clumps of mycelia, but no spores. They require nutrients similar to those of yeast, but are more versatile. Unlike yeasts, some of them can thrive on cellulose (*Trichoderma reesei*) or lignin (*Phanaerochaete chrysosporium*) (Ch. 6).

Fungi of the *Aspergillus* and *Penicillium* genera are used in fermentation, especially in the enzymatic degradation of starch and protein in barley, rice, and soy beans, while *Aspergillus niger* produces citric acid. Penicillin is obtained from *Penicillium chrysogenum* (Ch. 4). Other *Penicillium* species are used in the production of certain types of cheese, such as Camembert or Roquefort.

Amylases and proteases produced by fungi are also harvested for industrial enzyme preparations (Ch. 2).

Endomycopsis and *Mucor* also produce industrial enzymes, while *Fusarium* species are used to obtain protein for human consumption (see Quorn, Ch. 6)

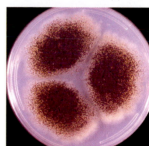

Aspergillus niger on agar.

Penicillin-producing mold (*Penicillium notatum*).

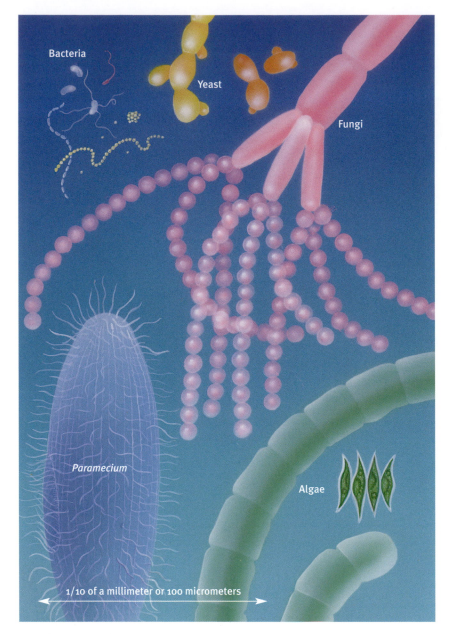

ose, glucose, and various other sugars by virtue of enzymes called amylases. Other enzymes (β-**cellulases**) break down the outer layers of the barley grain, thus making the starch inside available to α-**amylase**. In a next step, the solid components of the mash are filtered out and the sweet liquid transferred to the copper tanks (coppers) where hops are added. Hops give the mixture a bitter taste.

The mixture is now called **wort** and poured into a fermentation vessel. Brewers' yeast is added to set off the fermentation process.

Once the alcoholic fermentation process is completed, the beer is allowed to mature in large tanks. Eventually, the beer is briefly heated to kill off any noxious microbes and then filled into bottles, cans, or kegs.

While the basic processes in beer-brewing have remained virtually the same for thousands of years, our knowledge has expanded. Our ancestors were unaware of the fact that they were harnessing microbes for their purposes.

Similar discoveries to those of the Sumerians were made by people all over the world. **Winemaking** probably developed 6000 years ago in the Mount Ararat region. However, recent findings point to the Chinese as the first winemakers – 9000 years ago (Box 1.9).

The ancient Greeks and Romans were lovers of fermented grape juice or **wine** (Box 1.3), and it was the Romans who brought the art of wine-growing

Fig. 1.10 True to scale reproduction of biotechnologically relevant microorganisms. The length of a *Paramecium* is roughly equivalent of the diameter of a human hair – one tenth of a millimeter or 100 micrometers.

Fig. 1.11 right: Size ratio of eukaryotic and prokaryotic cells and viruses.

Fig. 1.12 far right: How many molecules does a single bacterial cell contain?

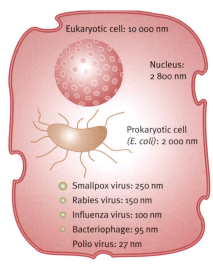

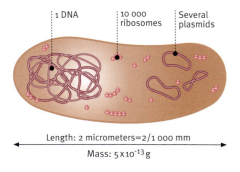

Number of Molecules and Proportion of Total Mass of the Cell

Water	10^{10}	80.0%
Proteins	10^6-10^7	10.0%
Sugars	10^7	2.0%
Sugars	10^8	2.0%
Amino and organic acids	10^6-10^7	1.3%
DNA	1	0.4%
RNA	10^5-10^6	3.0%
Inorganic matter	10^8	1.3%

Box 1.3 Wines And Spirits

In wine preparation, grapes are first crushed in a grape mill. For **white wine**, the grapes are immediately pressed and mashed, and the juice is separated from stalks, grape skin, and pips, also known as pomace. More and clearer juice can be extracted if **pectinases** (Ch. 2) are added. For **red wine**, the mash is fermented straightaway, as only the fermentation process makes the anthocyans in the skin of the grapes dissolve and color the juice red. The mash is only pressed after 4 or 5 days.

The fermentation process is brought on by yeasts adhering to the outside of the grapes, or – the safer bet – by inoculation with pure **yeast cultures** (*Saccharomyces cerevisiae* strains). Foam develops in the process. This frothy, cloudy mixture is very popular in many regions of Southern Germany, served with onion tart in the autumn.

During the **main fermentation** process which lasts four to eight days, nearly the whole sugar content is used up. Proteins and pectins do not dissolve, but precipitate to form sediment that needs to be separated from the wine. Over the first twelve months, the wine is stored in cool cellars where the remaining sugar undergoes slow fermentation. Again, sediment forms, and the wine develops its particular flavor or bouquet. At the end of the fermentation process, the young wine is transferred into firmly bunged-up storage casks that have been treated with sulfur. During the final **maturation** process, it may be exposed to air at regular intervals. This is the time for cellar treatment which involves enhancing the storability of wine, e.g., by using sulfur, as sulfur dioxide is more toxic to bacteria than to

yeast. Most wines have an ethanol content of 10 to 15%.

Wines can be distinguished by color (mostly red and white), origin and grape variety (e.g., Riesling, Pinot noir, Chardonnay, Syrah etc). To obtain some **residual sweetness**, the fermentation process can be interrupted or some must (juice of freshly pressed grapes, prior to fermentation) can be added. Dry wine contains up to 9 grams/liter, semi-dry wine up to 18 grams/liter, and anything above 18 grams/liter is considered sweet. There are also **fortified wines** to which sugar, ethanol, and sometimes herbs have been added, but no further microbial action is involved.

Vineyard in Stellenbosch near Cape Town (South-Africa).

Champagne and other sparkling wines

undergo two fermentation processes, and CO_2 is finally sealed into the bottle. At the beginning of the second fermentation, sugar syrup is added to the mix of selected wines (assemblage) which is filled into bottles inoculated with specific champagne yeast. During the slow process lasting for several months, the bottles are riddled – i.e., turned and gradually tilted until they stand on their heads and the yeast and sediment (lees) collect in the neck. The cap that has so far sealed the bottle is taken off, the sediment removed (disgorgement), and before it is replaced by a proper wire-hooded champagne cork, sugar syrup and brandy are added. This results in a noble and stable wine that will give off fine bubbles long after it has been opened, in contrast to those **sparkling wines** where CO_2 is added to a still wine under pressure.

Spirits include brandy, liqueur, punch, and cocktails. Cognac or Armagnac, rum, arrack, whisky, and fruit schnapps are sometimes called noble spirits. The aromatic byproducts of the fermentation process (esters, higher alcohols, aldehydes, acids, acetates etc) remain at least partially in the distillate to add to the taste. The fermentation of starch products does not produce a high level of fusel alcohols.

Ordinary **spirit drinks** are mostly prepared by mixing primary spirit with water and aromatizing

substances (e.g., aniseed, fennel, caraway or, juniper). They must contain at least 32% ethanol. **Grain spirits** are made from rye, wheat, buckwheat, oats, or barley. In **whisky** (at least 40% ethanol production) some peat is often added to the kiln when drying the malt to give it a distinctive smoky taste. **Vodka** has an ethanol content of 40-60% and is made from rye, potatoes, or other starchy plants. It is produced from mash through continuous distillation. For the preparation of **gin**, Steinhäger etc, juniper berries or alcoholic juniper extracts are added to the grain mash.

Fruit brandies contain at least 38% ethanol and are distilled from fermented fruit without the addition of sugar, more ethanol or coloring. They include calvados and grappa, whereas kirsch, plum (slivovitz), apricot (marillen), peach spirits are made from fresh fruit to which alcohol is added. Brandy (at least 38% alcohol) must be distilled from wine or fermented fruit juice.

Cognac is the name of distilled wine made from grapes originating from the French départements Charente Maritime, Charente, Dordogne, and Deux Sèvres. There is also Armagnac, and countries like Germany, Greece, Italy and Spain also produce wine-based brandies. **Rum** is distilled from fermented sugarcane debris and has an alcohol content of 38%. The basis for **arrak** is rice or the sap of coconut palm flowers.

To prepare **liqueurs**, spirits are spiked with sugar and aromatic fruit and plant extracts. **Punch** (derived from the Hindi word *punsha* – five) is a hot drink made of five ingredients: ethanol, spices, lemon juice, sugar, and some tea or water. **Cocktails** are ethanol-containing appetizers and owe their name to the American War of Independence when the skill was developed to fill a glass with several layers of different liquids without mixing them so that they formed layers like the feathers on a cock's tail.

The author exploring the (empty!) fermentation tanks in Stellenbosch.

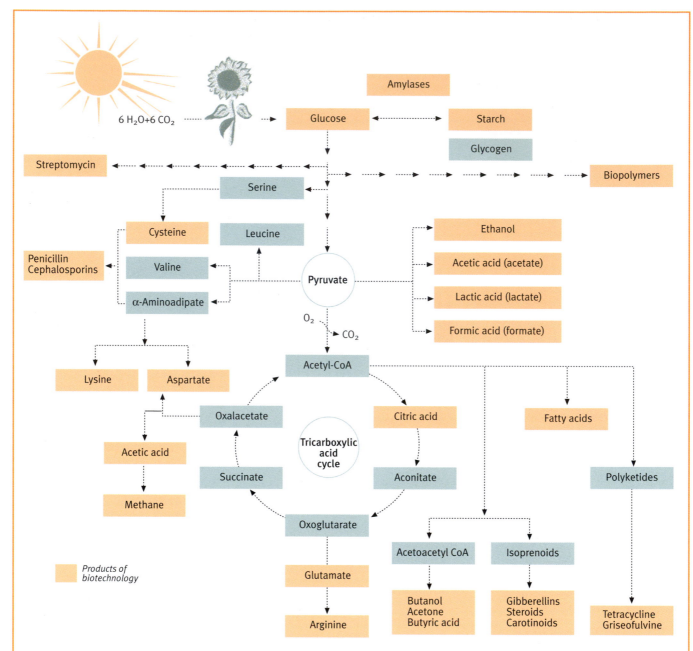

6 H₂O+6 CO₂

Amylases

Glucose

Starch

Glycogen

Streptomycin

Biopolymers

Serine

Cysteine

Leucine

Penicillin
Cephalosporins

Valine

Ethanol

Acetic acid (acetate)

Lactic acid (lactate)

Formic acid (formate)

α-Aminoadipate

Pyruvate

O₂

CO₂

Acetyl-CoA

Lysine

Aspartate

Oxalacetate

Citric acid

Fatty acids

Acetic acid

Tricarboxylic
acid
cycle

Succinate

Aconitate

Polyketides

Methane

Oxoglutarate

Acetoacetyl CoA

Isoprenoids

Products of
biotechnology

Glutamate

Butanol
Acetone
Butyric acid

Gibberellins
Steroids
Carotinoids

Tetracycline
Griseofulvine

Arginine

Box 1.4 The Transformation of Glucose

The breakdown of glucose (*glycolysis*) is initiated in the cell through enzymatic transfer of a phosphate group from the cofactor adenosine triphosphate (ATP) to the glucose molecule (Fig.1.15). While this process uses up energy, the subsequent steps of the metabolic process will continually liberate energy.

Cofactors nicotinamide adenine dinucleotide (NADH) and **ATP** are essential regenerative compounds that work in conjunction with enzymes (see Ch. 2). The original glucose molecule contains six carbon atoms and is enzy-

matically split into two 3-carbon molecules of **pyruvate** (pyruvic acid). Pyruvate plays an essential role in metabolic processes, as it is oxidized to a large extent to produce activated acetic acid or acetyl coenzyme A. The **CO₂** released in the process is the first metabolic end product, leaving two carbon atoms in coenzyme A. A small portion of pyruvate is turned into oxalacetate, which sets off the **tricarboxylic acid cycle**, also known as citric acid or Krebs cycle, named after Sir Hans Krebs who was awarded the Nobel Prize for his research into it. The two carbon atoms in acetyl coenzyme A enter the cycle to be oxidized to two CO₂ molecules. Intermediate products of the cycle, such

as citric acid, are only produced in low concentration and remain in a dynamic equilibrium (Ch. 4). During the breakdown of glucose, energy is released through dehydration, involving the oxidized cofactor NAD⁺ to which hydrogen is temporarily transferred.

The thus reduced cofactor NADH must get rid of the hydrogen in order to be available for energy-producing reactions. This happens in the mitochondria through enzyme reactions in the **respiratory chain**. The oxygen molecule in NADH is used by the enzyme **cytochrome oxidase** (Fig. 1.14 and Ch. 2) to produce water in a very energy-effective cold combustion process.

and winemaking to what is now Germany and Britain.

In Africa, **pombe** beer was produced using *Schizosaccharomyces pombe*, while tribes of Central Asia made **kumys**, a drink out of fermented horse milk, which they carried around in leather bags. The Japanese prepared **sake**, a rice-based alcoholic drink, and the Russians made **kvass** using *Lactobacillus* species and the mold *Aspergillus oryzae* for saccharification.

In **wine production** (Box 1.3), too, very little has changed. Red and white grapes are harvested and crushed, then pressed in order to obtain juice. The juice is then kept in sealed vessels – traditionally wooden casks, but nowadays, metal tanks with a capacity of up to 250,000 liters (66,000 US gallons). The world production of wine lies at half a billion hectoliters (13.2 billion US gallons) per annum.

■ 1.4 Cells Work On Solar Energy

Every day, the sun sends four million billion kWh of energy to the earth, delivering all the energy that all the inhabitants of the earth need in a year within 30 minutes for free. Only three thousandths of this energy are turned into chemical energy by green plants through **photosynthesis.**

In the chloroplasts, energy-charged photons cleave water into its components oxygen (100 billion metric tons/110 billion US tons per annum) and hydrogen. Only the **oxygen** molecules are released into the atmosphere and made available to all organisms relying on aerobic **respiration** or "cold combustion". The hydrogen released during photosynthesis is temporarily bound to carbon (220 billion US tons per annum) in carbohydrates (sugars).

Carbohydrates are the main product of photosynthesis and also the most important source of energy for most living organisms.

When hydrogen is severed again from its carbon bonds, it is only gradually released in what could be called a "biochemical detonating gas reaction", producing energy on its way through the respiratory chain. All this energy has its origins in the sun!

Cells need a constant supply of energy, whether they are growing or quiescent. This energy comes from the nutrients taken in and broken down by the cell's **metabolism** in a controlled cascade of enzymatic

reactions along specific metabolic pathways. The products of these reactions provide **building blocks** and **energy** for syntheses and other energy-intensive processes.

Nutrients are first broken down into smaller components and then transformed into low-molecular compounds that form the **building blocks of cells** – glucose, amino acids, nucleotides, (pyrimidine and purine bases and sugar phosphates), organic acids and fats.

These are the components that build the giant molecules of proteins, of nucleic acids (DNA and RNA), polysaccharide reserve materials and cell wall components.

Since photosynthesis produces mainly carbohydrates, providing the nutritional basis for most organisms, the primary starting material for metabolic processing is **glucose**. The various ways in which glucose can be broken down in a cell open up several pathways for biotechnological production.

■ 1.5 For Yeast, Alcohol Has Nothing to Do with Enjoyment, But All With Survival

Our biochemical knowledge tells us that yeasts are facultative aerobic organisms – they can live either aerobically **through respiration** or anaerobically **through fermentation**. Yeast thrives in the presence of oxygen, turning sugar into CO_2 and water through respiration while extracting the energy needed to grow and form new cells.

When the air supply is cut off, yeast cells switch their metabolism to an **anaerobic emergency mode**. Fermentation enables them to survive in a hostile environment, even though it is energetically inefficient. In 1861, **Louis Pasteur** discovered that in the absence of oxygen, yeast uses up more glucose than in its presence. This is known as the **Pasteur effect**. In an anaerobic environment, yeasts continue to process glucose molecules at an even higher rate than in an aerobic environment in order to compensate for the energy loss. In the absence of oxygen, the respiration chain cannot be completed to produce NADH + H+, but grinds to a halt at the pyruvate stage.

The cell transforms pyruvate into acetaldehyde, releasing CO_2. Due to the lack of oxygen, the accumulating NADH + H+, which would otherwise be

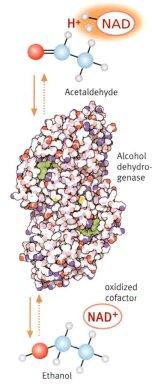

Fig. 1.13 Alcohol dehydrogenase from yeast converts acetaldehyde into ethanol. In bioanalysis (Ch. 10), the reverse reaction is used to detect ethanol levels in blood.

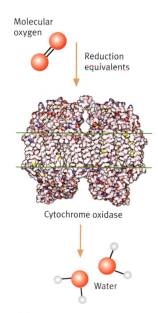

Fig. 1.14 Cytochrome oxidase, the respiratory enzyme discovered by Otto Warburg. In a cold combustion process, hydrogen is turned into water and energy released.

Fig. 1.15 Glycolysis and the enzymes involved in the process: Degradation of glucose into pyruvate (simplified).

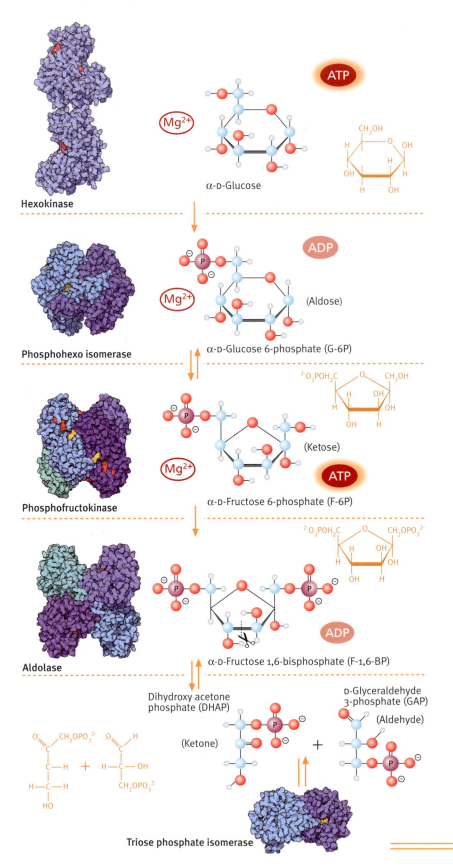

Hexokinase

ATP

Mg²⁺

α-D-Glucose

Phosphohexo isomerase

ADP

Mg²⁺

(Aldose)

α-D-Glucose 6-phosphate (G-6P)

Phosphofructokinase

Mg²⁺

(Ketose)

ATP

α-D-Fructose 6-phosphate (F-6P)

Aldolase

ADP

α-D-Fructose 1,6-bisphosphate (F-1,6-BP)

Dihydroxy acetone phosphate (DHAP)

D-Glyceraldehyde 3-phosphate (GAP)

(Ketone)

(Aldehyde)

Triose phosphate isomerase

re-oxidized into NAD^+, cannot be used up as usual in the citric acid cycle. The only alternative is to turn acetaldehyde into ethanol and NAD^+ by virtue of **alcohol dehydrogenase** (Fig. 1.13). In other words, glucose is degraded into alcohol and CO_2 in an incomplete combustion process. The fermentation of a glucose molecule yields only one to four ATP molecules, compared to up to 38 molecules resulting from respiration. However, it is efficient as a survival strategy.

Producing alcohol is therefore not as pleasurable an activity as humans like to think, but an emergency strategy for yeasts. Once the alcohol exceeds a certain level, the yeast will perish. In contrast to the traditional view, it is possible to produce ethanol aerobically (Crabtree effect) if the glucose content exceeds a glucose/nutrient medium ratio of 1:10,000. In the resulting overflow reaction, pyruvate is not oxidized via the citric acid cycle, but reduced to ethanol.

Modern **gene chips** (Ch. 10) for mRNA (a single-stranded nucleic acid that carries the protein building instructions from nuclear DNA to ribosomes, see Ch. 3) have been used to demonstrate that when a shortage of oxygen occurs, anaerobic yeasts suddenly switch to producing a set of enzymes that varies largely from that of aerobic yeasts. The oxygen supply determines which set of genes is read and what proteins will be expressed.

Other organisms produce **lactic acid** (*Lactobacillus*), **butyric acid** (*Clostridium butyricum*), **propionic acid** (*Propionibacterium*), **acetone** and **butanol** (*Clostridium acetobutylicum*) (Ch. 6) and other compounds as fermentation byproducts while synthesizing ATP.

These byproducts are usually **produced in large quantities** because only a large turnover of nutrients can ensure a viable supply of ATP for the cells in the absence of oxygen.

What seems to have been a drawback for the microorganisms proved to be an advantage in the human exploitation of fermentation processes – alcoholic and lactic fermentation yielded large quantities of human produce in a short space of time. From a human perspective, this was highly effective.

1.6 Highly Concentrated Alcohol Is Obtained by Distillation

If the alcohol produced by yeast exceeds a certain concentration, it will kill the yeast, which is why beer and wine have a fairly low alcohol content. Wine contains 12-13%, sake even 16-18% alcohol.

High-alcoholic drinks such as whisky or brandy were probably not known before the 12th century when wine was heated in closed kettles.

Alcohol distillation was possible because the temperature at which alcohol evaporates (172.4°F/78°C) is considerably lower than the boiling point of water (212°F/100°C). The alcohol vapors were collected in a pipe immersed in cold water where they cooled down and condensed into droplets. The highly concentrated alcohol was collected in a vessel.

Distilled drinks containing al least 64 proof (32 Vol%) ethanol (Box 1.3) are called spirits. They include brandy, whisky, schnapps, vodka, and gin.

Cognac was developed by vintners in the French region of Charente who realized that their wine could not compete with that of the neighboring region of Bordeaux, so they hit on the idea of distilling their wine. Later, it was even distilled twice in succession, and even nowadays, the young Cognac that is stored in casks of Limousin oak for several years has an alcohol content of 70%. Only when it has matured and developed the right color and flavor is it diluted to 40%.

Pure alcohol is extracted from grain and potato starch in modern distilleries. Amylases turn the starch into sugar which is then fermented into alcohol by yeast. The alcohol is heated and distilled to a concentration of 96%. Everybody has heard of the disinfecting powers of alcohol, and indeed 70% alcohol is used externally for medical disinfection.

Even in concentrations of 2-3%, alcohol increases the permeability of the cytoplasmic membrane in bacteria, thus hampering their growth. It is thus hardly surprising that highly concentrated alcohol has even better inhibitory properties. Fruit preserved in alcohol, for example, keeps a long time, demonstrating the microbe-containing effect of alcohol. Most microorganisms responsible for food poisoning or rot are alcohol-sensitive and will not proliferate when exposed to it.

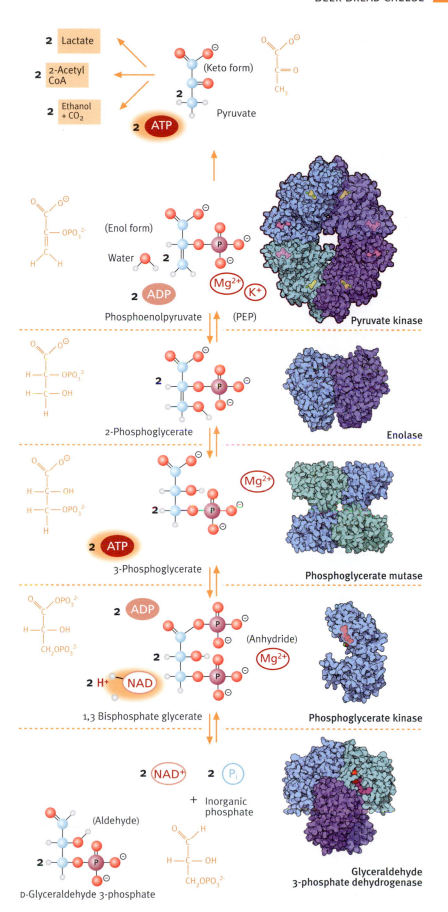

2 Lactate

2 2-Acetyl CoA

2 Ethanol + CO₂

(Keto form)

2 Pyruvate

2 ATP

Pyruvate kinase

(Enol form)

Water

2 ADP

Mg²⁺ K⁺

Phosphoenolpyruvate   (PEP)

Enolase

2-Phosphoglycerate

Mg²⁺

Phosphoglycerate mutase

2 ATP

3-Phosphoglycerate

2 ADP

(Anhydride)

Mg²⁺

2 H⁺ NAD

1,3 Bisphosphate glycerate

Phosphoglycerate kinase

2 NAD⁺   **2** Pᵢ

+ Inorganic phosphate

(Aldehyde)

Glyceraldehyde 3-phosphate dehydrogenase

2 D-Glyceraldehyde 3-phosphate

Box 1.5 Products Derived from Sour Milk

Sour milk products are the result of lactic fermentation.

Camembert preparation.

Sour milk (thick milk) is prepared from pasteurized milk inoculated with *Streptococcus cremoris*, *Str. lactis* and *Leuconostoc cremoris* to ensure a pleasant aroma develops during 16 hours of tank fermentation.

Camembert, the result of *Penicillium* species at work.

Yoghurt is made from goat's, sheep's, or cow's milk and contains a symbiotic culture of thermophilic lactic bacteria (*Streptococcus thermophilus* and *Lactobacillus bulgaricus*). *Lactobacillus* provides a cleavage product of casein that stimulates the growth of *Streptococcus*, while the formic acid produced by *Streptococcus* has preservative properties.

Kefir is a thick, slightly fizzy drink containing 0.8% to 1% lactic acid as well as 0.3 to 0.8% ethanol and carbon dioxide. In conjunction with small amounts of diacetyl, acetaldehyde, and acetone, these compounds all contribute to the refreshing taste of kefir. Kefir grains are cauli-

Local cheeses on the Mediterranean island of Corsica.

flower-like structures containing yeasts that ferment casein and lactose (*Saccharomyces kefir* and *Torula kefir*) as well as *Lactobacillus, Streptococcus, Leucostoc* species, and acetic acid-producing bacteria.

Kumys is made from horse milk fermented with a mixture of lactic acid bacteria and yeasts.

Sour cream butter (cultured butter) is obtained through microbial fermentation. The cream is skimmed off and heat-treated, then cooled and "matured", during which process the butter fat crystallizes. After inoculation with acid-forming bacteria (*Streptococcus lactis, Str. cremoris,* and *Leuconostoc cremoris*), lactose is fermented into lactic acid, diacetyl, and acetoin. The butter is then churned to remove surplus liquid or buttermilk.

Roquefort with air channels for mold growth.

In **cheese preparation**, pasteurized milk is inoculated with starter cultures of lactic acid bacteria and mold fungi. Rennet is added and makes the milk curdle within the hour. The thickened milk is then carefully cut into half-inch lumps, the whey is discarded and the curd filled into molds.

Emmental cheese is turned several times over two weeks, soaked in salt water, and dried before

being stored in a cellar for six to eight weeks, during which time Propionibacteria are fermenting lactic acid into CO_2 and propionic acid, giving the cheese its characteristic taste and "holey" structure. The cheese is then left for six months to mature.

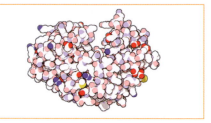
Rennin (chymosine) molecule.

After drying, **Limburg** cheese is brushed several times with a *Brevibacterium linens* culture to produce the red slimy layer on top of the cheese.

The famous holes in the cheese: product of propionic acid bacteria.

Camembert is inoculated with spores of either the fast-growing *Penicillium caseicolum* or the traditional *P. camemberti* and stored in a dry cellar where the mold grows within three to four days. After nine to eleven days, the cheese is ready to be packed and sold and may only reach its full maturity in the home of the end consumer. The protein-cleaving enzymes in the mold fungi soften the cheese, releasing aromatic compounds and pungent ammonia.

Real **Roquefort** is made from fresh unpasteurized sheep's milk. The cut curd is inoculated with *Penicillium roqueforti*. Once pressed into moulds, the cheeses are brought to Roquefort from all over France, where they are pierced with skewers to open up air channels for the mold to grow in. The cheese matures exclusively in the natural cellars or caves in a mountain in Roquefort. The ripening process takes 20 days under aerobic conditions, followed by three months under anaerobic conditions, wrapped in tin foil. The protein and fat-cleaving enzymes continue to be active.

Box 1.6 Sake, Soy Sauce And Other Fermented Asian Produce

The production of Japanese **sake** resembles the beer-brewing rather than the wine-making process because rice is rich in starch that needs to be broken down by amylases into fermentable sugars. Mold fungus spores are added to cooked rice which is then kept at a temperature of 95°F (35°C) for 35 days to produce *koji*. Koji is rich in amylases secreted by the fungi and is added to larger batches of rice and starter cultures (moto) of yeast strains such as *Saccharomyces cerevisiae*. Left to ferment for three months, the rice is turned into sake with an alcohol content of about 20%.

The production method for **soy sauce** (*shoyu* in Japan, *chiang-siu* in China, *siau* in Hong Kong) is very similar. *Moromi* is made from soy beans, wheat and koji plus large amounts of salt. It is left to ferment for eight to twelve months, using *Aspergillus soyae* and *A. oryzae*. Starter cultures of the bacterium *Pedicoccus soyae* and the yeasts *Saccaromyces rouxii*, *Hansenula* and *Torulopsis* are added to produce lactic acid and alcohol. At the end of the fermentation process, the soy sauce is strained, and the solid remains are used as cattle feed. Soy sauce contains not only 18% salt, but also 1% of flavor-enhancing amino acid salt **monosodium glutamate** (see Ch. 4) and 2% alcohol.

Miso is fermented soy paste, one of the major traditional Japanese sources of protein. Again, koji is involved in the preliminary production stages. **Tofu** (also called **sufu**) is obtained from acid-curdled soy protein fermented by *Mucor sufu*.

Sake barrels, piled up in front of a Japanese temple.

Natto with its pungent ammonia smell is made from steamed soy beans rolled into a slither of pine wood inoculated with koji (*Aspergillus oryzae*) and left to stand for several months before it undergoes a second fermentation with *Streptococcus* and *Pediococcus* bacteria.

Angkak (red rice) is made from steamed rice and the fungus *Monascus purpureus*. It is used as a hot spice and pigment in China, Indonesia, and on the Philippines.

Indonesian **Tempeh** is produced from cooked soy beans inoculated with *Rhizopus* mold and wrapped in banana leaves.

Soy beans fermented with mold fungi – natto.

Soy sauce with *wasabi* (grated horse radish containing peroxidase, s. Ch. 2).

In the early days of humankind, alcohol certainly had a part to play in preserving food and making it safe.

1.7 Bacterially Produced Acidic Preservatives

Bacteria are prokaryotes and about a tenth of the size of yeast cells. To give you some idea of the size of bacteria, imagine a tiny cube, the sides 1 mm long (1 cubic millimeter), filled with no less than a billion bacteria.

Bacteria (Fig. 1.8) are often rod-shaped *bacilli*. Some are spherical – **cocci** (Greek *kokkus*, spherical nucleus), some are comma-shaped and constantly vibrating **vibriones**, some spiral-shaped **spirilli**. Many bacteria carry flagella, long tails that act as propellers. Most bacteria reproduce by splitting in half. The daughter cells mostly detach themselves, but can also form chains. The chain-forming bacteria are called **streptococci** (Greek *streptos*,

chain). Others stick together in grape-like bundles and are known as **staphylococci** (Greek *staphyle*, grape).

Alongside yeast, there is a wide range of bacteria involved in the production of various foods. Some of these are even suitable for alcohol production. *Zymomonas mobilis* was used by native Mexicans for centuries to prepare **Pulque** (better known as **Tequila**, which is the local variety of the town of Tequila), by fermenting agave sap and palm wine. Only recently has it been discovered that the bacterium grows very well in high-sugar media, at about six to seven times the growing pace of the best yeast strains.

Bread-baking, as we know it, probably developed later than brewing. The first known bread was firm and flat. A fluffier type of bread was made from fermented dough (sourdough) by Egyptian bakers about 6000 years ago.

Fig. 1.16 Professor Robert Koch ensures his fungi will be 'cultured' (contemporary cartoon).

Fig. 1.17 Bread is the result of lactobacilli and yeasts at work.

Fig. 1.18 Medieval bread-baking.

Fig. 1.19 Modern bread-baking on an industrial scale. The basic processes remain unchanged.

Pasteurizing

It was Louis Pasteur who discovered that heating up wine very briefly was enough to kill off any bacteria that might spoil it. The method was also suitable for preventing milk from turning sour. In honor of its inventor, the process that kills off most microorganisms present in a product was named pasteurization. Even raw milk produced to high hygienic standards contains 250,000 to 500,000 microbes per ml (cubic centimeter). Milk sold in supermarkets is therefore usually briefly heated to between 160 and 165 °F (71 and 74 °C), i.e., it is pasteurized, killing off 98 to 99.5% of microorganisms contained in it. UHT milk, which can keep four weeks without refrigeration, has been briefly heated over steam to 248 °F (120 °C) and is then filled into pasteurized containers.

Sourdough like the world-famous San Francisco Sourdough contains a mixture of *Lactobacillus* varieties and acid-tolerant yeasts, i.e., yeasts that can live in a neutral as well as in an acidic environment. Species include *S. cerevisiae, Candida humilis,* and *Pichia saitoi.*

The **aroma and taste** of bread are the result of byproducts of the fermentation process, such as alcohol, acetic acid, lactic acid, 2-acetyl-1-pyrroline, dicarbonyl, and fusel alcohols (for details, see Box 1.7).

The desired product in bread-baking fermentation is not alcohol, but **carbon dioxide** (CO_2) to lighten the dough with its gas bubbles. The fermentation process is fed by the glucose and **maltose** produced by amylases from the starch contained in the flour, and by added free sugars such as **glucose, fructose,** and **saccharose** (**sucrose**). It comes to an end during the baking process when the heat kills off yeasts and bacteria. Any alcohol formed in the process evaporates, and all that remains in the baked bread are the honeycomb-like hollow spaces formed by the CO_2 bubbles.

Commercial sourdough bread is made from a mixture of flour, yeast, salt, and water plus ready-made sourdough. The dough is kneaded and left to rise for several hours before it is cut into loaf-size portions by a machine. These portions are left to rise again, then rolled up and put in tins.

Another rising process follows before the dough is put in the oven. After baking, the crisp loaves are taken out of the oven to cool down. White bread and cakes are leavened with yeast only, without any lactic acid fermentation.

Bakers' yeast (*Saccharomyces cerevisiae*) is often grown on molasses, a residue from sugar beet or sugar cane production, and 1.5 million metric tons (1.65 million US tons) worth $630 million, are sold worldwide every year.

When humans began to domesticate sheep, goats, and cows and use their **milk**, they discovered that milk (Box 1.5) seemed to turn sour by itself when left standing for some time. They also discovered that milk that had been boiled did not go off as quickly. In fresh milk straight from the animal, lactobacilli proliferate rapidly and ferment lactose into **lactic acid** (**lactate**). The result is a tasty, storable, and nutritious product. The acidic environment inhibits the growth of putrefactive and pathogenic microbes, such as staphylococci.

When milk turns sour, the protein **casein** precipitates into small flakes and becomes easier to digest. Lactic acid also undergoes a reaction with calcium in the stomach, resulting in calcium lactate, which can then easily be absorbed through the intestinal walls. This ensures that calcium, so essential for our bone structure, is not wasted. The differences in climate and milk properties all over the world have led to a wide variety of milk treatment traditions. In Europe, we have **curdled or thick milk**, and **curd, quark** or **fromage frais**, in the Balkans and the Middle East **yoghurt**, in the Caucasus **kefir**, in Central Asia **kumys**, in India **dahi** and in Egypt **laben**.

When **butter** is produced, milk is first pasteurized (see side column on the left and Ch. 4), the cream is separated and inoculated with a mixed culture of lactobacilli (cultured butter). The cream matures in special vessels for 16 to 30 hours during which the bacteria produce acids and acetoin. When acetoin oxidizes, the typical butter flavor (diacetyl) develops. When butter is churned, **buttermilk** forms as a byproduct.

According to German beliefs, sour food lifts up spirits – well, at least a certain amount of acidity in the food can be pleasant. **Lactic fermentation** of **cabbage** and **cucumbers** is an ancient preservation method. Gherkins are also often preserved together with other vegetables (mixed pickles) and are often recommended as a hangover cure. **Olives** that have been soaked in a weak sodium hydroxide solution to remove bitter-tasting oleuropin are then preserved with lactobacilli. Unripe olives are green, ripe olives black. The lactobacillus involved in these processes is mainly *Lactobacillus plantarum*.

In **sauerkraut** production, finely chopped white cabbage is layered with salt (spices optional) and crushed. The liquid emanating from the crushed plant cells must cover the cabbage completely to exclude air. Stored in a cool place, the cabbage will ferment and turn into delicious sauerkraut. Sauerkraut will keep because lactic and acetic acid ensure that the environment remains far too acidic for rot-promoting microbes to develop. Nowadays, sauerkraut is often produced in containers holding up to 100 US tons. It is ready to eat within seven to nine days and often briefly heated (blanched) to kill off the lactobacilli and end the fermentation process to make sure the sauerkraut will retain its mild taste.

Not only humans, but also cattle love acidic food – they adore **silage**. Fresh forage, corn, and turnip leaves are shredded and tightly packed into a silo for fermentation. This provides reliable and nutritious winter feed.

In the early stages of silaging, aerobic bacteria use up the available oxygen. This environment of diminishing oxygen supply and increasing acidity makes *Lactobacillus, Streptococcus,* and *Leuconostoc* species thrive.

Where the formation of lactic acid is insufficient, **butyric acid-producing bacteria** (*Clostridium butyricum*) will take over, giving off a tell-tale acrid smell. As a country boy I learned that such silage should never be fed to cattle if you want to spare them severe indigestion. They produce incredible amounts of gas then...

Dry and semi-dry sausages are another kind of food that may be preserved by lactic acid fermentation, e.g., salami, mettwurst, and cervelat sausages (Fig. 1.22). The sausage meat is inoculated with starter cultures of lactobacilli and bacteria of the micrococcus family. Sugar is added to the mixture to start off lactic fermentation, as the sugar content of meat is insufficient. Lactic acid not only adds to the flavor, but – in connection with pickling salt (mixture of NaCl and the controversial sodium nitrite) – also inhibits the growth of undesired microbes and adds to the firmness of the sausage. The sausage meat is stuffed into casings and hung up to mature for two weeks in maturing chambers where the sausages also undergo intermittent smoking.

When wine has been exposed to air over a period of time or has been kept in a vessel that has not been properly sealed, it turns sour and ends up as **vinegar**.

The transition from alcohol to vinegar is something many of us have observed when an uncorked wine or beer bottle was left in a warm room. Vinegar preparation also goes back a long way – it was known to the Sumerians who used palm sap and date syrup as starters, later also beer and wine. Greeks and Romans found diluted vinegar a refreshing drink. In medieval France, vinegar was produced using the Orléans method.

Modern industry prefers quick vinegar production, in which alcohol is oxidized by *Acetobacter suboxydans*, using the oxygen in the air. Vinegar fermentation is not a genuine fermentation process because it is not anaerobic.

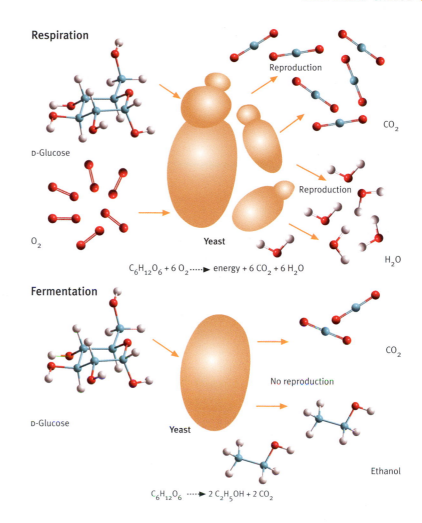

Respiration

D-Glucose

O_2

Yeast

Reproduction

Reproduction

CO_2

H_2O

$C_6H_{12}O_6 + 6\,O_2 \cdots\!\!\rightarrow energy + 6\,CO_2 + 6\,H_2O$

Fermentation

D-Glucose

Yeast

No reproduction

CO_2

Ethanol

$C_6H_{12}O_6 \cdots\!\!\rightarrow 2\,C_2H_5OH + 2\,CO_2$

■ 1.8 Coffee, Cocoa, Vanilla, Tobacco – Fermentation For Enhanced Pleasure

Coffee, cocoa, tea, tobacco, and vanilla are traditional fermentation products, i.e., they have always been modified by microorganisms or by the plant's own enzymes. In the **fermentation of coffee**, the flesh of the coffee cherry is broken down by bacteria containing pectin-cleaving enzymes (pectinases). All fruit contain pectin as their major constituent.

The flesh of **cocoa beans** is degraded by yeast, producing alcohol which is further processed by acetic acid-producing bacteria. This heat-releasing process is called 'sweating' and is crucial for the quality of the cocoa. As the seeds die, polyphenols develop and lend their flavor to the cocoa. The bitter tannins are broken down and the rich brown color develops. Eventually, the seeds are roasted (Fig. 1.23).

Vanilla pods (*Vanilla planifolia*) are the fruit of orchids and owe their flavor to the action of

Fig. 1.20 Comparison of respiration and fermentation of a glucose molecule by yeast cells.

Fig. 1.21 Genuine aceto balsamico or balsamic vinegar originates from Modena, Italy.

Fig. 1.22 Sausage preservation through lactic fermentation.

Box 1.7 San Francisco Sourdough Bread

Nothing beats the taste and texture of true San Francisco French style sourdough bread. From its hard crust to its soft interior, this bread has a unique flavor that is hard to resist. Although traditionally served with shellfish or seafood, this bread can be served with most dishes, including red meats and poultry. You can jet it even if you do not live in San Francisco as many US-supermarkets now carry prepackaged loaves as well as most major airports.

Colonies of slime-forming bacterium *Lactobacillus sanfranciscensis*.

A **sourdough culture** begins with a mixture of flour and water. Fresh flour naturally contains a wide variety of yeast and bacteria. When wheat flour contacts water, some of the gluten and starch is degraded by naturally-occurring enzymes in the flour, providing sugars and amino acids that yeasts and bacteria can metabolize. Initially, a wide variety of microorganisms starts to grow – the dough becomes sour and may even develop a bad smell. Only after repeated feedings with fresh flour and water does the mixture develop a **balanced, symbiotic culture**. All sourdough starters contain a stable symbiotic culture of yeast and lactic acid bacteria, most typically the yeast *Candida humilis* and the bacterium *Lactobacillus sanfranciscensis*, isolated first from San Francisco sourdough.

An active sourdough enables bread baking with only three ingredients – flour, water, and salt. Sourdough bread is made by using a small amount of "starter" dough with active yeasts and lactobacilli, and mixing it with new flour and water. Part of this resulting dough is then saved for use as starter next time. As long as the starter dough is fed flour and water daily, the sourdough remains healthy and usable almost indefi-

nitely. It is not uncommon to have a baker's starter dough that has had years of history, from many hundreds of previous batches.

Yeast and lactobacilli in the sourdough will metabolize sugars in the dough – mainly maltose and sucrose – to **produce the gas CO_2**, which leavens the dough. Obtaining a satisfactory rise from sourdough, however, is more difficult than with packaged baker's yeast. To leaven a dough with baker's yeast, a large number of yeast cells is added (usually more than 100 million cells per g of dough), which rapidly produce enough gas to leaven the dough. In sourdough, not as many yeasts cells are present (about 10 million cells per gram), and although they are supported by the lactobacilli, it takes a longer time to produce enough gas. Additionally, proteolysis also results in weaker gluten, and a denser finished product. An advantage of gluten breakdown by enzymes in the flour during sourdough fermentation is the liberation of amino acids: their transformation to flavor compounds by yeasts and lactobacilli as well as during baking contributes to the specific flavor of sourdough bread.

Gold rush in California: paving the way for sourdough bread.

History of sourdough

The history of sourdough probably began with cultivation of cereals – humans realized already in ancient times that milling, fermentation, and baking convert raw cereal grains to tasteful and nutritious food. Descriptions of sourdough baking were found in the tombs of **ancient Egypt**. Egyptians were also the first to virtually industrialize bread production, to feed the thousands of workers constructing the pyramids.

Bread made from **rye flour**, which is very popular in the northern and eastern parts of Europe, is leavened with sourdough. Rye flour has a high amylase activity. Unless the rye amylases are inhibited by the acidity of sourdough, they degrade the starch during baking, converting the bread crumb to a slimy mess. In those parts of Europe where bread is produced from wheat flour,

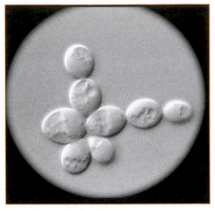

The yeast *Candida humilis*.

sourdough was replaced by baker's yeast in the last century. Only for specialty products such as **baguette in France** and **panettone in northern Italy**, the use of sourdough remained a must to achieve the right taste and aroma.

American pioneers heading west on their covered wagons also carried sourdough - an active culture enables bread baking with flour, water, and salt only and spared the need to carry other leavening agents. For the same reason, "sourdough" was used by the gold prospectors during the **1848 California Gold Rush**. In Northern California, sourdough bread was so common that sourdough became the general nickname for the gold diggers. In 1898, the sourdoughs and their "sourdough" moved to the Yukon in the Klondike Gold Rush but sourdough bread remains a major part of the culture of San Francisco. Baking was probably more lucrative than gold mining and a lot steadier...

Today, in **San Francisco** alone, almost a thou-

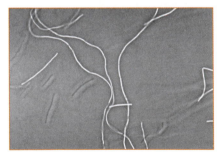

The bacterium *Lactobacillus*.

sand men and women work around the clock to produce sourdough bread. They annually produce 60,000,000 salable units a year (a unit can be anything from a loaf of bread to a bag of rolls), of which 70% is sourdough, 25% sweet French, and 5% specialty items like focaccia and ciabatta. Together they service about 4,000 retail outlets in Northern California.

Each bakery's mother dough, also known as the mother "sponge," has been in continuous use since the respective bakery's founding, carefully maintained and replenished by generations of bakers. It could be said, with some truthfulness, that San Francisco's sourdoughs are the oldest inhabitants of the city. The mother dough in one bakery will not be the same as the mother dough in another bakery across town – depending on the ambient temperature, humidity and how the baker propagates the mother sponge, different strains of *L. sanfranciscensis* and yeasts will develop. Their metabolic activities are decisive for bread flavor and thus each bakery produces different bread that has it adherents.

The biotechnology behind sourdough

What can modern biotechnology do today to improve the well-established sourdough procedure and to adapt the traditional fermentation to industrial baking?

The first questions is: **which bacteria and yeasts** proliferate in sourdough?

Yeast and bacteria are symbiotic partners and their cells occur in a ratio of about 1:100. Many different species of yeast and lactobacilli were isolated from sourdough but *Candida humilis* (previously *C. milleri*) and *L. sanfranciscensis* remain the most fascinating.

L. sanfranciscensis has to date only been found in sourdough. Moreover, it populates sourdoughs around the world and is used to make West-phalian Pumpernickel (in Northwest Germany), Italian Panettone and, of course, the famous San Francisco Sourdough Bread.

The second question: why is *L. sanfranciscensis* so widely successful in sourdough but fails to grow just about anywhere else? This organism has adopted a truly American way of life, making wasteful use of abundant resources. Dough abounds in maltose which is formed from starch through the action of amylase enzymes.

L. sanfranciscensis use maltose-phosphorylase to cleave maltose to glucose-1-phosphate and glucose without the expenditure of ATP – the energy rich glucose-1-phosphate is converted to lactate, CO_2 and ethanol, the glucose is thrown out. Fructose, the second most abundant sugar in dough, is not used as carbon source but reduced to a sugar alcohol, mannitol. This process allows *L. sanfranciscensis* to gain more energy from maltose and to produce acetate instead of ethanol. This and other metabolic features ensure that *L. sanfranciscensis* is the fastest growing bacterium in sourdoughs that are propagated daily, and thus outcompetes all other bacteria. The yeast can not use maltose but utilizes all other sugars present in dough (including the glucose thrown out by the lactobacilli). Thus the two critters do not compete for a carbon source. Moreover, *C. humilis* is much more resistant to lactic and acetic acid than baker's yeast.

Finally, which metabolites of yeasts and lactobacilli are responsible for the **distinctive flavor** of sourdough bread? Its taste is due mainly to lactic and acetic acids produced by the lactobacilli, but the flavor is a result of **teamwork between yeasts and lactobacilli**. There are about 20 important flavor compounds in sourdough bread. One example is **2-acetyl-1-pyrroline**, a compound generated during baking from the amino acid ornithine, which imparts the roasty odor to the fresh crumb of wheat bread (and contributes much to the smell of a bakery). Wheat flour contains very low levels of amino acids. During sourdough fermentation, wheat proteins are degraded to amino acids by flour enzymes. Sourdough lactobacilli convert arginine to ornithine which accumulates in the dough. Ornithine, which is not a proteinogenic amino acid, is converted during baking to 2-acetyl-1-pyrroline.

Enjoy your surdough bread!

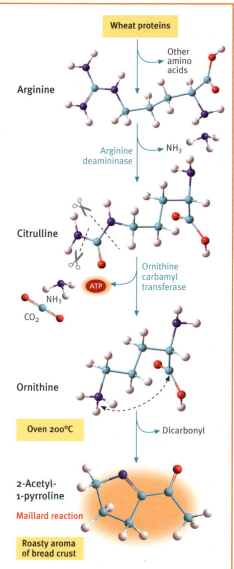

How the flavor is generated.

Prof Michael Gänzle is a leading international specialist on sourdough microbiology. He is at present teaching at the University of Alberta. Here, he is baking sourdough bread with son Markus.

Fig. 1.23 Cocoa fruits grow on the stem of the cocoa tree (*Theobroma cacao*) and take 9 months to ripen. Theobroma is Greek for "food of the Gods".

Fig. 1.24 Vanilla (*Vanilla planifolia*) is an orchid that has to undergo fermentation to release its aroma.

Fig. 1.25 Tea fermentation in China.

Fig. 1.26 Tobacco (*Nicotiana tabacum*).

enzymes. They are harvested unripe (Fig. 1.24) and dried in the sun, which gives them their typical dark-brown color. The glycosides contained in the pods are turned into vanillin through enzymatic activity.

Tea leaves are left to wilt for 24 hours before they are rolled up in order to crush the cells. The sap is spread over the leaf surfaces, and through the action of oxidizing plant enzymes, bacteria, and yeasts, the typical taste and smell of tea develops (Fig. 1.25).

The taste and smell are due to organic compounds in the leaves called polyphenols. As they develop, the color of the leaves changes, starting from the edges and progressing towards the middle. The leaves, once green, turn copper-colored and then brown or purple.

The tea drunk in the Western world is mostly fully fermented black tea. Green tea marks the other end of the spectrum, i.e., non-fermented tea. In-between, there are half or partially fermented teas such as oolong, yellow tea, or lightly fermented white tea.

When processing the various kinds of tea, the tea master must choose the right moment to stop the fermentation process. If it is stopped too early, the aroma will not develop, and the result is an insipid brew. If it is left too late, the tea is "burnt" and tastes bitter.

Similar enzymatic and microbial activity is involved in **tobacco** production (Fig. 1.26). Tobacco for cigars usually undergoes a natural fermentation process where cured (pre-dried) leaves are tied into bundles or "hands" and stacked into piles which heat up to 122°C in the fermentation process and need to be turned regularly. The fermentation process allowed to go on for three to four months to ensure a longer shelf life and enhance the coloring of the tobacco leaves.

While **food fermentation** may have been discovered by chance, its advantages – longer storage life, better digestibility, richer flavors, and the inebriating properties of alcohol – were so obvious that most civilizations adopted some fermentation process in their very early stages. Fermentation can thus be considered as a way of **enhancing the quality** of food.

The first human settlers knew only two ways of preserving food – **drying and pickling** in salt. The latter was often in short supply and very precious.

The introduction of fermentation gave therefore rise to the production of a wider and tastier variety of food while at the same time reducing the risk of food poisoning.

While in modern industrialized societies, fermented food is mainly produced for taste and enjoyment, it remains an invaluable food processing technology in developing countries where one third of all food is lost through decay. Compared to refrigeration, chemical preservation or freeze-drying, fermentation is cheap, does not require expensive machinery, and results in a widely accepted product. On top of that, it is an industry that provides jobs.

1.9 An Alliance of Molds and Bacteria in Cheese Production

Milk with its richness in nutrients, vitamins, and minerals has always been a major food source for the peasant population. **Jared Diamond**, professor in physiology and best-selling author, maintains that cattle-breeding marks a watershed in the development of civilizations across the continents (for details see Box 1.8). Animals suitable for domestication were only found in Eurasia where agriculture took off in leaps and bounds, due to the breeding of cows, goats, pigs, and above all, the use of oxen to pull plows. Horses provided the means of speedy transport. By contrast, the only animals domesticated in America were lamas and alpacas, and not a single animal was domesticated in Australia. Domestic animals provided **milk, meat, and manure** and pulled the plow. Milk-producing animals such as cows, sheep, goats, horses, reindeer, water buffalo, yak, or camels produced their body weight in protein several times over during their lifetime. The milk they produced was preserved through **lactic fermentation**.

The filtered solid components of sour milk, **curd**, were processed into a more storable product, i.e., cheese. The milk usually curdles within 30 minutes and solidifies, as casein precipitates and coagulates. Once the liquid part or whey has been separated, salt is added to the resulting curd which is cut into portions.

Soft cheeses, such as Camembert or Brie, have mold growing on the surface of the cheese mass, which is often specific to a certain location, e.g., the **Camembert** or the **Roquefort** mold fungus. This is reflected in the names of the cheeses.

Box 1.8 Biotech History:
The Historical Importance of Plant and Animal Domestication and the Storage of Food

In his book "Guns, germs, and steel – a short history of everybody for the last 13,000 years", Professor Jared Diamond inquires into the reasons why Europe became the cradle of modern societies and how Europeans could conquer the whole world. He gave his kind permission for the reprint of this condensed extract about the role of farming and preserving food (see also Ch. 5 for the role of microbes):

Europa holding the horn of Zeus (disguised as a bull) – Athenian vase found in northern Italy, (480 BC)

For most of the time since the ancestors of modern humans diverged from the ancestors of living great apes, around 7 million years ago, all humans on Earth fed themselves exclusively by hunting wild animals and gathering wild plants, as the Blackfeet still did in the 19th century. It was only within the last 11,000 years that some peoples turned to what is termed food production: that is, **domesticating wild animals and plants** and eating the resulting livestock and crops. Today, most people on Earth consume food that they produced themselves or that someone else produced for them. At current rates of change, within the next decade the few remaining bands of hunter-gatherers will abandon their ways, disintegrate, or die out, thereby ending our millions of years of commitment to the hunter-gatherer lifestyle.

Different people acquired food production at different times in prehistory. Some, such as **Aboriginal Australians**, never acquired it at all. Of those who did, some (for example, the **ancient Chinese**) developed it independently by themselves, while others (including **ancient Egyptians**) acquired it from neighbors. But, as we'll see, food production was indirectly a prerequisite for the development of guns, germs,

and steel. **Hence geographic variation in whether, or when, the peoples of different continents became farmers and herders explains to a large extent their subsequent contrasting fates**.

The first connection (through which food production led to all the advantages, RR) is the most direct one: availability of **more consumable calories means more people**. Among wild plant and animal species, only a small minority are edible to humans or worth hunting or gathering. Most species are useless to us as food, for one or more of the following reasons: they are indigestible (like bark), poisonous (jellyfish), tedious to prepare (very small nuts), difficult to gather (larvae of most insects), or dangerous to hunt (rhinoceroses). Most biomass (living biological matter) on land is in the form of wood and leaves, most of which we cannot digest.

By **selecting and growing those few species of plants and animals that we can eat**, so that they constitute 90 percent rather than 0.1 percent of the biomass on an acre of land, we obtain far more edible calories per acre. As a result, one acre can feed many more herders and farmers- typically, 10 to 100 times more- than hunter-gatherers. That strength of brute numbers was the first of many military advantages that food-producing tribes gained over hunter-gatherer tribes.

In human societies possessing domestic animals, **livestock fed more people in four distinct ways:** by furnishing meat, milk, and fertilizer and by pulling plows.

First and most directly, domestic animals became the societies' **major source of animal protein**, replacing wild game. Today, for instance, Americans tend to get most of their animal protein from cows, pigs, sheep, and chickens, with game such as venison just a rare delicacy.

In addition, some big domestic mammals served as **sources of milk and of milk products** such as butter, cheese, and yogurt. Milked mammals include the cow, sheep, goat horse, reindeer, water buffalo, yak, and Arabian and Bactrian camels. Those mammals thereby yield several times more calories over their lifetime than if they were just slaughtered and consumed as meat.

Big domestic mammals also interacted with domestic plants in two ways to increase crop production. First, as any modern gardener or farmer still knows by experience, crop yields can be greatly increased by **manure applied as fertilizer**. Even with the modern availability

of synthetic fertilizers produced by chemical factories, the major source of crop fertilizer today in most societies is still animal manure- especially of cows, but also of yaks and sheep. Manure has been valuable, too, as a **source of fuel for fires** in traditional societies (even today, see Ch. 6: biogas).

In addition, the largest domestic mammals interacted with domestic plants to increase food production by **pulling plows** and thereby making it possible for people to till land that had previously been uneconomical for farming. Those plow animals were the cow, horse, water buffalo, Bali cattle, and yak/cow hybrids.

Here is one example of their value: the first prehistoric farmers of central Europe, the so-called Linearbandkeramik culture that arose slightly before 5000 B.C., were initially confined to soils light enough to be tilled by means of hand-held digging sticks. Only over a thousand years later, with the introduction of the ox-drawn plow, were those farmers able to extend cultivation to a much wider ranger of heavy soils and tough sods. Similarly, Native American farmers of the North American Great Plains grew crops in the river valleys, but farming of the tough sods on the extensive uplands had to await 19th-century Europeans and their animal-drawn plows.

Even the strong EURO shows what we owe to farm animals...

All those are direct ways in which plant and animal domestication led to **denser human populations by yielding more food** than did the hunter-gathered lifestyle. A more indirect way involved the consequences of the sedentary lifestyle enforced by food production. People of many hunter-gatherer societies move frequently in search of wild foods, but **farmers must remain near their fields and orchards**.

The resulting fixed abode contributes to **denser human populations by permitting a shortened birth interval**. A hunter-gatherer mother who is shifting camp can carry only one child, along with her few possessions. She cannot afford to bear her next child until the previous toddler can walk fast enough to keep up with the tribe and not hold it back.

See next page

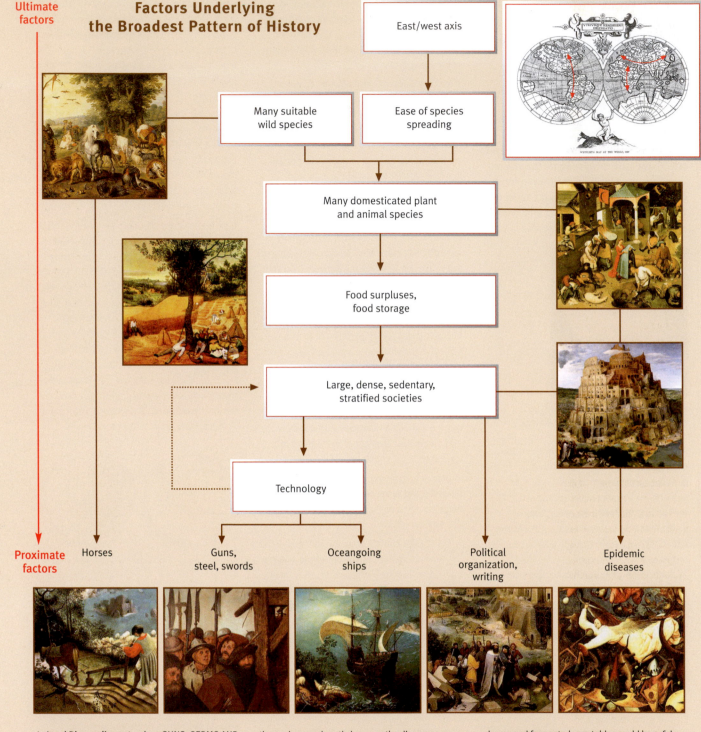

Factors Underlying the Broadest Pattern of History

Ultimate factors

East/west axis

Many suitable wild species

Ease of species spreading

Many domesticated plant and animal species

Food surpluses, food storage

Large, dense, sedentary, stratified societies

Technology

Proximate factors

Horses

Guns, steel, swords

Oceangoing ships

Political organization, writing

Epidemic diseases

In **Jared Diamond's** masterpiece GUNS, GERMS AND STEEL, Diamond shows the causal connection between factors (such us guns, horses, and diseases) that enabled some nations to conquer others, and more distant causes (such as the orientation of continental axes).

For example, diverse epidemic diseases of humans evolved in areas with many wild plant and animal species suitable for domestication, partly because the resulting crops and livestock helped feed denser populations in which epidemics could maintain

themselves, and partly because the diseases evolved from germs endemic to the domesticated animal population.

Animals kept for their milk (cow, sheep, goat, horse, reindeer, water buffalo, yak, and camel) yielded several times more calories over their lifetime than if they had been just slaughtered for their meat.

The manure from the animals could be used as fertilizers or fuel. Early biotechnology products like beer, wine or vinegar provided safe drinks, while

cheese and fermented vegetables could be safely stored for a longer period of time.

Diamond's scheme is illustrated here with some details from my favorite paintings by the Brueghel family. Pieter Brueghel (about 1525-69), usually known as Pieter Brueghel the Elder, is the greatest Flemish painter of the 16th century.

Jan Brueghel (1568-1625), called the "velvet Brueghel," was the second son of Pieter Bruegel the Elder and known for his still lives of flowers and for his landscapes.

When I was a kid I collected wooden farm animals.

In practice, nomadic hunter-gatherers space their **children** about four years apart by means of lactational amenorrhea, sexual abstinence, infanticide, and abortion. By contrast, sedentary people, unconstrained by problems of carrying young children on treks, can bear and raise as many children as they can feed. The birth interval for many farm peoples is around two years, half that of hunter-gatherers. That higher birthrate of food producers, together with their ability to feed more people per acre, lets them achieve **much higher population densities** than hunter-gatherers.

A separate consequence of a settled existence is that it permits one to **store food surpluses,** since storage would be pointless if one didn't remain nearby to guard the stored food. While some nomadic hunter-gatherers may occasionally bag more food than they can consume in a few days, such a bonanza is of little use to them because they cannot protect it. But **stored food is essential for feeding non-food-producing specialists,** and certainly for supporting whole towns of them. Hence nomadic hunter-gatherer societies have few or no such full-time specialists, who instead first appear in sedentary societies.

Two types of such **specialists are kings and bureaucrats.** Hunter-gatherer societies tend to be relatively egalitarian, to lack full-time bureaucrats and hereditary chiefs, and to have small-scale political organization at the level of the band or tribe. That's because all able-bodied hunter-gatherers are obliged to devote much of their time to acquiring food. In contrast, **once food can be stockpiled, a political elite can gain control of food produced by others**, assert the right of taxation, escape the need to feed itself, and engage full-time in political activities. Hence moderate-sized agricultural societies are often organized in chiefdoms, and kingdoms are confined to large agricultural societies. Those complex political units are much better able to mount a sustained war of conquest than is an egalitarian band of hunters. Some hunter-gatherers in especially rich environments, such as the Pacific Northwest coast of North America and the coast of Ecuador, also developed sedentary societies, food storage, and nascent chiefdoms, but they did not go farther on the road to kingdoms.

A **stored food surplus built up by taxation** can support other full-time specialists besides kings and bureaucrats. Of most direct relevance to wars of conquest, it can be used to feed professional soldiers. That was the decisive factor in the British Empire's eventual defeat of New Zealand's well-armed **indigenous Maori population.** While the Maori achieved some stunning temporary victories, they could not maintain an army constantly in the field and were in the end worn down by 18,000 full-time British troops. Stored food can also feed priests, who provide religious justification for wars of conquest; artisans such as metalworkers, who develop swords, guns, and other technologies; and scribes, who preserve far more information that can be remembered accurately.

Pieter Brueghel the Elder is often credited as being the first Western painter to paint landscapes for their own sake. His earthy, unsentimental vivid depiction of the rituals of peasant life including agriculture, hunts, meals, festivals, dances, and games are unique windows on a vanished folk culture and a prime source of evidence about both physical and social aspects of 16th century life.

Of equal importance in wars of conquest were the **germs that evolved in human** societies with domestic animals. Infectious diseases like smallpox, measles, and flu arose as specialized germs of humans, derived by mutations of every similar ancestral germs that had infected animals (see Ch. 5). The humans who domesticated animals were the first to fall victim to the newly evolved germs, but those humans then evolved substantial **resistance to the new diseases**. When such partly immune people came into contact with others who had had no previous exposure the population was killed. Germs thus acquired ultimately from domestic animals played decisive roles in the European conquests of Native Americans, Australians, South Africans, and Pacific islanders.

In short, plant and animal domestication meant much more food and hence much denser human populations. The resulting food surpluses, and (in some areas) the animal-based means of transporting those surpluses, were a prerequisite for the development of settled, politically centralized, socially stratified, economically complex, technologically innovative societies stratified, economically complex, technological innovative societies.

Hence the **availability of domestic plants and animals** ultimately explains why empires, literacy, and steel weapons developed earliest in Eurasia and later, or not at all, on other continents. The military uses of horses and camels, and the killing power of animal-derived germs, complete the list of major links between food production and conquest....

Jared Mason Diamond

Jared Mason Diamond (born 1937) is an American evolutionary biologist, physiologist, biogeographer, and nonfiction author. He is best known for the Pulitzer Prize-winning book, Guns, Germs, and Steel (1997). He also received the National Medal of Science in 1999. He became a professor of physiology at UCLA Medical School in 1966. While in his twenties, he also developed a second, parallel, career in the ecology and evolution of New Guinea birds, and has since led numerous trips to explore New Guinea and nearby islands. In his fifties, Diamond gradually developed a third career in environmental history, becoming a professor of geography and of environmental health sciences at UCLA, his current position. In his most recent book, Collapse: How Societies Choose to Fail or Succeed (2005), Diamond examines what caused some of the great civilizations of the past to collapse into ruin and considers what contemporary society can learn from their fates.

Cited Literature:

Diamond J (1998) *Guns, Germs, and Steel. A short history of everybody for the last 13,000 years.* Vintage, Random House, London, Pp 210-213 and 214

Box 1.9 Deciphering the Vintage Code

Patrick McGovern of the University of Pennsylvania has a dream job – he combines chemical analysis with archeology, analyzing traces of wine in ancient vessels. His claim to fame was the discovery of the purple imperial dye obtained from the mollusk *Murex trunculus*. This was followed by McGovern's **Noah hypothesis**: As described in the Bible, Noah landed on Mount Ararat in Eastern Turkey and soon started growing wine there. It seems that that area is the cradle of agriculture where einkorn wheat was grown for the first time.

Chinese wine vessels and oracle bones.

McGovern began by comparing the **DNA of wild vines** (*Vitis vinifera sylvestris*) (Ch. 10), which is found between Spain and Central Asia. It is the ancestor of the cultured vine varieties. He is now exploring the Turkish Taurus mountains where the river Tigris rises, hoping to find the original wild vine. José Voulilamoz at the Italian Istituto Agrario die San Michele all'Adige in Trento and Ali Ergül at Ankara University are part of the DNA team. Every type of grape they can lay their hands

on in the region is being collected in order to trace back the origins of viticulture.

Scientists are also looking for potsherd (bits of ancient crockery) carrying wine residue. Tartrate, investigated by Louis Pasteur in his day, is a reliable indicator for the presence of wine.

Ten years ago, Patrick McGovern thought he had finally found the oldest traces of wine and barley beer in the **Iranian village Haji Firuz Tepe**. They were about 7,400 years old.

However, in Jiahu in the Chinese province of Henan, oracle bones with very early writing, flutes made out of bird bones, and **potsherd with wine residue** have now been found. McGovern took 16 of the 9,000-year-old pieces of potsherd back to the Pennsylvania Museum of Archaeology and Anthropology in Philadelphia. He also took 90 sealed bronze vessels dating from the Shang Dynasty.

Using a combination of gas and liquid chromatography, infrared spectrometry, and isotope analysis, it was established that the wine consisted of a complex mixture of fermented rice, beeswax, hawthorn berries (very high in sugar and possibly the source of the fermenting yeast), and wild grapes.

During the Shang period much progress has been made on the wine production front. The wine of the Shang emperors, found in those bronze vessels, contained traces of chrysanthemum flowers, pine resin (reminiscent of modern-day Greek retsina), camphor, olives, tannins, and even wormwood, the absinthe plant that ruined the life of Parisians in Toulouse-Lautrec's Belle Epoque.

McGovern refrained from sampling the aromatic Shang wine, as the vessels contained up to 20% **lead**. This is an intriguing find. The acidity of wine helps dissolve metals, and so, just like the upper classes in Rome with their lead chalices, the Chinese emperors probably poisoned themselves with lead. The symptoms range from cramps to hallucinations. Like all heavy metals, lead inhibits the action of enzymes by breaking up the stabilizing disulfide bonds of proteins (Ch. 2).

Some of the 9,000-year-old Chinese drinks had been prepared from millet, and it is amazing to find similar traces of millet in 7500-year-old Iranian wine. There seems to have been a vivid exchange of ideas throughout Central Asia.

For Patrick McGovern, studying wine with all its social and economic implications opens a door to the old civilizations. "Even a good bottle of Merlot or Shiraz, enjoyed today, can recreate history, in a sense."

Animal-shaped bronze vessel from China, used in wine ceremonies.

Fig. 1.27 Chinese and Japanese scripts use a single sign for fungi and bacteria. The sign shows an ear of grain on which plants grow – an apt description of rice fermentation!

Hard cheeses like Swiss **Emmental** are hardened in a cheese press and keep better than soft cheese. Specific molds are added to the cheese mixture. These grow not only on the surface, but also within the cheese as long as there is a good supply of air. Thin aeration canals are made with skewers, and in the mature cheese, the fungal lawn can be seen growing along these channels.

Some hard cheeses owe their particular flavor and their holes to **propionic acid bacteria** that ferment sugar to propionic acid, acetic acid, and CO_2. The red smeary layer on top of some cheeses like **Munster, Limburg, or Romadur** is due to bacteria living on the surface of the cheese.

Mold fungi can be seen with the naked eye. In contrast to yeasts, they are multicellular organisms.

What we usually see are the spore carriers or fruit of the fungi, just like the mushrooms that grow in woodland.

The actual body of the fungus is not eye-catching at all, but consists of long thread-like structures called the **mycelium**. The mycelium sprouts spore carriers or sporangia that form thousands of spores to be carried by the wind or washed away by rain water. The spores will germinate on a substrate rich in nutrient and form new mycelia.

Some cheese molds with sporangia in the shape of little brushes are aptly named *Penicillium* (Latin, little brush).

Other molds (*Aspergillus*) grow on bread and jam. In contrast to the harmless molds used in cheese preparation, these molds, e.g., *Aspergillus flavus*,

Box 1.10 Biotech History:
Pasteur, Liebig and Traube or What Is Fermentation?

After Leeuwenhoek's discoveries, it took nearly 200 years until microbes grabbed the limelight again. By the middle of the nineteenth century, large industrial complexes had developed throughout Europe, and alcohol was also produced on an industrial scale rather than in small family-run companies. Accurate and detailed knowledge about the fermentation process was needed if one was to avoid costly failures.

Louis Pasteur (1822 – 1895), founding father of microbiology and biotechnology, in his laboratory.

In 1856 in the French city of Lille, a certain Monsieur Bigo, owner of an alcohol distillery came to see Louis Pasteur (1822-1895), professor of chemistry. His son was a student of Pasteur's. What troubled Bigo was the strange disease that had affected many of his fermentation vessels. Instead of turning sugar beet juice into alcohol, they produced some sour-smelling, slimy gray liquid. Pasteur took his microscope with him and went to the distillery. He took samples from diseased as well as healthy containers. The healthy samples contained small yellow blobs of yeast that stuck together like grapes. They produced side shoots much like germinating grain, so they must be alive and could be responsible for the transformation of sugar into alcohol. Then he investigated the slimy mass and could not find any yeast at all. Instead, he discovered small gray dots, each of which contained a tangle of oscillating rods – millions of them. The acidic substance these rods seemed to produce turned out to be lactic acid.

Pasteur added a few drops of rod-contaminated liquid to a flask containing a clear yeast-and-sugar solution. After a short while, here, too, all yeasts had disappeared and the rods had taken over. Again, lactic acid was produced instead of alcohol.

The rods were named bacteria, after their shape – the Greek word *bakterion* meaning rod. It would appear that bacteria turned sugar into lactic acid through fermentation, while yeasts turned sugar into alcohol and a gas, carbon dioxide.

Soon after his discovery of lactic acid bacteria, the vintners of his home town, Arbois, turned to Louis

Pasteur for help. They, too, had trouble with the alcoholic fermentation process. Even the juice of their best grapes turned thick, oily, and bitter. Again, Pasteur found bacteria instead of yeast in the spoilt wine, but this time, they formed chains. During his thorough investigations, Pasteur encountered a wide range of bacteria that could spoil wine. Eventually, he was able to predict the taste of the wine sample without even tasting it, which perplexed the vintners. All he did was look through the microscope and identify the species of yeast or bacteria. He also found out that heating the wine briefly was enough to kill off any unwanted bacteria.

The same technique was also effective in preventing milk from going sour. In honor of its inventor, this way of killing off the majority of microorganisms was called pasteurization.

Pasteur was not the only one looking for the actual cause of the fermentation process. In 1810, **Joseph Louis Gay-Lussac** (1778 - 1850) had shown that when grape juice is fermented, sugar (glucose) is transformed into ethanol and carbon dioxide.
In the middle of the nineteenth century, the eminent German chemist **Justus von Liebig** (1803 - 1873) developed his own theory, claiming that the production of alcohol was a purely chemical and not a biological process. Liebig found the idea of microscopic creatures being the cause of fermentation simply ludicrous. In his opinion, vibrations emanating from the degradation of organic matter would spread to the sugar and turn it into CO_2 and alcohol. However, it was found that yeast, a living organism, was involved in all alcoholic fermentation.

Louis Pasteur, then at the beginning of his scientific career, began a dispute with Justus von Liebig, an international authority, insisting that "there is no alcohol production without living yeasts"

Liebig retorted that "Those who believe that animalcules are the cause of fermentation resemble children who think the water in the river Rhine is caused to flow by the turning wheels of the water mills on its banks." They remained entrenched in their positions for years, and the dispute only came to a close after both protagonists had died.

In 1876, Pasteur published the results of two decades of research in a substantial volume in which he stated that "fermentation is respiration without oxygen", stimulating organisms to produce energy. All organisms obtain the energy they need to stay alive from their metabolism, mostly through the breakdown of sugars, fats, and proteins inside their cells. Sugar, for example, is turned into carbon dioxide and water. These are released from the cell,

while the energy liberated in the process enables the body to move its muscles, for example. This cold combustion process requires oxygen from the air just as hot combustion does when turning wood into ashes. Highly developed animals and plants cannot produce energy without access to oxygen.

Justus von Liebig (1803 - 1873), the most eminent chemist of the nineteenth century, in his laboratory in Gießen, Germany.

Some microorganisms, by contrast, can fall back on a kind of emergency respiration when oxygen is in short supply – the fermentation process. Fermentation probably has its origins in the beginnings of life when there was no oxygen in the earth atmosphere. Oxygen was only released by plants as a cleavage product of water, and before there was enough to go round, fermentation was just the normal way of energy production in microbes.

Was Pasteur right in claiming that fermentation was not possible without microbes? **Moritz Traube**, a student of Liebig's, predicted already in 1858 that fermentation was not necessarily the result of yeast activity, but of chemical processes catalyzed by oxidizing and reducing "ferments", or enzymes, as they were called later. Traube was the first to characterize enzymes as proteins acting as catalysts, i.e., as defined chemical compounds within an organism that, by their successive oxidation and reduction, are able to cause oxidation and reduction inside as well as outside living cells.

He went on to classify enzymes according to their reactive type and established that for a reaction to take place, direct molecular contact between enzyme and substrate was a prerequisite.

What has not been mentioned in the biochemical history books is that he was also thinking about the qualitative aspect of reaction kinetics and was the first to link reaction time to enzyme quantity.

It was not until 1897 that **Eduard Buchner** (1860-1917) carried out an experiment that clinched the argument in the dispute between Liebig and Pasteur (s. Ch. 2).

Fig. 1.28 In 1516, Duke William IV of Bavaria (1493-1550) established what became known as the purity requirements.
Bottom: facsimile of the document.

Here is an excerpt:

We stipulate in particular that the beer served in our towns, markets and country establishments must be brewed from nothing but barley, hops and water. Whoever knowingly flouts our ruling and disobeys, will have the offending barrel of beer confiscated by the court of justice. The full force of the law will hit the culprit whenever he reoffends."

Signed by William IV, Duke of Bavaria, Ingolstadt, St George's Day A.D. 1516.

Fig. 1.29 Soy sauce advertising.

can form **aflatoxins**, which are then transformed in the liver into active toxins due to oxygenation by the cytochrome P450 enzyme system, and cause liver cancer (Ch. 4).

■ 1.10 Sake and Soy Sauce

Mold fungi and their enzymes have been used for centuries in East Asia (Japan, China, Korea) to pre-pare protein-rich **soy beans and rice** for further processing by alcoholic or lactic fermentation. The enzymes involved are starch-degrading amylases and protein-degrading proteases. The preparation of **soy sauce** (*shoyu*) is a well-known example (Box 1.6 and Fig. 1.29). About 2.6 gal per person per year are produced and consumed in Japan. It is made from a mixture of soy and wheat, inoculated with *Aspergillus oryzae* or *Aspergillus soyae*.

The fungal enzymes cleave the soy bean protein and the starch molecules of the wheat, and large amounts of salt are added to prevent spoilage. Over a period of eight to ten months, yeasts and *Pedio-coccus* bacteria develop and finish off the fermenta-tion process. The soy sauce is ready to be filtered.

Rice wine (**sake**) is produced in a similar way. First, the rice starch must be degraded into fer-mentable sugars, which is done by enzymes (amy-lases) released by the fungi. The sugars are then fer-mented into alcohol by *Saccharomyces* strains.

Sake contains about 20 Vol% (40 proof) – not a trivial matter, keeping in mind that many Asians slightly differ from Caucasians in their genetic makeup of liver enzymes. They carry a molecular variant (isoenzyme) of acetaldehyde dehydrogenase which breaks down the acetaldehyde resulting from alcohol degradation (1.13) more slowly than the Caucasian isoenzyme. As a result, in Asians, smaller quantities of alcohol have the same intoxicating effect as larger quantities drunk by Caucasians. This may be very economical, but the hangover is worse!

■ 1.11 What Exactly is Fermentation?

All methods described so far have been used by humans over thousands of years. The knowledge was passed on from generation to generation. What was less clear, though, was the process of fermenta-tion itself, what causes it, and how it develops. **Louis Pasteur** (1822-1895) was the first person to shed some light on this matter and laid the founda-tion for the controlled industrial application of microbial fermentation. He is rightly considered one of the fathers of modern biotechnology (Box 1.10).

Alcoholic fermentation is the result of yeasts degrading sugars into alcohol and carbon dioxide in an anaerobic environment. Yeasts can function by respiration or fermentation, depending on the avail-ability of air. Fermentation, however, produces far less energy than respiration, and the reproduction rate slows down accordingly to just 5% of the aero-bic reproduction rate. It is a distress reaction that humans can bring about for alcohol or bread (CO_2) production.

In order to survive, yeasts must process much more sugar during fermentation, i.e., to increase their alcohol and CO_2 production. Conversely, if the desired result is an increase in yeast reproduction, e.g., to set off bioprocesses or to produce animal yeast feed, the nutrient solution may be boosted by additional oxygen. This, of course, will produce only very little alcohol.

While yeasts lead a dual existence as facultative aer-obic organisms (facultative anaerobes), other microbes are strictly anaerobic (obligate anaerobes) and will die from exposure even to low levels of oxygen. These include methane bacteria which only produce methane in the absence of air – or clostridia that live in airtight cans and can cause severe food poisoning.

Pasteur and the Germs

Omne vivum ex vivo.
(All life comes from life)

Louis Pasteur 1860

Cited and recommended Literature

- **Schmid, RD** (2003) *Pocket Guide to Biotechnology and Genetic Engineering*, Wiley-VCH A MUST-HAVE in your biotech bookshelf

- **Berg, JM, Tymoczko JL, Stryer L** (2007) *Biochemistry*, 6th edn., WH Freeman and Co, NY For me, "the Stryer" is THE modern Biochemistry textbook!

- **Nelson DL, Cox MM** (2005) *Principles of Biochemistry*. 4th edn., WH Freeman NY "The Lehninger" has educated generations of biochemists.

- **Barnum S R** (2006) *Biotechnology: An Introduction*, updated 2nd edn., with InfoTrac®) Brooks, Cole. The best available undergraduate textbook in biotechnology. White Biotech is not covered in detail though!

- **Ullmann's** *Encyclopedia of Industrial Chemistry* Wiley-VCH Apart from the latest (6th) print edition published in 2002, Ullmann's Encyclopedia is available now in two electronic versions – as a fully networkable DVD version and as an online database. A state-of-the-art reference work, detailing the science and technology in all areas of industrial chemistry

- **Thieman WJ, Palladino MA** (2004) *Introduction to Biotechnology*, Benjamin Cummings, San Francisco. Good textbook for students specifically interested in pursuing a career in biotechnology, clear illustrations, easy to read.

- **Bourgaize D, Jewell TR, Buiser RG** (1999) Biotechnology: *Demystifying the Concepts* Addison-Wesley-Longmann, San Francisco. Great college textbook, unfortunately not updated for 8 years.

- **Walker S** (2007) *Biotechnology Demystified*, McGraw-Hill, New York. The book promises to be "simple enough for beginners, but challenging enough for advanced students", using Multiple Choice as a learning tool. Having read it, I found myself more confused than enlightened about biotechnology.

- **Ratledge C, Kristiansen B** (2006) *Basic Biotechnology*, 3rd edn. Cambridge University Press; Multi-authored book, contributions vary widely in quality and style. Too detailed for a beginner, but a good reference book

- **Bains W** (2004) *Biotechnology from A to Z*, 3rd edn. Oxford University Press, USA; Great and concise biotech dictionary.

- **Diamond J M** (1999) *Guns, Germs, and Steel: The Fates of Human Societies* W. W. Norton & Company New York.

Useful Weblinks

- Biotechnology in general: a comprehensive site of biotechnology information maintained by the Biotechnology Industry Organization: *http://www.bio.org/*

- BioQUEST: a site for educators interested in the reform of undergraduate biology and contains curriculum and workshop information for secondary teachers: *http://bioquest.org/*

- Biotechnology Institute: an independent, national nonprofit organization for biotechnology education that publishes Your World: Biotechnology and You, a magazine for students in grades 7-12: *http://www.biotechinstitute.org/*

- World Wide Web Virtual Library: Biotechnology: catalog of WWW resources related to biotechnology: http://*www.cato.com/biotech/*

- A look through Leeuwenhoek's microscope: Brian J Ford found Leeuwenhoek's microscope and samples and confirmed that he saw indeed the bacteria: *http://www.brianjford.com/wav-mics.htm*

- All about microorganisms: *http://www.microbes.info/*

- Beer, bread, wine, cheese, fermented food. Extensive and (in most cases) reliable information by Wikipedia, the Free Encyclopedia. For example wine: *http://en.wikipedia.org/wiki/Wine*

Eight Self-Test Questions

1. What are the differences between prokaryotes and eukaryotes (name at least three criteria)?

2. In what way did alcohol nurture civilization? Who invented wine?

3. What are the biochemical reasons for the high productivity of fermentation within a short space of time?

4. In what way do bacteria differ from yeasts? Give two examples for each of the two types of organisms and their applications.

5. What food products owe their existence to fermentation?

6. Why does alcohol inhibit the growth of bacteria?

7. Given the choice, would yeasts produce alcohol?

8. What was Justus von Liebig's position in the debate about fermentation, and what was the position of Pasteur? Who was proved right eventually?

ENZYMES –
Molecular Supercatalysts
for Use at Home and in Industry

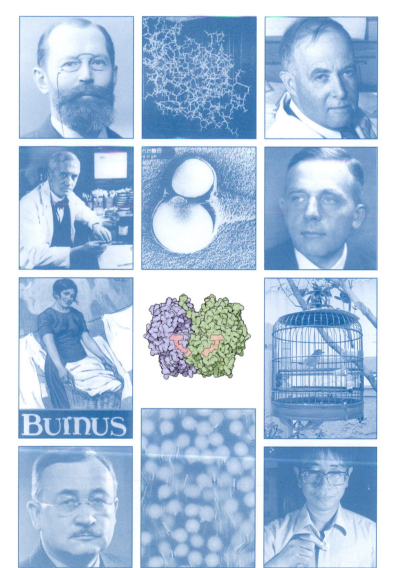

CHAPTER **2**

Fig. 2.1 In 1926, James B. Sumner (1887 - 1955) of Cornell University was the first to extract crystalline urease from jack beans and demonstrate that it had protein properties.

Fig. 2.2 John H. Northrop (1891 - 1975) succeeded in crystallizing the digestive enzymes trypsin and pepsin between 1930 and 1933. He also demonstrated that they consisted exclusively of proteins. Only after these findings did the science community accept Sumner's result, and both were awarded the Nobel Prize in 1946.

Fig. 2.3 Otto Heinrich Warburg (1883 - 1970) discovered the cofactor nicotinamide adenine dinucleotide (NAD) and respiratory enzymes containing iron, such as cytochrome oxidase (see Fig. 1.14, Ch. 1). He was awarded the Nobel Prize in 1941.

■ 2.1 Enzymes are High-performing and Highly Specific Biocatalysts

Nearly all chemical reactions within living cells are catalyzed and controlled by **enzymes**.

More than 3,000 different enzymes have been described in detail up to now. The total number of enzymes found in nature is estimated at around 10,000. Their presence in a cell varies in quantity from just a few molecules to 1,000 or even 100,000 molecules. All enzymes act as **biological catalysts** – turning substances into other products without undergoing any change themselves.

Due to enzymes, chemical reactions reach their equilibrium much faster and may be speeded up by a factor of up to 10^{12}. It is the activity of enzymes that makes life possible at all. Turning sugar into alcohol and carbon dioxide is a matter of seconds for the enzymes in yeast cells, but without them it would take thousands of years and be virtually impossible. Enzymes are highly effective, high-performing biocatalysts.

In cells ranging in size between a tenth and a thousandth of a millimeter, thousands of coordinated enzymatic reactions take place every second.

This can only work because each of the molecular catalysts involved recognizes its specific **substrate** among thousands of other compounds within a cell and turns it to a specific product. This process, called biocatalysis, takes place in the **active site** of the enzyme.

Nearly all biological catalysts are proteins. However, RNA (ribonucleic acid) can also act as biocatalysts (see Chapter 3). These **ribozymes** often break down other RNA molecules. RNA can also be used to build artificial **aptamers** that bind to designated compounds (see Chapter 10).

As early as 1894, the German chemist **Emil Fischer** (1852-1919), who was later awarded the Nobel Prize (Fig. 2.4), postulated that enzymes recognize their substrates on a **lock-and-key principle**. If the active site of an enzyme lies in a dimple (cavity, crevice) on the molecule's surface, the substrate molecule must fit accurately, just like a key into its lock. Even slightly modified molecules will no longer interact with the enzyme. Lock and key is a viable preliminary explanation for the high substrate specificity of enzymes. It also explains why the shape of **competing enzyme inhibitors** (e.g.,

penicillin) strongly resembles that of the original substrate. Like a skeleton key, they block the active site of an enzyme (competitive inhibition), thus preventing the substrate from locking on.

A simple biochemical experiment (Box 2.1) illustrates the **specificity** of enzymes. It is easy to see why enzymes such as glucose oxidase (Fig. 2.5) must be **highly substrate-specific** so as not to upset the finely tuned cellular mechanisms – this is a high-security level.

By contrast, **extracellular enzymes** such as protein-cleaving (proteases) or starch-degrading enzymes (amylases) are **far less specific** because it would be uneconomical to deploy a specific enzyme for each type of protein or polysaccharide to be degraded.

Let us focus on **two extracellular proteases** secreted by cells into their surroundings.

● The enzyme **trypsin** has the ability to break down all proteins found in a pig or human stomach into smaller fragments. Although it has low substrate specificity, its action specificity is high, as it cleaves protein compounds precisely at the binding sites of certain defined amino acids (at the carboxyl side of lysine or arginine residues in the peptide chain). This makes trypsin a suitable tool for specific syntheses, such as the transformation of pig insulin into human insulin (Chapter 3).

● **Subtilisin** is an enzyme produced by microbes which hardly differentiates at all between the side chains of amino acids next to which it cuts a peptide bond. It has broad substrate and action specificity. Subtilisin is released by *Bacillus subtilis* into a medium where it breaks down proteins into small fragments. Its biochemical shredder qualities make it a suitable candidate for technical applications, e.g., use in biological detergents (Fig. 2.17 and Box 2.9).

Enzymes have been **universally classified and named** (Box 2.3) in the IUPAC enzyme nomenclature on the basis of their action specificity.

■ 2.2 Lysozyme – The First Enzyme to be Understood in Structure And Function Down to Minute Molecular Detail

The first enzyme to be analyzed in its spatial structure was lysozyme. Its properties have been explored right down to the smallest atom. It was

Box 2.1 Glucose oxidase (GOD) – Highly Specific Recognition and Conversion of Sugar

A test tube contains a mix of carbohydrates – glucose, fructose, saccharose, maltose, and starch.

We now add the enzyme GOD to the mixture and give it some time to act. Subsequent chemical analysis shows that the glucose has all but disappeared and has been replaced by a new compound, gluconolactone, an oxidation product of glucose. All other carbohydrates, however, remain unchanged. Apparently, the glucose has been converted into gluconolactone by glucose oxidase. Thus, glucose is the substrate of glucose oxidase and gluconolactone, the oxidation product of the enzymatic reaction. β-D-Glucose has been singled out by glucose oxidase as a suitable substrate from among five different carbohydrates. Not even α-D-glucose has been recognized and is thus not converted. β-D-Glucose is characterized by the position of the hydroxyl group (-OH) on the C_1 atom above the ring plane, while in α-D-glucose, the hydroxyl group is positioned on the C_1 atom below the ring plane. Alongside glucose conversion, a second reaction takes place, involving oxygen as a substrate. It is reduced, producing hydrogen peroxide (H_2O_2).

Maltose, saccharose, and starch molecules are too big to fit into the glucose oxidase "keyhole". Fructose may be small enough, but it does not fit accurately and is thus not converted by the enzyme. The only precise fit is β-D-glucose.

Mix of various sugars

Fructose   Sucrose   Maltose

Starch

Oxygen   β-D-Glucose

Glucose oxidase

Gluconolactone

H_2O_2

Catalysis within the enzyme-substrate complex

Fig. 2.4 Lock-and-key: Nobel Prize winner Emil Fischer (1852–1919) postulated that enzymes and substrates work on a lock-and-key principle, which explains the complementary properties of both reagents.

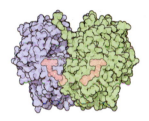

Fig. 2.5 Spatial structure of glucose oxidase (GOD). GOD is a dimeric molecule, an oxidoreductase that has been used with great success in biosensors to measure blood sugar levels in diabetics (see Ch. 10). The prosthetic group in the active centre of GOD, flavin adenine dinucleotide (FAD), is shown in pink.

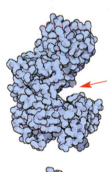

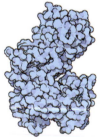

Fig. 2.6 Hexokinase, the first enzyme involved in glucose breakdown in a cell, shows a change in its conformation (below) after binding glucose (arrow).

discovered by Alexander Fleming some years before he discovered penicillin (Box 2.4).

Lysozyme provided an excellent model for studying **interactions between enzyme and substrate**. The lock-and-key model had its merits, but proved to be too rigid to describe the dynamic and flexible structure of proteins.

In 1958, sixty years after the Fischer postulate, **Daniel Koshland** (b. 1920) could confirm his new **"induced fit" theory** on the lysozyme model. He argued that substrate and enzyme were like hand-in-glove rather than key-in-lock. A hand can move, and a glove can be deformed, and hand (substrate) and glove (enzyme) affect one another. A glove is more than a 3-dimensional negative print of the hand – it could also be a mitten. Only through the

hand sliding into it can the exact fit be realized (Fig. 2.6).

Following active interaction and an **intermediate state**, enzyme and substrate fit perfectly. The enzymes bind not to the original configuration, but to the intermediate state of the substrate in their active site (Figs. 2.10 and 2.11).

In a first state of euphoria, it was assumed that the lysozyme model could be applied to all enzymes and that the substrate was always shaped by the enzyme. However, the reasons for the high catalytic performance of enzymes have proved to be far more complex (Box 2.5).

Due to their protein structure, enzymes have crevices or cavities which enable them to bind rapidly to the relevant substrate. Their active sites

Box 2.2 Biotech History:
The Discovery of Enzymes

Enzyme reactions have been observed and utilized since the dawn of mankind. **Homer** describes how milk curdles when fig juice is added (Fig. 2.7). Rennet from calves' stomachs was used in cheese preparation, and game that had been shot always had to hang for a while to make it palatable. Serious research into enzymes, however, only began at the end of the 18th century.

The degradation of any kind of compound due to the action of another was generally described as fermentation. In 1780, the Italian scientist **Lazzaro Spallanzani** (1729 - 1799) agreed with the Frenchman **Antoine Ferchault de Réaumur**, inventor of the first temperature scale, that the digestive juices in the stomachs of birds turned meat into liquid.

Lazzaro Spallanzani (1729 - 1799).

Antoine Ferchault de Réaumur (1683 - 1757).

In the early nineteenth century, it was generally assumed that fermentation is a chemical change brought about by a particular compound which was dubbed a ferment.

In 1814, a member of the St. Petersburg Academy of Science, the German **Gottlieb Sigismund Constantin Kirchhoff** (1764 - 1833) showed that germinating grain contains a substance that converts starch into sugar.

Anselme Payen (1795 - 1871), director of a Paris sugar factory who later was to discover cellulose, and his colleague **Jean François Persoz** (1805 - 1865) isolated what they called a starch-liquefying principle from germinated barley,

observing some properties that are now established to hold for all enzymes. Comparatively small amounts of the preparation were able to turn large quantities of starch into liquid. However, this ability was lost when exposed to heat. The active compound was separated from the solution as a powder. When rediluted in water, it became active again. The described compound was named **diastase** (Greek diastasis, separation) and was the first plant enzyme to be purified and studied in a laboratory.

Three years later, **Theodor Schwann** (1810 - 1882), co-founder of cytology, successfully isolated pure **pepsin**, an animal digestive juice enzyme, to study it.

Jöns Jakob Berzelius (1779 - 1848) conjectured in 1836 that thousands of catalytic processes took place in every living organism.

The great Swedish chemist, **Jöns Jakob Berzelius** (1779 - 1848) was ahead of his time when he stated that fermentation must involve catalytic processes (Greek katalysis, decay). He defined catalysts as bodies that, by their mere presence, induce chemical activity that could not take place without them.

In 1836, he wrote with quasi prophetic insight:

"We have reason to assume that in living plants and animals, thousands of catalytic processes take place among their tissues and fluids."

Other discoveries, however, muddied the waters a bit, and the term "**ferment**" led to some confusion when **Theodor Schwann** (1810 - 1882) found that rot, which is a decaying process of organic material and thus a fermentation process, is caused by microorganisms. This led to the distinction between two categories of fermentation, one caused by "genuine organized ferments" such as yeasts and other microorganisms and another caused by "non-organized soluble ferments" (e.g. diastase). It was thought that the non-organized ferments could be separated from vital processes. **Wilhelm Friedrich Kühne** (1800 - 1882) suggested the name **enzymes** for this type of ferment in order to

avoid misunderstandings or the need for long-winded explanations. The name was introduced in 1878.

Wilhelm Friedrich Kühne (1837 - 1900) coined the term "enzymes" for what was hitherto known as "non-organized ferments" in 1878.

Friedrich Wöhler (1800 - 1882).

Organized ferments were one of the last bastions of vitalism, a concept that put the force of life, a god-given *vis vitalis* at the heart of all life. It had first been thought that organic compounds could not be produced in the laboratory because they contain the force of life.

The latter hypothesis was proven wrong when **Friedrich Wöhler** (1800 - 1882) managed to synthesize urea from inorganic matter.

He wrote to Berzelius: "I can no longer, so to speak, hold my chemical water and must tell you that I can make urea without needing a kidney, whether of man or dog; the ammonium salt of cyanic acid is urea."

Moritz Traube (1826 - 1894) predicted in 1858 that "ferments" are catalytic proteins (see Chapter 1, Box 1.10).

It was not until 1897 that **Eduard Buchner** carried out the experiment that clinched the argument between Liebig and Pasteur. He wanted to find out if fermentation was possible without the involvement of living cells, crushing yeast with a pestle and adding quartz and kieselguhr to the large mortar. The resulting mass was wrapped in strong canvas and squashed in a hydraulic press. The run-off, which did not contain any cells, but all the chemicals found in yeast, was added to concentrated sugar solution and left to stand overnight. It did not take long for vigorous fermentation to set in, and carbon dioxide was produced!

Eduard Buchner (1860–1917), Nobel Prize for Chemistry in 1907.

For the first time, it was possible to observe alcoholic fermentation in the absence of living yeast cells. This ground-breaking discovery of the enzyme zymase earned Eduard Buchner the Nobel Prize for Chemistry in 1907.

Two years after the death of Pasteur, he had shown that **fermentation is possible in the absence of living cells**.

Pasteur and Liebig – both of them were right in their own way. Fermentation is caused by microbes, but it is the enzymes they contain that are actually responsible for the chemical conversions. Enzymes can produce the same results *in vitro*, i.e. outside living cells.

James B. Sumner (1887–1955), in spite of having lost one arm, was a very skillful experimenter. He succeeded in crystallizing urease.

The next finding was truly sensational: Enzymes can be crystallized. In 1926, **James B. Sumner** (1887–1955) of Cornell University was the first to extract crystalline urease from jack beans and demonstrate that it had protein properties.

Between 1930 and 1933, **John H. Northrop** (1891–1975) of the Rockefeller Institute for Medical Research in New York City crystallized digestive enzymes, and after long debates, it was established that **enzymes were proteins**.

Sumner and Northrop were jointly awarded the Nobel Prize for Chemistry in 1946.

contain clusters of **highly reactive chemical groups** (proximity effect).

Electrical charges "lure" the substrate to the centre in what has been somewhat poetically described as the Circe effect. In a matter of seconds, the conversion process is completed, and the enzyme is back to its original state, able to convert further substrate molecules.

Compact arrangement and concerted action of the reactive groups in the active site of the enzyme molecule help reduce the **activation energy** to trigger a chemical reaction (free activation enthalpy). Compared to the energy required in the absence of an enzyme, it is a substantial reduction. By enabling metabolites to go through transitory stages, enzymes help establish an equilibrium of reactions at a much faster rate (Box 2.5).

A **3D image of lysozyme** shows very clearly the activity of the groups and forces involved in shaping the structure of an active site (Figs. 2.9 to 2.11).

The **molecule is stabilized** through various amino acid side groups, arranged in the peptide chain to stick out in all directions like the bristles of a bottle brush. This facilitates interaction between them, should the chain curl or tangle.

The molecule owes further stability to the stable chemical bonds that form between adjacent sulfur-containing side groups (S-H) of two cysteine building blocks. The resulting four **disulfide bridges** (S-S) provide the main support of the 3D structure of the lysozyme. Disulfide bonds in proteins can be destabilized by heavy metal ions, which accounts for their toxicity. They act on enzymes through non-competitive inhibition.

In a watery medium, the spherical organization of the enzyme molecule (Figs. 2.8 and 2.10) is stabilized by polar and nonpolar amino acid side groups. While the **polar groups are hydrophilic** (literally: water-friendly) and arranged on the outside where they are in contact with water, the **nonpolar side groups** (Fig. 2.10) – which are hydrophobic (literally: afraid of water) turn inward to avoid contact with water. They hold the molecule together in much the same way as drops of oil stabilize in water. Apart from the interaction between side groups, loose **hydrogen bonds** develop between neighboring oxygen and hydrogen atoms in the backbone of the various amino acids.

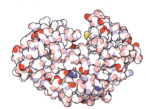

Fig. 2.7 Homer (top) describes how milk curdles when fig juice is added. Fig juice (middle) contains the protease ficin. Theodor Schwann was the first scientist to isolate a pure animal enzyme, pepsin (bottom).

Jöns Jacob Berzelius (1836)

"The catalytic force actually appears to consist in the ability of substances to arouse the affinities dormant …by their mere presence and not by their affinity and so as a result in a compound substance the elements become arranged in another way – such that a greater electrochemical neutralization is brought about."

Example Enzyme	Enzyme Class	Products

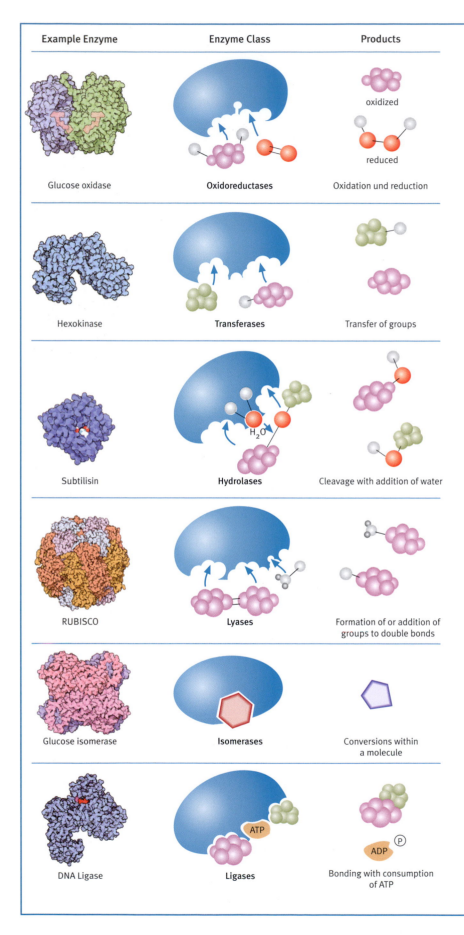

Glucose oxidase — **Oxidoreductases** — Oxidation und reduction (oxidized / reduced)

Hexokinase — **Transferases** — Transfer of groups

Subtilisin — **Hydrolases** (H₂O) — Cleavage with addition of water

RUBISCO — **Lyases** — Formation of or addition of groups to double bonds

Glucose isomerase — **Isomerases** — Conversions within a molecule

DNA Ligase — **Ligases** (ATP → ADP + P) — Bonding with consumption of ATP

Box 2.3 Naming and Classifying Enzymes

Emile Duclaux, a student of Louis Pasteur's, suggested in 1883 that enzymes should be characterized by adding the **suffix-ase to the name of their substrate**. All ester-degrading enzymes were to be called esterases, cellulose-degrading enzymes cellulases, and protein-degrading enzymes proteases. However, since many substances act as substrates to several enzymes, the reaction type was soon included in the name, for example glucose oxidase, glucose isomerase, and glucose dehydrogenase.

Alongside these names, a range of other, sometimes quite imaginative, names were also in use, such as respiratory ferment, pH5 enzyme, Old Yellow Ferment, which caused some confusion. Meanwhile, the number of known enzymes grew continuously. While in 1964, 900 enzymes had been described in detail, this figure grew to 1300 in 1968, and over 3000 today. This led the International Union of Biochemistry (IUB) to suggest that all enzymes should be classified into **six categories according to their specific action**. It is quite astonishing that the wide range of reactions in organisms can be categorized according to just six action principles.

Apart from its IUB classification, each enzyme is also given a code number, consisting of four digits separated by dots (class, group, subgroup, serial number) The name of the enzyme is made up of the names of the compounds involved in the enzyme reaction and the name of the major enzyme category.

Thus, the official name of glucose oxidase is β-D-glucose: O_2 oxidoreductase and the code number 1.1.3.4.

These exact enzyme names are very awkward and difficult to pronounce, so most enzymes are still called by their nicknames – glucose oxidase, pepsin etc. In scientific publications, however, their classification numbers and official names must be added.

Emile Duclaux (1840 - 1904) suggested using the suffix-ase to characterize enzymes.

2.3 The Role of Cofactors in Complex Enzymes

Not all enzymes consist exclusively of protein, as does lysozyme. Many include additional chemical components or **cofactors** which serve as tools. Such enzymes are known as qualified enzymes and have more complicated reaction mechanisms.

Cofactors can consist of one or more **inorganic ions** (such as Fe^{3+}, Mg^{2+}, Mn^{2+}, or Zn^{2+}) or more complex organic molecules, known as **coenzymes**. Some enzymes require both types of cofactors.

Coenzymes are organic compounds that bind to the active site of enzymes or near it. They modify the structure of the substrate or move electrons, protons and chemical groups back and forth between enzyme and substrate, negotiating considerable distances within the giant enzyme molecule. When used up, they separate from the molecule.

Many coenzymes are derived from **vitamin precursors**, which explains why we require a constant low-level supply of certain vitamins. One of the most essential coenzymes, NAD+ (**nicotinamide adenine dinucleotide**), is derived from niacin. Most water-soluble vitamins of the vitamin B group act as coenzyme precursors very much like niacin.

Otto Heinrich Warburg (1883-1970, Fig. 2.3) discovered the respiratory enzyme cytochrome oxidase (Ch. 1, Fig. 1.14) and NAD. His discovery and subsequent structural analysis was one of the shining hours of modern biochemistry. In the absence of niacin in the diet, certain enzymes (e.g., dehydrogenases) cannot work effectively in the body. The affected human will develop pellagra, a disease caused by vitamin B (niacin) deficiency. Otto Warburg developed an optical test making it possible to quantify reduced NADH at a wavelength of 340 nm (the oxidized NAD+ does not absorb light at this wavelength). It was now possible to measure essential enzyme reactions, such as the detection of glucose using glucose dehydrogenase (see Ch. 10)

Nowadays, **vitamins** like B_2 (riboflavin), B_{12} (cyanocobalamin), and **C** (ascorbic acid) are produced by the ton using biotechnological methods (Chapter 4).

Cofactors that are covalently bonded to the enzyme are called **prosthetic groups**. Flavin adenine dinucleotide (FAD) acts as a prosthetic group for glucose oxidase. Peroxidase and cytochrome P-450 contain

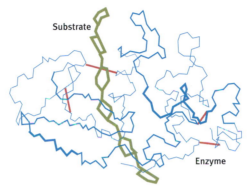

Substrate

Enzyme

a heme group, as found in myoglobin and hemoglobin. The heme group itself consists of a porphyrin ring incorporating an iron ion in its centre.

Coenzymes, by contrast, have only loose bonds, and, just like substrates, they undergo changes in the binding process and are used up. Unlike substrates, however, they bind to a whole host of enzymes (e.g., NADH and NADPH of nearly all dehydrogenases) and are regenerated and recycled inside the cells (see 2.13). Enzymes that bind to the same coenzyme usually resemble each other in their chemical mechanisms.

While I referred to the cofactors as "tools", the protein section of the enzyme is the "craftsman" using these tools, who is responsible for their effectiveness. As always, craftsmen and tools rely on each other to achieve the best possible result.

2.4 Animals, Plants, and Microorganisms As Enzyme Sources

Along with the discovery of digestive enzymes in the 19th century, methods were developed to obtain such **enzymes from slaughtered animals** (see Box 2.2). Even nowadays, **pepsin** is collected

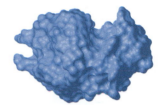

Fig. 2.8 Anatomy of an enzyme: Lysozyme
Above: Primary structure of lysozyme, the sequence of amino acids in the molecule (all 20 naturally occurring amino acids are present, abbreviated according to international nomenclature). The four disulfide (SS) bridges are shown in color. Below: Tertiary structure of lysozyme, the spatial arrangement of the peptide chain. In order to give a better overview, only the backbones of the peptide change and the sugar rings of the enzyme and its active centre are shown. (disulfide bridges in color).

Spatial structure of lysozyme from eggwhite.

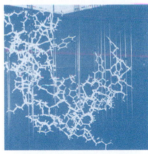

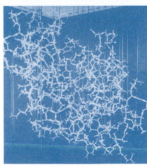

Fig. 2.9 Reconstruction of David Phillipps spatial model of lysozyme. Above: the first 38 amino acid residues. Centre: Residues 1-86. Below: The complete model of 129 residues.

Fig. 2.10 Tertiary structure of lysozyme and part of its substrate (the natural substrate is a polymer, and only a small part of it is shown here). All atoms of both molecules are shown, highlighting the side groups of the amino acids of the active site that are involved in / binding and converting the substrate. Asp52 and Glu35 stand for aspartic acid in position 52 of the peptide chain and glutamine in position 35 of a total of 129 amino acids.

Substrate
(six sugar rings)

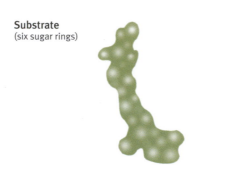

Lysozyme

Inside of the enzyme

Nonpolar hydrophobic amino acid side residues

Surface of the enzyme

Polar hydrophilic amino acid side residues

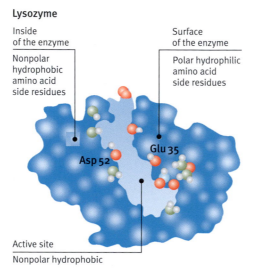

Asp 52

Glu 35

Active site
Nonpolar hydrophobic

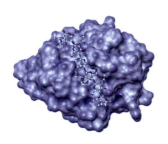

Fig. 2.11 Below: Temporal sequence of the breakdown of the substrate. Only sugar rings 4 and 5 are shown with the relevant atoms in the skeleton whose bonds undergo the cleavage.

1. The substrate binds to the active site.

2. Deformation of the 4th sugar ring.

3. A proton (H+) from glutamic acid attacks the bond between the 4th and 5th sugar rings.

4. A positively charged carbonium ion (C+) is formed on the 4th sugar ring and is stabilized by a negatively charged aspartic acid (Asp). The bond between the 4th and 5th ring is cleaved, and the first products (5th and 6th ring) are released from the active site. The carbonium ion bonds to a hydroxyl ion (OH-) from a water molecule. The proton of the water molecule fills the vacant place in the glutamic acid in the active site.

5. The second cleavage product is formed (1st to 4th ring) and detaches from the enzyme.

6. The enzyme has been totally regenerated, ready to bind to the next substrate molecule and convert it. Two substrate molecules per second are converted.

Lysozyme-substrate-complex

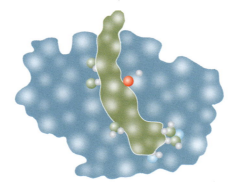

Substrate cleavage process as a time sequence

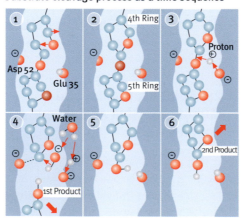

1 4th Ring

Asp 52

Glu 35 5th Ring

3 Proton

4 Water

1st Product

6 2nd Product

from the stomach lining of pigs and cattle, rennet from calf stomachs, and enzyme cocktails containing **trypsin, chymotrypsin, lipases,** and **amylases** from the pancreas of pigs. The pepsin wine bought in pharmacies usually contains pepsin from pigs' stomachs. Organs with a high metabolic turnover, such as muscle, liver, spleen, kidney, heart, and small intestine provide enzymes for analytical and medical purposes. They need to be purified to a high standard.

Not only animals, but also **plants** have been investigated as potential sources for industrial enzyme use. Grain, left to soak and germinate, turns into **malt** which contains amylases (starch-degrading enzymes) and proteases and has traditionally been used in beer brewing and spirit distillation.

Already in the 19th century, simple methods were in use to obtain high yields of proteases from the sap of tropical plants – **papain** and chymopapain from papaw trees (Fig. 2.19), **ficin** from fig trees, and **bromelain** from pineapple stalks. They are still in use for tenderizing meat to help digestion or for cleaning contact lenses (Fig. 2.18).

In Europe, however, it is rather difficult to obtain enzymes from plants, as supply and enzyme concentration vary with the seasons and the processing of large amounts of plants is very labor-intensive. While animal enzymes can be obtained as a byproduct of meat production, plant enzyme processing requires enormous quantities of plant material. However, both enzyme sources cannot satisfy the huge appetite of modern industry for enzymes. This is where **microorganisms** come in as an easily manageable, high-yielding source of enzyme production.

The industrial use of **microbially produced enzymes** began in 1894 when **Jokichi Takamine** (1854 - 1922) (Fig. 2.13), a Japanese-born biochemist working in Peoria (USA), patented his method of producing Takadiastase. His method of growing *fungi* in an **emersed culture** was both simple and ingenious. Nutrients and minerals were added to wheat straw which was then inoculated with sporangia of *Aspergillus oryzae*. The straw, kept in crates, was stored in incubating rooms. Once the fungi had grown, the straw was washed in a salt solution to extract the enzymes (amylases, proteases) secreted by the fungal cells. Right up to the end of World War II, there were enzyme factories in

Box 2.4 Biotech History:
Alexander Fleming's Cold and Its Implications for Enzymology

A cold would not stop **Alexander Fleming** (1881-1955) from coming to work in his London laboratory in 1922. Following his research instinct, he added some mucus from his nose to a bacterial culture and was perplexed when he found a few days later that something in the mucus seemed to have killed the bacteria. It could only be an enzyme that lyzed certain microbes, so he called it lysozyme. Soon, Fleming was able to detect lysozyme in all bodily fluids, such as tear fluid.

Alexander Fleming (1881-1955) discovered first lysozyme and later penicillin.

The story goes that all colleagues, students, and even visitors were expected to provide tear fluid for Fleming. Egg white is also very rich in lysozyme. This could be seen as a protective mechanism to avoid microbial invasion. However, it turned out that lysozyme, which was so effective in killing harmless microbes, had no effect whatsoever on pathogens.

It took two further years until Fleming discovered a very effective antibiotic, **penicillin**, in what seemed another hap-hazard experiment (see Chapter 4).

Lysozyme, however, was destined to take a very special place in the history of modern biology, as it was the first enzyme to be studied in its spatial structure. Its properties were explored in atom-level detail. The substrate of lysozyme is a molecule consisting of the sugar rings of acetylmuramic acid and acetyl glucosamine (Fig. 2.10), also called a mucopolysaccharide, which serves as a building component of cell walls.

When lysozyme cleaves its substrate, the cell becomes permeable, takes up liquid and finally bursts from the high osmotic pressure inside.

In 1963, the exact number and sequence of the amino acids (primary structure) of lysozyme isolated from egg white was established (Fig. 2.8).

The 20 amino acids differ in their side groups. It was not known, however, how the long protein thread consisting of 129 amino acids and a total of 1950 atoms could form an active site and bind and convert substrates.

Only through **X-ray structural analysis** was it possible to gain insights into the 3D structure of lysozyme. It was the third compound to be analyzed in that way, after the blood pigment hemoglobin and the muscle protein myoglobin (Nobel Prize in 1962 for **John C. Kendrew** (1917-1997) and **Max F. Perutz** (1914-2004), both from Cambridge, U.K).

David Philipps (1924-1999) and his colleagues at the Royal Institution in London already began in 1960 to work on the spatial structure of lysozyme. First, the protein was crystallized, and in spring 1965, the first 3D model of lysozyme was finished. It did not show a long stretched-out molecule, but a compact structure with a crevice. This remarkably deep crevice must contain the active site – it was the first time an active site in an enzyme could be shown by X-ray structural analysis (Figs. 2.8 - 2.11).

David Philipps (1924-1999) and lysozyme crystals.

What the 3D image did not show was the way substrates were bound and cleaved and where the energy for these processes came from. On the basis of the spatial structure, Philipps built a bit-by-bit wire-and-ball 3D model of lysozyme (Fig. 2.9).

Simultaneously, he built a 3D wire model of the part of its substrate that interacts with lysozyme. It consisted of six interconnected sugar rings, each of which was shaped like an armchair.

Three sugar rings were an exact key-and-lock fit in the upper part of the cleft. The "chair" conformation of the fourth one did not fit because several of its atoms clashed with the side groups of the amino acid side groups in the active site.

Philipps bent his wire model of the fourth sugar ring into an unnatural, "half-chair" conformation – and it was a perfect fit! What worked in the model must work for lysozyme in real life. After the fourth sugar ring had changed its shape to fit, the fifth and sixth ring fitted into the crevice without any further deformation. It was pre-

cisely between the taut and twisted fourth ring and the normal fifth ring that the substrate was cleaved (Fig. 2.11).

The cleft was mainly formed by nonpolar amino acids – i.e., their side groups did not carry an electric charge. Right next to the bond to be cleaved – between the fourth and fifth ring – there was the polar side group of aspartic acid (Asp) with a strongly negative charge. The opposite side of the crevice contained a side group of glutamic acid (amino acid no 35) (Fig. 2.11) Both side groups act like pliers on the bond they are supposed to break. For the first time, it was possible to show the mechanism of an enzyme reaction right down to the last atom. General rules about the structure of enzymes could be studied with the lysozyme model.

There was a further insight to be gained from the lysozyme model. According to **Emil Fischer's lock-and-key hypothesis**, one would have expected the substrate to fit precisely into the active site. However, the substrate only fitted after it had been re-shaped by the enzyme. Furthermore, it was not only the substrate that changed shape, but as was shown by X-ray structure analysis, the crevice in the lysozyme also deepened when it bound to the substrate. It resembled a rubber key and a rubber lock, or, to put it more elegantly, it was an **induced fit** like a hand in a glove.

Lysozyme in space!

Starting in 1985, NASA and ESA performed experiments in space with lysozyme as a model substance to quantify improvements in protein crystal growth in microgravity. The Space Shuttle Columbia carrying the International Microgravity Laboratory (IML-2) on board was launched on 8 July 1994. Columbia returned safely on 23 July after a very successful mission. It was also a milestone for microgravity research.

A flawless single crystal of lysozyme, grown in space, approximately 1.8 mm in length.

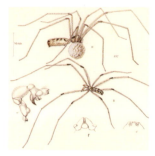

Fig. 2.12 The Daddy longlegs spider *Pholcus phalangioides* bites into the leg of a midge caught in its web to inject digestive juices. The victim is thus digested from within. The leg of the midge serves as a "straw" through which the whole body content of the midge is slurped up, taking up to 16 hours. It takes so long because the digestive enzyme has to reach the furthest ends of the midge body first before the meal is ready.

How fascinating! You, an alkaline protease, are a fellow omnivore?

Fig. 2.13 Jokichi Takamine (1854–1922) was one of the first industrial users of microbial enzymes to aid digestion. The product was called Takadiastase.

the US with a daily output of up to 10 tons of "mold fungus straw". They were replaced by **submerged culture** at the end of the 1950s.

Using **microorganisms** as sources of enzymes has obvious advantages, as it is possible to produce large amounts at comparatively low cost. Production is unaffected by seasonal changes other than location-linked restrictions. The use of suitable mutants, induction and selection processes can enhance production. It is even possible to produce tailor-made enzymes through genetic engineering and protein design.

Apart from the **high yields** required for industrial applications, the enzymes must also be **very stable**. As many microorganisms are able to survive under extreme conditions, the enzymes they produce must be extremely stable. The **thermophilic** (heat-loving) **microorganisms** found in the hot springs of the Yellowstone National Park or Kamchatka must produce heat-stable enzymes, or they would not be able to survive. The polymerase obtained from *Thermus aquaticus*, known as **Taq polymerase**, has become an indispensable tool in modern gene technology (s. Chapter 10). Thermostable enzymes are also found in microorganisms living in ordinary compost heaps which heat up considerably. Salty lakes are the habitat of **halophilic** (salt-loving) **bacteria**, whereas **psychrophilic** (cold-loving) **enzymes** are extracted from microorganisms found in Antarctica.

2.5 Extracellular Hydrolases Degrade Biopolymers into Smaller Manageable Units

Microbial enzyme production concentrates on **simple hydrolytic enzymes** (proteases, amylases, pectinases) that degrade natural polymers such as proteins, starches, or pectin. The microorganisms secrete the enzymes into their nutrient medium to make better use of it. These extracellular enzymes break up the giant molecules of the substrate into smaller ones that can feed the microorganisms. Similar processes are known in the animal kingdom, in spiders, for example, where they are referred to as extraintestinal digestion (Fig. 2.12).

It comes as no surprise that industry has so far mainly focused on extracellular enzymes, as they can be extracted easily and cheaply from a medium and do not require cost-and-labor-intensive cell disruption and purification procedures. A bacterial cell contains more than a thousand different enzymes, and any **intracellular enzyme** must be separable from other enzymes and cell structures. There are two distinct properties in proteins that can be used for separation purposes – molecular mass and electric charge.

All protein separation methods known to date utilize these techniques: salt precipitation, movement through an electric field (electrophoresis), binding to charged or non-charged carriers (chromatography), mass spectrometry, and other methods.

2.6 Amylases are Used For Brewing, Baking, and Desizing

Duke William IV of Bavaria (Fig. 1.28) issued the **purity law** for beer in 1516. It stated that "nothing but barley, hops, and water should be used for beer-brewing." This rule is still followed in Germany, whereas in most other countries, economic considerations have led to the **replacement of malt** by ungerminated grain, corn, or rice. As these do hardly contain any enzymes, ready-made enzyme preparations containing amylases, glucanases, and proteases obtained from mold fungi or bacteria, known as brewing enzymes, are added.

According to the Japanese brewery Kirin, the annual per capita consumption of beer (in 2004) was: Czech Republic 156.9 L, Ireland 131.1 L, Germany 115.8 L, Australia 109.9 L, followed by UK with 99 L. The US was in 13th position with 81.6 L., Japan at rank 32 with 31 L, and China, not a serious contender, with 22.1 L per capita. Worldwide, beer consumption in 2004 increased by approximately 6.1 million kL from the previous year, amounting to around 150.392 million kL. This is equivalent to 238 billion standard size bottles. The largest market is China. Having moved up one rank last year, China has secured its No. 1 position again this year. The volume consumed in China is 4.67 million kL more than that in the US, while Germany takes rank 3 behind the US.

The underlying processes, however, have remained the same as in ancient Egypt 2000 years ago.

The starch contained in grain is degraded enzymatically into sugars which then undergo alcoholic fermentation by yeasts. It is an energy storage compound in plants and one of the major nutrition sources for animals and humans. It is a polysaccha-

Box 2.5 How Do Enzymes Do It?

Enzymes speed up chemical reactions by a factor of 100 million to 1000 billion (10^8–10^{12}). Suppose an enzyme-catalyzed reaction is completed within one second, the same uncatalyzed reaction would, in theory, take 3000 years (factor of 10^{10}). The progress of such a reaction would be barely measurable. Most metabolic reactions in organisms would be so slow without enzymes that they would be completely pointless. Life is only possible due to enzymes.

The substrate (red) binds to the active site of lysozyme (blue).

For substances to interact with each other, they must first be activated, i.e., put into a reactive state. The energy needed to achieve this is called **activation energy**. The energy curve of a chemical reaction can be compared to a mountainous landscape (see Figure below) where the **starting compounds** lie like stones on the far side of the mountain. In order to roll into the valley on the near side (to be converted to **products**), they need to be pushed over the top with the help of activation energy. This energy could be supplied via a rise in temperature or pressure, which would kill any living cell. Broadly speaking, all

catalysts lower the activation energy (free or Gibbs activation enthalpy). In our image, they would flatten the mountain top to such an extent that only very little energy is needed to get across. This small amount of energy can be supplied fairly easily and frequently. The catalysts could also be compared to a mountain guide who, rather than climbing over the mountain tops, chooses a path via several lower passes, thereby saving energy. The stages of equilibrium are shown as a trough (initial stage) and the valley (end stage). The mountain top represents a labile **transition state** of the activated complex.

Enzymes do not change the equilibrium position of the reaction (that would mean interfering with the depth of the trough or the valley), but they make it possible to reach the **state of equilibrium much faster**. They only accelerate a process that would have taken place without them, albeit at a much slower – sometimes immeasurably slow – pace.

How this is achieved remains a matter of debate among enzymologists, and it seems that one explanation does not fit all enzyme reactions.

We have seen one possibility in lysozyme (Figs. 2.8 to 2.11) where the **substrate is deformed** by the enzyme and can only get out of its taut and twisted position (intermediate state) by product conversion.

A large amount of the energy required for the reaction is probably obtained during the **binding of the substrate** to the active site. In a solution, reactants are only brought together by chance, whereas enzymes bring them into **close proximity** in their active sites, thereby increasing the probability of a reaction.

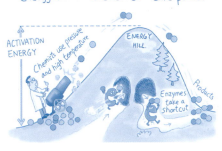

Only when the substances have climbed the "energy hill" a reaction can take place.

In the active site, **highly reactive functional groups** are concentrated in a very small space and arranged in a way that they are in direct contact with the bonds of the substrates they are going to modify, thus ensuring constant exposure. The active site contains mainly nonpolar groups, which makes it resemble a nonpolar organic solvent. Organic reactions are generally speeded up in a nonpolar organic solvent, compared to the reaction times in polar water. In the organic environment of the active site, the few existing charged **polar side groups of amino acids become super-reactive** in comparison to their behavior in a watery solution.

All the factors mentioned so far help explain why an enzymatic reaction is far superior to normal catalyzed chemical reactions, including those involving technical catalysts. This also explains why enzymes consist of such **enormous chain molecules** – it enables them to fold in order to arrange the required reactive groups in one place at a given time. An enzyme molecule is not a rigid structure, but flexible and deformable.

Mountain top

Activated complex

Uncatalyzed reaction

Activation energy of uncatalyzed reaction

Starting compounds

Trough

Enzyme-substrate complex

Enzyme reaction

Products

Valley

ΔG

Products

Energy required in an uncatalyzed reaction (back) and enzyme catalysis (front). The difference in activation energy between the reactions is striking.

Fig. 2.14 At the beginning of the 20th century, German industrialist and inventor of Plexiglas® Otto Röhm (1876-1939), former student of Eduard Buchner's, had the idea of cleaning laundry using diluted pancreatic enzyme extracts. His company produced a product called Burnus, containing pancreatic proteases, in which the laundry was soaked before washing.

Fig. 2.15 A packet of Burnus pancreatic enzyme pre-wash, as sold in 1914.

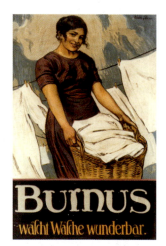

Fig. 2.16 Pectinases liquefy fruit.

ride exclusively composed of glucose components. **Malt** contains various starch-degrading amylases that cut the bonds between glucose residues in various ways, resulting in molecules of various sizes. Although malt contains no more than 0.5 to 1% amylases, it is the most significant supplier of enzymes worldwide. To obtain glucose, potato or corn starch must be degraded as completely as possible, which used to be done by exposing it to acids at raised temperatures (acid hydrolysis).

Over the last twenty years, however, amylases have gradually replaced acid hydrolysis. In a first step, starch is heated up to 176 to 221°F (80 to 105°C), broken down into short sections by **α-amylase** (see Fig. 4.43) and turned into a thin liquid. The resulting dextrin mix is then degraded by another amylase, **glucoamylase**, into its basic components, glucose. It is turned into pure dextrose through crystallization.

Heat-stable α-amylase is obtained from a *Bacillus* and remains active for two to three hours at a temperature of 203°F (95°C). Glucoamylase is derived from mold fungi (*Aspergillus* strains). Enzymatic saccharification results in a high yield of dextrose, shortened starch degradation times and does not require any acid treatment – a boon for the environment.

In **baking**, too, the addition of enzymes has become popular. Through the accelerated degradation of starch, the sugar content in the dough rises, which, in turn, accelerates the fermentation process. Proteases are added to degrade the "sticky proteins" (gluten) in the dough. Gluten partly binds water and has a gel-like consistency. It is degraded by proteases obtained from mold fungi to make the dough stretchier and increase its ability to retain air bubbles. The volume of enzyme bread rolls exceeds that of bread rolls prepared without enzymes.

In the **textile industry** (e.g., jeans production), the warp of the jeans fabric is treated with starch in order to make the fibers stick together and to enhance their robustness and elasticity during the weaving process.

Once the fabric has been woven, it must be desized, i.e., the starch must be removed. This used to be done by amylases originally obtained from malt or animal pancreases, but nowadays from bacteria. As these are fairly heat-stable, it is possible to work at high temperatures to speed up the process and bleach the material at the same time.

2.7 Pectinases Increase Fruit and Vegetable Juice Production

Fruit and vegetable juices have become part of a modern healthy lifestyle. When juice is pressed from fruit and vegetables, high-molecular pectins reduce the output. Every housewife knows pectin, obtained from apple cores, as a gelling aid when preparing jam and jelly. It is exactly this gelling effect that is not wanted when extracting juice.

Pectinases derived from *Aspergillus* and *Rhizopus* molds are produced in submerged cultures. Worldwide pectinases production is about 100 tons. Fruit and vegetables are chopped up and the pectinases added to degrade long-chain pectins. The viscosity of the juice is reduced, which helps filtration, and the yield is higher. Baby food is another major application area for pectinases. They macerate fruit and vegetables to make them softer and easier to eat.

Fruit yoghurt and cloudy fruit juices are also often the result of biotech production. In order to produce carrot juice or puree, not only pectinases, but also cellulases obtained from mold fungi, are added. Cellulases degrade cell walls.

2.8 Biological Detergents – the Most Important Application Area of Hydrolytic Enzymes

Stains that contain protein (e.g., milk, egg yolk, blood, or cocoa) are difficult to remove because proteins do not readily dissolve in water. At high temperatures, the protein curdles on the textile fibers and is even harder to remove. Dirty laundry contains a mixture of dust, soot, and organic matter such as fat, protein, carbohydrates, and pigments, and in bed linen and underwear in particular, **fat and proteins act as glue** that makes dirt stick.

During the laundering process, detergents detach fat from the fabric and distribute it in the washing liquid, while proteins are still stuck to the material. **Otto Röhm** (1876-1939), a former student of **Eduard Buchner**'s, pioneered the use of enzymes in laundering (Fig. 2.14).

However, the pancreatic enzymes used initially proved to be not very stable and very expensive.

The first biological detergents were unsuitable for mass production until 1960, when subtilisin was discovered in *Bacillus licheniformis*. The enzyme is active under alkaline conditions.

These days, biological laundry detergents are widely used. They contain proteases at a ratio of 1:50, which reach their optimal activity during the washing process. Their **low substrate specificity** makes them ideal omnivores, able to degrade sticky proteins into amino acids and short-chain peptides (Fig. 2.17). Proteins are thus detached from textile fibers and washed away.

From the mid-1960s, biological detergents have been widely sold in the US, Western Europe, and Japan. Initially, the dust arising from detergent production caused allergies in factory workers, but the problem was quickly solved by **granulation**. The particles were coated with wax to form granules. Liquid detergents and detergent powder in tablet form avoid the problem altogether (Box 2.6).

Raised energy awareness in recent years shifted the focus on a further property of biological detergents. As the enzymes involved reach their optimum action at 120 to 140°F (50 to 60°C), the laundry need not be boiled any more, thus saving precious energy. Often, amylases and lipases are added to the proteases in order to degrade starch residues and fat.

Cellulases are added to detergents to degrade microfibers sticking out from cotton fibers, giving the fabric a softer feel and making colors appear more vibrant. Dishwasher detergents also make more and more use of enzymes, mainly of **amylases and lipases**.

Enzymatic stain remover contains proteases, lipases, and amylases and dissolves not only stubborn stains of red wine, grass, fruit, coffee, and tea, but also the much more common mixed stains. Fruit ice cream or yoghurt or tomato sauce contain not just the pigment, but also protein and fat. Gravy, ketchup, and many ready meals contain a bit of everything: pigment, protein, fat, and starch.

Like detergents, stain removers also contain oxidizing bleaches that destroy enzymes. However, through successful **protein engineering**, it has been possible to fine-tune the *Bacillus* proteases to the laundering process. 1000 tons of genetically engineered proteases are produced every year. These are mainly

Soiled fiber

Fat | Dirt particle | Protein stain

Biological detergent

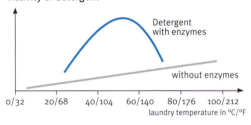

Enzyme | Peptides and amino acids

Activity of detergent

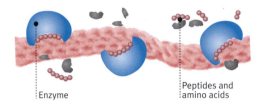

Detergent with enzymes

without enzymes

0/32   20/68   40/104   60/140   80/176   100/212
laundry temperature in °C/°F

subtilisin variants in which an amino acid sensitive to oxidation (methionine in position 222) has been substituted by a more stable amino acid (see Box 2.9 for details).

The enzymes retain their stability at pH 10 at 140°F (60°C) and are resistant to surfactants, complexing agents (water softeners) and oxidants.

■ 2.9 Proteases for Tenderizing Meat and Bating Leather

When the Spanish conquered Mexico, they noticed that the indigenous population wrapped leaves of the **papaya tree** (*Carica papaya*) around meat they wanted to cook or fry or they rubbed a slice of papaya fruit over it. There is method to the madness – the papaya contains high concentrations of the proteases papain and chymopapain. They degrade connective tissue in the meat, making it nice and tender.

Hundreds of tons of **papain** are used in the US every year for tenderizing meat. Other proteases – ficin from fig trees and bromelain from pineapple plants – are also suitable for the purpose. **Tenderizing powders** containing proteases are sold in many

Fig. 2.17 Diagram of biodetergent action (left). Serin proteases (below). A particularly reactive serin group in the active site lends its name to the proteases.

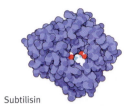

Subtilisin

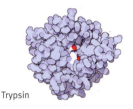

Trypsin

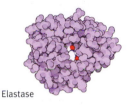

Elastase

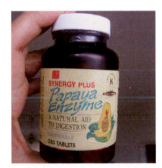

Fig. 2.18 Papain, obtained from the papaya fruit, sold as digestive aid.

Fig. 2.19 Papaya tree (*Carica papaya*).

Box 2.6 How Enzymes for Detergents Are Made

The enzyme-producing cells are cultured in 2,500 to 25,000 gal (10,000 to 100,000 liters) stirred vessels. The nutrient medium contains pre-degraded starch (5-15%) as carbohydrate and nutrient source, 2 - 3% soy flour, and 2% milk protein or gelatin as cheap sources of protein and nitrogen, as well as phosphate to stabilize the pH level. The medium is steam-sterilized for 20 to 30 minutes at 250°F (121°C) and then cooled down to a cultivation temperature of 86 to 140°F (30 to 60°C). The pH level is kept neutral and the culture well aerated.

Action of biological detergents.

The starting culture is a pure culture of a specifically chosen strain of *Bacillus licheniformis*. In a first step, nutrient medium is inoculated with the culture in a shaking flask. This becomes the inoculum for a smallish bioreactor (2.5 - 15 gallons or 10 - 50 liters). Often there is another intermediate production step at a volume of 130 - 260 gallons or 500 - 1000 liters before final production in the reactor goes ahead.

The bacteria use up the easily available nitrogen sources in the medium, and after 10 - 20 hours, subtilisin, a protease that they have been secreting, is detected in the medium. The presence of amylases for further degradation of the pre-degraded starches is only transient, while protease production will continue as long as the medium contains protein.

In order to prevent other microbes from settling in the culture and compromising the product, the contents of the bioreactor are cooled quickly to around 41°F (5°C). The cells (about 100 grams/liter) are centrifuged or filtered and then separated. The clear supernatant of the culture is concentrated through ultrafiltration. The membranes used in ultrafiltration have pores so small that even dissolved molecules like those of proteases are unable to pass, whereas water and dissolved salts and other small molecules pass unhindered. The process results in a ten-fold concentration of proteases.

Japanese washing machines do not come with a heating program because for decades, the Japanese have relied on biological detergents.

Dry enzyme preparations can be obtained through precipitation, but spray-drying is the method of choice. A solution containing enzyme is sprayed into a current of hot air in which water evaporates rapidly. The particles that form are sized between 0.5 and 2 millimeters. They are embedded in prills or coated with (marumerizer method) inert material and wax-like preparations, such as polyethylene glycols.

Fig 2.20 Otto Röhm (right) tanning leather.

Fig. 2.21 *Oropon* ® was the first industrially produced leather-bating product.

Fig. 2.22 Chicken and pigs benefit from phytase food additives. Farmers can reduce the phosphate content added to the feed and thus protect the environment from excessive phosphate contamination.

countries. These are rubbed into the meat which is then left to stand at room temperature for several hours. During this time, plant proteases degrade connective tissue proteins, such as collagen and elastin, speeding up natural maturing processes. We all know that **game has to hang** for a while to make it tastier. The animal's own proteases (cathepsins) are at work in the maturation process.

Microbial proteases have been also highly effective in **leather production**. After the hair was removed and prior to tanning, hides had to be softened by a process known as bating. Typically, this was carried out by immersing them in a mixture of water and fermented dog (or bird) dung.

Otto Röhm (Fig. 2.14) held the first German patent (1911) for the use of animal pancreatic enzymes in leather-bating in preparation for the tanning process (Figs. 2.20 and 2.21).

The enzymes replaced the traditional use of dogs' excrement which had not exactly added glamor to the image of the tanning profession. We know now, of course, why the dogs' turds worked – they are a breeding ground for bacteria that produce leather-softening proteases.

2.10 Immobilization – Reusing Enzymes

For a successful application of the aforementioned enzymatic processes, the following conditions must be met: The enzymes must be so **inexpensive** that they can be allowed to remain in the end product in an active or inactivated state or be discarded after use. This mainly applies to **extracellular hydrolases** (proteases, amylases, and lipases) which are stable and do not require cofactors.

It would be uneconomical to use intracellular enzymes that have been costly and labor-intensive to isolate in the same way. Methods must be found to **enhance enzyme stability and enable reusage**. In many applications, e.g., pharmaceutical use, the end product must be completely free from enzyme residues in order to avoid immune reactions. It is therefore necessary to find ways of separating the enzymes from the end product, for example, through fixation to the carrier material, also known as **immobilization** (Box 2.8).

When an intracellular enzyme is isolated from cells, it undergoes complex processing to free it from all impurities. Its properties and behavior are monitored *in vitro*, and it is often forgotten that this is a totally unnatural environment for the enzyme, so

Box 2.7 The Expert's View:
Phytase – Managing Phosphorus

Phosphorus is becoming a major focus of interest, as it is no longer available in abundance, but has become a rare and precious resource that needs to be managed with care on sustainability principles.

For decades, the massive impact of phosphorus used in industrial farming has been a major environmental concern. Phosphorus is used in **fertilizers** as well as **animal feed** fortified by inorganic phosphates. Both animals and plants need phosphorus – either to build bone substance or to produce the energy carrier adenosine triphosphate. Excessive amounts of phosphates and nitrates from agricultural run-off cause **eutrophication** (over-fertilization) in lakes and rivers, turning them into a smelly, slimy mess. In the Southern seas, such as the South China Sea in Hong Kong, eutrophication leads to what is known as Red Tide. Similar sights have been seen in the Baltic Sea Red Tide. They are caused by microscopic algae of the dinoflagellate family (dual-flagellated carapaced algae) that thrive on a combination of warm water and nutrients and produce toxins.

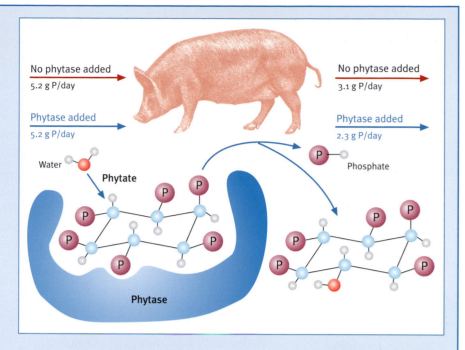

The **second global phosphorus rush** was brought about by the sudden dynamic growth of China and India, leading to uncontrolled surges and fluctuations of raw material prices. As a reaction, phosphorus management is beginning to get established, and its sharpest tool is phytase.

Monogastric animals such as chicken, pigs (Fig. 2.22), and humans cannot make use of a large proportion of the phosphates they take up in their food because the phosphate in plant seeds is stored as myo-inositolhexakis phosphate, also known as phytate, a compound that monogastric stomachs are unable to hydrolyze, due to insufficient enzyme activity. The multiple stomachs of ruminants, by contrast, are inhabited by microorganisms that produce phytase activity. Microbial **phytases** make phosphate groups bioavailable through hydrolytic cleavage. Phytase is produced by many microbes, including fungi, and excreted extracellularly.

Biotechnologists and animal nutritionists hit on the idea of producing microbial phytases (of fungal or bacterial origin) in bioreactors on an industrial scale in order to add it to pig and chicken feed. At the same time, the amount of inorganic phosphate feed additives was reduced. As **pigs** are notorious for their **phosphate-rich liquid manure**, they have been an environmental scourge.

Through phytase-catalyzed hydrolysis of phytate and reduction of phosphate feed additives, it was possible to reduce the phosphate excretion of the animals by a whopping 25-30%. In other words, the addition of inorganic phosphate to grain feed could be drastically reduced by liberating the phosphate contained in the grain itself. In countries with intensive animal farming, the use of phytase has already proved to be ecologically efficient, reducing phosphate emissions by several thousand tons a year as well as making the addition of inorganic phosphate almost obsolete. All this has contributed to the extraordinary global success of phytase.

One of the main global producers, the Danish company Novozymes, had a turnover of around US$ 100 million in 2003 from animal feed enzymes alone, a major part coming from phytase sales. The phytase is obtained from the fungus *Peniophora lycii* and is sold under the name *RONOZYME*® P by business partner DSM Nutritional Products. By far the most successful product has been temperature-stable *CT* granulate (*CT* stands for coated thermostable).

RONOZYME® *PCT* has been the only phytase product so far able to withstand the high temperatures of 176-185°F (80-85°C) at which animal feed pellets are processed.

While phytase products are booming, Green Biotechnology has been making progress by inserting phytase genes into corn, rice, and soy beans. Furthermore, plants with low phytate content and more bioavailable phosphorus are being developed by selective mutation.

However, as long as Green Biotechnology continues to struggle with technological problems and is not readily accepted in Europe, phytase looks like the more successful candidate in the short term.

Dr Frank Hatzack is Science Manager of Novozymes, Denmark.

Fig. 2.23 In the 1970s, Ichiro Chibata (above) and Tetsuyo Tosa, of Tanabe Seiyaku Co. Ltd (Osaka) developed a method using immobilized acylase – a world premiere.

Fig. 2.24 Atsuo Tanaka (Kyoto) immobilized microbes in customized gels. He gave the author of this book lessons on biotechnology in Japan.

Fig. 2.25 Miner with "immobilized" canary. The last canaries in British mines quit service in 1996.

its behavior could be equally unnatural. *In vitro* conditions are only normal for a few extracellular enzymes released from the cells into the surrounding medium. Inside the cell, most enzymes are in some way bound to the cell components or embedded in membranes. They are part of intricate complexes involving other proteins or fatty compounds called lipids. Efforts have been made to simulate a natural environment for the intracellular enzymes by bonding them to artificial carriers and membranes or enclose them in gel. Enzymes that have been bonded in that way have been largely immobilized and are called immobilized or carrier-bound enzymes.

Immobilized enzymes can be seen in analogy to the "immobilized nightingales" on the bird market in Hong Kong (Fig. 2.27). The birds are shut into narrow cages (polymer gels) from which they cannot escape (leakage) and are protected from cats (proteases, microbes). Food, water, and oxygen (substrate) enter freely, while metabolites are easily disposed of. Their activity can be seen and heard. Due to their long lifespan, they are re-usable. By the way, in the Western coal mining industry, **canaries** were kept in cages underground as **living biosensors** (see Chapter 10) to warn miners of carbon monoxide poisoning (Fig. 2.25). In the presence of carbon monoxide, the bird would drop dead.

With the production cost of enzymes remaining comparatively high, it makes sense to apply the same economical methods the cells use, i.e., to stabilize enzymes and use them several times over in substrate processing.

The success stories of immobilized glucose isomerase and aminoacylase demonstrate that it is possible to create the right working conditions for immobilized enzymes.

■ 2.11 Glucose Isomerase and Fructose Syrup – Boosting the Sweetness of Sugar

The worldwide consumption of sugar is continuously going up, while sugar cane and sugar beet have not changed their high demands on soil and climatic conditions. Furthermore, the raw material needs to be processed immediately after harvesting to avoid losses. Starch, by contrast, the natural storage product in plants, can be obtained from a wide range of plants grown under all sorts of condi-

tions (potatoes, grain, cassava, sweet potatoes). It is storable and can easily be broken down into sugar (glucose).

As we have seen in beer-brewing, starch can be degraded on an industrial scale using amylases. The resulting **glucose** has, however, one major drawback – it has only **75% of the sweetness of saccharose** (sucrose), so more glucose is needed to achieve the same sweetening effect.

Fructose, by contrast, contains about 80% more sweetness than saccharose, which makes it twice as sweet as glucose. As fructose is an **isomer** of glucose with the same empirical formula ($C_6H_{12}O_6$) and the same atoms, it should be possible to restructure glucose into fructose, thus **doubling its sweetening power**. Read about the development of HFCS (high fructose corn syrup) in Box 2.10.

Currently, the worldwide production of **glucose isomerase** amounts to roughly 100,000 tons, whereas fructose production is nine to ten million tons.

Fructose syrup is the preferred drink sweetener in the US. Glucose isomerase is mostly derived from *Streptomyces* spp. and then immobilized. Mostly killed-off and disrupted microbial cells with intact glucose isomerase are used and linked with glutardialdehyde to stabilize them.

Fructose is a highly attractive choice for food manufacturers, as it is absorbed more rapidly than other sugars – ideal for **sports drinks**. It enhances the taste of fruit or chocolate and masks the bitter taste of artificial sweeteners (Fig. 2.26). It softens ice cream by lowering its freezing point, making it creamier, and easy to eat. Clinical tests have shown that diabetics find it easier to control their glucose levels by eating food containing fructose rather than saccharose or starch.

Fructose is mainly metabolized by the liver, independent of insulin levels. It is therefore a major dietary nutrient, and, due to its higher sweetening power, fewer calories land on the plate of the consumer.

In the Western world, most fructose is produced from starch though amylase degradation and subsequent restructuring through glucose isomerase. Alternatively, **sucrose** can be used as a starting **product**. Through acid hydrolysis or cleavage using the enzyme invertase, invert sugar is produced, i.e., fructose plus glucose. The not so sweet glucose

Box 2.8 Immobilized Enzymes

The point of immobilized enzymes is that they are recyclable. They are bound to large carriers which are visible to the naked eye and can be separated mechanically from the reaction solution (e.g. by filtration).

A large range of immobilization techniques has been developed. Enzymes can be bound to their carriers by chemical (covalent bonds) or physical means (adsorption or electrostatic forces).

Special reagents are used to cross-link enzyme molecules, or the enzyme molecules are mechanically entrapped by gel or hollow fibers.

The production of immobilized enzymes for industrial processes must be simple and relatively cheap. Enzyme activity must be high in relation to the carrier mass, and the enzymes must be very stable within their action range. Immobilized enzymes are used in various types of enzyme reactors, most of which are either column reactors or stirrer vessels.

They have obvious advantages over soluble enzymes: they are recyclable, they have the desired chemical and physical properties and often also higher stability in a wider pH range and at higher temperatures. The end product is free of enzyme residues.

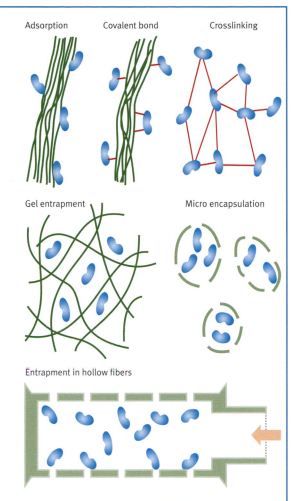

Adsorption Covalent bond Crosslinking

Gel entrapment Micro encapsulation

Entrapment in hollow fibers

Fig. 2.26 Fructose is used as a low-calory sweetener for diabetics.

Fig. 2.27 An analogy to immobilized enzymes – caged bird at the bird market in Hong Kong. Substrates (air, food) are delivered into the cage, while its products (CO_2, excrement) are disposed of. The cage prevents the bird from flying away and protects it from cats.

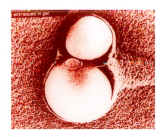

Fig. 2.28 Yeast cells entrapped (immobilized) in polymers retain their activity – even their ability to sprout.

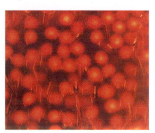

Fig. 2.29 Alginate beads containing immobilized yeast in a column reactor. The yeast is fed by a sugar solution running through the reactor.

which is also undesirable for diabetics is separated through chromatography.

The use of immobilized glucose isomerase reduces the total production cost by about 40%, compared to the use of soluble enzyme.

■ 2.12 Immobilized Enzymes in Human and Animal Food Production

The example of glucose isomerase showed that immobilized enzymes make the production of cheap products even cheaper, whereas the next example, **penicillin acylase**, used for tailor-made modifications of the penicillin molecule, will show how immobilized enzymes can also be used effectively in the production of small amounts of high-value products (Ch. 4).

Humans, pigs, and other non-ruminants do not synthesize amino acids such as lysine and methionine – or they synthesize them far too slowly. These essential compounds must therefore be taken up with food.

There is a fast-growing need for **essential amino acids** as animal food additives and for medical purposes (infusion solution). In industrial production, fermentation and chemical methods have replaced the conventional isolation of L-amino acids from protein hydrolysates (s. Ch. 4). Chemically synthesized amino acids are optically inactive mixtures (racemates) of D and L-isomers. Only the optical L-version (exception: methionine) is physiologically active and in demand in medicine, sport (Fig. 2.30), and animal food production.

In Japan, **Ichiro Chibata** (Fig. 2.23) and his group developed a process by which chemically produced racemates of acetyl-D-, L-amino acids are enzymatically cleaved into L-amino acids and unhydrolyzed acetyl-D amino acid (Fig. 2.34). Due to its reduced solubility, the desired L-amino acid can be easily

Box 2.9 The Expert's View:
Protein Engineering: Tailor-Made Enzymes

Protein Engineering has held a fascination for me ever since I began to work in this field as a post-doc almost 20 years ago. As a chemist by training, I had the opportunity to discover the weird and wonderful world of 3D protein structures. During my studies, I began to realize that 3D protein structures can tell us something about the relationship between structure and function of proteins. My initial fascination never tired, and I am always keen to find out more through reading and my own research.

My first encounter with the world of 3D protein structures came when building a model of the *Bacillus*-derived protease subtilisin as a postdoc researcher at the GBF in Braunschweig, Germany.

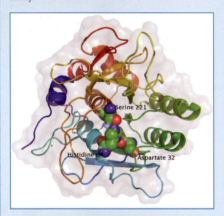

Cartoon of subtilisin Carlsberg (PDB code 1CSE) displaying the secondary and tertiary structure and the water-accessible surface of the enzyme. The residues of the catalytic triad are shown as spheres. The key to the function of this enzyme lies in just three of the 275 amino acids. The atoms of these three residues are highlighted as spheres. The colors indicate the chemical elements, which make up the amino acids (red=oxygen, green=carbon, blue=nitrogen, yellow=sulfur, hydrogen is invisible). The three residues, aspartate number 32, histidine 64, and serine 221 are known as the catalytic triad of subtilisin. The side chain oxygen of serine 221 actively breaks the peptide bond of the proteinaceous substrate, while the two other residues of the catalytic triad assist this bond breaking-process by shuffling electrons.

Subtilisin is widely used in the detergent industry for the removal of protein-based stains like egg, milk, or blood. The figure above gives an overview of the subtilisin 3D-structure. The picture shows a cartoon of this 275 amino acid protein. It contains a central beta-sheet system, flanked by alpha-helices on either side. In this view, we are looking into the active site of the enzyme. As can be seen, it is wide open and easily accessible for the substrates. This feature might be the reason why subtilisin is so effective against the various proteins normally found in laundry stains.

A solid knowledge of the structure of proteins is essential to much of the protein engineering work. "Protein Engineering" is the combined use of several techniques of modern biotechnology to modify the properties of natural (wild-type) proteins, mainly enzymes.

Parallels can be drawn between protein engineering and a traditional mechanical engineering process like the design and production of cars. Car designers propose variations on a prototype car. Some engineers produce the cars, while other engineers test the new prototype: is it suitable for racing, the transport of goods, or as a family car?

Protein designers propose modifications of a wildtype enzyme. Molecular biologists then instruct a host cell to produce the modified enzyme, which is subsequently tested by biochemists. This is done for several variants, until the most suitable enzymes are selected for mass production on an industrial scale.

Protein engineering has been shaped by several technologies, most importantly by those developed in molecular biology. The discovery of **DNA-polymerases**, which catalyze the polymerase chain reaction (PCR), was the biggest step forward in protein engineering. PCR enables protein engineers to modify genes to the specifications of protein designers that will yield the desired proteins.

In parallel, rapid changes took place in a completely unrelated field that had a bearing on protein engineering – the **IT revolution**. Faster and ever faster computer systems enabled more and more scientists to elucidate 3D protein structures and evaluate them on screen or with computerized data analysis tools in order to plan changes in a protein.

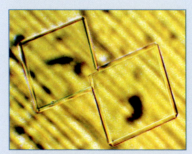

Subtilisin crystals.

Two main approaches are used in modern protein engineering: The random generation of multiple mutations in a gene, which are then screened for desirable changes in the properties of the protein. The alternative approach is called "Rational Protein Engineering". Structural and functional information about the protein in question is analyzed to predict successful mutations.

The use of random mutation technologies has been dubbed "Directed Evolution".

Directed evolution techniques in protein engineering can be described as the man-made version of natural evolution. In this process, the protein engineering team generates a library of variants of the target protein by using molecular biology tools. This library is then screened for variants showing the desired properties. One of the improved variants is taken as the basis for a new cycle of mutagenesis, screening, and selection. This process of variant generation, screening, and selection is repeated until a variant of the protein has been identified that fulfills the criteria for the successful use of the enzyme.

Rational Protein Engineering uses a knowledge-based approach, i.e., scientists in a protein engineering project determine the three-dimensional structure of the enzyme (with X-ray crystallography or NMR techniques). The 3D-structure sheds light on the function of the various amino acids of a protein. The 3D-structure often reveals a protein-substrate or protein-inhibitor complex. Thorough biochemical investigation further clarifies the relation between structure and function of the investigated enzyme. This information is used to produce targeted mutations in the enzyme. These are then painstakingly analyzed to identify the most successful mutations and develop strategies for further mutations.

In practice, many protein engineering projects rely on a combination of both approaches, as they complement each other.

The groundwork for protein engineering has been done through academic research. Once the

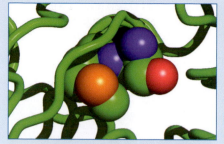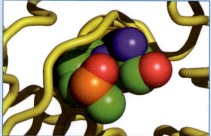

Active site of the native (PDB code 1CSE, left) and the oxidized (PDB code 1ST2, right) form of subtilisin. The residues 221 and 222 are shown as spheres.

fundamental techniques had been developed, academic researchers began to use protein engineering to further study and analyze the catalytic mechanisms of various enzymes. They were thus able to explain and identify general principles for the stability of proteins and get insight into the mystery of the surprisingly fast protein-folding process.

While I was still finding my way into protein engineering, my future colleagues at Genencor (**Scott Power, Tom Graycar,** and **Rick Bott**), then working for Genentech in South San Francisco, CA, were able to demonstrate the effectiveness of rational protein engineering by improving subtilisin for **detergent applications**.

A loss of enzyme activity had been observed in the presence of **bleaching agents** in a detergent formulation. The bleaches used in detergents are oxidizing reagents like hydrogen peroxide or peracetic acid. The enzyme lost about 80 percent of its original activity during the washing process. Apparently, this partial loss of enzyme activity was caused by the presence of bleach. This would explain why it underperformed when it came to removing protein stains. To support this hypothesis, the enzyme was incubated with hydrogen peroxide. Again, it lost 80% of its activity. Finally, the researcher isolated the oxidized subtilisin, crystallized it, determined the 3D-structure, and compared it with the crystal structure of the native (non-oxidized) subtilisin.

The figure (above) shows the small, but significant differences around the active site in the two enzyme structures. On the right-hand side of the picture, an extra oxygen atom is visible on the sulfur atom of methionine at position 222, which is next to the active site serine 221.

An experienced observer will notice that the side chain (indicated by the red sphere of the oxygen atom) of the serine adopts a slightly different position in the oxidized enzyme.

This minor displacement, which is caused by a hydrogen bond between the sulfoxide (oxidized sulfur) of methionine 222 and the alcohol func-

tion in the side chain of serine 221, is the reason for the loss of catalytic efficiency.

When methionine was replaced by one of the other 19 amino acids, the resulting variants turned out to be **bleach-resistant** and produced better stain removal results in bleach-containing detergents than the wildtype enzyme. The procedure was repeated for other detergent enzymes and led e.g. to the development of bleach-resistant α-amylases (for degrading starch).

As the example about bleach stabilization of enzymes suggests, protein engineering efforts in industry focus very much on the applicability of enzymes in technical processes. In many cases the wildtype enzymes are not suited to the often harsh conditions in technical applications.

The **synthesis of intermediates** is another area of major interest, mainly for the production of drugs in the pharmaceutical industry. The interest of the industry focuses mainly on the capability of enzymes to catalyze enantioselective synthesis resulting in enantiopure compounds.

Enzymes are not enantioselective towards all substrates, especially if the substrates are not the "natural substrates" of the enzymes. Protein engineering techniques are used to make these enzymes enantioselective towards the non-natural substrates. Thus, the synthesis is directed towards the desired version of the two enantiomers of a compound.

Protein engineering has been used to improve the efficiency of many enzymes for industrial applications. These improvements were mainly driven by the wish to make enzyme usage more cost-effective.

An underestimated side effect of protein engineering in industry is a **tremendous saving in energy and raw materials** due to enzyme production.

In a life cycle assessment, it has been shown that improved performance of a detergent protease through protein engineering in combination with an optimized microbial production system led to

a more than 10-fold reduction of all relevant input parameters like energy and raw materials. Consequently, the waste stream consisting of carbon dioxide, water, inorganic salts, and cell mass was reduced more than ten-fold as well.

Protein engineering is a modern research discipline of biotechnology. It has deepened our understanding of the relationship between the structure and function of proteins.

This work that was initiated in a university context yielded results that were soon adapted for commercial purposes. Protein engineering has certainly paved the way for the use of more and more enzymes in industrial processes. I believe that we are only at the beginning of enzyme usage in industry. Other industrial sectors, like the pulp and paper industry or the cosmetic industry, need still to be convinced of the economic and other advantages that enzymatic catalysts have to offer.

The most convincing argument for the introduction of a new technology is the **demonstration of benefits of a new technology** in comparison with traditional methods. The coming years will see an increasing use of enzymes tailor-made by protein engineers.

Wolfgang Aehle is a senior scientist for Protein Engineering at Genencor International, a Danisco Company. After finishing his Ph.D. in Organic Chemistry, he discovered the beauty and excitement of working with protein 3D-structures during his postdoc period on molecular modeling of subtilisin at the GBF in Braunschweig (Germany). Aehle further pursued his career in industry as a protein engineer. He has focused his work on a significant number of industrial enzymes. The main goal of his work is to improve enzyme function in the industrial environment. His work has been documented in numerous scientific publications and patent applications. He is the editor of the handbook "Enzymes in Industry".

Box 2.10 Fructose Syrup

The first purely chemical attempts to isomerize fructose using technical catalysts at a high pH level were unsuccessful, resulting in dark and nasty-tasting byproducts. These would have been very expensive to remove.

In 1957, xylose isomerase was discovered, which has not only the ability to isomerize xylose into xylulose, but also glucose into fructose. As this additional activity of the enzyme has become the focus of commercial interest, it is nowadays often called glucose isomerase. It is an intracellular enzyme obtained from various microorganisms, e.g. *Streptomyces* spp.

Transport of fructose syrup in the US.

The first enzymatic glucose isomerization process was patented in the US in 1960, and in 1966, Japanese researchers in Chiba City described an industrial process that uses soluble glucose isomerase. The industrial isomerization process of glucose results in a very sweet mix of glucose and fructose which can be used as syrup instead of crystalline sucrose.

In the US, the Clinton Corn Processing Company started producing glucose/fructose syrup using glucose isomerase in 1967. Initially, the syrup contained only 15% fructose, and it

became clear that the glucose isomerase process could only be economically viable if the expensive enzyme could be reused. Fortunately, glucose isomerase is an ideal enzyme to be immobilized, as it is stable at high temperatures, and the substrate (glucose) as well as the product (fructose) consist of very small molecules, so when the immobilized enzyme is packed into columns, diffusion problems are negligible. As glucose and fructose molecules do not carry electric charges, glucose isomerase could be bonded to charged cellulose derivatives as carrier material. Otherwise, both substrate and product would bind to the carrier.

In 1968, the Clinton Corn Processing Company introduced a discontinuous fructose production method using immobilized enzymes which yielded 42% fructose. In 1972, a continuous process using immobilized glucose isomerase was developed.

However, while it was important to have found a technological solution, its breakthrough depended on the market price of sugar.

In the 1960s, the sugar price was 15 to 20 US cents per kilogram, and it was impossible to produce fructose syrup more cheaply than that. At that time, the enzymatic production process was still fraught with problems and drawbacks. Not only was it necessary to overcome preconceptions within the industry, but a whole new production method had to be developed. A complex system of pressure filters and an apparatus for the removal of cobalt, a heavy metal used as an enzyme stabilizer, had to be developed.

When in November 1974 the sugar prices climbed to $1.25 per kilogram, isomerization suddenly became a very attractive alternative. At that time, the Danish firm Novo Industry A/S

was developing a more cost-effective method of producing glucose isomerase that required no addition of cobalt and could withstand the high pressures of industrial reactors. The columns of these reactors could be over seven yards high.

In 1976, 750 tons of immobilized glucose isomerase were produced in the US alone, enabling the production of 800,000 tons of 42% fructose syrup. When sugar prices fell again to 15 cents per kilogram at the end of 1976, the new production method was already well established, and fructose syrup was being produced at a lower cost than that of sucrose. In 1978, further progress was made and due to a new separation procedure, the fructose content went up to at least 55%. It was only 15 to 25% more expensive than the 42% syrup and was suitable for acidic drinks such as Coca Cola (pH value 4.0) that required a syrup of at least 55% fructose to replace sucrose. This opened the doors to the use of fructose syrup in a huge market.

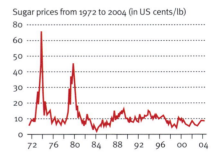

Sugar prices from 1972 to 2004 (in US cents/lb)

Fig. 2.30 In Japan, amino acids are added to fitness drinks.

separated from the acetylated D-amino acid. The undesired D-amino acid is reused in the synthesis of the racemate. This process was first carried out using soluble aminoacylase from kidneys or *Aspergillus oryzae*, and the enzyme was not recycled.

However, in certain cases (e.g., infusion solutions containing amino acids), it is crucial to **remove all traces of enzymes** to avoid immune reactions in the patient. Such removal procedures used to be costly and reduce productivity, so immobilization offered a viable alternative. In a straightforward low-cost procedure, the enzyme binds through adsorption to a DEAE cellulose carrier (Fig. 2.33) and remains stable – losing only half its activity after 65 days. Tanabe Seiyaku (Osaka, Japan) was the

first company to use immobilized enzymes in industry.

From 1969 on, this method was applied to produce L-phenylalanine, L-valine, and L-alanine on an industrial scale (Fig. 2.33 and Ch. 4). The total cost of the process using immobilized aminoacylase is about 40% below that of using soluble enzymes (Fig. 2.34). As the process has been largely automated, labor cost has been reduced, the cost of catalysts has gone down, and productivity has increased. Using immobilized enzymes is arguably more effective.

The same rate of success has not been achieved in the production of other enzymes, although some of them, e.g., soluble glucoamylase has been used in industry for 20 years. Many successful attempts

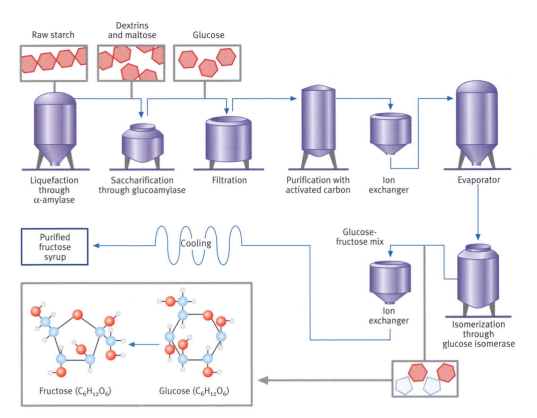

α-Amylase

Glucoamylase

Glucose isomerase

Liquefaction through α-amylase | Saccharification through glucoamylase | Filtration | Purification with activated carbon | Ion exchanger | Evaporator

Raw starch | Dextrins and maltose | Glucose

Purified fructose syrup

Cooling

Glucose-fructose mix

Ion exchanger

Isomerization through glucose isomerase

Fructose (C₆H₁₂O₆) | Glucose (C₆H₁₂O₆)

have been made to immobilize glucoamylase, but in spite of all the technological advantages, immobilized enzyme production could not compete with the soluble enzyme which is so easy and cheap to produce. In the dairy industry, both immobilized and soluble lactase are used to produce lactose-free, easily digestible milk.

Another soluble enzyme is making inroads on the animal food market. Phytase (Box 2.7) cleaves phosphate groups from forage grain, helping to reduce the need for expensive phosphate additives and thereby also reducing the phosphate rate in chicken and pig excrements to a more environmentally compatible level.

■ 2.13 Making Use of Cofactor Regeneration – Enzyme Membrane Reactors

The development of **enzyme membrane reactors** which use cofactor-dependent enzymes opened up a new era in enzyme application. The used-up cofactors, which would be very expensive to replace, are enzymatically regenerated in a continuous process (Box 2.11).

In membrane reactors, cofactor-dependant (NADH) amino acid dehydrogenase and **formate dehydrogenase** (FDH, used in the regeneration of cofac-

tors) are placed side by side between ultrafiltration membranes through which substrate solution is passed. The enzymes cannot penetrate through the membranes. To prevent the low-molecular, weight cofactor from leaving the reaction chamber, **polymer carriers** (polyethylene glycol, PEG) are bound to it, retaining it by virtue of their much larger molecular size. With the participation of the cofactor, amino acid dehydrogenase converts keto acids

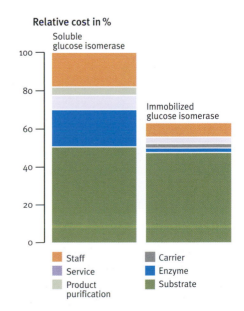

Relative cost in %

Soluble glucose isomerase

Immobilized glucose isomerase

Staff | Carrier
Service | Enzyme
Product purification | Substrate

Practical tip: Fructose as a hangover remedy

Some experts recommend taking a spoonful of fructose after a good night out. Some people, however, have a fructose intolerance that could even be fatal. It is safer to take in fructose in honey or jam or fresh fruit. In the liver, fructose is broken down by fructokinase ten times faster than glucose by hexokinase. In the process, cofactor NAD⁺ is produced, which is indispensable for the degradation of alcohol (by alcohol dehydrogenase and acetaldehyde dehydrogenase). Alcohol and hangover-causing acetaldehyde are broken down much faster.

Fig. 2.32 Production cost of fructose syrup using soluble glucose isomerase compared with the cost of using immobilized glucose isomerase.

Fig. 2.33 The first industrial-scale process using immobilized enzyme: Amino acylase is used to separate synthesized D,L-acetyl amino acid into physiologically active L-amino acid which, not being easily soluble, is crystallized out. Inactive acetyl α-amino acid is converted back into the starting mixture by chemical or thermal means. Gradually, all of the starting material can be used up in the process if it is appropriately carried out.

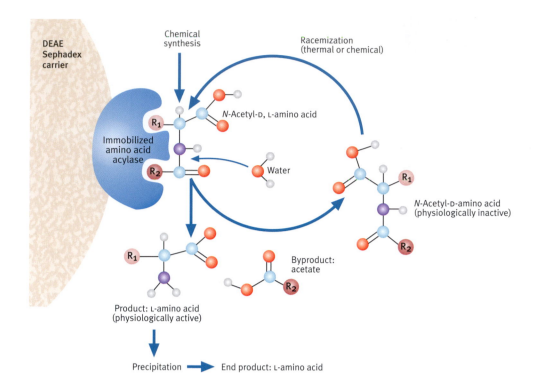

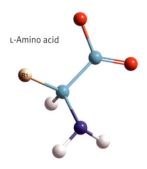

L-Amino acid

into L-amino acids (Fig. 2.37) which diffuse through the membranes. The once so expensive cofactor NADH is oxidized to NAD⁺ and thus no longer useful to the enzyme. However, the other enzyme trapped in the reactor chamber, formate dehydrogenase, has the ability to turn cheap formic acid into CO_2 and reduce NAD⁺ to NADH in the process. NADH, the expensive cofactor, has been restored and can be reused. The concept of the regeneration of cofactors in an enzyme membrane reactor, brainchild of **Christian Wandrey** (Box 2.11), **Maria-Regina Kula,** and **Frits Bückmann** from Jülich and Braunschweig in Germany, proved successful. Over a period of 90 days, each of the molecules in the bioreactor was regenerated 700,000 to 900,000 times from NAD⁺ (Fig. 2.37).

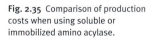

Fig. 2.34 Fortune, the kitten, enjoying a bowlful of lactose-free milk, produced using immobilized lactase (β-galactosidase). Lactase cleaves lactose into glucose and galactose.

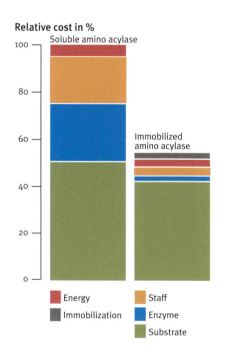

Relative cost in %

Fig. 2.35 Comparison of production costs when using soluble or immobilized amino acylase.

■ 2.14 Immobilized Cells

Not only enzymes, but entire cells can be immobilized. Japan has been pioneering research in the field. **Ichiro Chibata** and **Tetsuya Tosa**, who worked for Tanabe Seiyaku in Osaka and were in the vanguard of enzyme immobilization, developed a cell immobilization application in 1973. It enables killed-off cells of *Escherichia coli* enclosed in gel to synthesize 600 tons of the amino acid **aspartate from fumaric acid** (see also Ch. 4). It took 120 days for *E. coli* bacteria activity to lose half of their activity, compared to free cells that had a half-life of just ten days. Using immobilized cells lowers production costs to 60% of the cost of using free cells. The cost of using catalysts drops from 30% in the use of free cells to 3%, the cost of labor and energy by 15%. A 1000 liter column reactor yielded nearly two tons (!) of L-aspartate per day.

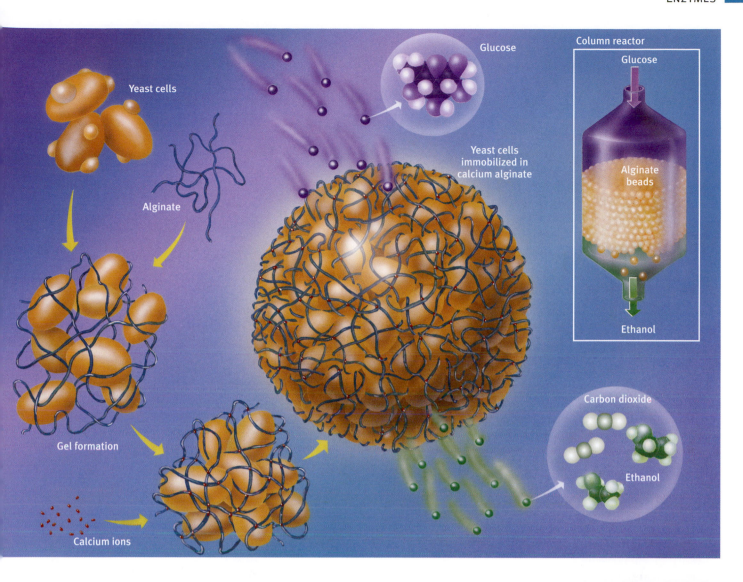

Glucose

Yeast cells

Alginate

Yeast cells immobilized in calcium alginate

Column reactor

Glucose

Alginate beads

Ethanol

Gel formation

Calcium ions

Carbon dioxide

Ethanol

Fig. 2.36. Ethanol production in immobilized yeast.

The immobilization, i.e., the fixation of cells onto a carrier, in ethanol production has several advantages. The cells can be re-used and have an extended lifespan. As the desired end product is largely free of biological compounds and cells, some of the conventional purification steps are not needed.

One of the most common immobilization methods is the inclusion of cells in gels where a cell suspension is mixed with gel-forming substances.

Alginate is a product of marine algae that is used by the food industry for the preparation of gels. In the presence of calcium ions, it forms a firm stable gel structure. Alginate beads are produced by adding drops of an alginate-cell mixture to a calcium chloride solution. The penetrating calcium ions cause the alginate to gel and entrap cells such as yeast cells in the process.

Small molecules such as glucose can pass through the gel pores to reach the cells while their metabolic products (alcohol and carbon dioxide) can exit the beads. The living yeast cells remain intact.

In order to tightly pack the alginate beads with cells, the yeast cells are first encouraged to grow by the ample provision of broth and oxygen. Once the available space in the beads is filled with cells, the beads are layered into a column bioreactor. Glucose solution flows through the reactor, providing the immobilized cells with substrate. As oxygen is only in short supply now, the yeast cells ferment the glucose into alcohol.

The alcohol and gaseous CO_2 exit the beads. Pilot plants working with immobilized cells have been able to produce alcohol from glucose continuously for months on end.

Plant and animal cells can also be immobilized in a similar way.

Not only beads, but also films containing immobilized cells can be made and used, for example, as membranes in biosensors.

The *Arxula* yeast is immobilized and used to measure the biochemical oxygen demand (BOD) in wastewater biosensors (see Chapter 10). The respiration of yeast cells immobilized in a polymer membrane is measured by an oxygen sensor. If the water does not contain any biodegradable compounds (clean water), the yeast will not ingest any nutrients and thus not use up any oxygen. The biosensor is able to deliver results within five minutes, whereas conventional monitoring takes five days.

Immobilized yeast cells.

Box 2.11 Biotech History:
Christian Wandrey, the Enzyme Membrane Reactor and Designer Bugs

Hermann Sahm, Maria-Regina Kula, and Christian Wandrey.

In summer 1973, Prof **Karl Schügerl**, doyen of chemical engineering in Germany, gave me the best advice I was ever given: "Wandrey, why don't you do something totally different and study biotechnology with Professor Kula?"

Maria-Regina Kula (b. 1937), who was an authority in biotechnology already, agreed to meet me in a motorway restaurant somewhere between Hanover and Braunschweig. Timewise, this seemed to be the most economical arrangement. I went because I had been told to do so, but did not expect much to come out of it.

However, the meeting did have far-reaching consequences – the fact that nowadays the Japanese, the world champions of amino acid production, import L-methionine from Germany is just one of them.

When I first visited Kula's lab, I asked all the naive questions a chemical engineer might ask a biochemist.
"Do these – what-do-you-call-them – enzymes really survive ultrafiltration?"

Her reply could not have been more to the point: "Well, why do you think we are doing it?"

"And can you run water over them to wash out impurities?

"Of course you can!"

The first EMR

I see. And when you run substrate solution over the enzyme you should get a product. If you introduced a membrane, the product could be kept separate from the enzymes which could then be reused in the reactor.

At that time, the immobilization of enzymes was all the rage. Enzymes were cross-linked, covalently bound or embedded in gels (see main text, page 45). The only trouble was that quite often, they lost their activity in the process. This is where our totally different approach came in. Due to the membrane, there was no need to interfere with enzyme activity, and yet the enzymes were separated from the product. The enzyme membrane reactor (EMR) was born.

The EMR had the additional advantage that it could be topped up with enzyme, something that could not be done with immobilized enzymes.

Only later did I find out that Alan Michaels (after whom AMICON has been named), had published a similar idea in 1968, but never put it into practice. Unfortunately, our wonderful idea was not patentable.

The Degussa EMR in Nanning.

We first tried it out with amino acylase, an enzyme that had been successfully immobilized by **Ichiro Chibata** at Tanabe Seiyaku in Japan on ion-exchange particles. It worked and the product in the ultrafiltered solution was quantified then with a flow polarimeter.

Today, our first EMR prototype can be admired in the Deutsches Museum among the "100 most important technological inventions". The notice says that the reactor produces 75% of the world supply of L-methionine. This, of course, does not refer to the 10 milliliter prototype reactor!

In 1981, the first EMR was launched in Konstanz, Germany. The German company **Degussa** opened an EMR plant with an annual production capacity of 500 tons of L-methionine in Nanning, China, in 2005.

So, why can't L-methionine be produced through fermentation? Well, the biosynthesis of S-containing amino acids is (as yet) too complicated.

Coffee on the train and cofactor regeneration

In 1976, when I had just been awarded my PhD, I sat on the train with Prof Kula, sipping coffee on the way back from a conference on biotechnology. Somebody at the conference had suggested producing gluconic acid (see Ch. 6, p. 189) using glucose dehydrogenase. The only snag was that this required cofactor NAD⁺, which cost a fortune then and would have driven the cost of industrial gluconic acid through the roof (read about Warburg's discovery of NAD in the main text, page 30).

While I was stirring my coffee, my thoughts went round in circles too – or should I say cycles? What about regenerating the cofactor?

Well, it would be possible to regenerate the expensive cofactor NADH from NAD⁺ using a dehydrogenase, Prof Kula said, but there was something even better – formate dehydrogenase (FDH). Her colleague **Hermann Sahm** had found it in a yeast, *Candida boidinii*, and Maria-Regina Kula had purified it.

FDH produces active hydrogen in the shape of NADH and CO_2. The gaseous CO_2 would be released from the reaction mix anyway, and formic acid was dirt cheap. The reaction was quickly scribbled on a napkin.

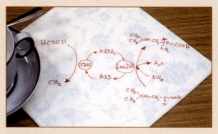

The historical cup of coffee with the basic idea for cofactor regeneration scribbled on a paper napkin.

The continuous release of CO_2 would tip the balance towards an increased production.

Our enthusiasm grew by the minute: Better still, if we took ammonium formate, that would provide the nitrogen source for the reductive amination of alpha keto-acids (2-oxoacids in modern nomenclature).

At that time, L-amino acids were produced by separating racemates (p. 50), which yielded only 50% and required further recycling. Our method, at least in theory, would yield 100%. It looked like what is nowadays called a win-win option.

α-Keto acids, the starting material, are easily chemically synthesized.

The Degussa team in Nanning (China).

No sooner said than done – starting with leucine dehydrogenase of which considerable quantities were sitting in Maria Kula's refrigerator.

To top it all, we produced not just natural essential leucine, but an artificial version called *tert-leucine*. This "fake leucine" could be inserted into artificial peptides without infringing on their action. As they were not easily degraded by proteases in the blood, they would be more stable than natural peptides. This looked like a clever little money-spinner.

So far so good, but what about those small NAD^+ and NADH molecules that passed through the membrane along with the product and needed to be replaced rather than regenerated? One would need to augment their molecular weight considerably without affecting their solubility in water, we thought. A carrier was needed onto which NAD could bind, but polystyrol or sephadex globules were no use because they did not dissolve in water.

Coenzymes are transport metabolites, i.e., they function as shuttles between different enzymes

and are highly mobile. They can be compared to a pendulum (see the cartoon).

The solution was to **bind NAD^+ to polyethylene glycols (PEGs)** which are completely soluble in water. The late **Frits Andreas Bückmann** of GBF Braunschweig knew the art of synthesis inside out. He bound NAD^+ to PEG, and it worked!

The dehydrogenases coped with the carrier-bound, high-molecular-weight NAD^+ of approximately 20,000 daltons because it was easily water-soluble. Like dogs chained to their huts, they behaved and remained highly reactive in their immobilized state in the EMR.

Nobel Prize winner Otto Heinrich Warburg on a German stamp: Warburg found the structures of NAD and NADP.

So, to keep the picture of the pendulum: the NAD-pendulum is loaded with two hydrogen atoms by the FDH. It swings to the amino acid dehydrogenase which picks the two hydrogens and uses it for synthesis. The "empty" pendulum swings back to the FDH and so on…

Soon, my students in Jülich reported that they were able to use the coenzyme pendulum 10,000 times and later, they could even run 30,000 cycles.

I promised a bottle of German champagne to the student who would manage 100,000 cycles.

"What about a bottle of real champagne for 200,000 cycles?" asked a smart female doctoral student. Okay, I went as far as paying for three bottles of champagne for 600,000 cycles. What a bright bunch of students I had!

The Future – the Cell-Membrane Reactor

What does the future hold for EMRs?

Ultrafiltration membranes have been developed to such a high standard that they can retain cofactors without the need for PEG. Besides, the price of NAD has fallen to 1 Euro per gram, so some loss of NAD will no longer hurt.

However, microbial cells are still unsurpassed as enzyme membrane reactors. Some of them have been genetically engineered to express higher levels of the desired enzymes as well as regenerating enzymes for their cofactors. In one of our startup companies in Jülich, they produce chiral alcohols. These designer bugs could also be described as cell membrane reactors that work highly efficiently within technical membrane reactors.

Back to Nature!

Christian Wandrey began his career as a doctoral student at the University of Hanover, Germany, and received his PhD in 1973. He was Assistant Professor in Hanover from 1974 to 1977 and then changed to the University of Clausthal. In 1979, he was given a chair in Biotechnology/Chemical Engineering at the University of Bonn and became Director at the Institute of Biotechnology at the Research Centre Jülich – posts he still holds today. Prof Wandrey has published more than 300 scientific papers, holds 100 patents and helped several of his former students to set up their own biotech companies. He received prestigious awards like the Philip Morris Award, the Gauss Medal, and the Wöhler Award.

Fig. 2.37 Working principle of an Enzyme Membrane Reactor (EMR) for the production of L-amino acids. The oxidized cofactor NAD⁺ is bound to the large water-soluble polyethylene glycol (PEG) molecule and is thus retained in the reactor. It is continuously reduced by formate dehydrogenase (FDH) and thus regenerated to be used by L-amino acid dehydrogenase.

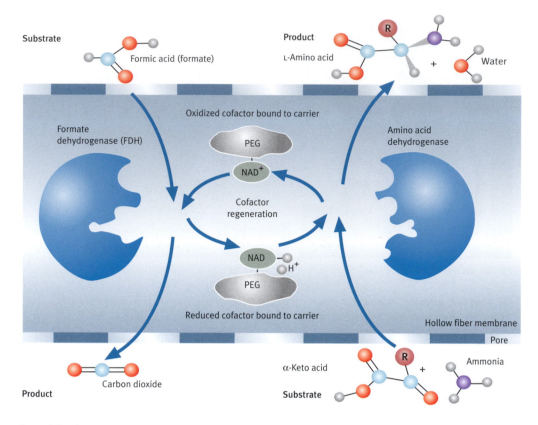

Substrate — Formic acid (formate)

Product — L-Amino acid + Water

Oxidized cofactor bound to carrier

Formate dehydrogenase (FDH)

PEG / NAD⁺

Cofactor regeneration

NAD / H⁺

PEG

Reduced cofactor bound to carrier

Amino acid dehydrogenase

Hollow fiber membrane

Pore

Product — Carbon dioxide

Substrate — α-Keto acid + Ammonia

Fig. 2.38 Time-honored fermentation traditions combined with modern technology: yeast cells entrapped and immobilized in alginate beads provide a constant supply of alcohol for use in Japanese households (cartoon from the 1980ies).

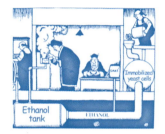

Fig. 2.39 An alcohol distillery in 1900 where amylases were used to break down starch.

Fig. 2.40 Contemporary ethanol plant in Japan where immobilized yeasts are used.

Immobilized microorganisms are an obvious choice for processes involving several steps such as **alcohol production** with yeast. The ethanol is obtained anaerobically from glucose in a multi-step reaction involving cofactor-regenerating systems in yeast cells.

The Japanese firm **Kyowa Hakko Kogyo Co.** (Fig. 2.40) started a pilot project in 1982, running five reactor columns, 4 cubic meters capacity each, to continuously produce 8.5% ethanol from sugar cane molasses. The reactor used living yeast cells enclosed in **alginate beads** (Fig. 2.36). Alginate is obtained from seaweed (Chapter 7) and is also used as a gelling agent in the food industry. The pilot plant produced 2,400 liters of pure alcohol per day.

The advantages of cell immobilization become apparent when comparing the **productivity** of their applications to conventional procedures. It is about 20 times higher than that of conventional production. Since the process can be run continuously, it is automated and computer-controlled, thus reducing labor costs. The ethanol plant stands here as a model for more sophisticated industrial synthesizing plants.

I think that enzymes are molecules that are complementary in structure to the activated complexes of the reactions that they catalyze, that is, the molecular configuration that is intermediate between the reacting substances and the products of the reaction.

The attraction of the enzyme molecule for the activated complex would thus lead to a decrease in its energy and hence a decrease in the energy of activation of the reaction and to an increase in the rate of the reaction.

Linus Pauling (1948)

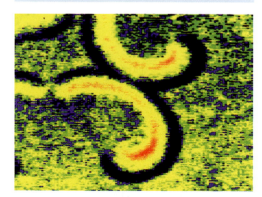

Fig. 2.41 Rotating spiral waves of NADH during the glycolysis in cytoplasm extracted of yeast cells.

Cited and recommended Literature

- **Aehle W** (2007) *Enzymes in Industry*, 3rd edn. Wiley-VCH, Weinheim. The complete and updated book on industrial applications of enzymes.

- **Buchholz K, Kasche V, Bornscheuer UT** (2005) *Biocatalysts and Enzyme Technology*. Wiley-VCH, Weinheim. Very good overview of our current knowledge of biocatalysis and enzyme technology.

- **Tanaka, A, Tosa T, Kobayashi T** (1993) *Industrial Applications of Immobilized Biocatalysts*. Marcel Dekker, NY. The classical monograph on immobilized biocatalysts.

- **Fischer E** (1894). *"Einfluss der Configuration auf die Wirkung der Enzyme"*. Ber. Dt. Chem. Ges. 27: 2985-2993. The classical paper on lock-and-key theory.

- **Koshland DE** (1958). *"Application of a Theory of Enzyme Specificity to Protein Synthesis"*. Proc. Natl. Acad. Sci. 44 (2): 98-104. The induced-fit hypothesis.

- **Blake CC, Koenig DF, Mair GA, North AC, Phillips DC, Sarma VR** (1965). *Structure of hen egg-white lysozyme.* "*A three-dimensional Fourier synthesis at 2 Angstrom resolution*". Nature 22 (206): 757-761. Lysozyme 3D structure by D Phillips.

Useful weblinks

- A starting point, the Wikipedia: *http://en.wikipedia.org/wiki/Enzyme*

- British website for enzyme technology, based on a textbook: *http://www.lsbu.ac.uk/biology/enztech/*

- Professor David Goodsell (Scripps Institute, La Jolla) who contributed most of the molecular graphics to this book, is running a homepage "Molecule of the Month" giving structures and details of biomolecules in a great way. *http://mgl.scripps.edu/people/goodsell/pdb*

- Web tutorial on enzyme structure and function: *http://tutor.lscf.ucsb.edu/instdev/sears/biochemistry/tw-enz/tabs-enzymes-frames.htm*

- BRENDA database, created by Dietmar Schomburg (University of Braunschweig). A comprehensive compilation of information and literature references about all known enzymes: *http://www.brenda.uni-koeln.de/*

- How lysozyme binds the substrate, computer animation: *http://www.locuspharma.com/structuremovies/lysozyme.php*

- Enzyme Structures Database links to the known 3-D structure data of enzymes in the Protein Data Bank. *http://www.ebi.ac.uk/thornton-srv/databases/enzymes/*

- World Wide Web Virtual Library: Biotechnology: catalog of WWW resources related to biotechnology; contains over 3500 links to companies, research institutes, universities, sources of information, and other directories specific to biotechnology. Emphasis on product development and the delivery of products and services. *http://www.cato.com/biotech/*

- Microbes-A look through Leeuwenhoek's microscope: Brian J Ford found Leeuwenhoek's microscope and samples and confirmed that he saw indeed the bacteria. *http://www.brianjford.com/wav-mics.htm*

- All about microorganisms: *http://www.microbes.info/*

- Beer, bread, wine, cheese, fermented food. Extensive and mostly reliable information by Wikipedia, the Free Encyclopedia For example Wine *http://en.wikipedia.org/wiki/Wine*

Eight Self-Test Questions

1. Why is it next to impossible for the metabolism of a cell to run without enzymes?

2. Are biocatalysts always proteins?

3. What strategies do enzymes use to lower the activation energy of reactions?

4. In what way did David Phillips's lysozyme model modify Emil Fischer's lock-and-key hypothesis? What is meant by Koshland's idea of an "induced fit"?

5. What are the advantages of immobilizing enzymes? Where are immobilized enzymes used?

6. What are the most important industrial applications of enzymes?

7. Why do Japanese washing machines not have a heating program?

8. Why and through what process have cofactors in enzyme membrane reactors been regenerated to produce high-value amino acids?

THE WONDERS OF GENE TECHNOLOGY

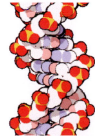

CHAPTER **3**

Fig. 3.1 James D. Watson (b. 1928)

Fig. 3.2 Francis C. Crick (1916-2004)

Fig. 3.3 DNA double helix. Sydney Brenner described his first impression: "The moment I saw the DNA model...I realized it is the key of understanding to all problems of biology."

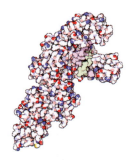

Fig. 3.4 DNA polymerase. The template strand is shown in purple, the newly formed strand in green.

■ 3.1 DNA – the Double Helix as a Physical Carrier of Genetic Material

At the heart of the biological revolution is the helix of life, or DNA – short for deoxyribonucleic acid. When in 1953, the journal "Nature" published an article written by two young scientists, **James Dewey Watson** (b. 1928) and **Francis Compton Crick** (1916-2004), it was the end of a long search for the structure of the physical carrier of genetic material (Figs. 3.1 and 3.2). A simple, but ingenious diagram of a double helix (Fig. 3.3) was included in the article.

To put it simply, DNA could be compared to a twisted zipper that uses four different kinds of teeth – the **four bases adenine (A), cytosine (C), guanine (G), and thymine (T)**. These bases are the components of nucleotides, the actual building elements of DNA. The nucleotides, in turn, consist of a sugar, a base and a phosphate residue (Fig. 3.8). This is not only a space-saving arrangement, but it also makes the information accessible from all directions.

Deoxyribose is the sugar carried by nucleotides. In contrast to ribose, the sugar of ribonucleic acid or RNA, deoxyribose lacks an oxygen atom at the 2' carbon. The backbone consists of a mix of deoxyribose and phosphate units. Thus, the sugars are interconnected via phosphodiester bridges. The four bases sit on the supporting backbone like teeth on the fabric strip of a zipper. What matters for genetic information is the **sequence of the bases**.

The teeth in a zipper are held together mechanically, while the bases on the DNA strands are held in place by molecular interaction. **Hydrogen bonds** form between bases lying opposite each other (Fig. 3.8).

In 1950, **Erwin Chargaff** (1905-2002, Fig. 3.5) used chromatography to show that in all organisms, adenine and thymine were always present in the same amounts, and the same applied to guanine and cytosine (**Chargaff's rule**) (Fig. 3.6). Now the Watson and Crick DNA model offered an explanation: A bases and T bases, on the one hand, and C bases and G bases, on the other are accurate spatial fits. G and C are held together by three, and A and T by two hydrogen bonds (Fig. 3.8). This base pairrule or **Watson-Crick rule** is a prerequisite for correct transfer of genetic information.

■ 3.2 DNA Polymerases Catalyze the Replication of the DNA Double Strand

The reproduction of cells is the basis of any transfer of hereditary traits. Two cells develop from one cell, each of them carrying identical genetic information. In order to achieve this, the DNA must **replicate**, i.e., create an exact copy of itself before the cell undergoes mitotic division. This is done by opening the DNA zipper. The two single strands separate, and the enzyme **DNA polymerase** (Fig. 3.4) locks on to the free ends to synthesize new DNA strands, thereby forming two new double helixes.

DNA polymerase is an enzyme of the transferase category (Ch. 2) and has the ability to transfer entire chemical groups. It was discovered by **Arthur Kornberg** (b. 1918, Nobel Prize 1959) who isolated it from *E. coli*. During the replication process, a liberated A-nucleotide attaches to a T-nucleotide provided by the cell, a liberated C-nucleotide attaches to a G-nucleotide etc.

While the "front" of the newly attached nucleotides is connected to the corresponding base via hydrogen bonds, the polymerase binds their deoxyribose and phosphate elements on their "back" into a firm, supportive "backbone structure".

DNA polymerase (Fig. 3.4) cleaves off pyrophosphate and catalyzes the formation of a phosphodiester bond between the 3'-OH terminus of the existing sequence and the 5'-triphosphate terminus of the newly attached nucleotide.

To do this, the enzyme needs a **template**, which can be a single DNA strand or a partially separated double strand.

Furthermore, the polymerase needs all components as activated precursor compounds, i.e., deoxynucleoside-5-triphosphates (**dNTPs**) and a double-stranded starting point or primer with a free 3'-hydroxyl group (see also Ch. 10).

The DNA polymerase shown in Fig. 3.4 displays dual activity. It attaches to a short single-strand section (nick) of an otherwise double-stranded DNA molecule, synthesizing a whole new strand (polymerase activity) by progressively breaking down the existing strand (nuclease activity). Degradation and synthesis are happening simultaneously, while the nick travels along the DNA. This is a troubleshooting device – wrongly placed nucleotides are being replaced, resulting in an incredibly accurate DNA replication at an error ratio of one error per 100,000,000 copied base pairs. Imagine retyping several thousands of novels and just making one single mistake!

The polymerase molecule is shaped like a human right hand. The space between the fingers and the right thumb is taken up by DNA, while the polymerase activity is happening on the "index and middle finger". In the center of the molecule, nuclease activity is going on, the proof-reading of the newly attached nucleotides. Thus, two accurate DNA copies arise from one double helix, each of them a complete molecule. When the mother cell divides, these will be passed on to the two daughter cells.

The polymerases used in the **polymerase chain reaction** (PCR, see Ch. 10), an indispensable technique in gene technology, are usually thermostable DNA polymerases obtained from the microorganism *Thermus aquaticus*, also called **taq-polymerase**. *Thermus aquaticus*, a member of the archaea family, is found in places like the boiling hot springs of Yellowstone National Park.

■ 3.3 Not All Genes Are Encrypted in DNA. RNA Viruses Use Single-Stranded RNA

The genomes of nearly all organisms consist of DNA. Some viruses, however, use **RNA** (**ribonucleic acid**) enclosed in a protein envelope. The tobacco mosaic virus (TMV), a tobacco leaf pathogen, consists of a single-stranded RNA molecule containing 6395 nucleotides and enclosed by a protein envelope consisting of 2130 identical subunits (Fig. 3.7). Its replication is catalyzed by an **RNA polymerase**.

Another important category of RNA viruses are known as **retroviruses**. Their genetic information does not follow the usual flow from RNA to DNA, but travels "backwards" (retro). Retroviruses include the AIDS pathogen HIV and some cancer viruses (Ch. 5).

They contain two copies of a single-stranded RNA. When entering the host cell, the RNA is transcribed into DNA by a viral enzyme called **reverse transcriptase** (**revertase**) (see 3.13 and Fig. 3.35). The obtained double-stranded DNA is integrated into the host genome. The host cell then produces new viral

RNA and new viral envelope proteins that assemble to produce infectious viral particles (see Ch. 5).

■ 3.4 Deciphering the Genetic Code

After the four bases had been identified as carriers of genetic information, the question arose as to how the information is coded in the base sequence.

The first clue came in the 1940s when US geneticists **George W. Beadle** (1903-1989) and **Edward L. Tatum** (1909-1975) from Stanford University in collaboration with **Joshua Lederberg** (b. 1925) from Yale, founder of phage genetics, postulated that **the production of an enzyme is controlled by a gene**.

Enzymes are protein molecules, and it was thought that, for example, the gene for the blue color of flowers controlled the production of the enzyme in control of blue flower pigment. Beadle, Tatum, and Lederberg were awarded the Nobel Prize in Physiology in 1959.

However, the key question remained to be answered: in what way did the DNA assembly instructions control the structure of proteins? Protein molecules come in various sizes and are all composed of just **20 amino acids**. The properties of a protein depend on the properties, number, and sequence of the amino acids they contain (as shown in the lysozyme example in Ch. 2).

The instructions for the integration of amino acids into the protein must be contained in the DNA sequence and its familiar ACGT base "language". The language of proteins is totally different, it is the language of amino acids, and the key that translates DNA into amino acids is called the **genetic code**.

In the 1950s, the Watson-Crick double helix model gave rise to a lot of speculation about the nature of the genetic code. Four different bases encode the makeup of 20 different amino acids. It had to be a **combination of several bases** that carried the instructions for the integration of a particular amino acid into a protein. A combination of two bases would allow for 4x4=16 different combinations – not enough to encode 20 different amino acids. Thus, a combination of at least three bases was required (**triplets** or **codons**) – 4x4x4=64. As only 20 and not 64 sets of instructions are required, it was assumed that several triplets control the inte-

Fig 3.5 Erwin Chargaff (1905-2002)

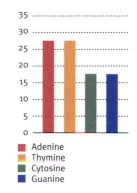

Fig. 3.6 Composition of human DNA. The distribution of the bases A, T, C, and G is given in percentages, demonstrating Chargaff's rule.

Fig. 3.7 Structure of the tobacco mosaic virus (TMV) with its protein subunits and its single-stranded RNA molecule.

Fig. 3.8 From DNA to protein. Top right: structure of the DNA double helix. Bottom right: transcription of the DNA base sequence into mRNA and subsequent translation into an amino acid sequence on the ribosome.

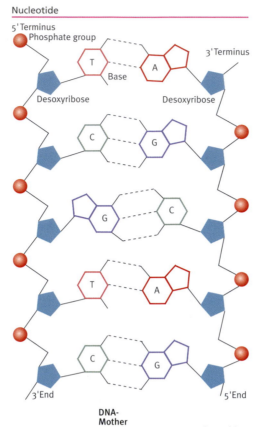

Nucleotide

5' Terminus
Phosphate group
Base
Desoxyribose
3' Terminus
Desoxyribose

T — A

C — G

G — C

T — A

C — G

3' End
5' End

Fig. 3.9 RNA polymerase with newly synthesized mRNA.

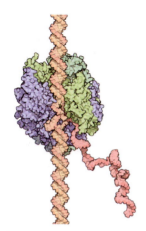

Fig. 3.10 Marshall Nirenberg (right, b. 1927) and Heinrich Matthaei (b. 1929).

Fig. 3.11 Har Gobind Khorana (b. 1922), Nobel Prize winner 1968 with Nirenberg and Holley.

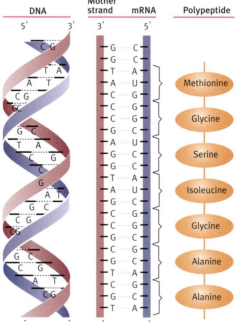

DNA	DNA-Mother strand	mRNA	Polypeptide
5' 3'	3'	5'	
C G	G	C	
	G	C	
T A	T	A	
A T	A	U	Methionine
C G	A	C	
G C	C	G	
	C	G	Glycine
G C	G	C	
	A	U	
T A	G	C	Serine
C G	G	C	
	T	A	
G A T	A	U	Isoleucine
G C	G	C	
C G	C	G	
C G	C	G	Glycine
	C	G	
G C	G	C	
C G	G	C	Alanine
A T	T	A	
C G	C	G	
G C	G	C	Alanine
	T	A	

gration of one single amino acid. This code is referred to as "**degenerate**".

At the end of the 1950s, it was also known that proteins are not produced directly on the DNA. The genomic DNA of prokaryotic bacteria floats freely in the plasma, while eukaryotic DNA is lined up in the nucleus (Fig. 3.16). Eukaryotes contain not only genomic DNA, but also mitochondrial and, in the case of plants, chloroplast DNA (ctDNA).

Proteins are formed in the cell plasma outside the nucleus in specific protein factories known as **ribosomes**. There must be a nucleus-to-ribosome transfer.

How does the information contained in nuclear DNA get to the ribosome in the cell plasma? It is done very economically – via a messenger.

■ 3.5 The Human Genome – A Giant 23-Volume Encyclopedia

3,080,000,000 characters, the equivalent of 750 megabytes in digital information terms, make up the human genome. It would fill about 5000 books the size of this textbook, but would fit on one single DVD.

Humans possess 29,000 **protein-coding genes**, and each of the 23 **human chromosomes** carries an average of 1000 different genes.

The human genome could fill an extra-fat 23-volume encyclopedia or 2000 New York phone directories (Ch. 10).

Each volume (chromosome) of the encyclopedia (genome) contains about 1000 protein assembly instructions (genes) of varying length. They are written in a strange three-letter word (codon) language in an alphabet that has no more than four letters. Some instructions only take up a few lines, others spread over several pages.

Each of the protein assembly instructions (**exons**) is interspersed with so far incomprehensible sections (**introns**). Sometimes, these seemingly pointless insertions and repeats take up pages and pages.

Whenever there is demand for a certain instruction (gene), the administration of the library (i.e., the cell) produces a copy rather than handing over the enormously heavy volume, and rightly so – it is invaluable!

The single-stranded gene copy is called **messenger ribonucleic acid** or **mRNA**.

The mRNA copy of a gene is chemically very close to the original. Like DNA, it has a backbone of various sugar and phosphate units, and it binds to the same bases as DNA – A, C, and G. However, the **place of thymine (T) is taken by uracil (U)** wherever it appears in the DNA (Fig. 3.8).

As the name suggests, the sugar in RNA is **ribose** which differs from deoxyribose in DNA in that ribose has an OH group on its second carbon atom.

When a cell requires a certain protein, it first produces an mRNA copy of the relevant gene DNA. In eukaryotes, the copy travels from the nucleus into the cell plasma to protein-producing ribosomes where the composition of the protein (Fig. 3.16) is controlled by mRNA.

The process of copying a DNA section into mRNA is called **transcription** and largely resembles the copying process (replication) preceding cell division in the mitotic S phase.

DNA polymerase is replaced by a different enzyme that binds to the openings of the RNA "zipper" around the start of the gene. **RNA polymerase** (Fig. 3.9) provides the relevant prefab building blocks:

- Uracil for DNA adenine
- Guanine for DNA cytosine
- Cytosine for DNA guanine
- Adenine for DNA thymine.

Incidentally, it is the irreversible inhibition of RNA polymerases through α-amanitin (Fig. 3.13) that makes *Amanita phalloides* ("death cap") such a deadly fungus that claims the lives of around 100 people every year. The cells are unable to produce mRNA and thus cannot make proteins.

■ 3.6 The DNA Code Deciphered – Artificial RNA Decodes the Codons

The working principle of the genetic code was known since the 1960s, but deciphering it was a longer process. Which of the 64 possible combinations of three bases (codon) was responsible for the integration of which of the 20 existing amino acids to a protein?

In 1961, **Marshall Nirenberg** (b. 1927) from the University of Michigan and the German **Heinrich Matthaei** (b.1929) went public with their breakthrough at the International Congress of Biochemistry in Moscow.

They had succeeded in synthesizing an mRNA molecule consisting entirely of uracil bases (U-U-U etc) and transferring it to a reaction mix they had prepared in advance. The mix contained all the ele-

ments a living cell uses to make proteins, except that the only available DNA or RNA was the synthetic U-U-U mRNA, and lo and behold, the "dead" reaction mix started producing proteins just like any other living cell. The artificial protein molecule consisted of a monotonous chain, repeating the amino acid phenylalanine sequence: Phe-Phe-Phe.

Thus, Nirenberg and Matthaei had been able to decode the first of 64 possible triplets in the genetic code. Codon U-U-U in the RNA controls the integration of phenylalanine into a protein molecule.

Soon afterwards, AAA was identified as the codon for lysine, CCC for proline. And so it went on until 1966 when **Har Gobind Khorana** (b. 1922), an Indian researcher working in the US (Fig. 3.11), finished by deciphering the last of the 64 possible triplet combinations.

It had thus been shown that amino acids were coded by **groups of three bases** (**triplets**), starting at a defined point of origin. 61 of the codons define amino acids while the remaining three (UAA, UAG, and UGA) are signals to break off the chain. In eukaryotes, AUG works as a starting signal as well as coding for methionine.

61 possible codes are used to code for just 20 amino acids – in other words, some amino acids are coded by several code words. This came to be known as a **degenerate code**. Valine and alanine are coded by four different codons each, leucine even by six. Due to this code degeneration, it is not possible to derive a protein from the precise nucleotide sequence in the DNA. Although the principle of the **genetic code is universal**, not all DNA follows the same rules. It is thought that eukaryotic cell organelles may have evolved from bacteria once ingested by eukaryotes that became symbiotic. This could

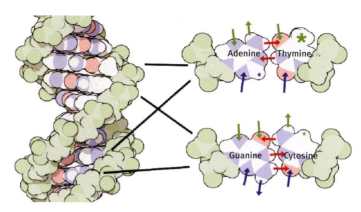

Fig. 3.12 Chemical pairing of bases (Watson-Crick rule): two hydrogen bonds for adenine – thymine; three hydrogen bonds for guanine – cytosine.

α-Amanitin

Fig. 3.13 Eukaryotes contain three types of RNA polymerase. The Death cap fungus (*Amanita phalloides*) produces α-amantin which inhibits human RNA polymerases II and III often with fatal consequences.

Box 3.1 Biotech-History: DNA

"While the abbot was one of the most popular clergymen in Brünn, nobody ever thought that his experiments were more than some nice leisure activity and his theories more than a charming conversation topic", writes a contemporary of Gregor Mendel.

In 1865, **Mendel** (1822-1884) published the results of his experiments with peas which he had begun in 1856, for the first time putting into writing the rules of heredity. The observed characteristics (e.g., color of flowers, shape of seeds and pods) are determined by factors that are passed on independently from each other. These factors were later called genes.

Gregor Mendel (1822-1884)

It was not until around 1900 that Mendel's laws were rediscovered by **C. E. Correns**, **E. von Tschermak**, and **H. de Vries**.

In 1869, **Johann Friedrich Miescher** (1844-1895) in Basel isolated a compound he called nuclein from pus found on discarded bandages. This compound, which could not be broken down by protein-cleaving enzymes, had acidic properties. It was therefore renamed "Nucleinsäure", i.e., nucleic acid. Miescher found nuclein also in yeast, liver, and kidney cells as well as in red blood cells.

However, it took almost fifty years to find out that DNA contains six components – phosphoric acid, the sugar deoxyribose, and the four organic bases adenine, guanine, thymine, and cytosine. This simple structure was deemed incapable of dealing with such complex matters as heredity transmission.

Rudolf Signer in Bern (1903-1990) was the first to synthesize pure DNA in 1938. With his statement that DNA bases must be flat rings positioned perpendicular to the axis of a chain molecule, he was ahead of his time. The 15 grams of highly purified DNA he brought to England was later dubbed "manna from Bern".

Friedrich Miescher (1844-1895) who discovered nucleic acids in 1869.

In 1928, British bacteriologist **Frederick Griffith** (1877-1941) carried out an experiment with the pneumonia pathogen *Pneumococcus*. He used two different strains, one did not form capsules and was called R for rough, while the other formed capsules and was therefore named the S-strain, i.e., S for smooth. In the experiment, the S strain proved to be fatal to mice, while the R strain was innocuous. Mice injected with killed bacteria of the dangerous S-strain were not harmed, while those injected with a mixture of live R-strain and killed S-bacteria died. This suggests that the heretofore innocuous R-pneumococci had incorporated the deadly factor from the S-strain.

Griffith was thus the first to describe the possibility of genetic transfer (transformation) between bacteria. We now know that the DNA of the deadly S-bacteria had survived the heating process and was taken in by the usually innocuous R-pneumococci. The S-strain DNA contained a crucial gene that protects the bacteria from the host's immune system.

After a period of working as a general practitioner, the Canadian scientist **Theodore Avery** (1877-1955) did research at the Rockefeller Institute of Medical Research between 1913 and 1947. 75 years after Friedrich Miescher's discovery, he and his colleagues **MacLeod** and **McCarty** succeeded in showing that nucleic acids, and not proteins, were carriers of genetic information, as had previously been assumed. What is the underlying "transforming principle" of gene transfer? What is the chemical makeup of a gene? Avery went a step further than Griffith, using not heat-inactivated S-pneumococci,

but cell extract that he purified further and further. All fractions (cell wall components, various protein fractions and fractions containing nucleic acid) were tested for their ability to turn R-strains into S-strains. Adding ethanol led to precipitation, and transformation was no longer possible. However, carbohydrates such as the bacterial capsules do not precipitate in the presence of alcohol, while nucleic acids do. Proteases did not seem to have any effect on transformation. This left the fractions containing nucleic acids as the culprit.

Theodore Oswald Avery (1877-1955) proved in 1944 that DNA was the physical carrier of heredity transfer.

However, when Avery added the RNA-cleaving enzyme ribonuclease, transformation still took place although both protein and RNA were out of action. The chemical test for the detection of the DNA sugar deoxyribose involved the formation of a blue color. DNA was responsible. Only fractions containing DNA and finally pure DNA turned innocuous R-recipient cells permanently into the pathogenic S-phenotype; the transforming principle had been identified. Griffith and Avery, whose experiments seem remarkably contemporary from today's perspective, could be considered the first genetic engineers in history.

The third part of the DNA saga was mainly shaped by Erwin Chargaff, Maurice Wilkins, Francis C. Crick, and James D. Watson.

Shown in Avery's article that made history: Small colonies of the non-encapsulated pneumococcal R-strain (top) and the encapsulated S-strain (bottom).

Erwin Chargaff (Fig. 3.5), coming from a Jewish background, left Germany after the Nazis came to power and went to the Pasteur Institute in Paris. He emigrated to the U.S. in 1935 and worked at Columbia University, New York, from 1938. His research into DNA began in 1944.

After Chargaff had found with the DNA of every organism he investigated that the ratio of adenine and thymine content as well as that of cytosine and guanine was always 1:1, he laid down what became known as Chargaff's rule (Fig. 3.6). In his later life, he became one of the most eminent figures to warn against the misuse of gene technology.

Rosalind Franklin (1920 - 1958)

A highly intelligent youth and enthusiastic ornithologist, James Dewey Watson began to study zoology at the University of Chicago at the age of 15. Increasingly interested in modern genetics, he wrote a doctoral thesis, supervised by **Salvador Luria**, one of the founding fathers of phage research. Phages are viruses that infect bacteria and have long been used in gene models.

Maurice Wilkins (1916 - 2004), who shared the Nobel Prize with Watson and Crick.

In 1950, Watson went to Europe to study the biochemical foundations of phage replication. While in Cambridge he turned out to be the right man, at the right place, at the right time. Within a short time, he and the physicist and crystallographer **Francis Crick** unveiled the double helix structure of DNA which was published in 1953.

The best report by far about the discovery of the double helix is by Watson himself.

They were spurred on by competition – including London-based DNA crystallographer **Rosalind Franklin** (1920 - 1958) who had identified the helical structure of DNA, and **Maurice Wilkins** (1916 - 2004) who was to share the Nobel Prize with Watson and Crick later. U.S chemist **Linus Pauling** had identified the helix as a key structural element of proteins in 1951 and then went on to investigate the structure of DNA. However, his assumption was a triple helix structure, as Watson and Crick, "full of academic glee", were told later by Pauling's son Peter.

Once the DNA race had been won, Watson turned to other promising subjects, but his attempt to identify the structure of RNA failed.

In the years that followed, young researchers formed what was known as the RNA Tie Club, which included Watson and Crick, **Sydney Brenner** (Ch.10), **Richard Feynman**, and **Alexander Rich**, as well as **Erwin Chargaff**, **Gunther Stent**, and **Edward Teller** who had been exiled from Germany. Their chairman was the Russian-American scientist **Georgi Gamow**, and their ambition was to solve the most pressing questions in molecular biology, the genetic code in particular, through informal and interdisciplinary cooperation.

Linus Pauling (1901 - 1994) who won the Nobel Prize twice

Watson also revolutionized academic teaching, stating that in order to achieve success in science, it was essential that professors allow their students to become independent as soon as possible. He began to practise what he preached at Harvard in the 1960s when he and **Walter Gilbert** broke with the long-standing academic tradition of publishing their students' work under their own name even if they had not been involved in the research.

Watson played an important part in the Human Genome Project (Ch. 10) and to this day has been sparking off debates and controversies by his pointed remarks and comments.

How to write down a DNA sequence

The 5'hydroxy group of a nucleotide is linked to the 3'hydroxy of the neighboring nucleotide via a phosphate group (Fig. 3.8). All phosphodiester bonds in DNA and RNA strands have the same orientation along the chain. This gives every nucleic acid strand a definite polarity and distinct 5' and 3' termini. By convention, sequences are written down in the 5' → 3' direction. Thus, ACCGGT stands for 5'-ACCGGT-3'. A has a free 5' phosphate group at its "head", while T has an unbound 3-OH group at its "tail". Due to their polarity, ACCGGT and the complementary TGGCCA are two totally different compounds!

What is a kilobase (kb)?

A kilobase stands for 1000 bases in a double or single DNA strand. 1 kb of a double-stranded DNA has a length of 0.34 micrometers and a mass of approximately 660,000 daltons.

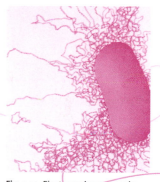

Fig. 3.14 Electron microscope view of an *E. coli* cell which has burst (arrow points out plasmid).

A burst *E. coli* cell with its ring-shaped main DNA (about a millimeter long) and some plasmids. The cell itself, however, measures no more than two micrometers.

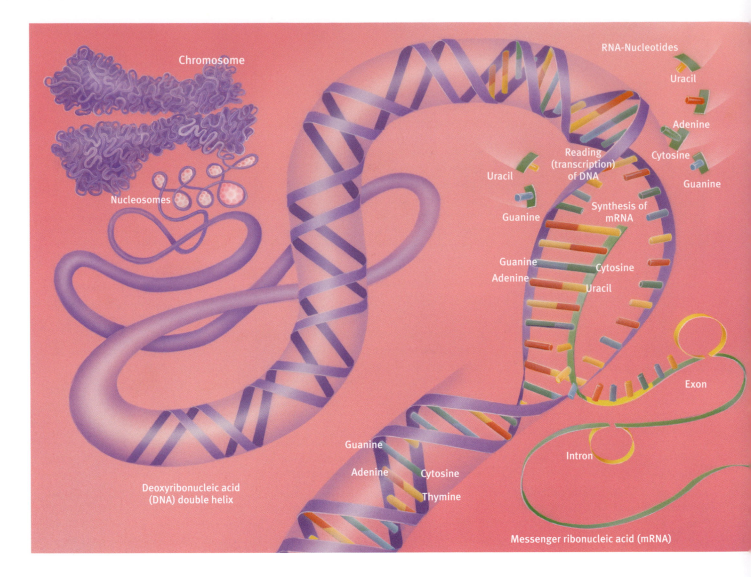

Chromosome

Nucleosomes

Deoxyribonucleic acid (DNA) double helix

Guanine
Adenine · Cytosine
Thymine

RNA-Nucleotides
Uracil
Adenine
Cytosine
Guanine

Uracil
Reading (transcription) of DNA
Guanine
Synthesis of mRNA
Guanine
Adenine
Cytosine
Uracil

Exon

Intron

Messenger ribonucleic acid (mRNA)

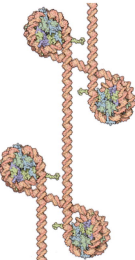

Fig. 3.15 DNA in eukaryotes is not naked, but closely bound to small proteins called histones. These units are called nucleosomes.

Figs. 3.16 und 3.17 Every living cell produces proteins. A protein molecule consists of amino acids. The directions determining the precise sequence of the amino acids in the protein are found in the deoxyribonucleic acid, long double-stranded molecules.

Every DNA strand, in turn, consists of building elements called **nucleotides** which are named after the bases they contain – adenine (A), cytosine (C), guanine (G), and thymine (T).

Nucleotides form complementary pairs, i.e., on a dual DNA strand; A/T and C/G are always found opposite each other.

Every amino acid in a protein has a corresponding **nucleotide triplet** in the DNA. The entirety of nucleotide triplets specifying a protein molecule is called the **structural gene of a protein**.

In eukaryotes, such a structural gene can contain several information-carrying sections (**exons**), interspersed with

introns that do not contain any information about the protein structure.

Nucleotides do not only specify amino acids, but also work as signals that **start or stop** the protein synthesis process.

In order to become effective in the cell cytoplasm, the information contained in the DNA must be transcribed into **messenger ribonucleic acid (mRNA)**. mRNA also consists of nucleotides, but the base thymine (T) has been replaced by uracil (U).

During the transcription process, the polymerase first copies the exons and introns from the DNA. Then the nucleotides representing introns are eliminated from the sequence (splicing) to produce a shorter RNA molecule, known as **mature mRNA**. It travels from the nucleus into the cytoplasm, carrying the assembly instructions for the protein molecule.

In the cytoplasm of the cell, the mRNA nucleotide sequence is translated into the amino acid sequence of the protein, using **ribosomes** and 20 different tRNAs carrying the different amino acids. The synthesized polypeptide chains then fold to form the protein.

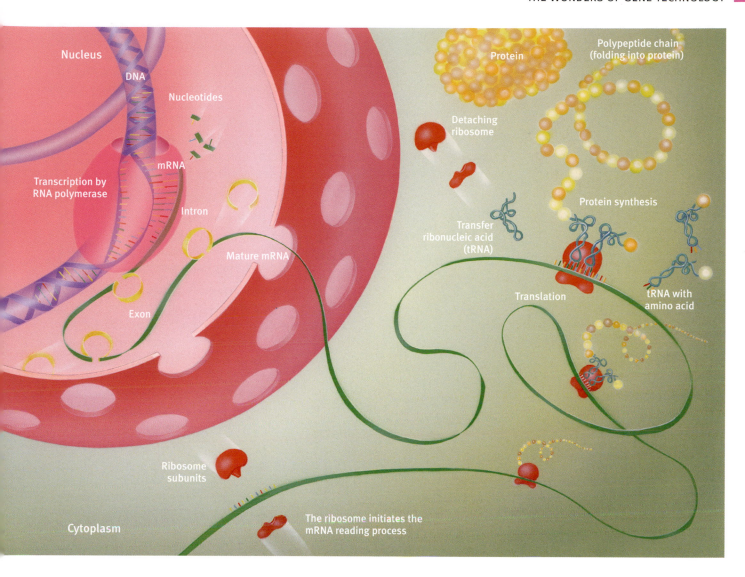

Nucleus

DNA

Nucleotides

mRNA

Transcription by RNA polymerase

Intron

Mature mRNA

Exon

Ribosome subunits

Cytoplasm

Protein

Polypeptide chain (folding into protein)

Detaching ribosome

Protein synthesis

Transfer ribonucleic acid (tRNA)

tRNA with amino acid

Translation

The ribosome initiates the mRNA reading process

explain why the mitochondrial code differs from that of the rest of the cell in some cases. In human mitochondrial DNA, UGA is not a stop signal, but codes for tryptophan.

Ciliates are another group that went their own way early in evolution, which led to slight differences in their genetic code. They have just one stop signal, UGA, while UAA and UAG, usually chain break-off signals – code for amino acids in ciliates.

Why has the code remained more or less unchanged over billions of years? Any mutation that would cause drastic changes to the reading pattern of mRNA would upset the amino acid sequences of nearly all proteins. Such mutations would be fatal and thus soon be eliminated by natural selection.

◼ 3.7 DNA Sites Around the Structural Genes Control the Expression of Genes

Since bacteria are prokaryotes, they do not have a nucleus. Their DNA is contained in a ring of at least one millimeter in circumference. The whole structure is extremely tightly packed and entwined to fit into a bacterial cell which is often only a thousandth of a millimeter thick (see p. 63).

In *Escherichia coli* K12, the "biosafety strain", precisely 4,639,221 base pairs are lined up in that 1 millimeter of DNA space, carrying almost 4300 protein-coding genes. **Structural genes**, each of them about 1000 base pairs long, are responsible for the structure of single proteins, in most cases an enzyme. The structural gene determines the correct arrangement of several hundred amino acids in the

Exons and Introns – A Bizarre Manuscript

The English science writer Steve Prentis aptly compared eukaryotic genes with a bizarre manuscript. Some pages of exquisite poetry are followed by nonsensical letters and words (introns) before the next "proper" passages begins.

Cut it down!

The editor must cut out (splice) those nonsensical units from the copy (mRNA) and paste the useful parts together. The ready manuscript (mature mRNA) is then sent to the printers (ribosome) and turned into a readable book (protein).

Fig. 3.18 Right: Translation of mRNA into a protein in a ribosome.

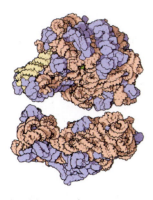

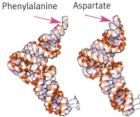

Phenylalanine   Aspartate

Fig. 3.19 A ribosome contains two subunits attaching to mRNA during protein synthesis. Top: A bacterial ribosome. Eukaryotic ribosomes are slightly bigger. Ribosomes consist mainly of RNAs (orange and yellow) in protein (blue) complexes. The small subunit matches tRNA anticodons for the relevant amino acids with mRNA codons.
The larger subunit catalyzes the synthetic process by transferring the amino acid from the tRNA to the growing polypeptide chain. Bottom: Two of 20 tRNAs. The aspartate tRNA carries aspartate, the phenylalanine tRNA carries phenylalanine. The anticodons are also clearly recognizable.

Fig. 3.20 Robert W. Holley (1922-1993) investigated tRNA and was awarded the Nobel Prize in 1968, together with Nirenberg and Khorana.

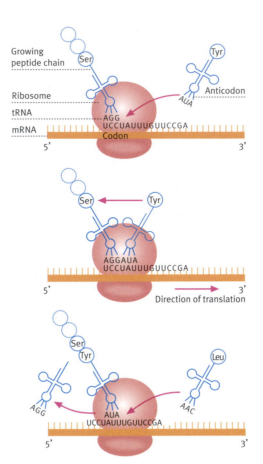

ribosome so that they can form peptide chains for a specific protein.

Not all DNA sequences code for proteins. Next to the structural gene are specific sequences controlling its **expression**, i.e., they initiate transcription and subsequent translation of the structural gene into a protein.

Transcription, the first of the two processes, is directed by two specific DNA sequences. One of them is the **starting point** or **promoter** – a short sequence that enables the enzyme RNA polymerase to bind to DNA, travel along the strands, and transcribe the DNA into RNA from that specific point before the structural gene.

The other relevant sequence is a **stop sequence** found somewhat behind the structural gene. It gives the signal to end transcription. In *E. coli*, the stop signal produces a hairpin structure in the newly synthesized mRNA strand which, in turn, causes the immediate separation of RNA polymerase from the newly formed mRNA.

Eukaryotes have other types of promoters (such as the metallothionein promoter that reacts to heavy

metals – more about it in Ch. 8), as well as enhancer sequences and stop sequences.

In eukaryotes, the **mRNA is modified after transcription**. A cap structure is added to the 5′ terminus, and a tail consisting of adenine components (poly-A tail) is attached to the 3′ terminus (see insulin expression in Fig. 3.36). Genes whose activity depends on the concentration of certain metabolites rely on further regulator regions.

As we will see when looking at the regulation of lactose processing (more in Ch.4 using the example of the *lac* operator), an **operator region** between promoter and structural gene may bind to what is known as a **repressor protein**.

When an **inducer** (e.g., a sugar) binds to a repressor protein, the spatial structure of the protein changes, causing it to leave the operator region. Then, the structural gene can be read for the relevant sugar-degrading enzyme.

■ 3.8 Ribosomes – Protein Production Plants Inside the Cell: Giant RNA and Protein Molecules

The entire information for protein synthesis – including start and stop signals – is contained in the mRNA which is read in the **ribosome**.

Ribosomes consist of a large and a small subunit (Fig. 3.19). They are protein and RNA complexes. **Ribosomal RNAs** (**rRNAs**) are single-stranded molecules that form part of the ribosome. The 20 different tRNAs are synthesized by 20 **aminoacyl synthetases** (Fig. 3.21) that vary widely in their structure. Each of them binds an amino acid to the relevant rRNA.

Apart from mRNA and rRNA, there is a third type of RNA – **transfer RNAs** (tRNAs) transfer amino acids in their activated form to the ribosome (Fig. 3.18). There is at least one tRNA for each of the 20 amino acids. tRNAs consist of RNA with an anticodon at one terminus and the activated amino acid at the other (Fig. 3.19).

Other DNA sequences that have been copied into mRNA alongside the actual structural gene supervise the **translation** into a peptide chain inside the ribosome (Fig. 3.18). The mRNA is bound to the ribosome by a specific binding site. The translation process begins at the start signal, the first codon of

Box 3.2 Biotech History:
James D. Watson about the discovery of the DNA structure

James Watson, co-discoverer of the structure of DNA tells the story of the amazing molecule since its discovery 50 years ago in his new book "DNA. The Secret of Life". E.O. Wilson calls the book "an immediate classic". The following excerpts give just a glimpse. Warning: reading Watson's book is addictive!

I got hooked on the gene during my third year at the **University of Chicago**. Until then, I had planned to be a naturalist and looked forward to a career far removed from the urban bustle of Chicago's South Side, where I grew up. My change of heart was inspired not by an unforgettable teacher but a little book that appeared in 1944, *What Is Life?*, by the Austrian-born father of wave mechanics, **Erwin Schrödinger** (see Plate 7). It grew out of several lectures he had given the year before at the Institute for Advanced Study in Dublin. That a great physicist had taken the time to write about biology caught my fancy. In those days, like most people, I considered chemistry and physics to be the "real" sciences, and theoretical physicists were science's top dogs.

Erwin Schrödinger (1887 - 1961), Austrian physicist, wrote this little book in 1944, which was to revolutionize biology.

Schrödinger argued that life could be thought of in terms of storing and passing on biological information. Chromosomes were thus simply information bearers. Because so much information had to be packed into every cell, it must be compressed into what Schrödinger called a "hereditary code-script" embedded in the molecular fabric of chromosomes. To understand life, then, we would have to identify these molecules, and crack their code. He even speculated that understanding life – which would involve finding the gene – might take us beyond the laws of physics as we then understood them. Schrödinger's book was tremendously influential. Many of those who would become major players in Act 1 of molecu-

lar biology's great drama, read *What Is Life?* and been impressed.

In my own case, Schrödinger struck a chord because I too was intrigued by the essence of life. A small minority of scientists still thought life depended upon a vital force emanating from an all-powerful god. But like most of my teachers, I disdained the very idea of vitalism. If such a

"vital" force were calling the shots in nature's game, there was little hope would ever be understood through the methods of science. On the other hand, the notion that life might be perpetuated by means of an instruction book inscribed in a secret code appealed to me. What sort of molecular code could be so elaborate as to convey all the multitudinous wonder of the living world? And what sort of molecular trick could ensure that the code is exactly copied every time a chromosome duplicates?

Years later....

Upon finishing my thesis, I saw no alternative but to move to lab where I could study DNA chemistry. Unfortunately, however, knowing almost no pure chemistry, I would have been out of my depth in any lab attempting difficult experiments in organic or physical chemistry. I therefore took a postdoctoral fellowship in the **Copenhagen lab** of the biochemist **Herman Kalckar** in the fall of 1950. he was studying the synthesis of the small molecules that make up DNA, but I figured out quickly that his biochemical approach would never lead to an understanding of the essence of the gene. Every day spent in his lab would be one more day's delay in learning how DNA carried genetic information.

My Copenhagen year nonetheless ended productively. To escape the cold Danish spring, I went to the Zoological Station at Naples during April and May. During my last week there, I attended a small conference on X ray diffraction methods for determining the 3-D structure of molecules. X ray diffraction is a way of studying trh atomic structure of any molecule that can be crystallized. The crystal is bombarded with X rays, which bounce off its atoms and are scattered. The scatter pattern gives information about the structure of the molecule but, taken alone, is not enough to solve the phase problem was not easy, and at that time only the most audacious scientists were willing to take it on. Most of the successes of the diffraction method had been achieved with relatively simple molecules.

My expectations for the conference were low. I believed that a three-dimensional understanding of protein structure, or for that matter of DNA, was more than a decade away. Disappointing earlier X-ray photos suggested that DNA was particu-

larly unlikely to yield up its secrets via the X-ray approach. These results were not surprising since the exact sequences of DNA were expected to differ from one individual molecule to another. The resulting irregularity of surface configurations would understandably prevent the long thin DNA chains from lying neatly side by side in the regular repeating patterns required for X-ray analysis to be successful.

It was therefore a surprise and a delight to hear the last-minute talk on DNA by a thirty-four-year-old Englishman named **Maurice Wilkins** from **the Biophysics Lab of King's College, London**. Wilkins was a physicist who during the war had worked on the Manhattan Project. For him, as for many of the other scientists involved, the actual deployment of the bomb on Hiroshima and Nagasaki, supposedly the culmination of all their work, was profoundly disillusioning. He considered forsaking science altogether to become a painter in Paris, but biology intervened. He too had read Schrödinger's book, and was now tacking DNA with X-ray diffraction.

He displayed a photograph of an X-ray diffraction pattern he had recently obtained, and its many precise reflections indicated a highly regular crystalline packing. DNA, one had to conclude, must have a regular structure, the elucidation of which might well reveal the nature of the gene. Instantly I saw myself moving to London to help Wilkins fond the structure. My attempts to converse with him after his talk, however, went nowhere. All I got for my efforts was a declaration of his conviction that much hard work lay ahead.

While I was hitting consecutive dead ends, back in America the world's triumph: he had found the exact arrangement in which chains of amino acids (called polypeptides) fold up in proteins, and called his structure the α-helix (alpha helix). That it was Pauling who made his breakthrough was no surprise: he was a scientific superstar.

As soon as I returned to Copenhagen I read about **Pauling**'s α-helix. To my surprise, his model was not based on a deductive leap from experimental X-ray diffraction data. Instead, it was Pauling's long experience as a structural chemist that had emboldened him to infer which type of helical fold would be most compatible with the underlying chemical features of the polypeptide chain. Pauling made scale models of the different parts of the protein molecule, working out plausible schemes in three-dimensional jigsaw puzzle in a way that was simple yet brilliant.

Whether the α-helix was correct – in addition to being pretty – was now the question. Only a

See next page

week later, I got the answer. **Sir Lawrence Bragg**, the English inventor of X-ray crystallography and 1915 Nobel laureate in Physics, came to Copenhagen and excitedly reported that his junior colleague, the Austrian-born chemist **Max Perutz**, had ingeniously used synthetic ploypeptides to confirm the correctness of Pauling's α-helix. It was a bittersweet triumph for Bragg's Cavendish Laboratory. The year before, they had completely missed the boat in their paper outlining possible helical folds for polypeptide chains.

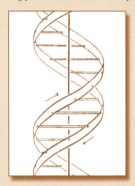

By then **Salvador Luria** had tentatively arranged for me to take up a research position at the Cavendish. Located at Cambridge University, this was the most famous laboratory in all of science. Here Ernest Rutherford first described the structure of the atom. Now it was Bragg's own domain, and I was to work as apprentice to the English chemist **John Kendrew**, who was interested in determining the 3-D structure of the protein myoglobin. Luria advised me to visit the Cavendish as soon as possible. With Kendrew in the States, Max Perutz would check me out. Together, Kendrew and Perutz had earlier established the Medical Research Council (MRC) Unit for the Study of the Structure of Biological Systems.

A month later in Cambridge, Perutz assured me that I could quickly master the necessary X-ray diffraction theory and should have no difficulty fitting in with the others in their tiny MRC Unit. To my relief, he was not put off by my biology background. Nor was **Lawrence Bragg**, who briefly came down from his office to look me over.

I was twenty-three when I arrived back at the MRC Unit in Cambridge in early October. I found myself sharing space in the biochemistry room with a thirty-five-year-old ex-physicist, **Francis Crick**, who has spent the war working on magnetic mines for the Admiralty. When the war ended, Crick had planned to stay on in military research, but, on reading Schrödinger's *What Is Life?*, he had moved toward biology. Now he was at the Cavendish to pursue the 3-D structure of proteins for his Ph.D.

Crick was always fascinated by the intricacies of important problems. His endless questions as a child compelled his weary parents to buy him a children's encyclopedia, hoping that it would satisfy his curiosity. But it only made him insecure: he confided to his mother his fear that everything would have been discovered by the time he grew up, leaving him nothing to do. His mother reassured him (correctly, as it happened) that there would still be a thing or two for him to figure out.

A great talker, Crick was invariable the center of attention in any gathering. His booming laugh was forever echoing down the hallways of the Cavendish. As the MRC Unit's resident theoretician, he used to come up with a novel insight at least once a month, and he would explain his latest idea at great length to anyone willing to listen. The morning we met he lit up when he learned that my objective in coming to Cambridge was to learn enough crystallography to have a go at the DNA structure. Soon I was asking Crick's opinion about using Pauling's model-building approach to go directly for the structure. Would we need many more years of diffraction experimentation before modeling would be practicable? To bring us up to speed on the status of DNA structural studies, Crick invited Maurice Wilkins, a friend since the end of the war, up from London for Sunday lunch. Then we could learn what progress Wilkins had made since his talk in Naples.

Wilkins expressed his belief that DNA's structure was a helix, formed by several chains of linked nucleotides twisted around each other. All that remained to be settled was the number of chains. At the time, Wilkins favored three on the basis of his density measurements of DNA fibers. He was keen to start model-building, but he had run into a roadblock in the form of a new addition to the King's College Biophysics Unit, **Rosalind Franklin**.

Just back form a four-year X-ray crystallographic investigation of graphite in Paris, Franklin had been assigned to the DNA project while Wilkins was away from King's. Unfortunately, the pair soon proved incompatible. Franklin, direct and data-focused, and Wilkins, retiring and speculative, were destined never to collaborate. Shortly before Wilkins accepted our lunch invitation, the two had had a big blowup in which Franklin had insisted that no model-building could commence before she collected much more extensive diffraction data. Now they effectively didn't communicate, and Wilkins would have no chance to learn of her progress until Franklin presented her lab seminar scheduled for the beginning of November. If we wanted to listen, Crick and I were welcome to go as Wilkin's guests.

Crick was unable to make the seminar, so I attended alone and briefed him later on what I believed to be its key take-home messages on crystalline DNA. In particular, I described from memory Franklin's measurements of the crystallographic repeats and the water content. This prompted Crick to begin sketching helical grids on a sheet of paper, explaining that the new helical X-ray theory he had devised with **Bill Cochran** and **Vladimir Vand** would permit even me, a former bird-watcher, to predict correctly the diffraction patterns expected from the molecular models we would soon be building at the Cavendish.

As soon as we got back to Cambridge, I arranged for the Cavendish machine shop to construct the phosphorous atom models needed for short sections of the sugar phosphate backbone found in DNA. Once these became available, we tested different ways the backbones might twist around each other in the center of the DNA molecule. Their regular repeating atomic structure should allow the atoms to come together in a consistent, repeated confirmation. Following Wilkin's hunch, we focused on three-chain models. When one of these appeared to be almost plausible, Crick made a phone call to Wilkins to announce we had a model we though might be DNA.

The next day both Wilkins and Franklin came up to see what we had done. The threat of unanticipated competition briefly united them in common purpose. Franklin wasted no time in faulting our basic concept. My memory was that she had reported almost no water present in crystalline DNA. In fact, the opposite was true. Being a present in crystallographic novice, I had confused the terms "unit cell" and "asymmetric unit." Crystalline DNA was in fact water-rich. Consequently, Franklin pointed out, the backbone had to be on the outside and not, as we had it, in the center, if only to accommodate all the water molecules she had observed in her crystals.

That unfortunate November day cast a very long shadow. Franklin's opposition to model-building was reinforced. Doing experiments, not playing with Tinkertoy representations of atoms, was the way she intended to proceed. Even worse, Sir Lawrence Bragg passed down the word that Crick and I should desist from all further attempts at building a DNA model. It was further decreed that DNA research should be left to the King's lab, with Cambridge continuing to focus solely on proteins. There was no sense in two MRC-funded labs competing against each other. With no more bright ideas up our sleeves, Crick and I were reluctantly forced to back off, at least for the time being.

It was not a good moment to be condemned to the DNA sidelines. Linus Pauling had written Wilkins to request a copy of the crystalline DNA diffraction pattern. Though Wilkins had declined, saying he wanted more time to interpret it himself, Pauling was hardly obliged to depend upon data from King's. If he wished, he could easily start serious X-ray diffraction studies at Caltech.

The following spring, I duly turned away from DNA and set about extending prewar studies on the pencil-shaped tobacco mosaic virus using the Cavendish's powerful new X-ray beam. This light experimental workload gave me plenty of time to wander through various Cambridge libraries. In the zoology building, I read Erwin Chargaff's paper describing his finding that the DNA bases adenine and thymine occurred in roughly equal amounts, as did the bases guanine and cytosine. Hearing of these one-to-one ratios Crick wondered whether, during DNA duplication, adenine residues might be attracted to thymine and vice versa, and whether a corresponding attraction might exist between guanine and cytosine. If so, base sequences on the "parental" chains (e.g., ATGC) would have to be complementary to those on "daughter" strands (yielding in this case TACG).

These remained idle thoughts until **Erwin Chargaff** came through Cambridge in the summer of 1952 on his way to the International Biochemical Congress in Paris. Chargaff expressed annoyance that neither Crick nor I saw the need to know the chemical structures of the four bases. He was even more upset when we told him that we could simply look up the structures in textbooks as the need arose. I was left hoping that Chargaff's data would prove irrelevant. Crick, however, was energized to do several experiments looking for molecular "sandwiches" that might form when adenine and thymine (or alternatively, guanine and cytosine) were mixed together in solution. But his experiments went nowhere.

Like Chargaff, Linus Pauling also attended the International Biochemical Congress, where the big news was the latest result from the Phage Group. **Alfred Hershey** and **Martha Chase** at Cold Spring Harbor had just confirmed Avery's transforming principle: DNA was the hereditary material! Hershey and Chase proved that only the DNA of the phage virus enters bacterial cells; it protein coat remains on the outside. It was more obvious than ever that DNA must be understood at the molecular level if we were to uncover the essence of the gene. With Hershey and Chase's result the talk of the town, I was sure that Pauling would now bring his formidable intellect and chemical wisdom to bear on the problem of DNA.

No. 4356   April 25, 1953   NATURE

MOLECULAR STRUCTURE OF NUCLEIC ACIDS

A Structure for Deoxyribose Nucleic Acid

WE wish to suggest a structure for the salt of deoxyribose nucleic acid (D.N.A.). This structure has novel features which are of considerable biological interest.

It has not escaped our notice that the specific pairing we have postulated immediately suggests a possible copying mechanism for the genetic material.

J. D. WATSON
F. H. C. CRICK
Medical Research Council Unit for the Study of the Molecular Structure of Biological Systems,
Cavendish Laboratory, Cambridge.
April 2.

The historic article in *Nature* – a masterpiece of science and understatement!

Early in 1953, Pauling did indeed publish a paper outlining the structure of DNA. Reading it anxiously I saw that he was proposing a three-chain model with sugar phosphate backbones forming a dense central core. Superficially it was similar to our botched model of fifteen months earlier. But instead of using positively charged atoms (e.g., Mg^{2+}) to stabilize the negatively charged backbones, Pauling made the unorthodox suggestion that the phosphates were held together by hydrogen bonds. But it seemed to me, the biologist, that such hydrogen bonds required extremely acidic conditions never found in cells. With a mad dash to **Alexander Todd's** nearby organic chemistry lab my belief was confirmed: The impossible had happened. The world's best-known, if not best, chemist had gotten his chemistry wrong. In effect, Pauling had knocked the A off of DNA. Our quarry was deoxyribonucleic acid, but the structure he was proposing was not even acidic.

Hurriedly I took the manuscript to London to inform Wilkins and Franklin they were still in the game. Convinced that DNA was not a helix, Franklin had no wish even to read the article and deal with the distraction of Pauling's helical ideas, wven when I offered Crick's arguments for helices. Wilkins, however, was very interested indeed in the news I brought; he was now more certain than ever that DNA was helical. To prove the point, he showed me a photograph obtained more than six months earlier by Franklin's graduate student **Raymond Gosling**, who had X-rayed the so-called B form of DNA. Until that

moment, I didn't know a B form even existed. Franklin had put this picture aside, preferring to concentrate on the A form, which she thought would more likely yield useful data. The X-ray pattern of this B form was a distinct cross. Since Crick and others had already deduced that such a pattern of reflections would be created by a helix, this evidence made it clear that DNA had to be a helix! In fact, despite Franklin's reservations, this was no surprise. Geometry itself suggested that a helix was the most logical arrangement for a long string of repeating units such as the nucleotides of DNA. But we still did not know what that helix looked like, nor how many chains it contained.

The time had come to resume building helical models of DNA. Pauling was bound to realize soon enough that his brainchild was wrong. I urged Wilkins to waste no time. But he wanted to wait until Franklin had completed her scheduled departure for another lab later that spring. She had decided to move on to avoid the unpleasantness at King's. Before leaving, she had been ordered to stop further work with DNA and had already passed on many of her diffraction images to Wilkins.

When I returned to Cambridge and broke the news of the DNA B form, Bragg no longer saw any reason for Crick and me to avoid DNA. He very much wanted the DNA structure to be found on his side of the Atlantic. So we went back to model-building, looking for a way the known basic components of DNA – the backbone of the molecule and the four different bases, adenine, thymine, guanine, and cytosine – could fit together to make a helix. I commissioned the shop at the Cavendish to make us a set of tin bases, but they couldn't produce them fast enough for me: I ended up cutting out rough approximations from stiff cardboard.

By this time I realized the DNA density-measurement evidence actually slightly favored a two-chain, rather than three-chain, model. So I decided to search out plausible double helices. As a biologist, I preferred the idea of a genetic molecule made of two, rather than three, components. After all, chromosomes, like cells, increase in number by duplicating, not triplicating.

I knew that our previous model with the backbone on the inside and the bases hanging out was wrong. Chemical evidence from the University of Nottingham, which I had too long ignore indicated that the bases must be hydrogen-bonded to each other. They could only form bonds like this in the regular manner implied by the X-ray diffraction data if they were in the center of the molecule. But how could they come together in pairs? For two weeks I got nowhere,

See next page

misled by an error in my nucleic acid chemistry textbook. Happily, on February 27, **Jerry Donahue**, a theoretical chemist visiting the Cavendish from Caltech, pointed out that the textbook was wrong. So I changed the locations of the hydrogen atoms on my cardboard cutouts of the molecules.

Replica of the original DNA double helix model in the Science Museum, London.

The next morning, February 28, 1953, the key features of the DNA model all fell into place. The two chains were held together by strong hydrogen bonds between adenine-thymine and guanine-cytosine base pairs. The inferences Crick had drawn the year before based on Chargaff's research had indeed been correct. Adenine does bond to thymine and guanine does bond to cytosine, but not through flat surfaces to form molecular sandwiched. When Crick arrived, he took it all in rapidly, and gave my base-pairing scheme his blessing. He realized right away that it would result in the two strands of the double helix running in opposite directions.

It was quite a moment. We felt sure that this was it. Anything that simple, that elegant just had to be right. What got us most excited was the complementarity of the base sequences along the two chains. If you knew the sequence – the order of bases – along one chain, you automatically knew the sequence along the other. It was immediately apparent that this must be how the genetic messages of genes are copied so exactly when chromosomes duplicate prior to cell division. The molecule would "unzip" to form two separate strands. Each separate strand then could serve as the template for the synthesis of a new strand, one double helix becoming two.

In *What Is Life?* Schrödinger had suggested that the language of life might be like Morse code, a series of dots and dashes. He wasn't far off. The language of DNA is a linear series of As, Ts, Gs, and Cs. And just as transcribing a page out of a book can result in the odd typo, the rare mistake creeps in when all these As, Ts, Gs, and Cs are being copied along a chromosome. These errors are the mutations geneticists had talked about for almost fifty years. Change an "i" to an "a" and "Jim" becomes "Jam" in English; change a T to a

C and "ATG" becomes "ACG" in DNA.

The double helix made sense chemically and it made sense biologically. Now there was no need to be concerned abut Schrödinger suggestion that new laws of physics might be necessary for an understanding of how the hereditary code-script is duplicated: genes in fact were no different from the rest of chemistry. Later that day, during lunch at the Eagle, the pub virtually adjacent to the Cavendish Lab, Crick, ever the talker, could not help but tell everyone we had just found the "secret of life." I myself, though no less electrified by the thought, would have waited until we had a pretty three-dimensional model to show off.

Among the first to see our demonstration model was the chemist Alexander Todd. That the nature of the gene was so simple both surprised and pleased him. Later, however, he must have asked himself why his OWN lab, having established the general chemical structure of DNA chains, had not moved on to asking how the chains folded up in three dimensions. Instead the essence of the molecule was left to be discovered by a two-man team, a biologist and a physicist, neither of whom possessed a detailed command even of undergraduate chemistry. But paradoxically, this was, at least in part, the key to our success: Crick and I arrived at the double helix first precisely because most chemists at that time thought DNA too big a molecule to understand by chemical analysis.

At the same time, the only two chemists with the vision to seek DNA's 3-D structure made major tactical mistakes: Rosalind Franklin's was her resistance to model-building; Linus Pauling's was a matter of simply neglecting to read the existing literature on DNA, particularly the data on its base composition published by Chargaff. Ironically, Pauling and Chargaff sailed across the Atlantic in the same ship following the Paris Biochemical Congress in 1952, but failed to hit it off. Pauling was long accustomed to being right. And he believed there was no chemical problem he could not work out from first principles by himself. Usually this confidence was not misplaced. During the Cold War, as a prominent critic of the American nuclear weapons development program, he was questioned by the FBI after giving a talk. How did he know how much plutonium there is in an atomic bomb? Pauling's response was "Nobody told me. I figured it out."

Over the next several months Crick and (to a lesser extent) I relished showing off our model to an endless stream of curious scientists. However, the Cambridge biochemists did not invite us to give a formal talk in the biochemistry building. They started to refer to it as the "WC", punning our initials with those used in Britain for the toilet or water closet. That we had found the double helix without doing experiments irked them.

The manuscript that we submitted to Nature in early April was published just over three weeks later, on April 25, 1953. Accompanying it were two longer papers by Franklin and Wilkins, both supporting the general correctness of our model. In June, I gave the first presentation of our model at the Cold Spring Harbor symposium on viruses. **Max Delbrück** saw to it that I was offered, at the last minute, an invitation to speak. To this intellectually high-powered meeting I brought a three-dimensional model built in the Cavendish, the adenine-thymine base pairs in red and the guanine-cytosine base pairs in green.

Nine years later....

It was not until 1962 that Francis Crick, Maurice Wilkins, and I were to receive our own Nobel Prize in Physiology or Medicine. Four years earlier, Rosalind Franklin had died of ovarian cancer at the tragically young age of thirty-seven. Before then Crick had become a close colleague and a real friend of Franklin's. Following the two operations that would fail to stem the advance of her cancer, Franklin convalesced with Crick and his wife, Odile, in Cambridge.

A 1962 photo shows Nobel Prize winners (from left): Maurice Wilkins (medicine), Max F. Perutz (chemistry), Francis Crick (medicine), John Steinbeck (literature), James Watson (medicine), and John C. Kendrew (chemistry).

It was and remains a long-standing rule of the Nobel Committee never to split a single prize more than three ways. Had Franklin lived, the problem would have arisen whether to bestow the award upon her or Maurice Wilkins. The Swedes might have resolved the dilemma by awarding them both the Nobel Prize in Chemistry that year. Instead, it went to Max Perutz and John Kendrew, who had elucidated the three-dimensional structures of hemoglobin and myoglobin respectively.

The discovery of the double helix sounded the death knell for vitalism. Serious scientists, even those religiously inclined, realized that a complete understanding of life would not require the revelation of new laws of nature. Life was just a matter of physics and chemistry, albeit exquisitely organized physics and chemistry.

the structural gene, and goes on until it reaches the stop signal at the end of the gene; then the ribosome releases the complete protein chain.

If you want to be able to control DNA programming, you must have a clear understanding of all the processes involved. Changes to the structural gene, for example, may result in a change of the amino acid sequence of an enzyme and thereby alter its activity. Even after a minor modification of its sequence, the promoter may bind more easily to RNA polymerase, which makes copying DNA to mRNA more efficient (Ch. 4). Also, **mutations** in the operator region or a regulator gene may prevent the repressor protein from binding to the correct site, leaving transcription running continuously at top speed. Furthermore, **foreign genes** that have been inserted are only translated into proteins if promoter and ribosomal binding sites of host and donor are sufficiently similar.

Eukaryotes – which include all higher organisms from yeasts and algae to humans – use **control signals** that differ from those used by prokaryotes such as bacteria. This is not the only difference. In eukaryotic cells, there is no "naked DNA" (Fig. 3.22). Eukaryotic DNA is wrapped up in proteins (histones) and distributed over individual chromosomes which are all contained in a cell nucleus.

A fungal cell contains ten times more DNA than a bacterium. Higher plants and animals contain even several thousand times as much, although that does not enlarge their genetic repertoire accordingly. Humans have 20,000 to 25,000 genes, which is only five times as much as what is found in an intestinal bacterium. This can be explained by the existence of what are known as **mosaic genes**, which could be described as structural genes that have been cut up. Their coding sections (**exons**) are interspersed with non-coding sections (**introns or junk DNA**). Furthermore, there are long sections with repetitive sequences whose function is as yet unknown.

Whatever the function of those introns (Ch. 10) which seem to have been shaped by evolution (probably including viral attacks), they do not contain information related to the formation of peptide chains. In eukaryotes, the introns are copied into mRNA where they form loops that are then **spliced off**, and only the exons are retained (Fig. 3.16). The resulting shortened mRNA, carrying exclu-

sively protein-coding information, is called **mature mRNA**.

■ 3.9 Recombination – A Genetic Reshuffling of Cards

Mutations modify the genes of an organism (more in Chapter 4).

Recombination – the geneticist's second most important tool – means genetically reshuffling the cards. Entire genes or parts of genes originating from two or more organisms are rearranged.

In **homologous recombination**, for example, two chromosomes containing identical or similar DNA sequences come together in a cell to exchange corresponding (homologous) parts. The exchange in which DNA is broken up and reunited is catalyzed by specific enzymes.

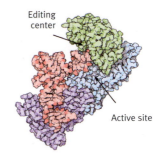

Fig. 3.21 Twenty different aminoacyl-tRNA-synthetases join the twenty amino acids accurately with tRNA. Isoleucine bears a strong resemblance to valine. To avoid errors, isoleucyl tRNA synthetase features an additional active center for editing and proofreading purposes. This ensures that only one error occurs in every 3000 syntheses, compared to one in 150 without this corrective function.

Fig. 3.22 Differences in protein biosynthesis between prokaryotes (top) and eukaryotes (bottom)

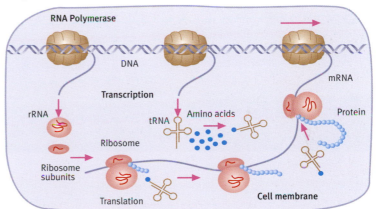

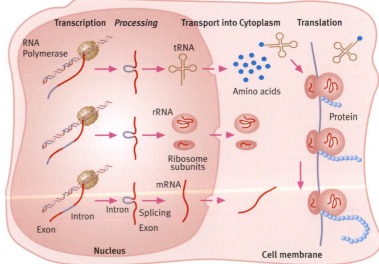

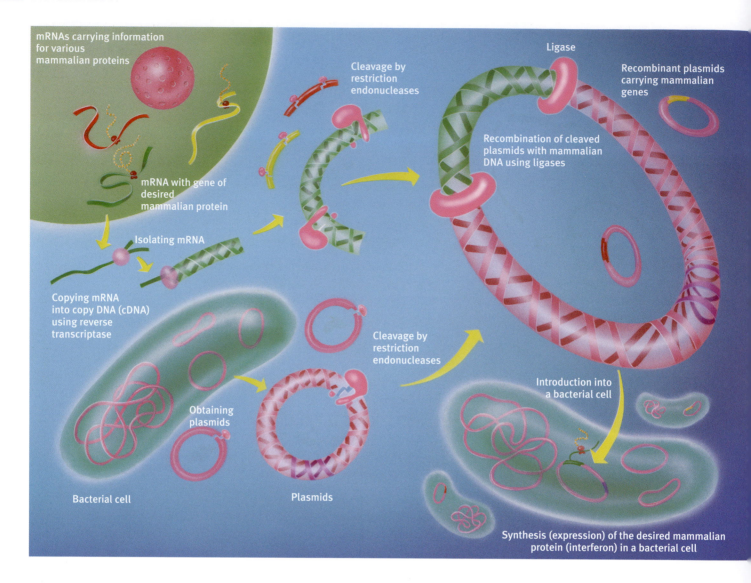

mRNAs carrying information for various mammalian proteins

mRNA with gene of desired mammalian protein

Isolating mRNA

Copying mRNA into copy DNA (cDNA) using reverse transcriptase

Cleavage by restriction endonucleases

Ligase

Recombinant plasmids carrying mammalian genes

Recombination of cleaved plasmids with mammalian DNA using ligases

Cleavage by restriction endonucleases

Obtaining plasmids

Introduction into a bacterial cell

Bacterial cell

Plasmids

Synthesis (expression) of the desired mammalian protein (interferon) in a bacterial cell

Fig. 3.23 Genetic modification of bacterial cells in the production of human interferon in bacteria.

The entire mRNA is isolated from human cells and enzymatically converted into double-stranded DNA, using reverse transcriptase.

The DNA is then specifically cut by restriction endonucleases and pasted into bacterial plasmids that have been cut by the same restriction endonucleases. These recombinant plasmids are introduced into bacteria and reproduced (cloned).

In a final step, those bacterial colonies that produce human interferon are isolated from tens of thousands of colonies.

In eukaryotes, homologous recombination mainly happens during **meiosis** (maturing of reproductive cells). The accidental distribution of maternal and paternal chromosomes (**homologues**) in the newly formed reproductive cells accounts for a further reshuffling. Two **haploid** (single) **sets** of parental **chromosomes** are brought together in the offspring to form a **diploid** (double) set.

The recombination between homologous DNA sections is a very efficient process indeed. The recombination of genes from two individuals differing in their genes or part of their genes results in two different **genotypes**, i.e., genetically different individuals.

If, for example, two strains of bacteria differ by a dozen base pairs, 2^{12} or nearly 5000 new genotypes can develop. In most cases, however, the differences comprise more than just a dozen base pairs,

making astronomical numbers of combinations possible.

Although probably all microorganisms are capable of exchanging genes with related strains, the possibility of natural genetic recombination in order to produce an industrially exploitable strain has not been used much.

The simplest cell cycle is found in **haploid yeasts**. During the major part of their cell cycle, their set of chromosomes is haploid, in contrast to most other animals and plants with diploid chromosome sets. A normal yeast cell can only reproduce sexually if in contact with a related cell of the opposite sex or mating type. Both cells merge into a transient diploid cell producing haploid sexual spores whose gene combinations differ from those of their parent cells. Many diversions from this simple pattern are found in industrial yeasts. These sometimes contain several sets of chromosomes or replicate without going through a mating process.

Cross-breeding (**hybridization**) of different strains – sometimes even species – plays an important part in the development of industrial yeasts. This applies in particular for yeasts that speed up hi-tech bread-making processes or yeast that increase alcohol production in distilleries, enabling the production of special beers in which almost the entire carbohydrate content has been metabolized (Ch. 1).

Natural recombination occurs not only in meiosis, but also under other circumstances. It is paramount in the production of the many shapes of **antibodies** and other molecules in the immune system (Ch. 5). Some **viruses** use recombination to integrate their genetic material into their host's DNA (Ch. 5). Recombination can be used to manipulate genes, e.g., in **knockout mice** (Ch. 8).

■ 3.10 Plasmids – Ideal Vectors for Genetic Material

In 1955, during a dysentery epidemic, Japanese microbiologists isolated a strain of the bacterium *Shigella* that defeated three kinds of antibiotics. The bacterium had become resistant. Over the following years, there was mounting evidence that the increasing use of antibiotics led to a rise in **antibiotic resistance**. Not only were the bacteria able to resist the attacks from antibiotics, but they also passed on their resistance to other bacteria, enabling them to produce enzymes that inactivate antibiotics in various ways.

In 1960, Japanese researcher **Watanabe** solved the puzzle. **Plasmids** are small ring-shaped DNA elements (comprising 3,000 to 100,000 base pairs) that are found outside the much larger main DNA in the bacterial cell (Fig. 3.29). A cell contains 50-100 smaller and one to two larger plasmids on average, most of which can reproduce independently within the cell. If two bacterial cells establish a contact between them, they can exchange large plasmids via a bridge (**conjugation**).

Smaller plasmids, however, are not transferable. The plasmid DNA itself does not directly confer antibiotic resistance to bacteria, but regulates production of enzymes that inactivate antibiotics (e.g., penicillinases or enzymes that inactivate tetracycline).

Stanley N. Cohen (b. 1935, Fig. 3.25), a plasmid specialist at Stanford University, California, was the first to understand how plasmid DNA could be harnessed as transporters for genetic material or **vectors**. If it were possible to smuggle in some foreign DNA, a plasmid could act like a cuckoo that lays its egg (foreign DNA) into a nest (bacterial cell). What was needed was a method of opening up the DNA rings and inserting foreign DNA.

This is not as easy as it sounds. Although if stretched out lengthwise, the main bacterial DNA would be about a millimeter long (s. p. 63-64), it is coiled up so tightly within the cell that it has a diameter of a thousandth of a millimeter. Plasmids, however, are yet a hundred times smaller. A gene for insertion into a plasmid is about a ten thousandth of a millimeter wide, and the DNA helix is only two millionths of a millimeter thick. Using mechanical **scissors and scalpels** at this range would be utterly hopeless, and the "scissors" would have to find the exact cutting location by themselves.

■ 3.11 Scissors and Glue at a Molecular Level – Restriction Endonucleases and DNA Ligases

In the 1960s, **Werner Arber** (Fig. 3.28), born in Geneva in 1929, along with US researchers **Hamilton D. Smith** (born in 1931), and **Daniel Nathans** (1928-1999), both at Johns Hopkins University, Baltimore, discovered a mechanism in bacteria that protected them from deadly viral invaders called **bacteriophages** – a discovery that earned them the Nobel Prize in 1978. Bacteriophages inject their DNA into the bacterial cell.

The bacteria use enzymes called **restriction endonucleases** (restrictases for short) to cut up the viral DNA and put it out of action. The bacterial DNA, however, can be protected from restrictase action by some additional methyl (CH_3-) groups.

In 1970, it was found that restriction endonucleases do not cut DNA just anywhere, but very precisely at certain base pairs. **Herbert W. Boyer** (b. 1936, Fig. 3.24) from the University of California in San Francisco investigated the restriction endonuclease *Eco*RI (named after the *Escherichia coli* strain RY 13). It cuts DNA only within the combination GAATTC, and only between bases G and

Fig. 3.24 Herbert W. Boyer (b. 1938), co-founder of Genentech. "*Wonder is what sets us apart from other life forms. No other species wonders about the meaning of existence or the complexity of the universe or themselves.*"

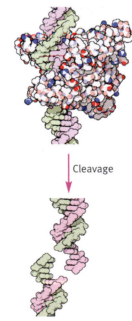

Fig. 3.25 Stanley N. Cohen (b. 1935), a plasmid specialist, was one of the founding fathers of genetic engineering.

Cleavage

Fig. 3.26 Top: A restriction endonuclease cutting DNA specifically and producing sticky ends (bottom).

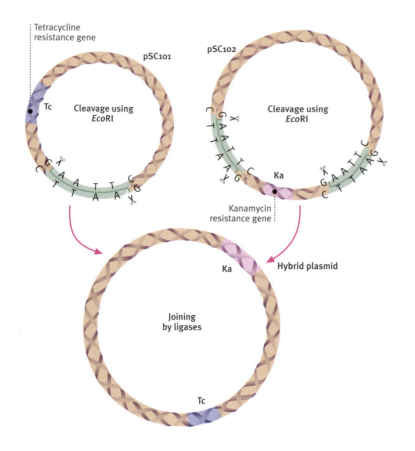

Fig. 3.27 The first experiment in gene technology by Cohen and Boyer.

Fig. 3.28 Werner Arber (b. 1929)

Fig. 3.29 Plasmid pSC101, also called *plasmid necklace* by Stanley Cohen.

A, and in the complementary strand CTTAAG, again it cuts between bases A and G.

3'-XXXXXXXXG/AATTCXXXXXXXX-5'

5'-XXXXXXXXCTTAA/GXXXXXXXX-3'

This does not produce a clean cut, but two fragments with ends that stick out.

3'-XXXXXXXXG   AATTCXXXXXXXX-5'

5'-XXXXXXXXCTTAA   GXXXXXXXX-3'

At low temperatures, the separated DNA does not fall apart, but its ends are loosely attached to each other. **Janet Mertz**, a collaborator of **Paul Berg's** (3.47) Stanford University lab, found that the protruding A and T bases or G and C bases, called **sticky ends**, are held in place by electrostatic attraction. It is even possible to re-attach them with an enzyme, the ATP-consuming **DNA ligase**.

Such **scissors and glue for DNA** had been found independently by several researchers, so it made sense to combine the results of their efforts.

More than 1200 restriction endonucleases, divided into three subcategories, are known today. Only a few of them, however, are relevant for genetic engineering. There are also restriction endonucleases that produce a clean cut and smooth ends which are called blunt ends. *Pvu*II from *Proteus vulgaris* and *Alu*I from *Arthrobacter luteus* are cases in point.

There are also several enzymatic scissors that produce identical ends – restriction endonucleases *Bam*HI (recognition sequence GGATCC) from *Bacillus amyloliquefaciens* and *Bgl*II (recognition sequence AGATCT) from *Bacillus globigii* give rise to the same sequence GATC, at both sticky ends. The gene fragments produced by one of these enzymes can thus be combined with those produced by the other.

■ 3.12 First Experiments in Gene Technology – Croaking Bacteria?

Early in 1973, not even a year after the discovery of the appropriate tools, **Stanley N. Cohen** and his collaborator **Annie C. Y. Chang** in Stanford joined forces with their colleagues from the University of San Francisco, **Herbert W. Boyer** (Fig. 3.24) and **Robert H. Helling**, to carry out the **first experiment in gene technology**.

The researchers chose a small nontransferable plasmid, named after Cohen's initials **pSC 101** (Fig. 3.29) which occurs in large quantities in bacterial cells. It carries the *tc* gene that confers resistance to the antibiotic tetracycline.

The plasmid was chosen because it contains just one GAATTC base sequence, to be cleaved by the restriction endonuclease *Eco*RI between bases G and A. If there had been more than just this one sequence, the plasmid ring would have been cut into several pieces rather than only once.

In order to identify the bacteria containing the foreign gene, it was crucial that the restriction process did not disable the antibiotic resistance properties. *Eco*RI originates from *Escherichia coli*, and the cutting process turns the plasmidic DNA ring into a linear structure with sticky ends. Cohen and Boyer used *Eco*RI to cut another plasmid derived from

E. coli (**pSC 102**) that carries a resistance gene to **kanamycin** (**Ka-gene**, Fig. 3.27). Again, this plasmid had no more than one restriction site, and the Ka-gene was left intact by *Eco*RI.

As both plasmids had been cut at the same site - G/AATTC-, they had **corresponding sticky ends**. Electrostatic attraction forces between the restriction sites made differing fragments loosely hang together. The researchers now added "glue", **DNA ligase**, (Fig. 3.33) to the mixture and thus produced new, larger, **recombinant plasmids**.

In a last step called **transformation**, the recombinant DNA was transferred into bacteria by adding calcium chloride ($CaCl_2$) to a solution containing *E. coli* bacteria. Calcium chloride is a salt that renders cell walls permeable for DNA (Fig. 3.30). This artificial process, found nowhere in nature, enabled the new plasmids to be introduced into the bacteria (see also Fig. 3.23). Finally, the crunch came when the bacteria solution was spread over nutrient plates containing tetracycline as well as kanamycin.

As expected, **most of the bacteria died**, but a few survived. They were bound to contain the artificially created plasmid conferring **dual resistance**. These bacteria were able to produce enzymes that inactivated both kanamycin and tetracycline. The surviving bacteria proliferated to form cell colonies of about 100 million identical offspring, all carrying the new recombinant DNA. A **clone** had been formed, a **group of genetically identical organisms**.

This success encouraged Cohen and his collaborators to go ahead with their next experiment. They recombined DNA segments of plasmids taken from different types of bacteria – pSC 101 from *E. coli* and a plasmid from a penicillin-resistant *Staphylococcus aureus* strain. The *S. aureus* plasmid contained 4 restriction sites for *Eco*RI and was thus divided into four pieces. Only one of them contained the penicillin resistance gene, so the number of possible combinations was higher than in the first experiment. The moment of truth came when the engineered bacteria were spread on nutrient plates containing tetracycline as well as penicillin.

There was an outbreak of joy in Cohen's lab – the recombinant plasmid worked in *E. coli* cells! For

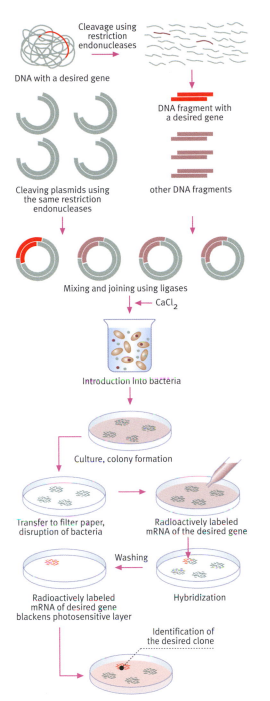

Cleavage using restriction endonucleases

DNA with a desired gene

DNA fragment with a desired gene

other DNA fragments

Cleaving plasmids using the same restriction endonucleases

Mixing and joining using ligases

$CaCl_2$

Introduction into bacteria

Culture, colony formation

Transfer to filter paper, disruption of bacteria

Radioactively labeled mRNA of the desired gene

Washing

Radioactively labeled mRNA of desired gene blackens photosensitive layer

Hybridization

Identification of the desired clone

Fig. 3.30 Diagram of the cloning process of a mammalian gene.

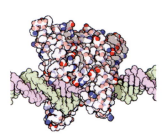

Restriction endonuclease

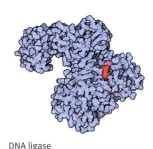

DNA ligase

the first time, genetic material from diverse species had been successfully transferred. However, the "crossing of barriers that normally separate biological species from each other", as Cohen described the experiment rather prematurely, had not yet been achieved. We now know that this is the way diverse microorganisms, including viruses, can "naturally" exchange genetic material. A virus infection could thus be described as a gene transfer!

What can be read from a gene?

"Genes are not like engineering blueprints; they are more like recipes in a cookbook.

They tell us what ingredients to use, in what quantities, and in what order – but they do not provide a complete, accurate plan of the final result".

Ian Stewart, Life's Other Secret, New York 1998

Box 3.3 Useful Gene Transporters

Plasmid pBR322

Plasmid **pBR322**, a very popular tool in genetic engineering, was developed from *E. coli* towards the end of the 1970s. **p** stands for plasmid, **BR** are the initials of the scientists involved, Bolivar and Rodriguez. The figure **322** distinguishes this particular plasmid from others created in the same lab, such as pBR 325, pBR 327 etc.

pBR322 contains genes for tetracycline and ampicillin resistance enzymes and sites where it can be cut by specific restriction endonucleases and matching DNA sequences inserted. If the restriction endonuclease is *Eco*RI, both antibiotic resistance genes remain intact. If, however, *Bam*HI is used, the tetracycline resistance (*tc*) gene is broken up and the foreign gene inserted into it.

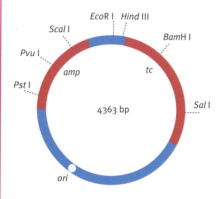

The genetic and physical map of pBR322 shows the positions of the two resistance genes, –(*amp* for ampicillin and *tc* for tetracycline), the replication starting point *ori* and the restriction sites for important restriction endonucleases.

These cells are resistant to ampicillin, but not to tetracycline, and can thus be easily selected.

Cells that have not taken in the vector are sensitive to both antibiotics, while cells that contain pBR322 and no insertion are resistant to both.

λ-Phage

λ-Phage (phage lambda) is a versatile creature. It can either destroy its host or become part of it. In the first case, it will lyse the host cell while rapidly producing its viral DNA and packaging it into viral particles (**lytic cycle**).

The other option is a **lysogenic cycle** which involves integrating the viral DNA into the host genome where it can be reproduced and carried

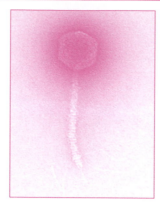

The λ-Phage

from generation to generation without damaging the host.

Certain environmental changes can suddenly activate dormant viral DNA. It separates from the host genome, starting a lytic cycle. Phage DNA contains 48 kb, and large parts of it are not needed for the infection process. The genes for the lysogenic cycle can thus be replaced by foreign DNA, making it an ideal vector.

Phage mutants have been developed specifically for the cloning of DNA. Their DNA can be cut by *Eco*RI in only two sites (compared to five in the wildtype). This results in three sections of which the middle one is removed.

The middle section is replaced by a matching long DNA sequence (about 10 kb), using ligases. The modified phage has not lost its infectiousness, but has lost its ability to run a lysogenic cycle, i.e., it cannot lie dormant. This makes it a very desirable cloning vector. Phage λ finds it easier than plasmids to enter a bacterial cell and it can introduce larger DNA sequences.

It has also been possible to create hybrids of λ-phages and plasmids, known as **cosmids**. The name is derived from DNA sequences called *cos* which originate from the phage λ genome.

These sequences enable cosmids to carry larger genes (up to 45kb). They are then packaged into phages that introduce the foreign genes into bacteria. Cosmids containing an ampicillin resistance gene convey to the bacteria the ability to survive in a culture medium laced with ampicillin. These bacteria will then reproduce and thereby replicate the foreign DNA.

The M13-Phage

M13 is a filamentous phage, looking completely different from phage λ. The threadlike virus is 900 nm long and has a diameter of only 9 nm. Its single-stranded DNA is ring-shaped, (some-

times also called + strand) and enclosed in a protein envelope consisting of 2710 identical protein subunits (see Ch. 5).

Intriguingly, the virus enters *E. coli* through their F pilus that enables the exchange of DNA between cells. M13 is a very attractive vector option because the cloned DNA can be obtained as single-stranded DNA. Especially in DNA-sequencing and in vitro mutagenesis, cloned genes are required as single strands (Ch. 10).

M13 is used in the **phage display technique** (Ch. 5). M13 DNA is not inserted into the bacterial genome, as is done with phage λ, and the cell is not lysed. The bacterium grows and divides, albeit at a slower rate than non-infected cells. The daughter cells continue to release M13 phages.

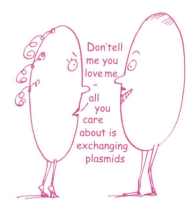

About 1000 new M13 phages are produced in every new generation. As M13 does not kill its host, it can be grown in large quantities and easily harvested.

M 13 also grabbed the headlines in a totally different field: **Batteries!**

In 2006, **Angela Belcher**, MIT Professor of Materials Science, Engineering and Biological Engineering, has demonstrated that genetically engineered M 13 viruses can work as positive electrodes in lithium ion batteries. The insertion of further nucleotides into the viral DNA led to the expression of an additional peptide that binds to cobalt ions. When brought into contact with a solution containing cobalt salts, the virus develops a cobalt oxide coating.

Used in batteries, cobalt has a much higher storage capacity than the conventionally used carbon-based materials in lithium ion batteries. It is a first step toward high-capacity, self-assembling batteries.

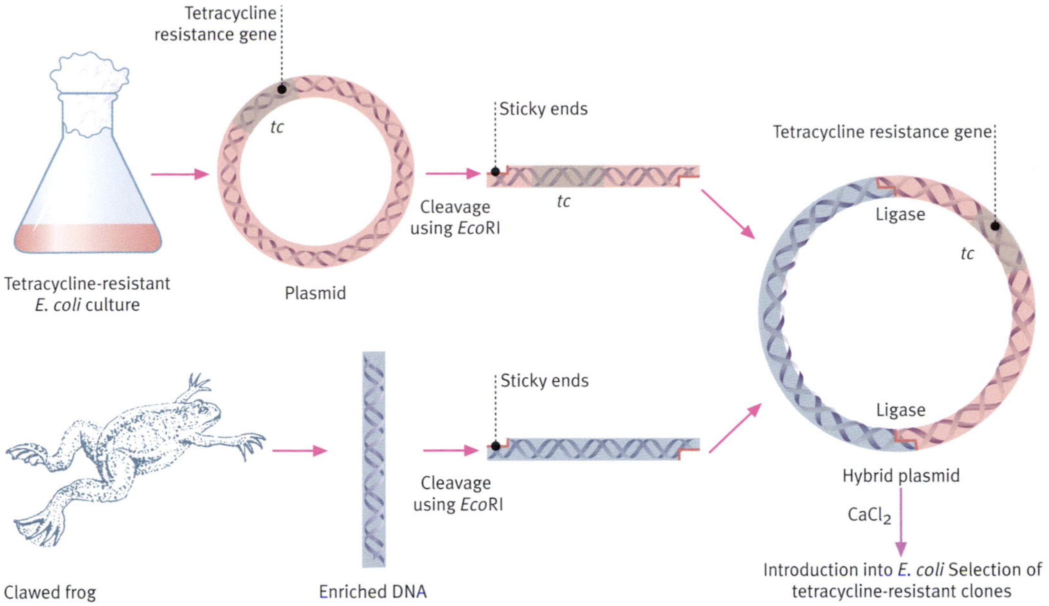

Fig 3.31 Insertion and cloning of clawed frog DNA into bacteria.

Tetracycline resistance gene

tc

Tetracycline-resistant *E. coli* culture

Plasmid

Cleavage using *Eco*RI

Sticky ends

tc

Tetracycline resistance gene

Ligase

tc

Ligase

Hybrid plasmid

$CaCl_2$

Clawed frog

Enriched DNA

Cleavage using *Eco*RI

Sticky ends

Introduction into *E. coli* Selection of tetracycline-resistant clones

I always knew we were born to higher things!

Spurred on by these successes, it was now a matter of raising the stakes and crossing even higher barriers – the species barrier between bacteria and frogs, the African clawed frog (*Xenopus laevis*), to be precise (Fig. 3.32)

Genomic DNA was isolated from clawed frogs and cut with *Eco*RI restriction enzymes. *Eco*RI was also used to cut the bacterial plasmid pSC 101 (Fig. 3.31).

The frog DNA attached to bacterial DNA and was glued with ligases, then introduced into *E. coli* cells and replicated. Those cells containing the new recombinant frog/bacterium plasmid could be identified due to their tetracycline resistance and – frog DNA not containing any antibiotics resistance gene – by chemically analyzing the amino acids.

On July 27th 1973, it was established that frog DNA had been "accepted" by bacteria.

The new plasmid was reproduced during the 1000 cell divisions – identical copies were made, i.e., the **frog-bacterium plasmid** had been cloned.

This was not yet a new species – after all, less than a thousandth of the bacterial DNA originated from the frog, and the manipulated bacteria did not croak like frogs, as researchers joked. However, something far more important had been achieved – a universal method of genetic engineering had been developed that could be used to produce and analyze large amounts of the heretofore completely inaccessible genetic material of higher organisms. The **technology for cloning DNA** was available.

3.13 How To Obtain Genes

Unfortunately, the production of higher organism proteins was not as easy as the "chopped-up frog DNA" experiment might suggest, because the DNA of higher organisms is interspersed with introns, containing non-protein-coding information.

Cleaving such intron-riddled DNA of higher eukaryotic organisms with restriction endonucleases would be futile. It would be possible to clone thousands of foreign DNA sequences, but what proteins they might produce is anyone's guess. In a best-case scenario, the result would be a protein containing all relevant amino acids coded in the exons, but interspersed with superfluous amino acids derived from introns.

Gene technologists opted therefore not to use the DNA of higher organisms which is loaded with introns, but to obtain **mature mRNA** which contains only the coded structural designs of proteins and no introns. The problem is how to bring together single-stranded mRNA with double-stranded plasmidic DNA. Luckily, an enzyme was found in retroviruses, **reverse transcriptase** (**revertase**) which can re-transcribe single-stranded RNA into double-stranded DNA (Fig. 3.35). The genetic material of retroviruses does not consist of DNA, but of a single-stranded RNA molecule.

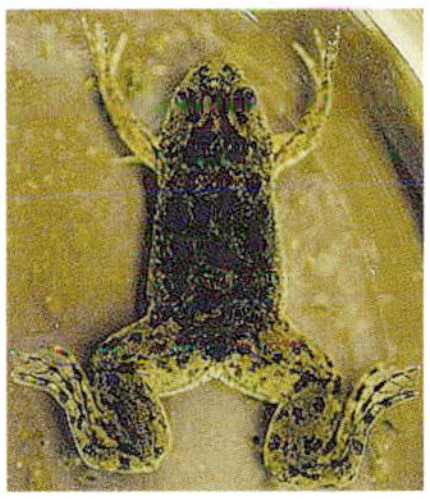

Fig. 3.32 The African clawed frog (*Xenopus laevis*), a new lab inhabitant alongside mice and rats is very easy to keep and is long-lived. Its name (strange foot) derives from its black claws. This extremely greedy amphibian, more than four inches long, has become quite a pest since it was introduced into California.

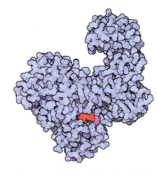

Fig. 3.33 DNA ligase, the glue that sticks the cut DNA sections together, is an ATP-dependent enzyme (see Ch. 2). Ligase is an indispensable tool in genetic engineering. Cofactor ATP (red) and a lysine residue (magenta) are essential to the process.

Fig. 3.34 Synthesis of rat proinsulin in bacteria.

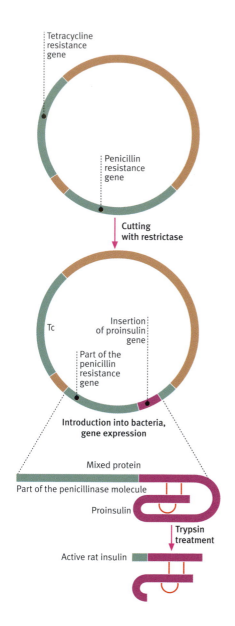

Tetracycline resistance gene

Penicillin resistance gene

Cutting with restrictase

Tc

Insertion of proinsulin gene

Part of the penicillin resistance gene

Introduction into bacteria, gene expression

Mixed protein

Part of the penicillinase molecule

Proinsulin

Trypsin treatment

Active rat insulin

Fig. 3.35 Top: Reverse transcriptase and its claw-like structure. Bottom: Two activities take place at the reverse transcriptase molecule. Polymerase activity produces an RNA-DNA hybrid molecule, while nuclease activity degrades the now redundant RNA and the DNA strand is complemented to form a double strand.

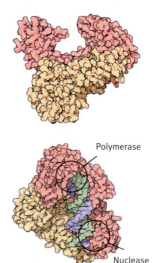

Polymerase

Nuclease

When such viruses infect host cells containing DNA (see Chapter 5), their revertase transcribes their single-stranded viral RNA into double-stranded DNA that can be inserted into the host DNA (Fig. 3.23).

Reverse transcriptase was used to synthesize a DNA strand along the mRNA strand. This DNA resulting from copying RNA is called **copy DNA** (**cDNA**).

If the amino acid sequence of a protein is known and the corresponding mRNA cannot be isolated from cells, there is another way of chemically synthesizing the gene. This is, of course, also a way of producing DNA that does not occur naturally. Nowadays, there are **DNA synthesizers** (Box 3.5) which use a start nucleotide bound to carrier material (such as silica gel or glass beads). Nucleotides

are attached to these in the desired sequence. By 1988, it was possible to synthesize 30 base-long DNA fragments in a day. In 1979, this would still have required six months of team work. Nowadays, it only takes a few hours to synthesize complete "tailor-made genes".

3.14 Human Insulin From Bacteria?

In July 1980, 17 volunteers at Guy's Hospital, London, were given insulin injections that made the headlines in the newspapers. What was so sensational about them? After all, millions of **diabetics** worldwide were treated with insulin every day (see Box 3.3). The insulin was obtained from the pancreases of cattle and pigs. The substitute insulin is supposed to regulate the blood sugar level in diabetics and avert the serious effects of a disease that has become one of the three top three causes of death in the industrial world (see Box 3.4).

Type I diabetes sets in before the 20th year and is caused by autoimmune destruction of the insulin-forming cells in the pancreas. Type II diabetes, by contrast, tends to develop in middle age (90% of all diabetics), especially in overweight people (Ch. 10). The 17 volunteers mentioned above were the first humans in medical history that were treated with a mammalian hormone that was not derived from the organ of a mammal, but from bacteria. This was the first test of a genetically engineered substance on humans. Two years later, an official license for the medical application of genetically engineered insulin was granted.

Demand for insulin is incredibly high. One single diabetic patient would require the pancreases of 50 pigs to cover a year's supply. Hoechst, one of the major insulin suppliers at the time, processed eleven tons of pigs' pancreases per day, provided by over 100,000 slaughtered animals.

In the US, Eli Lilly and Co. discontinued their sales of animal-based insulin Iletin in 2005. By then, the product was bought by a mere 2,000 customers who, it was suspected, might have bought it for their diabetic cats or dogs. Iletin as a brand had been introduced by Lilly in the 1920s and was the last insulin product that was still made from mashed animal pancreases.

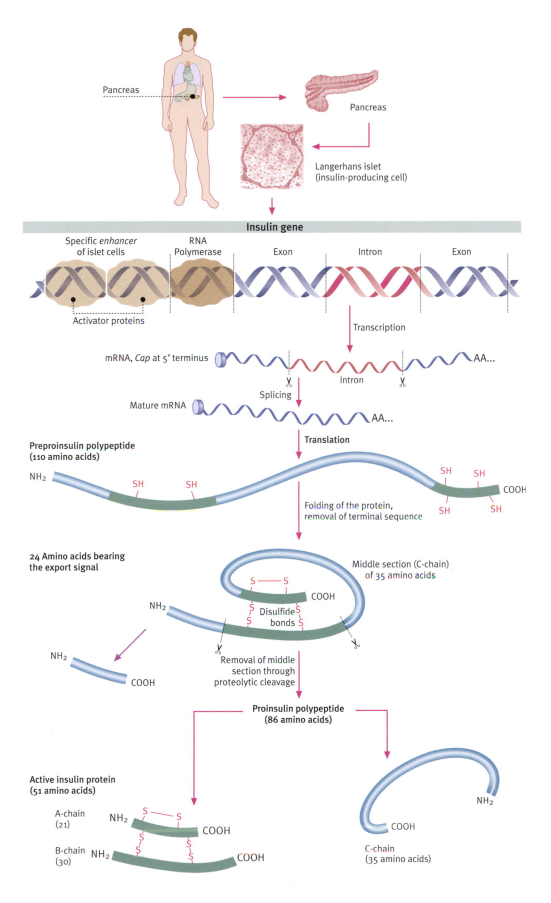

Pancreas

Pancreas

Langerhans islet
(insulin-producing cell)

Insulin gene

Specific *enhancer*
of islet cells

RNA
Polymerase

Exon

Intron

Exon

Activator proteins

Transcription

mRNA, *Cap* at 5' terminus

Intron

AA...

Splicing

Mature mRNA

AA...

Translation

**Preproinsulin polypeptide
(110 amino acids)**

NH₂

SH   SH

SH   SH

COOH

SH   SH

Folding of the protein,
removal of terminal sequence

**24 Amino acids bearing
the export signal**

Middle section (C-chain)
of 35 amino acids

NH₂

COOH

Disulfide
bonds

NH₂

COOH

Removal of middle
section through
proteolytic cleavage

**Proinsulin polypeptide
(86 amino acids)**

**Active insulin protein
(51 amino acids)**

A-chain
(21)

NH₂

COOH

B-chain
(30)

NH₂

COOH

NH₂

COOH

C-chain
(35 amino acids)

Fig. 3.36 Insulin synthesis in the
islet cells of the pancreas.

The diagram shows a typical insulin
gene in a mammalian cell, including
introns, coding sequences (exons)
and regulatory sequences needed
for transcription.

Right in front of the insulin gene (in
the flanking region at the 5'terminus) lie several sequence elements
that are crucial for the production of
insulin. They bind to several regulatory proteins that turn on their activity. In cells that do not produce
insulin, these DNA sections are
blocked by other proteins.

RNA polymerase begins its transcription process immediately
behind these sequences. In eukaryotes, the termini of the transcribed
mRNA are modified by a 5' cap and a
3'poly(A) tail. The cap protects the
mRNA from degradation by phos-
phatases and nucleases and
enhances the translation process,
while the poly(A) tail is not coded in
the DNA and seems to stabilize the
mRNA.

Even after the translation process in
the ribosome, the speed of insulin
synthesis can still be regulated. The
preproinsulin molecule (128 amino
acids) is longer than the active
hormone. At the amino terminus,
specific enzymes cleave off a short
section (24 amino acids), i.e., the
signaling sequence that enabled its
passage through the membrane at
the endoplasmic reticulum.

The resulting proinsulin has a middle section in the peptide chain,
called C-peptide, which is then
removed in order to produce insulin
from the short A and B chains, held
together by disulfide bridges (SS,
shown in red) between the cysteines.

The amount of active protein available is often determined only after
its synthesis by a bodily process
involving feedback regulation.

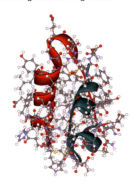

Spatial structure of an insulin molecule with disulfide bridges (yellow)
and α-helices of the A and B chains.

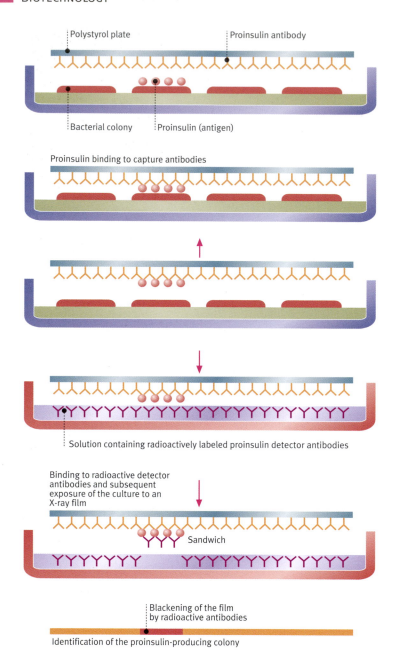

Fig. 3.37 Finding proinsulin-producing bacterial colonies using antibodies in a radioimmunoassay.

Fig. 3.38 Rosalyn Yalow (b. 1921) developed the radioimmunoassay (RIA) method and was awarded the Nobel Prize in 1977.

In the figure: Polystyrol plate; Proinsulin antibody; Bacterial colony; Proinsulin (antigen); Proinsulin binding to capture antibodies; Solution containing radioactively labeled proinsulin detector antibodies; Binding to radioactive detector antibodies and subsequent exposure of the culture to an X-ray film; Sandwich; Blackening of the film by radioactive antibodies; Identification of the proinsulin-producing colony.

According to the World Health Organization, in 2006, at least 171 million people worldwide suffered from diabetes. The incidence is increasing rapidly, and it is estimated that by the year 2030, this number will double.

Diabetes rates in North America have been increasing substantially for at least 20 years. In 2005, there were about 20.8 million diabetics in the United States alone. About 5-10% of diabetes cases in North America are type 1, the rest is type 2.

The American Diabetes Association pointed out that one in three Americans born after 2000 will develop diabetes in their lifetime. The Centers for Disease Control and Prevention and Prevention have termed this development an epidemic.

3.15 Insulin Synthesis in Humans

The hormone insulin is a small molecule consisting of two protein chains, one made up of 21 (A-chain), the other (B-chain) of 30 amino acids. **Fred Sanger** (Box 3.4) had begun deciphering its primary structure in 1945, and it took 10 long years of hard work in the cellars of the Biochemical Institute in Cambridge (England) to accomplish it.

At the time, trying to analyze insulin was considered a foolhardy undertaking. Sanger used crystallized insulin from 120 cows as raw material. He was awarded the Nobel Prize in 1957, only three years after decoding the sequences of all 51 amino acids in the two insulin chains.

Both chains are first synthesized in the B-cells of the Langerhans islets of the pancreas as parts of a longer chain of 110 amino acids. The long-chained version is known as **preproinsulin** (Fig. 3.36).

Like precursors of other peptide hormones, it is secreted into the endoplasmic reticulum. Its first 24 amino acids are the signal for the cell membrane to release the insulin from the cell. During its passage through the membrane, these 24 amino acids are separated from the chain by enzymes (peptidases) and remain in the cell. The remaining 86 amino acids are called **proinsulin**, consisting of B-chain, C-peptide, and A-chain.

The beginning and end of the molecule (B and A) interact with each other, binding via two disulfide bonds (-S-S-). Then the central part of the molecule (C-chain or C-peptide, consisting of 35 amino acids), is separated by membrane-bound enzymes (proteases) in the Golgi complex of the cell.

The C-chain gives the A and B-chain the correct orientation in relation to each other, a prerequisite for the correct spatial folding of the fully functioning insulin.

Active insulin and C-peptide are stored in vesicles and released together upon a stimulus signal. The daily "consumption" of insulin in a human is about 1.8 milligrams.

3.16 Rat Proinsulin – The Beginnings of Genetic Engineering

In order to produce genetically engineered insulin using bacterial cells, **Walter Gilbert** (Fig. 3.39) and **Lydia Villa-Komaroff** from Harvard University used animal cancer cells – β-cells of a rat pancreatic tumor. When they were beginning their research, experimenting with human genes was still prohibited.

The mRNA of the cancer cells was transcribed into cDNA using reverse transcriptase and then cut with restriction endonucleases. The resulting sticky ends were inserted into the DNA of bacterial plasmids carrying resistance to penicillin as well as to tetracycline (Fig. 3.34). These plasmids were then introduced into bacterial cells. A clone was grown from each of these cells, and they were spread on plates with nutrient agar containing penicillin and tetracycline.

Would any of these clones be able to produce rat proinsulin? In order to find out, Gilbert and his collaborators put large amounts of **proinsulin antibody** (produced by mice injected with the protein) on **plastic plates**. Antibodies bind easily by adsorption to plastic such as polystyrol (PS). The researchers then brought the antibody-coated plates into contact with the bacterial colonies. By adding lysozyme, the cells were lyzed and their content could be analyzed.

If the cells contain any protein from the proinsulin amino acid sequence, that protein is bound to the antibodies on the polystyrol plate (Fig. 3.37). The antibodies are also called **capture antibodies**.

How can we know if and where proinsulin has bound? A solution containing **radioactively labeled detector antibodies** that also recognize proinsulin and binds to it is incubated on the PS plate. If the result is positive, a **sandwich structure** forms. The plate binds the capture antibody, the antibody binds proinsulin, and the proinsulin molecule binds, with its other end, to the radioactively labeled detector antibody.

In the absence of proinsulin, the detector antibodies will not bind and are washed away in a subsequent step. In this case, there is no sandwich formation

and no bound radioactivity. This method is called a **radioimmunoassay** (RIA, see also Ch. 10).

In 1959, **Rosalyn Yalow** (b. 1921, Fig. 3.38) in New York developed the first RIAs to detect insulin in blood (Nobel Prize 1977 together with **Andrew Guillemin** and **Andrew V. Schally**). Since the antibodies carry radioactive labeling, they are detected when the plate is placed on a sensitive X-ray film (**autoradiography**).

One clone that elicited a positive reaction and blackened the film produced the same effect on a polystyrol plate that carried not proinsulin, but penicillinase antibodies. In other words, the synthesis product detected by the antibodies appeared to be a **mixed penicillinase-proinsulin product**. The proinsulin gene had been inserted right into the penicillinase gene.

When the clone was isolated and the cells were grown in a liquid nutrient medium, it was possible to isolate the mixed protein from the culture solution without having to destroy the cells. Since the penicillinase part of the molecule had enabled it to penetrate the cell membrane, it was found in the medium.

The secretion of the protein into the medium was a major step forward in relation to the rat proinsulin that had been confined to the inside of the cell.

The penicillinase had been preceded by its **signal sequence** which enabled the export of the "protein thread" from the cell. Otherwise, the proinsulin molecule would have remained within the cell.

The subsequent nucleotide sequence analysis of the foreign DNA contained in the bacterial plasmid showed that it indeed only contained the sequence for proinsulin.

In order to obtain pure insulin from the penicillinase-proinsulin mixture, Gilbert used the digestive enzyme trypsin (Ch. 2 and Fig. 3.47) to remove the major part of the penicillinase segments as well as the middle (C) segment of the proinsulin molecule.

This resulted in **active insulin** which, as researchers in Boston found, had the same effect on the sugar metabolism of fat cells as normal rat insulin. The bacteria had been able to produce "genuine" rat insulin!

Fig. 3.39 Walter Gilbert (b. 1932), Nobel Prize for Chemistry in 1980.

Fig. 3.40 Frederick G. Banting (1891-1941) studied medicine and began to practice medicine in London, Ontario. He gave lectures on endocrinology at the University of Ontario. On the night of 31 October 1920, during his routine reading of articles in a medical journal, he wrote down an idea for a method to isolate the internal secretion of the pancreas. Dissatisfied with his practice and fascinated by his idea, Banting left London and moved to Toronto. There, in May 1921 he began his research at the University of Toronto, under the supervision of Professor John Macleod (1876-1935). He was assigned a single assistant to help him, the student Charles H. Best (1899-1978). During a summer of intense work, Banting tested his idea, performing operations on dogs to tie up their pancreatic ducts. A pancreas extract would then be made from it and administered to dogs with diabetes.
The method worked!
Banting could keep dogs with diabetes alive with the extract. He enthusiastically reported his findings to Macleod, who had been away on his summer holidays. The use of dog pancreases was soon abandoned in favour of using the pancreas from calves and cows. All the efforts concentrated on extracting a substance from a normal pancreas, named insulin. Insulin was put into mass production in a matter of months. In 1923 Banting and Macleod received the Nobel Prize in Medicine. Banting shared the award money with Best, believing that Best deserved the Prize more than Macleod.

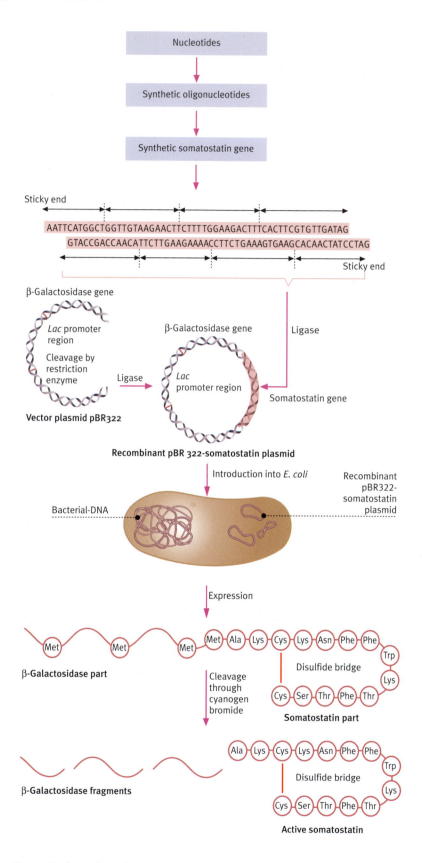

Nucleotides

Synthetic oligonucleotides

Synthetic somatostatin gene

Sticky end

AATTCATGGCTGGTTGTAAGAACTTCTTTTGGAAGACTTTCACTTCGTGTTGATAG
GTACCGACCAACATTCTTGAAGAAAACCTTCTGAAAGTGAAGCACAACTATCCTAG

Sticky end

β-Galactosidase gene

Lac promoter region

Cleavage by restriction enzyme

Vector plasmid pBR322

Ligase

β-Galactosidase gene

Lac promoter region

Ligase

Somatostatin gene

Recombinant pBR 322-somatostatin plasmid

Introduction into *E. coli*

Recombinant pBR322-somatostatin plasmid

Bacterial-DNA

Expression

β-Galactosidase part

Met — Met — Met — Met — Ala — Lys — Cys — Lys — Asn — Phe — Phe — Trp

Lys

Disulfide bridge

Cys — Ser — Thr — Phe — Thr

Somatostatin part

Cleavage through cyanogen bromide

β-Galactosidase fragments

Ala — Lys — Cys — Lys — Asn — Phe — Phe — Trp

Lys

Disulfide bridge

Cys — Ser — Thr — Phe — Thr

Active somatostatin

Fig. 3.41 The first synthesis of a human peptide, the hormone somatostatin, using genetically engineered bacteria.

3.17 DNA-Hybridization – How to Find Bacteria Using DNA Probes

In the case of rat proinsulin, the recombinant clone was found by detecting the gene product with the help of antibodies in a sandwich radioimmunoassay. It is, however, more common to detect bacteria (engineered or not) by using **DNA probes**.

This is done by utilizing the **DNA hybridization** process, the formation of double-stranded DNA from complementary single strands of diverse origins, resulting in a hybrid molecule.

When searching for a specific DNA sequence in bacteria, a complementary single-stranded oligonucleotide is used, also known as a DNA probe.

Colonies of cells are grown on a nutrient medium (Fig. 3.44) and marked (usually using numbers clockwise on the lid of the Petri dish). A mirror image of the culture is fixed to a **nitrocellulose filter**, and the cell membranes are disrupted by a **detergent** such as Tween to release the cell contents. The bacterial DNA binds firmly to the nitrocellulose. **Sodium hydroxide** solution is added and destroys the hydrogen bonds that hold the bacterial double helix together. The denatured DNA separates into two single strands that continue to bind to nitrocellulose.

Now the DNA probe is added, a single-stranded oligonucleotide made up of about 20 nucleotides. Such a probe can be manufactured in a **DNA synthesizer** (see Box 3.5) if the structure (sequence) of the wanted gene is known. The probe is then marked either by a radioactive or – more recently – fluorescent molecules. After **hybridization**, labeled probes that have not bound are washed out. Only the wanted DNA sequences have formed hybrids with the DNA probes. When an X-ray film is laid over the nitrocellulose filter, it is blackened at the sites of the hybridized probes (autoradiography). Where fluorescent molecules are used, light of a specific wavelength is used to excite them, so they shine. Once the site of the hybridization has been found, the relevant colony on the agar plate can easily be identified.

This method is suitable for identifying genetically-modified cells as well as detecting natural bacterial pathogens for diagnostic purposes, e.g., cholera pathogens (Ch. 6). The latter, however, is only possible if part of the DNA sequence of the pathogen is known.

3.18 A Slight Diversion – Somatostatin – The First Human Protein Obtained From Bacteria

The next step was to produce the first human protein, **somatostatin**, in bacterial cells (Fig. 3.40). In the process, crucial techniques for the later production of insulin were learned.

Somatostatin is made up of only 14 amino acids and is one of several hormones produced in the hypothalamus, a region in the diencephalon. The bloodstream carries it to the pituitary gland where it is responsible for triggering the release of insulin and slowing down the release of the human growth hormone.

This makes somatostatin a potentially valuable compound in medical treatment, which is why (in 1977) **Herbert Boyer** and his collaborators at the City of Hope National Medical Center and the University of California chose it for their research. Until then, it had not been possible to isolate the somatostatin gene from human cells, but its base sequence could be derived through the knowledge of the genetic code for the known sequence of the 14 amino acids that made up the peptide.

Thus, an **artificial gene** was built from groups consisting of three bases each (Fig. 3.41). The gene contained 52 base pairs, 14 x 3 = 42 of which carried the encoded structure of somatostatin. The remaining ten base pairs were supposed to ensure the expression of the gene in the ribosome (realization of the structural plan), facilitate the isolation of the hormone and provide the sticky ends required to insert the double-stranded DNA fragment into a plasmid vector. *E. coli* was chosen as the host organism. In order to facilitate the transfer, the synthetic gene was combined with a plasmid named **pBR322** (see Box 3.2) and a section of the *lac*-operon of *E. coli* (more about this in Ch. 4).

For an effective expression of a cloned gene, clear signals that the host cell can understand are required. These are called **promoters** and are often derived from a gene that is expressed at a high level in the host. The simplest way to harness them is by docking the cloned DNA onto the DNA of a cell gene and hitching a lift on its very efficient promoters. The somatostatin gene was inserted at the end

of the bacterial gene that codes for the enzyme β-**galactosidase**. β-galactosidase is an enzyme found in large quantities in *E. coli* that degrades lactose. It is an inducible enzyme (Ch. 4).

Once the plasmid has been successfully transferred into an *E. coli* cell, somatostatin is formed as part of the fusion protein somatostatin-β-galactosidase. It is no more than a short peptide (14 amino acids) tail attached to the β-galactosidase enzyme.

By adding **cyanogen bromide** (**CNBr**), a chemical compound that only cleaves proteins at sites carrying the amino acid **methionine**, somatostatin is then separated from β-galactosidase.

Since the gene had been produced synthetically, it was easy to insert the required methionine at the beginning of the somatostatin molecule – just a case of inserting the codon ATG for methionine. This was necessary because if somatostatin had entered the bacterial cell on its own, it would have been rapidly degraded by bacterial proteases. If, however, it enters under the cover of the familiar galactosidase molecule, the proteases do not "notice" the somatostatin tail and will not attack it. In other – more scientific – terms: by being expressed as a fusion protein containing the cell's own protein, the foreign somatostatin part is protected from degradation by bacterial proteases.

The somatostatin synthetically produced in *E. coli* turned out to be identical with natural human somatostatin. Each cell produced 10,000 somatostatin molecules. This remarkable yield encouraged researchers to try and obtain other peptides in a similar fashion.

After all, it had taken **Roger Guillemin** (b. 1924, Nobel Prize 1977 together with **Andrew V. Schally** and **Rosalyn Yalow**, Fig. 3.38), the discoverer of animal somatostatin, 500,000 hypothalami to obtain about five milligrams of pure animal somatostatin.

As little as eight liters of *E. coli* suspension yielded the same amount of human somatostatin.

3.19 How Enzymes Turn Porcine Insulin Into Human Insulin

The insulin that had heretofore been used in diabetes treatment had been extracted from the pancreas of cattle and pigs. The amino acid sequence was not completely identical with human insulin.

Fig. 3.42 27-year-old Robert Swanson (1947 - 1999) met Herbert Boyer in 1975. Together, they founded the firm Genentech: *"All the academics I called said commercial application of gene splicing was ten years away. Herb didn't."*

Fig. 3.43 The first genetic engineering firm was not Genentech, as is often assumed, but the Cetus Corporation, founded in Berkeley in 1972 by physician Peter Farley, biochemist Ronald Cape, and Nobel Prize winner Donald Glaser. MIT professor Arnold Demain (top left then and bottom, today), Joshua Lederberg, and Stanley Cohen of Stanford University were consultants.

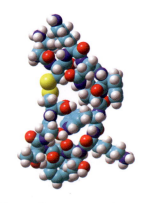

Fig. 3.44 Human somatostatin

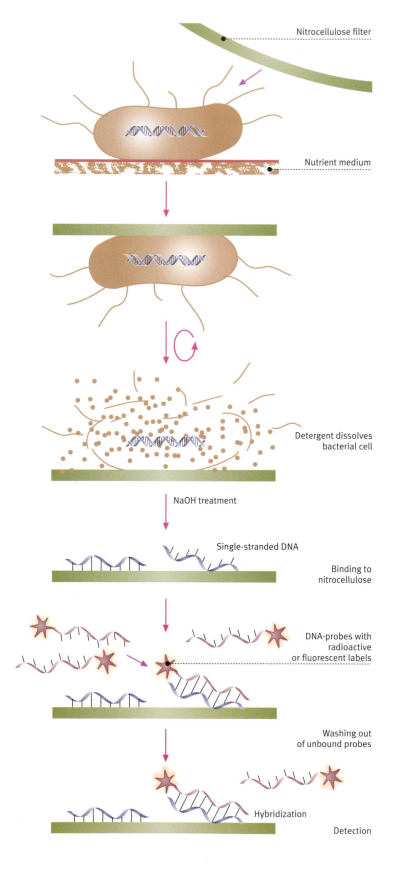

Nitrocellulose filter

Nutrient medium

Detergent dissolves
bacterial cell

NaOH treatment

Single-stranded DNA

Binding to
nitrocellulose

DNA-probes with
radioactive
or fluorescent labels

Washing out
of unbound probes

Hybridization

Detection

Fig. 3.45 Detection of bacteria using
DNA probes and DNA hybridization.

Insulin from pigs (porcine insulin) has one different amino acid, bovine insulin (derived from cows) three.

Two years after its discovery, **porcine insulin** was introduced into diabetes treatment. Although it brought the major symptoms of the disease under control, it had **unpleasant side effects**. Some diabetic patients had allergic reactions because their immune system recognized the animal hormones as foreign proteins and developed antibodies.

If it was possible to turn **porcine insulin into human insulin** using enzymes, this problem could be dealt with. Researchers at what was then Hoechst AG found an elegant solution in the early 1980s. The C-terminal amino acid (alanine) of porcine insulin would be enzymatically exchanged for threonine.

Trypsin (Fig.3.48), a hydrolase usually acting as a universal protein shredder in the stomach (Ch. 2) has an amazing ability of specifically hydrolyzing proteins and peptides next to the amino acids lysine and arginine. In porcine insulin, alanine happens to be preceded by lysine at the end of the B-chain – what could be better?

It was thus possible to cleave off the terminal alanine (Fig. 3.48). The procedure looks quite "normal" to begin with – porcine insulin is treated with trypsin while a threonine ester (threonine tertiary butyl ester) is added at the same time. The reaction takes place in 55% organic solvent.

The **reverse reaction of the enzyme** is truly astonishing: proteases are not only able to cleave peptides with the help of water, but they can also reconstitute peptides (and amino acid esters) in an anhydrous environment. In this reaction called **transpeptidation**, the threonine ester is attached to the molecule to replace the porcine alanine. All that needs to be done now is to add water in order to hydrolyze the ester and cleave off butanol – and the result is active human insulin, produced through enzyme catalysis.

This elegant method takes about six hours. It still requires, however, huge amounts of porcine insulin as raw material.

As mentioned before, Hoechst used to process nearly a dozen tons of porcine pancreases per day from over 100,000 slaughtered animals.

3.20 Eureka! The First Genetically Engineered Insulin Is Produced

In 1979, the first bacterial synthesis of human insulin was successfully accomplished. Yielding 100,000 molecules per cell, bacterial insulin production was even more efficient than somatostatin production, and what is more, insulin has a much larger role to play in medicine. How was the synthesis brought about?

As mentioned before, insulin consists of two peptide chains that comprise 21 and 30 amino acids, respectively. The necessary **processing** steps to turn it into active insulin, however, are not contained in the gene sequence. First, the **signal peptide** and the C-peptide sequence must be removed and the cysteines must be connected accurately to **disulfide bridges** – which can only be done by the Langerhans islets in the pancreas.

E. coli cells or even eukaryotic yeasts do not have this ability (see Sect. 3.25). At least, for the proper functioning of insulin, it was not necessary to attach a sugar residue (glycosylation). Isolating insulin mRNA was not an option, so the same approach as with somatostatin was required. It took the researchers of the City of Hope National Medical Center three months of painstakingly synthesizing the genes for the two peptide chains. Due to ten years of research by Frederick Sanger, the structure of insulin was known and it was possible to engineer the appropriate gene sequences.

18 fragments, each consisting of several nucleotides, were put together to form the longer chain, while the shorter one was made up of eleven fragments. In order to be on the safe side, the A-chain and the B-chain were synthesized in separate microbial cultures (Fig. 3.52). Thus, the insulin could not become active within the microbial cell.

3.21 Asilomar – How Dangerous Is Modern Gene Technology?

Is gene technology something to be afraid of? The risks involved in gene technology were discussed by worried scientists such as **Paul Berg** (Fig. 3.48), **David Baltimore**, **Maxine Singer**, and **Sydney Brenner** among many others at a conference in Asilomar in 1975. Would it be possible, for example, to create antibiotic-resistant carcinogenic intes-

tinal bacteria and thus cause a cancer epidemic? What the scientists did not know, however, was that their own concerns were blown into huge horror stories about the dangers of genetic engineering by the media.

What would happen **if bacteria formed active insulin** and, by some mistake, managed to enter the human gut and survive there? Well, they could cause a fatal insulin shock, which is why soon after the conference, the National Institute of Health (NIH) issued strict gene technology guidelines.

Debilitated *E. coli* **biosafety strains** were created that are unable to survive outside a lab, and a **biosafety level hierarchy** (**P1 to P3**) was introduced for labs to suit their various purposes.

Although the artificial genes for the A- and B-chain of insulin may not have carried the exact same sequence as the natural gene sections (which were unknown), they coded for the correct polypeptides. Each of them was inserted into their respective plasmid vectors directly behind the β-galactosidase gene promoter and the β-galactosidase gene as it had been done with somatostatin.

Between the insulin chains and the β-galactosidase gene, **a cutting site** – a codon for the amino acid methionine (ATG) – was inserted so that once the A- and B-chains were expressed, they carried an **additional methionine** molecule at their N-terminus and could thus be separated from the β-galactosidase (Fig. 3.52).

After the transfer into two different *E. coli* strains and the successful production of polypeptides, **cyanogen bromide** was used to cleave off galac-

Primary structure (sequence) of human insulin

Fig. 3.46 Enzymatic conversion of porcine to human insulin.

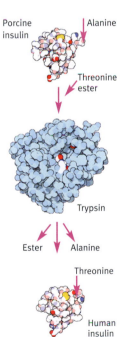

Fig. 3.47 Spatial structure of porcine and human insulins as well as trypsin (center).

Box 3.4 Insulin und Diabetes

Insulin is one of the most important mammalian protein hormones. When Sanger identified its structure in 1953, he demonstrated for the first time that a protein is characterized by a precise amino acid sequence (primary structure) and insulin consists exclusively of L-amino acids interconnected through peptide bonds between α-amino and α-carboxyl groups.

Insulin is produced in the pancreas and released into the bloodstream after meals when glucose levels rise. A signal is sent through the whole body, to liver, muscles and adipose cells to take up glucose from the bloodstream to synthesize storage compounds such as glycogen, triglycerides and protein. Insulin is a very small protein. Small protein molecules pose a particular challenge to the cell, as it is difficult to produce a small protein that folds into a stable structure. The cell solves the problem by forming longer peptide chains, folding them correctly and then cutting out the superfluous section (in the case of insulin, the C-peptide) (see. Fig. 3.36)

When the insulin function is impaired (through a diseased pancreas or old age or an unhealthy lifestyle), the glucose level in the bloodstream can reach dangerous levels, leading to dehydra-

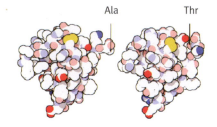

Ala   Thr

Human insulin (top right) differs from porcine insulin by just one amino acid. In position 30, humans carry threonine (Thr) instead of the porcine alanine (Ala).
In the lower part of the diagram, the A- and B-chains of porcine insulin are shown in green and blue. They are connected via three disulfide bridges (one of them is shown here in brown).

tion, as the body is trying to wash out surplus glucose via the urinary tract. The blood pH changes dramatically, and cells are affected.

Diabetes has become one of the major diseases in industrialized countries (Chapter 10) and is turning into a worldwide epidemic. Type I diabetes or insulin-dependent diabetes mellitus (lat. mellitus – sweetened with honey) is caused by

the destruction of Langerhans islets in the pancreas through the immune system. The patient needs insulin to survive. Most diabetics have normal or even elevated blood insulin levels but do not respond well to the hormone (type II diabetes, see in more detail Ch. 10).

Fred Sanger (b. 1918) worked from 1943 to 1953 to solve the structure of insulin, which earned him his first Nobel Prize. When working in his Cambridge University basement lab, he often noticed a crazy-looking pair of scientists engaged in lively, intense discussion. They would walk up and down and then get more and more worked up about something. They were Crick and Watson, and it was obvious that sooner or later, they would inspire Sanger to work on DNA. In fact, he managed to pick up a second Nobel Prize for his DNA sequencing method.

Fig. 3.48 Paul Berg (b. 1926) became a professor at Stanford at the age of 33. He wrote the much-quoted *Berg letter*, pleading for a moratorium on risky DNA experiments. He organized the meeting in Asilomar in 1975 where guidelines regarding the work with DNA were initiated. He shared the Nobel Prize for Chemistry in 1980 with Walter Gilbert and Frederick Sanger.

tosidase from both peptides at the predetermined site. Figure 3.50 shows how insulin chains can be chromatographically separated from a mix of various compounds using insulin antibodies. The underlying principle of the separation is the affinity of insulin to specific insulin antibodies – hence the name **affinity chromatography**.

The last chemical step is called **oxidative sulfitolysis**. Sulfite, oxygen, and a strongly basic pH lead to the oxidation of cysteines which carry an SH group. They are now highly reactive, linking neighboring oxidized SH groups via a disulfide (-S-S-) bond to form active insulin.

However, the yield of active insulin is somewhat reduced, as some of the A- and B-chains may not have been correctly linked and need to be separated. The disulfide bridges determine the spatial structure of insulin and thereby its activity.

Correctly formed insulin must contain three disulfide bridges, one within the A-chain and two between the A- and B-chains. **Recombinant insulin**, produced as described here, began to revolutionize the insulin market in 1982.

The **safety concerns** that led to the separate production of A- and B-chains were indeed quite considerable. From its first application in 1984 to build a production plant for recombinant human insulin, it took Hoechst fourteen years to battle through bureaucracy and public hearings before the Hoechst Marion Roussel plant could be finally opened in Germany in 1998.

■ 3.22 Human Proinsulin Obtained From a Single *E. coli* Strain

The next step ahead was the production of proinsulin in a single bacterial strain. The entire proinsulin gene, except for the sequence of the signal peptide, was synthesized. Again, the first amino acid was preceded by an inserted methionine codon (ATG).

Thus, only one single plasmid and one single *E. coli* strain were required to produce human proinsulin. The proinsulin gene was controlled by a strong tryptophan synthase promoter. Lined up on the plasmid were: the tryptophan synthase promoter, the tryp-

Box 3.5 Test Tube Genes – DNA Synthesizers

Nowadays, machines are available that automatically synthesize sequences of single-stranded DNA at a very fast rate. The whole process, including the exact base sequence, is controlled by microprocessors.

DNA synthesizer

In the synthesizer shown here, the desired DNA sequence is entered into the machine, and microprocessors ensure that nucleotides, reagents and solvents are correctly pumped into the synthesizing column at every step of the process. The column contains silica gel granules, about the size of fine sand grains. These granules are there for bonding by the DNA chains.

To synthesize a certain sequence, e.g., TACG, the first nucleotide (e.g, thymidylic acid) should already be attached to the column granules by its 3'terminus. Prompted by microprocessors, millions of molecules of the next nucleotide are pumped through the column. The 5'terminus of the A molecule is chemically masked to ensure the correct orientation when binding to T. The protecting group is then removed, and a new cycle begins.

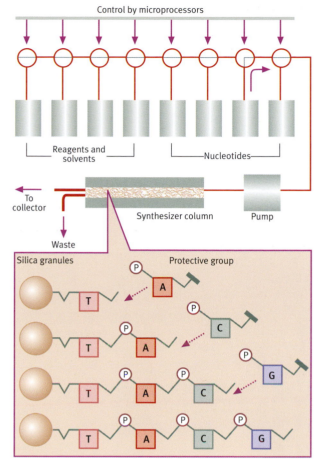

Operating principle of a gene synthesizer

Short chains of up to 50 nucleotides can be synthesized in this way. The finished oligonucleotide chain is cleaved off the granule and washed out of the column. Through continuous improvement of the automatic process, the synthesis of nucleotides can be carried out ever more rapidly. All that is required is to type in the sequence, and the ready-made oligonucleotide is delivered shortly. There are three reasons for the key role of oligonucleotides in biotechnology. They can be combined to produce completely synthetic genes (as shown with insulin and somatostatin). They can also be used as DNA probes (Fig. 3.45) or as primers for a polymerase chain reaction (PCR) and for sequencing purposes (Ch. 10).

---

tophan synthase gene, the ATG-sequence (coding for methionine, as a cutting site), and the proinsulin sequence B-C-A.

A **fusion protein** was synthesized and the methionine plus tryptophan synthase cleaved off using **CNBr** in order to obtain proinsulin. Through oxidative sulfitolysis, the correct cysteine residue links could be conserved, and the C-peptide in proinsulin made for a much better orientation of A- and B-chains. The C-peptide was then cut out enzymatically by tryposin to obtain active, correctly folded insulin which was 100% identical with human

insulin. This is how pharmaceutical firms such as Berlin-Chemie, Sanofi-Aventis, and Eli Lilly currently produce insulin.

■ 3.23 Bakers' Yeast For Proinsulin Production

The Novo Nordisk company chose to use eukaryotic bakers' yeast rather than prokaryotic bacteria for their insulin production. Like their competitors, they use an artificial gene. However, after failing to produce proinsulin containing a complete C-peptide, they are using a C-peptide containing only nine

Fig. 3.49 Bakers' yeast (*Sacharomyces cerevisiae*) are minute eukaryotic single-cell organisms. Shown here are live yeasts expressing green fluorescent protein (GFP, see Chapter 7).

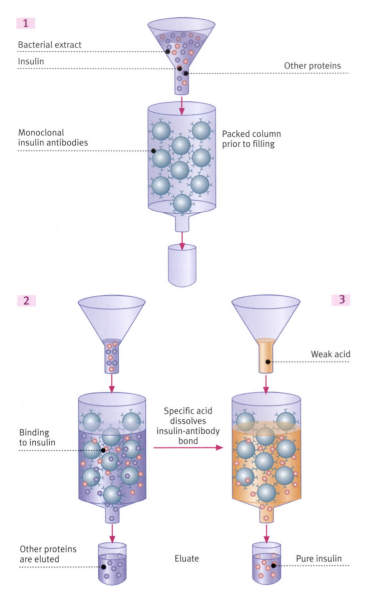

1

Bacterial extract

Insulin

Other proteins

Monoclonal insulin antibodies

Packed column prior to filling

2

Binding to insulin

Specific acid dissolves insulin-antibody bond

Other proteins are eluted

Eluate

3

Weak acid

Pure insulin

Fig.3.50 Purification of insulin by affinity chromatography.

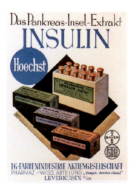

Fig.3.51 Above: Advert for pig insulin from Hoechst in the 30ies; right: Insulin variants.

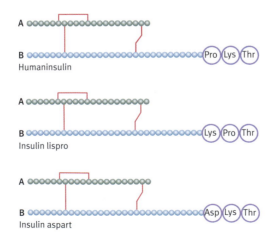

A

B Pro Lys Thr
Humaninsulin

A

B Lys Pro Thr
Insulin lispro

A

B Asp Lys Thr
Insulin aspart

nucleotides, i.e., it consists of just three amino acids. This peptide is too short to be hydrolyzed by yeast proteases. Moreover, it has been ingeniously composed to bring A-and B-chains so close together that disulfide bonds between them can be established while still inside the yeast cell rather than at the end of the biosynthetic process. The **mini-proinsulin gene** is controlled by the promoter of the triose phosphate isomerase (a glycolytic enzyme, see Ch. 2) gene, but the key point is that a **signal sequence** has been inserted between promoter and gene (B-chain, shortened C-peptide, A-chain) which transports the mini-proinsulin out of the cell. The signal peptide itself is cleaved off by yeast proteases during the process.

Thus, correctly folded proinsulin is secreted into the medium. The mini C-peptide is removed in the same way as porcine insulin is turned into the human form. In an anhydrous (organic) medium and in the presence of threonine tertiary butyl ester, trypsin cleaves behind the two lysine residues, lengthening the B-chain by a threonine ester residue during the reverse reaction. Enzymatic hydrolysis results in active human insulin.

■ 3.24 Artificial Insulin Variants (Muteins) Obtained by Protein Engineering

As all the methods mentioned so far produce authentic human insulin, the next logical step would be to produce **insulin variants that don't exist in nature**. This is done by recombinant technology.

The two major properties desired in insulin are rapid action on the one hand and effectiveness for 24 hours on the other.

When insulin is injected subcutaneously (under the skin), there is a brief surge in insulin concentration under the skin. When highly concentrated, the insulin molecules form hexamers, i.e., they form groups of six. However, in order to be absorbed into the bloodstream and transported to their target cells, they must first dissociate into dimers and monomers. In other words, the injected insulin reaches the bloodstream too slowly and, once in the blood, the insulin level remains high for far too long – a serious problem for diabetics.

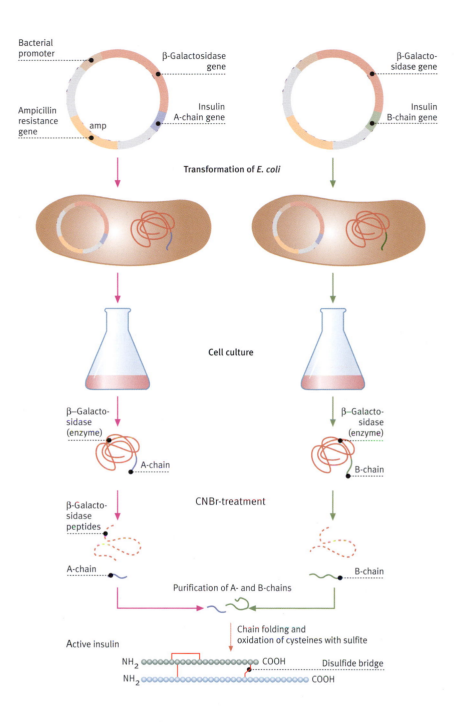

Bacterial promoter

β-Galactosidase gene

Ampicillin resistance gene

amp

Insulin A-chain gene

β-Galactosidase gene

Insulin B-chain gene

Transformation of *E. coli*

Cell culture

β–Galactosidase (enzyme)

A-chain

β–Galactosidase (enzyme)

B-chain

CNBr-treatment

β-Galactosidase peptides

A-chain

B-chain

Purification of A- and B-chains

Chain folding and oxidation of cysteines with sulfite

Active insulin

NH₂ — COOH **Disulfide bridge**
NH₂ — COOH

Fig. 3.52 Insulin expression in two separate *Escherichia coli* strains:

Two oligonucleotides (A- and B-chain) are chemically synthesized in a first step. Each of the DNA fragments codes for one of the insulin chains. Once they have been cut by restriction endonucleases, they are inserted into separate plasmids.

The insulin DNA is preceded by – galactosidase-encoding DNA (β-Ga) and the bacterial promoter of β-Gal.

The resulting fusion proteins accumulate in separate places in *E. coli* as β-galactosidase A-chain insulin and β-galactosidase B-chain insulin.

The cells are harvested and both fusion proteins are purified.

The insulin-coding DNA has been engineered to start with a methionine codon. Cyanogen bromide (CNBr) always splits the peptide binding after a methionine. Thus, the fusion protein can be cleaved by CNBr right behind methionine. The natural insulin chains are thereby produced separately. The produced β-galactosidase contains several methionine residues and is cut therefore into small peptide pieces. One can get rid of β-Gal by this trick.

The ingenious idea: Since the insulin chains themselves do not contain methionine internally, they remain intact!

After purification, they undergo oxidative sulfitolysis, thus activating the cysteines in the A- and B-chain to produce functional human insulin.

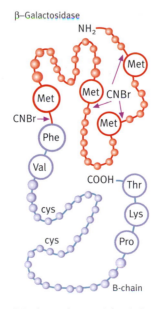

β–Galactosidase

CNBr cleaves the peptide bond after methionine.

Ways had to be found to enhance the rapid availability of insulin. In 1996, a product named Humalog® was developed by swapping proline and lysine at positions 28 and 29. The product is also known as insulin lispro (Fig. 3.51), a name derived from "lysine-proline". Another variant, insulin aspart®, was created by replacing proline by aspartate.

Both are fast-acting variants, reaching maximum plasma insulin concentrations in 60 rather than 90 minutes, and as with natural insulin, plasma con-

centration also goes down faster, which makes meticulous planning for meals and the need to carry around snacks a thing of the past.

Intense research is also going into insulins with delayed response. Insulin glargine (HOE901, Lantus®) is a recombinant human insulin analogue which makes use of the reduced solubility of insulin hexamers in order to delay the action – just the opposite of what the fast-acting insulins are aiming for.

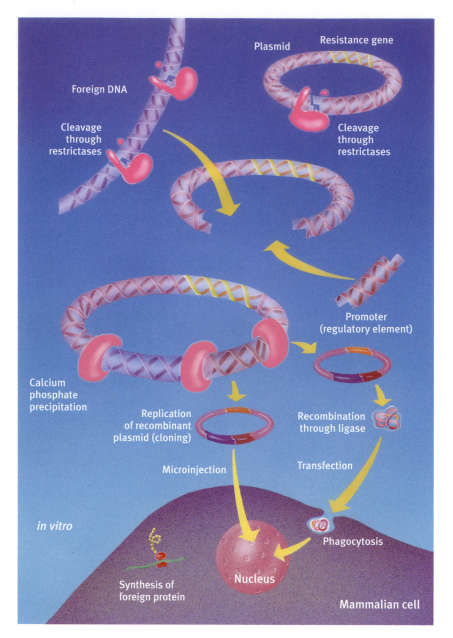

Fig. 3.53 Gene modification in mammalian cells.

In a first step, the foreign gene is isolated and cloned into bacterial cells (or replicated through PCR) and introduced into a vector, using restriction endonucleases and ligases (shown here: bacterial plasmids carrying a specific antibiotic resistance gene). If the foreign gene is to be read (expressed) in mammalian cells, an animal promoter (the recognition sequence for the binding of RNA polymerase) must be inserted right before the foreign gene.

The recombinant plasmid is then introduced into bacteria. The cells containing the plasmids are resist-

ant to specific antibiotics and can thus be selected. The plasmid containing the foreign gene is cloned in the bacteria and subsequently isolated. At the end of the processes, it will be available in large amounts.

It is introduced (transfected) into the mammalian cell through microinjection into the nucleus or through precipitation of the plasmids and direct uptake by the cell. A small proportion of the introduced DNA is stably inserted into the genetic information of the mammalian cell and converted into a foreign protein. The mammalian cell types used here have been cultured to facilitate the isolation and purification of the foreign protein.

The B-chain was lengthened by two arginines at the C-terminal and asparagine in position 21 substituted by glycine (hence the gl in the name). This makes insulin glargine difficult to dissolve within the physiological pH range (pH 7.4). Injected as a clear solution, the insulin forms stable hexamer associates that dissolve slowly, releasing insulin over 24 hours. It can thus be used for one-shot injections.

In all these cases, targeted genetic engineering was used to swap around specific amino acids, and the insulin was produced by *Escherichia coli* – a concrete example of successful **protein engineering**.

■ 3.25 Genetically Modified Mammalian Cells For the Production of Modified Complex Proteins

After the first euphoria about the ability to produce human insulin from bacteria, there was a – perhaps not unexpected – disappointment in store for researchers. Not all proteins could be produced via modified microbes. By a fortunate coincidence, insulin is a small molecule consisting exclusively of amino acids.

The amino acid sequence is not the only important characteristic of a protein. **Complex animal and human proteins** also require modifications such as sugar residues, phosphate, or nucleotide groups in order to function. These cannot be produced by prokaryotic bacteria or at least, when produced, they are inactive. **Eukaryotic yeasts** were an obvious alternative with their more sophisticated structure and metabolism. They were used successfully in the production of proinsulin, hirudin (Ch. 9), and the hepatitis B vaccine where *E. coli* failed miserably.

However, the possibilities of yeast are also limited, as they often lack enzymes and cofactors needed for certain protein modifications, or they may not have the cell organelle structures to enable the newly synthesized proteins to fold into their correct spatial structure and form stabilizing disulfide bridges.

The remaining alternative is to genetically alter **mammalian cells** (Fig. 3.53) and grow them in large quantities in bioreactors. What first looked quite unpromising has now become technologically viable.

Genetic engineering of mammalian cells is much more labor-intensive than that of bacteria because their DNA is organized in a more complex way, packed away in chromosomes within a nucleus. Not only must the foreign gene be introduced into the nucleus (**transfection**), but it must be expressed within the cell.

In a first step, the desired **mammalian gene** is isolated and cloned either in bacteria or by using the **polymerase chain reaction** (PCR) (Ch. 10) in order to obtain sufficient quantities. For the transfection process, a **vector system** is required (see Box 3.3).

The most common vectors are bacterial **plasmids** containing an antibiotic resistance gene for the selection of those bacterial cells that have been successfully transformed and can propagate the plasmid along with their own reproduction.

The antibiotic would be one that is normally toxic to animal cells. Alternatively, the plasmid could possess a gene encoding an enzyme that would compensate for a metabolic defect in the mammalian cells used for the purpose. This ensures that only cells containing the plasmid will survive in selective media.

Using molecular glue and scissors as mentioned above, i.e., restriction endonucleases and DNA ligases, the human gene is inserted into the plasmid. In order to enable its subsequent expression in the mammalian cell, an animal promoter must be inserted just before it.

The hybrid plasmid is mixed with bacteria. All cells that have taken it up can be easily identified, as they carry a resistance gene to a certain antibiotic, which enables them to grow and form a colony of identical cells or clones. Alternatively, they contain a gene compensating for a genetic defect present in the mammalian cells. These cells all contain copies of the same plasmid which can easily be extracted from the bacterial cells.

There are several possible **transfection** methods, i.e., the introduction of the hybrid plasmids into mammalian cells:

The vector can be introduced through **microinjection**, using a very fine injection needle, the tip of which is pushed right into the nucleus. If this is successful, the plasmid integrates into the chromosomal DNA of the mammalian cell.

The **calcium phosphate precipitation** method involves adding plasmids to a phosphate buffer and adding calcium ions. The poorly soluble calcium phosphate precipitates, forming calcium-plasmid aggregate granules. The calcium ions speed up the intake of the DNA granules through the cell membrane into the mammalian cell. A small proportion is thus transported into the cells and are stably integrated into the chromosomal DNA

In **electroporation**, the mammalian cells are exposed to strong electric impulses which briefly open up large pores in the cell membrane through which the plasmids can be taken in.

The transfected mammalian cells are grown in a suitable **selective medium** (Fig. 3.54). Just as in the case of genetically cloned *E. coli* bacteria, only genetically engineered mammalian cells will survive, as the inserted plasmid carries either a resistance gene to an antibiotic that would otherwise be toxic to the mammalian cell, or an enzyme gene compensating for a metabolic defect of the mammalian cells used in the experiment.

There are several methods of making the desired product available. If the products remain inside the cell, the cells are harvested and solubilized. The product is then isolated and purified (Fig. 3.50).

If the products have been secreted by cells (e.g., antibodies), they can be obtained from the drained medium while the cells remain in the reactor.

The genetically modified cells form **colonies** on suitable nutrient media, secreting large quantities of the desired protein into the medium.

Some mammalian cells need firm surfaces to grow on naturally, such as glass or plastic, where they form cell lawns (Fig. 3.55). However, most mammalian cells are able to grow as single cells in a liquid nutrient medium in bioreactors.

Mammalian cells have also been enclosed in gels. Antibody-forming hybridoma cells (see Ch. 5) are grown in alginate capsules, secreting the antibodies through the capsule membrane into the medium.

Pharmaceutical products obtained from genetically manipulated cells rang in a new era in the development of biotechnology.

The first major milestone in biotechnology, however, was reached in the 1940s with the industrial production of penicillin (see Sect. 4.14).

Fig. 3.54 Culture of mammalian cells in special media containing nutrients and special antibiotics.

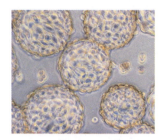

Fig. 3.55 Mammalian cells growing on spheric microcarriers.

Fig. 3.56 Bioreactor for mammalian cell cultures.

Fig. 3.57 In this mammalian cell reactor, the cells are provided with nutrients through hollow fibers.

Possible modifications of a protein after its synthesis in eukarytic cells

- Disulfide groups between cysteine residues.
- Attachment of sugars (glycosylation).
- Blockage of an N-terminus.
- Farnesylation or myristinylation in order to associate proteins with the membrane.
- Special hydrolyses (proteolytic cleavage).

Box 3.6 The Magic of DNA

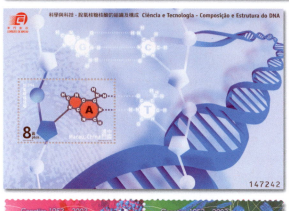

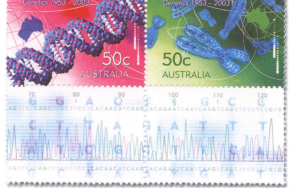

The five stamps were issued by the Royal Mail in 2003 to commemorate the discovery of the DNA double helix by Francis Crick and James Watson in Cambridge in 1953, and the subsequent work on the Human Genome Project. The cartoons are the work of The Times newspaper's political cartoonist Peter Brookes.

Norvic Philatelics issued a First Day Cover designed by John Billings, reprinted here with permission.

Also in 2003, the Special Administrative Region Macao, part of China since 1999, issued four stamps in presentation packs, showing the structures of DNA nucleotides.

The Human Genome Project was the subject of an Australian stamp series.

Points to Ponder about Gene Technology

I think we must take the latest developments seriously.

All progress in the past, from the taming of fire to the development of supercomputers, had an impact on our relationship with nature.

We were the active party, and nature was our subject that was acted upon.

Our aim was to free human beings from the constraints imposed by nature.

However, what we are about to do now is something entirely different. We are no longer trying to tame living and inanimate nature. Instead, we are beginning to design ourselves, our bodies, our brains, turning biology into technical design.

We are interfering with the basics of our being, no longer content with designing our environment. We want to perfect therapies, improve our bodies or even optimize them.

Will we be successful?

And if we are, will we then still be human?

Jens Reich, in: Es wird ein Mensch gemacht (A Man is in the Making), Rowohlt Berlin 2003

Cited and recommended Literature

- **Watson JD** (2004) DNA. *The Secret of Life*. Arrow Books, London
  This book you MUST have, an immediate classic!

- **Watson JD, Baker TA, Bell SP, Gann A, Levine M, Losick R** (2003) *Molecular Biology of the Gene* (5th edn.) The long-awaited new edition of James D. Watson's classic text, Molecular Biology of the Gene, co-authored by five highly respected molecular biologists, provide current, authoritative coverage of a fast-changing discipline, giving both historical and basic chemical context. Unfortunately, students may not be able to afford this quite expensive book ($130!!!).

- **Brown TA** (2001) *Gene Cloning and DNA Analysis-An Introduction*. 4th edn. Blackwell Science Ltd, Oxford.
  A beginner's book, complementary to this book here. A good start!

- **Thieman WJ**, Palladino MA (2004) *Introduction to Biotechnology*, Benjamin Cummings, San Francisco.
  Good textbook for students specifically interested in pursuing a career in biotechnology, clear illustrations, easy to read.

- **Micklos D** (2003) DNA *Science: A First Course*. 3rd edn., ASM Press
  Well-written textbook for students in biotechnology and genetic engineering

- **Lewin B** (2004) *Genes VIII*. 8th edn. Pearson Prentice Hall, Upper Saddle River Complete textbook for the advanced student and researcher.

- **Berg, JM, Tymoczko JL, Stryer L** (2007) *Biochemistry*, 6th edn., WH Freeman, N.Y. Basic dynamic understanding about DNA and genes

- **Judson, HF** (1996) *The Eighth Day of Creation: Makers of the Revolution in Biology*, Cold Spring Harbor Laboratory Press. The best history of DNA research to date!

- **Clayton, J** (Ed.) (2003) *50 Years of DNA*, Palgrave MacMillan Press.
  A reader containing up-to-date DNA history contributions

Useful weblinks

- A starting point, the Wikipedia, many useful weblinks:
  http://en.wikipedia.org/wiki/DNA

- The original papers of the DNA pioneers as downloads:
  http://www.nature.com/nature/dna50/archive.html

- Website of the textbook Genes VIII, updated permanently:
  www.ergito.com

- DNA interactive, mainly for teachers :
  http://www.dnai.org/

- World Wide Web Virtual Library: Biotechnology, catalog of WWW resources:
  http://www.cato.com/biotech/

Eight Self-Test Questions

1. What are the differences between DNA and RNA? Name at least three!

2. This is the sequence of a DNA strand.
   5′-AATTCGTCGGTCAGCC-3′
   What is the complementary sequence?

3. You discovered a new bacterial strain with a DNA consisting of 17 % adenine. What is its guanine percentage?

4. A eukaryotic mRNA has the following sequence:
   5′-AUGCCCCGAACCUCAAAGUGA-3′
   How many codons does it contain, and how many amino acids have been coded? (Normally, mRNAs are much longer. Note that there are not only protein-coding codons).

5. What sort of protein would result from cutting human genome DNA directly with restriction endonucleases, inserting it into plasmids, and having it expressed? How could this result be avoided?

6. What makes bacterial plasmids ideal vectors for foreign DNA? Name at least one other vector that is based on a virus.

7. Why are DNA probes always single, stranded molecules?

8. Why were the A- and B-chains of human insulin first expressed in separate strains of *E. coli*?

WHITE BIOTECHNOLOGY
Cells As Synthetic Factories

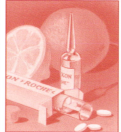

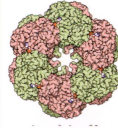

CHAPTER **4**

Fig. 4.1 Ludwig von Bertalanffy (1901-1972), co-founder of Theoretical Biology, sometimes applied his theory in surprising practical ways.

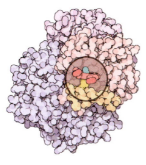

Fig. 4.2 Hemoglobin (with oxygen bound to the iron-porphyrin complex of the heme group) was the first protein in which allosteric effects could be shown. The binding of oxygen changes the shape of the molecule, thus making it easier for more oxygen to bind.

Fig. 4.3 Jacques Monod (1910-1976), Nobel Prize winner and director of l'Institut Pasteur, Paris, from 1971 until his death in 1976. Monod wrote a much-debated philosophical treatise, titled "Chance and Necessity".

Fig. 4.4 The metabolic network of a cell. Complex sugars (blue, top left) are turned into glucose, which is converted during glycolysis (dark blue vertical line down the middle), ending in the citric acid cycle (dark blue, center at the bottom). Amino acids are formed at the junctions.

■ 4.1 An Overview

A cell is not an isolated unit, but an open system through which there is a constant flow of substances that are immediately converted. This process which never reaches a stationary equilibrium is called a steady state. **Ludwig von Bertalanffy** (1901 - 1972, Fig. 4.1), co-founder of Theoretical Biology, expressed the concept thus: "If you have a dog, you may believe that in five years time, the dog that comes running to you when you call his name is still the same dog you are seeing today. In fact, there will be hardly anything left of it in terms of material components. In five years time, your dog will contain perhaps just a molecule and a few cells of the dog you used to know. Apply the same principle to humans, your wife, for example, and I leave it to you to draw your own conclusions."

A **human weighing 70 kilograms** (154 lbs) takes in 50 - 100 grams (1.75 - 3.5 oz) of protein, 300 grams (11 oz) of carbohydrates, and 40 to 90 grams (1.4 - 3.2 oz) of fat per day, along with water, vitamins, minerals, and about 500 liters (18 cubic feet) of oxygen. The human body converts this into approximately 70 kg of "energy currency" ATP and two kilograms of citric acid (see later in this chapter), which are released and degraded, producing energy and non-recyclable waste.

Only the constant provision of raw material can help maintain the **steady state of the cell system.** Our metabolism is further assisted by up to two kilo-

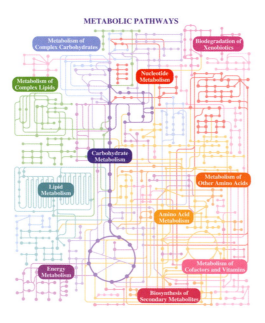

METABOLIC PATHWAYS

grams of intestinal bacteria that also produce some essential vitamins.

In an **average-sized single cell** (between 1/1000 to 1/100 nm), myriads of metabolic reactions take place simultaneously, in parallel, or in sequence. They are all kept in sync, maintaining their equilibrium, yet flexible enough to adapt to changing circumstances such as rest or activity, hunger, growth, reproduction, heat, or cold. All organisms have a close network of regulatory and control mechanisms working on a wide range of principles. At the cell level, enzymes have an important part to play. Most of them are organized like production processes in a factory, in multiple chains and cycles (Figs. 4.4 and 4.6), whereas in some simple organisms, the enzymes float freely in the **cytoplasm**.

The cytoplasm of a single cell (Fig. 4.5) contains between 50,000 and 100,000 molecules of the most important glucose-degrading (glycolytic) **enzymes** (see Ch. 1). Their end products are small molecules that move with ease through the cytoplasm on their way from one enzyme to another. However, distances that seem minimal to us can be quite a challenge to overcome in a metabolic reaction, especially where activators or inhibitors of enzymes and enzymes involved in other metabolic pathways could interfere.

In more highly organized enzymes, such interferences are avoided and enzymatic effectiveness is enhanced by the formation of enzyme complexes. These are structures that consist of several enzymes, known as **multi-enzyme systems.**

German biochemist and Nobel Prize winner **Feodor Lynen** (1911 - 1979) carried out long-term studies on the synthesis of fatty acids within a cell. He found that the enzymes involved in a cell's synthetic processes are part of a large multi-enzyme system. In yeasts, the fatty acid synthetase complex consists of seven different enzymes that are stably interlinked. A multi-enzyme system helps overcome distances, handing the substrate from one enzyme to the next in a closed circuit. Only the end product is allowed to exit the system, thus avoiding interference.

The first electron microscopes showed that a **eukaryotic cell** is not just a sack containing a medley of components, but a sophisticated ensemble of **channels and membranes**. Figure 4.8 shows the complex interior of a eukaryotic cell.

Just like a factory, a eukaryotic cell is divided into several functional units or **compartments**. **Membranes** provide walls, pipes, or wires between the compartments (Fig. 4.5). The **nucleus** is the headquarters, the **mitochondria** are the power stations, and the **ribosomes** are protein assembly plants. Just as in a factory, you would find certain machines and specialists only in the relevant functional units and certain enzymes pertaining to specific compartments. Thus, otherwise incompatible processes can run simultaneously in separate compartments. For example, fatty acids are assembled in the cytoplasm, but degraded in mitochondria, which are separated from the cytoplasm by a membrane.

Membranes are not just barriers for enzymes, but also provide sites for substrates to accumulate, as well as combining and coordinating enzyme reactions. There is increasing evidence that most enzymes do not occur freely in the cell, but are bound to various structures, with a large proportion of enzymes floating in the lipids of membranes.

A cell has **two basic ways** of regulating its metabolism through enzymes:

Tactical adaptation uses the enzymes available, activating or inhibiting them chemically or physically. These processes permit short-term reactions, sometimes within a fraction of a second.

Strategic adaptation, on the other hand, is a response to general changes in living conditions and genetically controlled.

■ 4.2 Tactical Adaptation

"Eureka, I have discovered the second secret of life!" exclaimed Professor **Jacques Monod** (1910 – 1976) (Fig. 4.3), having worked over two decades at the Institut Pasteur in Paris. From the mid-1950s onwards, several research teams had been investigating the mechanisms of metabolic regulation in their "pet organism", the intestinal bacterium *Escherichia coli*.

The research focused on enzyme chains involved in the step-by-step conversion of the original substrate into an end product. The precision of the control mechanisms acting on these pathways was simply beyond belief. In order to stop the synthesizing process immediately, it sufficed to add the end product. However, when the first enzyme of the chain

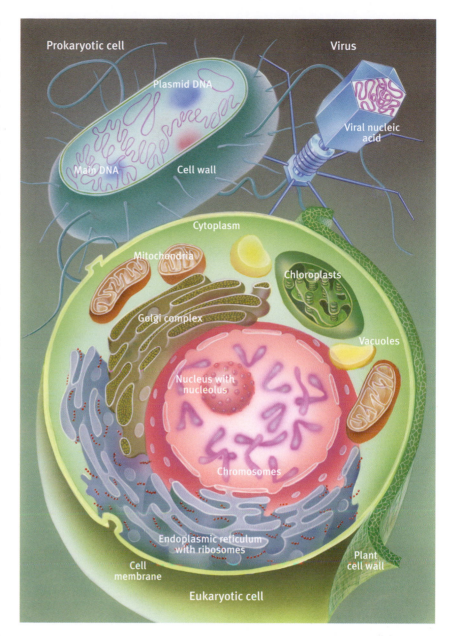

Fig. 4.5 Eukaryotic (below, featuring both plant and animal components) and prokaryotic cell (top left), and a virus (a bacteriophage).

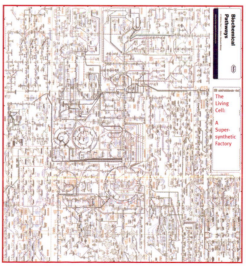

Fig. 4.6 Roadmap of a cell – a fascinating view.
This is a section of Gerhard Michal's famous map of *metabolic pathways* (Roche Applied Scienes), showing all known enzymes, substrates, products, co-factors, interconnections, and feedback.
A clear picture of the map is shown at the end of this book.

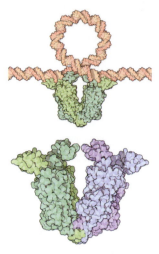

Fig. 4.7 Induction of lactose-degrading enzymes in a bacterial cell by lactose (or more precisely, the real inducer allolactose) (right). The Lac repressor binds to the operator region in the DNA.

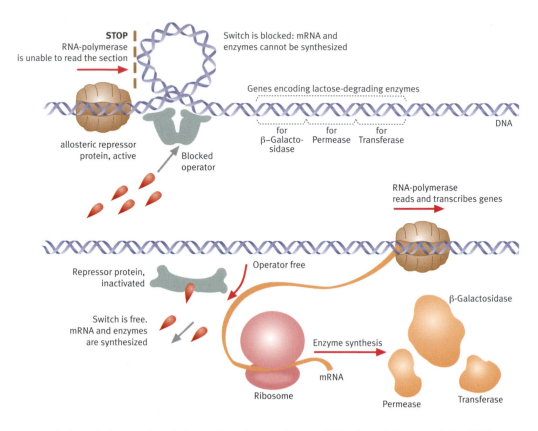

STOP
RNA-polymerase is unable to read the section

Switch is blocked: mRNA and enzymes cannot be synthesized

Genes encoding lactose-degrading enzymes

DNA

for β–Galacto-sidase

for Permease

for Transferase

allosteric repressor protein, active

Blocked operator

RNA-polymerase reads and transcribes genes

Repressor protein, inactivated

Operator free

Switch is free. mRNA and enzymes are synthesized

β-Galactosidase

Enzyme synthesis

mRNA

Ribosome

Permease

Transferase

Fig. 4.8 David Goodsell's panoramic section through a eukaryotic cell.

Below left: plan view of the section. Below right: The panoramic view begins on the left with the cell surface, carrying on through part of the cytoplasm. In the endoplasmic reticulum, ribosomes are shown during protein synthesis. Then comes the Golgi apparatus and a vesicle. The center of the diagram shows a mitochondrion, and the nucleus can be seen on the right. All macromolecules are shown: proteins are blue, DNA and RNA red and orange, lipids yellow, and carbohydrates green. The ribosomes, consisting of RNA and protein, are colored in magenta. In an existing cell, the spaces between macromolecules are filled with small molecules, ions, and water.

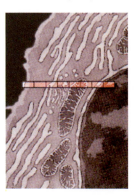

was gently heated, the reaction chain could not be switched off. Although it still showed some activity, the switch on the enzyme had been destroyed.

How could the end product switch off the first enzyme of the chain? Competition between substrate and end product could be ruled out, as there was no similarity between the two.

After carrying out further experiments, Monod came to the conclusion that the switch must be a binding site for the end product which is sterically (spatially) separated from the active center – a phenomenon that became known as **"allostery"** (Greek *allos*, different).

The allosteric site seemed to have some impact on the active site. Incidentally, it was around that time that **John Kendrew** (b. 1917, Nobel Prize shared with **Max Perutz** in 1962) revealed the structure of hemoglobin through X-ray analysis. While at a first glance, these results seemed to bear no relevance to the solution of allostery problems, it turned out that the entire hemoglobin molecule changed its shape when one of its four subunits had bound an oxygen molecule (Fig. 4.2).

Could it be that in allosteric enzymes, the binding of an inhibitor modified the entire spatial structure of the enzyme and thus the shape of its active site in a way that inhibited the conversion of further substrates? Conversely, the binding of activators to the allosteric site should optimize the spatial structure of the enzyme and its active site.

It was found that **allosteric enzyme**s consist of several **subunits**, all of them containing one active site and at least one allosteric site. The subunits are so closely linked that the binding of an allosteric

Cell membrane

Ribosomes with mRNA

Endoplasmic reticulum

Golgi complex

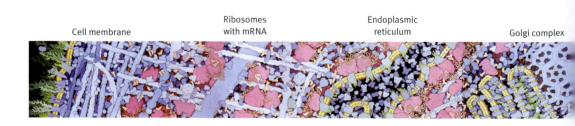

inhibitor molecule not only changes the spatial structure and thus the activity of the binding sub-unit, but also inhibits the other subunits. Cooperation between the subunits thus enhances the regulatory effect of the inhibitors or activators.

In other words, allosteric regulatory enzymes are **controlled by comparatively weak signals** (inhibitors or activators) that act like electrical relays and have a strong impact within the cell. They are placed in key positions in the metabolic process. Sometimes, a single **allosteric pacemaker enzyme** at the start of a metabolic pathway is enough to set the pace for the whole pathway. This is true for the synthesis of glutamine in bacteria (Sect. 4.4). The accumulation of a metabolic product at the end of a pathway triggers **negative feedback**.

The end product has a remote effect by binding to the pacemaker enzyme of the chain and inhibiting it. The whole conveyor belt system is thus brought to a halt, and the starting substrate can be used for other purposes until overproduction has been dealt with through degradation or use of the end product.

Not only enzymes, but other proteins as well can be regulated by allosteric mechanisms. Allosteric proteins mediate interactions between compounds with no chemical affinity, thus regulating the flow of matter and energy through the whole system with a minimum of energy.

Monod and **François Jacob** (b. 1920) shared the Nobel Prize with **André Lwoff** (1902-1999) in 1965.

We will have a closer look at the allosteric regulation of the lac repressor and glutamine synthesis.

■ 4.3 Strategic Adaptation: Enzyme Production on Demand

All cells in an organism carry the same complete set of genetic information encoded in their DNA.

A highly specialized neural or intestinal cell, however, does not need the whole genetic range. Producing acetylcholine esterase in an intestinal cell would not make any sense and producing digestive enzymes in a neural cell could even be fatal. A large part of the genetic information is therefore **suppressed** or blocked, depending on the cell type. Surely, there must be a control system that decides what information is released where and when.

Jacques Monod and his colleague **François Jacob** had been working on this riddle since the early fifties. They transferred bacteria from a nutrient medium that contained their favorite food glucose to another medium containing only lactose. Immediately, their growth was stopped, as the bacteria did not have enough of the enzymes required to break down lactose. After about 20 minutes of standstill, they began to grow furiously again, having now the right enzymes for lactose degradation.

This phenomenon had already been known since the 1930s and was known as **induction**. It was then assumed that there were two categories of enzymes – **constitutive and inducible enzymes**. The constitutive enzymes were the essential metabolic enzymes, indispensable for the formation and survival of a cell and produced at an approximately steady rate. Inducible enzymes are only produced when required. Their production depends on the presence of certain inducing compounds. These enable a cell to adapt to varying environmental conditions. The concentration of these enzymes in cells varies greatly.

In the aforementioned experiment, the addition of lactose had induced the production of three **lactose-degrading enzymes**. Previous experiments had established that these three enzymes were coded by **three gene**s that lay next to each other in the DNA.

How is it possible that a simple compound such as lactose can switch on the complex genetic apparatus of a bacterial cell? At this point, Monod's clear

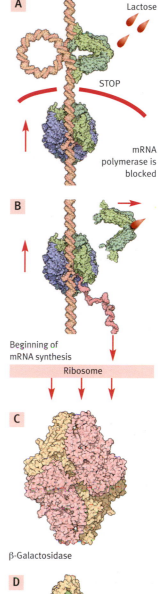

A

Lactose

STOP

mRNA polymerase is blocked

B

Beginning of mRNA synthesis

Ribosome

C

β-Galactosidase

D

Galactoside acetyltransferase

E

Lactose permease

Fig. 4.9 The Lac induction process (top). Molecular structure of the three lactose-degrading enzymes (bottom). β-galactosidase catalyzes the breakdown of lactose to simple sugars – glucose and galactose. The second enzyme is galactoside acetyltransferase, and the third is lactose permease, a protein in the bacterial membrane that transports lactose into the cell.

Mitochondrion   Nuclear membrane   Nucleus with nucleosomes

Box 4.1 Biotech History: *Aspergillus niger* – The End of the Italian Monopoly

The commercial production of citric acid began under **John** and **Edward Sturge** in Selby, England in 1826. The starting material was the juice of imported Italian citrus fruit (lemons and limes) from which calcium citrate was obtained. This was then easily converted into citric acid.

Within a very short time, Italy had a monopoly on the production of citrus fruit, and prices rose. During the First World War, however, Italy neglected her citrus plantations, and the citric acid price soared to dizzying heights as a consequence. The importing countries began to look for cheaper alternatives.

In 1917, **J. N. Currie** published an article in the Journal of Biological Chemistry that was to end Italy's long-held monopoly within a few years. The article dealt with a particular metabolic activity of the fungus *Aspergillus niger*. Currie had discovered that it produces far more citric acid than other fungi and was studying the conditions under which its yields were highest.

With the help of John Currie, the Pfizer corporation, New York, began the first large-scale production of citric acid in 1923, with subsidiary plants in Germany, Czechoslovakia, the U.K, and Belgium.

Aspergillus was cultured on the surface of a liquid medium. Glucose, sucrose, and molasses from sugar beet were the starting substrates.

100 years after its foundation, the English firm John & E. Sturge (Citric) Ltd, which was based in York, adopted Currie's method and combined it with the traditional chemical production of calcium citrate from lemon juice. After the Second World War, a new submerged culture was developed that was easier to control.

In the 1960s and 70s when the oil price was low, attempts were made to synthesize citric acid using yeasts (*Candida* strains) from crude oil fractions, but once oil prices began to rise, it was rejected as uneconomical (see also Ch. 7, Single Cell Protein).

Nowadays, the previous biotechnological methods are being used again, with not only *Aspergillus niger*, but various yeasts as citric acid producers.

With a world production of 800,000 tons of citric acid per year, even with the best will in the world, it would not be possible to revert to production from citrus fruit unless one wanted to cover every square mile in Italy with lemon trees!

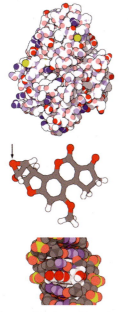

Fig. 4.10 Cytochrome P-450, here from the camphor-degrading bacterium *Pseudomonas putida*. The enzyme detoxifies foreign substances by inserting molecular oxygen. Some compounds such as benzopyrene and aflatoxins, however, are "toxified", as the epoxide (arrow) of aflatoxin (center) binds to DNA (bottom), causing mutations and cancer.

idea of allostery connected to the unsolved problem of induction suggested an ingenious solution: The **inducer** must bind to an allosteric protein that blocks the genes for lactose-degrading enzymes. Through the binding process, the **allosteric repressor protein** would undergo spatial deformation. This would make it inactive and cause it to detach from DNA. This idea proved to be a viable concept, although Jacob and Monod never succeeded in experimentally detecting a repressor protein. It was not until 1967 that **Walter Gilbert** (see Ch. 3) and **Benno Müller-Hill** were able to isolate and purify the repressor protein of lactose-degrading enzymes. Only a few repressor molecules were found in each cell. Thus, Jacob and Monod's hypothesis was experimentally confirmed two years after they had been awarded the Nobel Prize.

The induction process through lactose in bacteria is shown in Figures 4.7 and 4.9.

The repressor is a tetramer consisting of four identical subunits (Fig. 4.7). It binds stably to a specific region of the bacterial DNA, the operator, which lies right next to the region coding the three lactose-degrading enzymes.

When lactose and similar sugars bind to the repressor protein, it changes its shape and can no longer remain attached to the DNA. RNA polymerase can now wander along the DNA to transcribe the genes, i.e. the enzymes are going into their production phase. When lactose becomes scarce, the enzymes are no longer needed, and the Lac repressor reassumes its original shape, blocking the DNA.

Right next to the genes coding for the degrading enzymes, there is a DNA region called an **operator** which serves as an on-and-off switch for the genes. It is blocked by the four subunits of the repressor protein which exactly fits the DNA region. The repressor literally bends the DNA to form a loop. Thus, the RNA polymerase is physically prohibited from starting the reading process. The three lactose-degrading enzymes cannot be read.

When the inducer, it this case lactose, is present, it binds to the active allosteric repressor protein, changing its spatial structure. However, the substrate to be degraded is not always the inducer. In our case, it is **allolactose**, a modified compound that is formed at the beginning of the induction process and inactivates the repressor. The repressor, not fitting the operator any longer, is repelled, the switch is released, and the DNA loop is straightened.

Now **mRNA polymerase** can spring into action (Fig. 4.9), transcribing the information for the three genes into a long mRNA sequence which is immediately read by the ribosomes. All three enzymes are produced one by one. The freshly produced enzymes speed up the uptake of lactose and its degradation, causing even more inducer molecules to enter the cell. Within a relatively short time, the bacterial cell has formed enough enzymes to be able to feed largely on lactose.

When these adapted bacteria are then retransferred to a medium containing pure glucose, the inducer disappears from one moment to the next, and the repressor protein that had been inactivated by the inducer reassumes its active form, binding to the operator and disrupting the mRNA polymerase. Thus, the synthesis of lactose-degrading enzymes comes to an end. The lactose-degrading enzymes still present in the cell become redundant and are broken down. To summarize – in the induction process, the inducer temporarily removes the inhibition of enzyme synthesis. This is called **negative control**.

This mechanism allows the cell **to use its resources efficiently**, as it would be highly uneconomical if a bacterium had to produce the thousands of enzymes it might need at a certain point of its life.

There is another mechanism called **repression** which stops the further production of a compound when large amounts of the end product are already available.

In **higher organisms**, the regulation of induction and repression is more complicated than in bacteria, and their effects are slower and less dramatic. Not only nutrients, but also the organism's own hormones or foreign compounds may act as inducers. In humans, the **induction of liver enzyme production** through a drug (the cytochrome P 450 system, Fig. 4.10) and alcohol (alcohol dehydrogenase and acetaldehyde dehydrogenase, Ch. 1) is an example. It is well known that the regular intake of sleeping pills reduces their effect in the long run because the cytochrome P-450 system is induced. The sleeping tablet is degraded into water-soluble

compounds for **detoxification**. However, when the induced P-450 is brought into contact with α-pyrenes (cigarettes, barbecued meat), it turns them into carcinogenic epoxides. This could be called a **toxification** process. A similar mechanism works in the case of aflatoxins (Fig. 4.10). Obviously, the toxification effect depends on the previous induction of P-450 (e.g. drug abuse).

By contrast, the regular consumption of ethanol induces alcohol dehydrogenase and acetaldehyde dehydrogenase in the liver, which enhances alcohol tolerance at the potential price of damaging the liver.

▬ 4.4 An Allosteric Molecular Computer – Glutamine Synthetase

An organism must be able to react to a variety of nutrients and have the necessary enzymes ready. This is also true for bacteria that, unlike us, often have no choice in what they take in. They must take in the nutrients available wherever they are and mobilize the relevant enzymes.

Glutamine synthetase (GS) is a key enzyme in the control of nitrogen in a cell (Fig. 4.11). Glutamine is an amino acid that serves not only as a protein building block, but also as a delivery system for nitrogen to the enzymes that assemble DNA bases and amino acids. GS must therefore be carefully controlled. When nitrogen is needed, it must be switched on so as not to starve the cell. When sufficient nitrogen is available it must be switched off to prevent "overeating".

GS functions like a molecular computer, recording the availability of nitrogen-rich molecules – amino acids such as glycine, alanine, histidine, and tryptophan, as well as the nucleotides adenosine monophosphate (AMP) and cytidine phosphate (CTP). Any surplus of these is registered by GS and the production reduced accordingly. If there is a dramatic surge in nitrogen-rich products, the end products slow down GS. Eventually, when enough of all compounds have been produced, the enzyme activity grinds to a halt.

The basis on which glutamine synthetase works is called **cumulative feedback**. There are eight end products, each of which has an inhibiting effect on

Glutamine synthetase (top view)

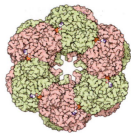

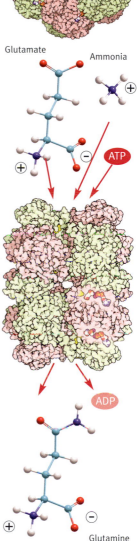

Glutamate

Ammonia

Glutamine

Fig. 4.11 Glutamine synthetase consists of twelve subunits, each of which features an active center for the production of glutamine. Glutamate and ammonia bind to them, and energy is provided by an ATP molecule. All subunits communicate amongst each other. When the concentration of a relevant compound rises, the total activity is reduced accordingly. It is also possible for the cell to shut down the enzyme completely from one moment to the next. This happens if certain amino acids, AMP, or CTP bind to the active centers and reduce their action.

Box 4.2 Biotech History:
Boyd Woodruff and the Second Important Antibiotic

A course in soil microbiology, taught by Professor Selman A. Waksman (1888-1973) became the turning point in H. Boyd Woodruff's life. He writes in his memoirs about his life with soil microbes. Here is a drastically shortened excerpt authorized by him, it starts at this turning point. (RR)

Boyd Woodruff and his teacher Selman A. Waksman in the laboratory in 1940.

"For the first time I realized the unity of biology and chemistry, that each biological observation has an underlying chemical cause, that in unraveling the latter, one could understand the other.

Graduation from Rutgers in 1939 brought me to the first great decision point in my life, the choice of a graduate career. All preparation had been for chemistry. Letters were written soliciting graduate school appointments. I was not prepared for, and did not expect, an offer received from **Dr Waksman** to join his student group on a University fellowship permitting full-time research at the then unusual stipend of $900 a year, fully 20 percent more than fellowships available elsewhere, without the need to perform time-requiring teaching assistantship duties.

Still the decision was difficult. Should I stay at Rutgers? Should I leave chemistry for biology? **Dr Jacob Lipman**, Dean of the Agricultural School, the directors of various departments were consulted, and all pointed to Dr Waksman's chemical approach to microbiology. Had he not been author of a book entitled *Enzymes*? It was recommended that I examine his publications to see the extent to which chemistry was utilized. I was convinced and decided to stay. The die was cast.

With the enormous financial support for research presently enjoyed by American universities, and the well equipped and staffed laboratories, it is difficult to remember the privations existing in the 1940s. It may be difficult also to recognize my good fortune at the time in joining the soil microbiology laboratory at Rutgers.

Dr Waksman was able to spend half of each day working in the laboratory. As his laboratory assistant, I learned directly by example and by discussion. The other member of the departmental staff was **Dr Robert Starkey**. He also worked in the same laboratory. His influence, the enthusiasm with which he attacked each research endeavor, his willingness to spend hours in explanation, to teach on a one-to-one basis, the effort he gave to planning group experimentation, all must be considered as having major influence on my development as a scientist.

I deserved the envy of my associates. Who else could report that their entire graduate study program was spent working at the same laboratory bench with two senior scientists, one a future Nobel Prize winner, and both later to serve terms as President of the American Society for Microbiology? After my brief probes in soil microbiology, Dr Waksman appeared in the laboratory one day. He was highly agitated. "Woodruff," he said, "drop everything. See what these Englishmen have discovered a mold can do. I know the actinomycetes will do better!"

He had just learned of the extension of **Fleming's** work on penicillin in **Florey's** Oxford University laboratory, proving its chemotherapeutic potential.

Thus was initiated the first search in Waksman's laboratory for antibiotics from actinomycetes. The work was quickly successful. At first, experience was gained reisolating the old bacterial product pyocyanase, but soon we had **actinomycin**, a new product to study and evaluate.

Fortunately, actinomycin was easy to produce and extract, was sufficiently stable to be studied chemically, and was so highly active that we could dream of practical applications. But all our dreams were defeated by **toxicity**. It was not until much later that the usefulness of the actinomycins in treatment of rare forms of cancer was recognized.

Actinomycin did, however, provide my first introduction to **Merck & Co.**, Inc., the place of my eventual employment. Merck microbiologist **Dr Jack Stokes**, a former Waksman student, offered to provide facilities for large-scale production of actinomycin. The hundreds of flasks incubated at Merck were inoculated directly by Waksman himself. I served as assistant and general handyman. Merck chemist **Dr Max Tishler**, later to become president of the Merck Sharp & Dohme Research Laboratories, joined with Dr Waksman in bringing actinomycin to the final state of purity and investigating its structure.

As a result of the early successes, incoming graduate students were assigned to antibiotic research. Many were engaged in screening. A culture isolated by **Dr Walter Kocholaty** was assigned to me for further research and from it was isolated **streptothricin, the first of the basic water soluble antibiotics**.

No facilities for animal study existed in the soil microbiology department, but they were available in the dairy department under **Dr H. J. Metzger**, who was studying contagious abortion, caused by *Brucella abortus*, in laboratory animals. The disease still was a serious human infection at that time. Our streptothricin concentrate was tested for toxicity and proven safe and then was shown to produce significant cure of experimental infections.

In his memoirs, H. Boyd Woodruff gives a humble understatement of his contribution. To be fair, I cite here The American Chemical Society about Woodruff's pioneering work (RR):

The first agent isolated by Boyd Woodruff in 1940 under the initial screening program

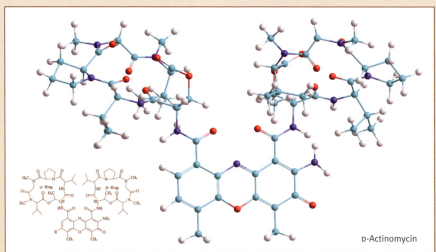

D-Actinomycin

was actinomycin. Woodruff demonstrated that the methodology would yield antibiotic-producing cultures. Actinomycin was active against a broad range of bacteria and even showed promise for attacking a tuberculosis strain, but it proved too toxic for therapeutic use in humans. Two years later, Woodruff isolated streptothricin, an antibiotic which was active against both Gram-positive and Gram-negative bacteria. The researchers were initially excited about streptothricin because, as Waksman said in his Nobel acceptance speech, it "gave promise of filling the gap left by penicillin in the treatment of infectious diseases due to Gram-negative bacteria." In addition to being active against Gram-positive and Gram-negative bacteria, initial tests of streptothricin showed that it was not toxic to animals. However, pharmacological studies demonstrated that streptothricin had a delayed toxic effect in animals and thus could not be used for therapeutic purposes in humans. The partial success of streptothricin indicated that Waksman and his students were on the right track.

Boyd Woodruff again: "The excitement in our group was intense. From that time on Dr Waksman's department specialized in antibiotics. The advantage of variety in research experience was lost to its students. We talked of antibacterial substances, nothing else. There was much debate whether the substances were natural soil products or products of artificial culture conditions. At one lunchtime discussion, Dr **Waksman** coined the word *antibiotic* and stated its definition. The word entered the scientific vocabulary."

What happened after that?

A couple of years later, in 1944, Waksman, with **Albert Schatz** and **Elizabeth Bugie**, isolated the first aminoglycoside, streptomycin, from *Streptomyces griseus*.

The Merck company immediately started manufacturing streptomycin with the help of Woodruff. Waksman sent Woodruff to Merck to help with the project, and after finishing his thesis, Woodruff continued working there. Simultaneously, studies at the Mayo Clinic confirmed streptomycin's efficacy against tuberculosis in guinea pigs and relatively low toxicity. On November 20, 1944, doctors administered streptomycin for the first time to a seriously ill tuberculosis patient and observed a rapid, impressive recovery.

Merck had just developed streptomycin and moved it into the marketplace when the com-

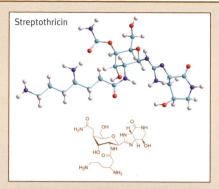

Streptothricin

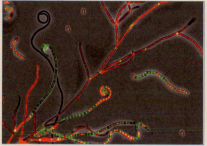

Streptomyces

pany stumbled upon another great discovery. At the time, doctors treated patients with **pernicious anemia** by injecting them with liver extracts, which contained a factor required for curing and controlling the disease. When patients stopped receiving injections, the disease redeveloped. The Merck chemists had been working on isolating what was called the pernicious anemia factor from liver extracts, and they decided to look at the cultures grown by Boyd Woodruff and other microbiologists at Merck, to see if one of the cultures might produce the pernicious anemia factor. They found a strain of *Streptomyces griseus* similar to the streptomycin-producing strain that made the pernicious anemia factor. The factor turned out to be a vitamin, and it was later named vitamin B_{12}.

Merck struck gold three times in a row. As Boyd Woodruff describes the period, "We jumped from penicillin to streptomycin to vitamin B_{12}. We got them 1-2-3, bang-bang-bang!"

Epilogue

Boyd Woodruff concludes his Review:

"Like Ulysses, the final return from my odyssey is now in sight and seemingly my ship speeds more rapidly as home approaches, but also like Ulysses when home is reached, life will not be at an end. Opportunity for creative activities will continue. It is soil microbiology that I see today as the great challenge, as the area where an industrial or academic laboratories are not required. Thus, this odyssey has brought me

from an initial introduction (in Waksman's department at Rutgers University) to problems in soil microbiology, to the many opportunities available in applied industrial research at Merck & Co., Inc., returning again at this later stage to soil microbiology as a fertile, largely uncharted field.

Starting in the ancient buildings on the banks of the Raritan, where I first became aware that the soil is not simply dirt but a teeming mass of microbial life, who could have envisioned the challenges, the frustrations, the thrill of accomplishments that have been the lot of this industrial microbiologist."

In a letter to me (RR) he wrote:

At time of retirement at age 65, I decided I would like to continue research in my own laboratory, so formally established a new company, Soil Microbiology Associates, which in fact consisted of my wife and myself. I constructed a laboratory in the basement of our home in Watchung. My objective was to continue a cooperative program I had established at the University of Western Australia, hopefully obtaining unusual actinomycetes from heavily stressed soils. We undertook extensive travels worldwide, carrying necessary equipment with us for plating soils, then carried the inoculated plates home with us for incubation and for isolating as many morphologically distinct actiomycetes as possible from the soils, sending them to Boehringer Mannheim in Germany, to Merck in the USA and to the Kitasato Institute in Japan for entry into screens they had available. The second paper, Natural Products From Microorganisms An Odyssey Revised, covers 17 years of my retirement period and emphasizes the relatively large number of new natural products discovered, with structures established, obtained from the approximately 5,000 cultures isolated.

Boyd Woodruff

Cited Literature:

Woodruff HB (1981) A soil microbiologist's Odyssey. Ann Rev.Microbiol. 35:1-28

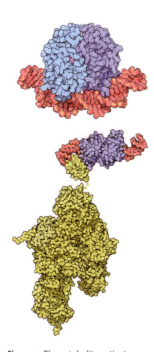

Fig. 4.12 The catabolite activator protein CAP facilitates the work of RNA polymerase by "fishing" for genes. When cAMP (top of the diagram, ruby dot in light blue section) binds to it, CAP changes its conformation slightly to bind perfectly to the relevant operator of the DNA, right next to the gene of a degrading enzyme. CAP has an iron grip on the DNA, bending it by nearly 90 degrees. Then it "lures" RNA polymerase towards the DNA, thus stimulating the transcription of neighboring genes. The RNA polymerase (yellow) carries a small subunit that interacts with CAP and DNA - similar to a fishing rod. The lower diagram shows how one half of the polymerase subunit binds to DNA (red) and CAP (purple). Both halves of the subunit are connected by a flexible linker. Thus, RNA polymerase is able to "fish" actively for the desired genes.

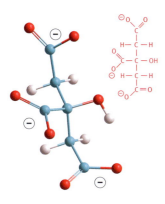

Fig. 4.13 Citric acid

the enzyme. Their effect is additive, i.e. each of these inhibitors can add to the overall activity inhibition even if all the other inhibitors have already been bound to saturation level.

The amino acid glutamine is an important nutrient in animal cell cultures. It is also often added to sports drinks because it provides energy. Glutamine and arginine are also popular with body builders.

■ 4.5 Catabolite Repression or Fishing for Polymerase

Bacteria love sweet things, especially glucose which is easily digested and converted into chemical energy. Where glucose is abundant, bacteria usually ignore other sources of carbohydrate.

Bacteria use an unusual modification of ATP to let the cell system know what they are ingesting at the moment. When glucose levels fall, a membrane-bound enzyme is activated as a glucose sensor – **adenylate cyclase**. It cleaves two phosphate residues off ATP and binds the free ends to a small molecule, cyclic AMP (cAMP). **Cyclic AMP** has also been called a second messenger.

The reactions that convert fuel into energy are called catabolic reactions or **catabolism**. The reverse, an energy-consuming reaction, is called **anabolism**.

The **catabolite activator protein** (CAP) is activated by cAMP, then activates and stimulates enzymes that degrade non-glucose nutrients. The binding of cAMP causes a slight change in the CAP conformation so that it perfectly fits the relevant operator in the DNA, right next to the gene for the degrading enzyme. The figure shows how RNA polymerase actively fishes out the relevant genes (Fig. 4.12).

In *E. coli*, there are numerous binding sites for CAPs that attach to promoters for multiple catabolic enzymes. As long as *E. coli* grows on glucose, cAMP concentration remains low, hardly any other sugar-degrading enzymes are produced, as it would be a waste to synthesize them. Only a lack of glucose and the presence of other sugars will cause the cell to switch over its enzyme production.

In contrast to the negative lactose induction process, **catabolite repression is a positive control mechanism**. The cAMP/CAP complex enhances the reading of DNA by polymerase 50-fold. At the lac operon, the gene expression of lactose degradation and transport enzymes is strongest when lactose or allolactose removes the lac repressor inhibition while the CAP/cAMP complex stimulates the binding of RNA polymerase.

■ 4.6 Mold Replacing Lemons

It was **Justus von Liebig** (Ch. 1) who identified the structure of citric acid in 1838, and in 1893, microbiologist **C. Wehmer** at the University of Hanover observed that fungi growing on sugar secreted citric acid (structure in Fig. 4.13) into the medium.

As the demand in citric acid rose after World War I, citric acid production from microbes offered a viable alternative to its labor-intensive isolation from **expensive imported citrus fruit** (Box 4.1). An intense search for the most productive mold fungus began and the crown went to *Aspergillus niger*, the **bread mold** (Fig. 4.14 and Ch. 1, Fig. 1.9).

It was soon found that the fungus' productivity largely depends on the proton (H$^+$) content, i.e. the acidity (the pH value is the negative logarithm to base 10 of the proton concentration) of the nutrient medium. A very acidic medium enhances the citric acid production of *Aspergillus niger*. A further increase in production can be achieved by reducing the number of free iron ions in the nutrient medium to 0.5mg/liter. It is thought that a low pH and a lack of iron inhibit the citric acid-degrading enzyme aconitase and citric acid is not broken down as a consequence. It is also thought that the higher proton content in the medium causes changes in the membrane structure of the fungal cells that facilitate the secretion of citric acid. A low pH has the added effect of inhibiting the growth of unwanted microbes during the fermentation process.

The iron ions are bound by adding potassium ferrocyanide to the fermenting solution, resulting in the poorly soluble Prussian Blue or ferric ferrocyanide. Blue is also the color of residues in citric acid production plants.

In the 1920s, the successful industrial production of citric acid devastated many small farmers in Italy whose livelihood depended on lemon plantations – an early example of collateral damage caused by the biotech industry (Box 4.1).

Nowadays, stirring reactors or reactor towers of 130 to 650 cubic yards (100 to 500 cubic meters) capacity turn 85% of raw material into citric acid. As the harvested solution is very acidic, it must be kept in anticorrosive steel tanks. Microbiologically produced citric acid is chemically identical to citric acid obtained from lemons and is used for flavoring sweets, lemonade, and other food. Citric acid is also under consideration as an alternative to environmentally harmful polyphosphates in detergents (Fig. 4.15) because it forms calcium and magnesium complexes. Citric acid is also used as emergency treatment for heavy-metal poisoning because of its ability to bind heavy metals.

The worldwide yearly production of citric acid amounts to approximately one million tons at a market value of around $ 1.2 billion. The industrial *Aspergillus* strains are among the most coveted and guarded possessions in the fermentation industry. The potential of cells as synthetic production plants for a wide range of products, including production methods and applications, have been discussed in more detail in other parts of this book (Ch. 1 and 2). We are now looking at the direct connection that exists between the citric acid cycle and the synthesis of amino acids.

■ 4.7 Overproduction of Lysine – How Mutants Outwit the Feedback Inhibition of Aspartate Kinase

Lysine is one of the **eight essential amino acids** (Figs. 4.17 map, 4.16 and 4.18 structures) that humans and many domestic animals are unable to synthesize. They must be taken up with food, just as **phenylalanine (Phe)**, **isoleucine (Ile)**, **tryptophan (Trp)**, **methionine (Met)**, **leucine (Leu)**, **valine (Val)**, and **threonine (Thr)** (see mnemonic aid in the margin of p. 108). Alongside methionine and threonine, lysine is one of those amino acids that are only found in small quantities in grain (wheat, corn, and rice), making it an important additive to animal feed. This is where most of the 550,000 tons of microbial lysine annually produced worldwide went. Lysine is also produced for human consumption.

Lysine is obtained from *Corynebacterium* strains (Fig. 4.22, for history see Box 4.4) – cudgel-shaped (Greek *koryne*, cudgel) bacteria. In order to produce the amino acid lysine in *Corynebacterium glutamicum*, enzyme **feedback inhibition** had to be overcome.

In wildtype strains, lysine is produced from oxalacetate in the citric acid cycle, using pyruvate, and following an enzymatic reaction pathway via aspartate. The starting point is the **allosteric pacemaker enzyme aspartate kinase** (Fig. 4.21). The enzymatic reactions are controlled by feedback inhibition. Aspartate kinase activity is inhibited by excessive amounts of lysine and threonine. Both must bind to aspartate kinase to slow down its production.

When more lysine and threonine are produced than actually needed, the cell inhibits aspartate kinase, and conversely, when there is a lack of lysine and threonine, it encourages their synthesis. This makes sense in nature, but can be rather annoying if you are after an overproduction of lysine. There is little hope of making the bacterium produce large amounts of lysine unless it is possible to outwit the regulation mechanism by switching off the allosteric controls.

The search was on for **defective mutants** of the bacteria, and two of them were eventually found. One of them lacked threonine formation (thr-) (Fig. 4.21) because the enzyme **homoserine dehydrogenase** had been inactivated by spontaneous mutation. The affected cell could no longer produce threonine, one of the two inhibitors of aspartate kinase. This mutant is grown in a medium containing just enough threonine to maintain cell growth, but too little to interact with lysine to switch off aspartate

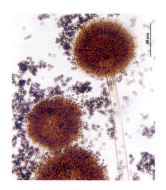

Fig. 4.14 *Aspergillus niger*, the fungus that made lemons obsolete.

Fig. 4.15 Citric acid is used in washing liquid. The complexes it forms with calcium and magnesium make the water soft.
Citric acid is added to blood samples in the laboratory to prevent them from clotting.

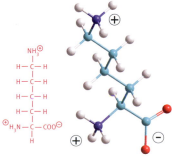

Fig. 4.16 Lysine (Lys,K), an essential amino acid which is vital in animal feed production.

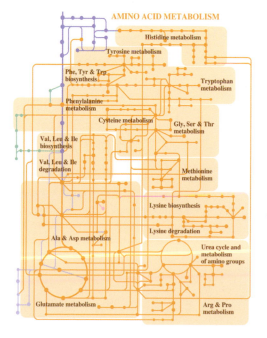

AMINO ACID METABOLISM

Histidine metabolism

Tyrosine metabolism

Phe, Tyr & Trp biosynthesis

Tryptophan metabolism

Phenylalanine metabolism

Cysteine metabolism

Gly, Ser & Thr metabolism

Val, Leu & Ile biosynthesis

Val, Leu & Ile degradation

Methionine metabolism

Lysine biosynthesis

Lysine degradation

Ala & Asp metabolism

Urea cycle and metabolism of amino groups

Glutamate metabolism

Arg & Pro metabolism

Fig. 4.17 Refined metabolism map of amino acids (see also Fig. 4.4).

Box 4.3 The Expert's View:
High-grade Cysteine No Longer Has to Be Extracted From Hair

Cysteine is a sulfur-containing amino acid that is unique because of its chemically very reactive sulfhydryl group.

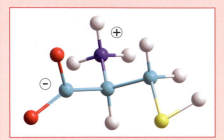

By building disulfide bridges, it contributes to the **stability of proteins**. Therefore, it enables the formation of strong fiber strands such as hair, wool and feathers, as well as horns, hooves and nails, all of which are made of proteins containing large amounts of cysteine.

Cysteine is used by **bakers** to break up the gluten in flour, thus reducing its stickiness and facilitating the kneading of the dough. Due to their high reactivity, cysteine and its derivates such as acetylcysteine are also used in **expectorants** (cough mixtures). Acetylcysteine breaks up the mucoproteins in the bronchia, liquefying the mucus.

Mao-jai, the kitten, can't get enough of the cysteine used in perms and shouts 'Yummy!'

Japanese **hairdressers** use cysteine in preparation for a perm rather than the pungent thioglycolic acid mostly used in Europe.

Somewhat surprisingly, given the versatility of cysteine, until very recently it has been one of the few amino acids that could only be extracted from animal and human material – **hair, feathers, bristles, or hooves**. Extracting cysteine has become big business in Asia. In China's hair salons, tens of thousands of tons of hair are swept up by professional collectors and delivered to cysteine manufacturers, who then extract cysteine using activated charcoal and concentrated hydrochloric acid. One ton of hair yields 100 kg (approx. 220 lb) of cysteine.

As the global pharmaceutical, cosmetic, and food industries are currently using about 4000 tons of cysteine per year and the demand is rising at an annual rate of 4 percent, huge quantities of raw material are required to meet the demand. Being able to produce cysteine on an industrial scale would be very welcome indeed, not only in terms of efficiency, but also for environmental reasons. However, an even more powerful drive to find an alternative to the traditional extraction process is the pharmaceutical industry's need for raw material guaranteed to be free from contaminants such as BSE, SARS, or avian flu. These materials are now obtained by biotechnological means alone.

The author at his Chinese barbershop. If you know what the hair is used for, you can look at yourself as a source of raw material – strange feeling!

Through targeted mutagenesis and selection, researchers at **Wacker Fine Chemicals** were able to switch off regulatory proteins in *E. coli*, which would normally close down cysteine production at the saturation point. Bacteria, in which these proteins have been switched off, keep on producing cysteine, secreting it through their cell membranes into a medium in fermentation vessels holding 50,000 liters (over 13,000 gal).

There are several advantages to this **biotechnological approach**. It is highly efficient, as 90 percent of cysteine ends up in the final product, compared to only 60 percent when extracted from hair. The extraction process also uses far less hydrochloric acid – just one kilogram per kilogram of extracted cysteine. Extracting cysteine from hair uses 27 kg of hydrochloric acid. The

Bioreactor containing cysteine-producing *E. coli*.

use of sugar, salts, and trace elements as raw material avoids contamination of the end product by pathogens and other unwanted material.

The product contains 98.5% of pure cysteine which fulfills all the food and pharmaceutical quality standards. It is also fit for use in **vegetarian food** to produce an artificial meat flavor. When cysteine binds to a sugar such as ribose, meat-like aromatic compounds develop during the heating process. This imitation of natural processes occurs, for example, when frying a chicken (i.e., the Maillard reaction). The cysteine present in the meat, reacts with the sugar to form characteristic aromatic compounds.

In 2004, 500 tons of cysteine – an eighth of global demand – had already been produced by this biotechnological method. The annual growth rate of biotechnologically-produced cysteine is over ten percent.

However, this will not put Asian hair collectors out of work straightaway. Several thousand tons of cysteine will still be extracted in the conventional way every year, but there will be a shift in market shares. Customers for whom a low price rather than the origin of the product is a consideration will continue to buy it – for example, to enhance the various meat aromas in cat and dog food. These will probably be the longest-lasting markets for the conventional production of cysteine.

Dr Christoph Winterhalter is project manager at WACKER FINE CHEMICALS.

kinase. As a result, lysine production continues unabated.

The other useful mutant carries a mutation in the aspartate kinase gene itself. The enzyme that has been slightly modified in its structure remains functional, but can no longer be inhibited by lysine, not even by a vast surplus. The production of lysine shows that the study of the enzyme reaction and the selection of suitable mutants are prerequisites for an economically efficient industrial production process. In 500 m³ bioreactors, mutants of *Corynebacterium glutamicum* and *Brevibacterium flavum* are currently responsible for the conversion of over a third of the sugar contained in the medium to lysine, achieving concentrations of 120 g lysine per liter of medium within 60 hours.

A specialist branch of gene technology called Metabolic Engineering is involved in developing industrially relevant strains. **Alfred Pühler's** research group at the University of Bielefeld (Germany) and two major amino acid producers in Japan have deciphered the 3.3 megabase genome of *Corynebacterium glutamicum* (Fig. 4.23).

Methionine is another essential amino acid with a particular role among amino acids, as animal organisms have the ability to convert D-methionine into L-methionine. This permits the synthetic production of methionine (from acrolein, methanethiol, and hydrogen cyanide) as racemate (D,L mix) (see Sect. 4.9). As much as 370,000 tons of methionine were produced this way in 2001.

Moreover, 70,000 tons of **L-threonine** per year have been synthesized mainly by high-performance mutants of *Escherichia coli* – partly at concentrations of 100 grams per liter in only 30 hours.

With the price of all three amino acids hovering around $ 5,000 per ton, their market volume is approximately one billion dollars. China in particular has a dynamically growing industry. In the future, **transgenic plants** (Ch. 7) containing higher proportions of essential amino acids might compete with amino acids produced by fermentation.

> Never underestimate
> the power of the microbe.
>
> *Jackson W. Foster (1914-1966)*

■ 4.8 L-Glutamate – "Levorotatory" Soup Seasoning in Abundance

Alongside the four Western qualities of taste – bitter, sweet, sour and salty – there is a **fifth taste** called *umami*. It has been described as "spicy", "meaty" or "savory" and was virtually unknown in Central Europe. Mediterranean cuisine, however, has always had a lot of tasty *umami*.

Three different compounds are responsible for the *umami taste* – monosodium glutamate (MSG), disodium inosinate (IMP) and disodiumguanylate (GMP). Glutamate has a key role in this.

As early as 1908, Japanese researcher **Kikunae Ikeda** (1864-1936) (Fig. 4.20) had identified the umami taste compound in *kombu* (*Laminaria japonica*), a seaweed from the Pacific, as glutamate. The salt of L-glutamic acid (L-glutamate) (Fig. 4.19) enhances the taste of soups and sauces considerably, as the human body has specific receptors for L-glutamate, but not for D-glutamate (Fig. 4.26).

In the 1920s and to a certain extent still today, L-glutamate was obtained from wheat. The world's largest producer was the Japanese firm Ajinomoto Co. (*aji-no-moto* means flavor enhancer in Japanese) (Fig. 4.20). The demand in glutamate increased dramatically after the Second World War with the introduction of ready meals, gravy powder, and ready seasoning.

In the early 1950s, the Japanese researcher **Shukuo Kinoshita** (born 1915) from the Kyowa Hakko Kogyo Co. tested bacteria and found one that accumulated glutamate when growing on glucose (see Box 4.4). The bacterium – we have come across it as a source for lysine – was given the name *Corynebacterium glutamicum*. Further glutamate-producing microorganisms, particularly from the *Corynebacterium, Brevibacterium, Arthrobacter,* and *Microbacterium* genera, were discovered later.

The microbial production of glutamate, mainly in China and Japan, now exceeds **1.6 million** tons and is worth over **one billion dollars per year**.

The example of glutamate shows how the fundamental knowledge of metabolic enzyme reactions can be converted into production technology. Like most of the 20 amino acids required for protein synthesis, L-glutamate originates from precursors pro-

Fig. 4.18 The eight essential amino acids (see also next page)

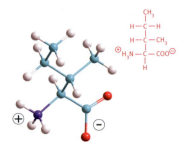

Isoleucine (Ile, I)

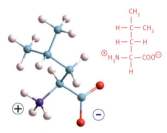

Leucine (Leu, L)

Threonine (Thr, T)

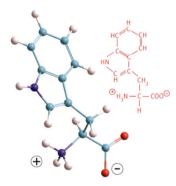

Tryptophan (Trp, W)

Methionine (Met, M)

Valine (Val, V)

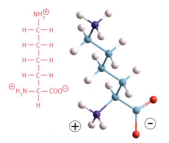

Lysine (Lys, K)

Phenylalanine (Phe, F)

Mnemonic for stressed-out students before an exam!

I Like To Teach My Vets Lumbar Puncture

Ile Leu Thr Trp Met Val Lys Phe

Fig. 4.19 Glutamate (Glu, E)

Fig. 4.20 Lysine production in gigantic bioreactors in Japan.

Fig. 4.18 (Fortsetzung)

duced from glucose degradation (glycolysis) and the citric acid cycle.

For genetic reasons, the oxoglutarate dehydrogenase in the citric acid cycle of *Corynebacterium glutamicum* is not very active. As this is the enzyme responsible for the further conversion of 2-oxoglutarate in the citric acid cycle, 2-oxoglutarate accumulates as a result. Excessive amounts might accumulate, were it not for glutamate dehydrogenase, another very active enzyme derived from the cell's citric acid cycle, which converts 2-oxoglutarate to L-glutamate by inserting ammonium ions (NH_4^+).

In order to produce large amounts of glutamate, a sufficient supply of ammonium ions must keep 2-oxoglutarate conversion going. Ammonia gas (NH_3) is bubbled into water where it is protonated into NH_4^+.

How is the excess glutamate brought from inside the cell into the culture liquid? Having to destroy the cells first and then to isolate glutamate from thousands of cell fragments would be very uneconomical. A way had to be found to make the living cells excrete glutamate into the medium.

Here, a characteristic of the *Corynebacterium* strain comes in handy – it is unable to produce the **cofactor biotin** (a vitamin) which is essential for building cell walls. It relies therefore on the biotin supplied by the nutrient medium. In a medium (e.g. molasses) containing minimal amounts of biotin, cell growth is still possible, but the cell walls become permeable for glutamate. Other ways of releasing glutamate include adding penicillin or using detergents.

How much **glutamate in a meal** can humans tolerate? In Asia, it is said that bad cooks make up for their lack in skill by adding glutamate. Fairly recent double-blind studies about glutamate consumption and "Chinese restaurant syndrome" showed that non-Asian consumers often had strong allergic reactions. This may hint at a psychological component of the problem.

Another amino acid has become increasingly interesting for the pharmaceutical, cosmetics, and food industries. Up to 4,000 tons of **L-cysteine** are used worldwide every year (Box 4.3). So far, it has mainly come from hair, but the German firm Wacker Fine Chemicals has been producing it biotechnologically from *E. coli*. Through targeted mutation and selection, the regulator proteins that

normally slow down cysteine production in *E. coli* have been switched off. The mutant bacteria keep up conveyor belt-style cysteine production, secreting the surplus amino acid through the cell membrane into a nutrient medium in 50,000 liter (over 13,000 gallon) fermenter tanks.

■ 4.9 Chemical Synthesis Versus Microbial Production

Why does L-glutamate and L-lysine production rely on microorganisms? Could they not be produced as cheaply through chemical synthesis?

Actually, in Japan, glutamate was first chemically synthesized from acrylonitrile, using cobalt carbonyl catalysts in what is known as Strecker synthesis. Lysine was produced chemically from aminocaprolactam. The products were mixtures (racemates) of two steric mirror images of glutamate or lysine atoms (Fig. 4.26).

The two mirror images are distinguished by their optical dextro- or levorotation and are called accordingly **D and L-forms**. Chemical synthesis results in equal amounts of D-and L-forms, but only L-glutamate provides the desired taste.

Our taste cells have **receptors** similar to those of enzymes in their active site. These only recognize the spatial arrangement of their reaction partner in about the same way as we would notice if somebody gives us their right or left hand.

As proteins are almost exclusively synthesized from L-amino acids, the synthetic apparatus of a cell is also exclusively laid out to produce L-amino acids. Incidentally, the opposite applies to **sugars** where only the **D-forms** are synthesized and recognized, for example β-D-glucose (dextrose).

The underlying causes for these phenomena are as yet unclear. It is probably due to some coincidence, and in theory, it would be perfectly possible to find organisms containing L-sugars and D-amino acids on some other planet. They might even look very similar. However, the two forms cannot be mixed, and L-glucose or D-glutamate on such a planet would be of no nutritional value to us because our enzymes could not recognize and convert them. Conversely, we would not be an attractive prey for the inhabitants of the other planet…

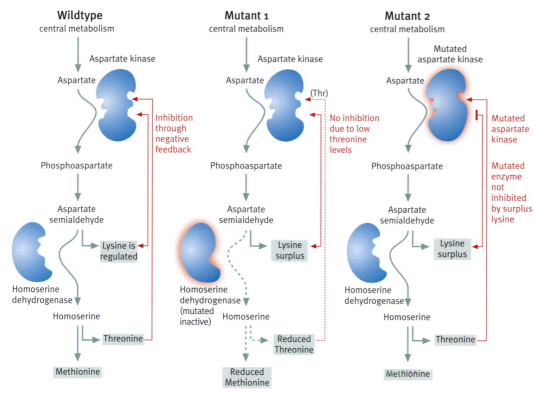

Wildtype
central metabolism

Aspartate kinase

Aspartate

Inhibition through negative feedback

Phosphoaspartate

Aspartate semialdehyde

Lysine is regulated

Homoserine dehydrogenase

Homoserine

Threonine

Methionine

Mutant 1
central metabolism

Aspartate kinase

Aspartate

(Thr)

No inhibition due to low threonine levels

Phosphoaspartate

Aspartate semialdehyde

Lysine surplus

Homoserine dehydrogenase (mutated inactive)

Homoserine

Reduced Threonine

Reduced Methionine

Mutant 2
central metabolism

Mutated aspartate kinase

Aspartate

Mutated aspartate kinase

Mutated enzyme not inhibited by surplus lysine

Phosphoaspartate

Aspartate semialdehyde

Lysine surplus

Homoserine dehydrogenase

Homoserine

Threonine

Methionine

Corynebacterium glutamicum

Fig. 4.21 Negative feedback during the synthesis of lysine in wildtype *Corynebacterium* (electron microscopic view, top) and two mutants. In the first mutant, the enzyme homoserine dehydrogenase has been inactivated. In the second mutant, aspartate kinase has been modified so that it cannot be inhibited even by excess production of lysine.

Goethe – Advocating Bioreactors Before His Time?

That is the property of matter:
For what is natural the All has place;
What's artificial needs restricted space.

J.W. Goethe „Faust. II"

Top: Scene with creating a homunculus in "Faust II" illustrated by the artist Paul Struck (1975/76)

Lewis Carroll followed this line of thought in *Through the Looking-Glass*: How would you like to live in Looking-glass House, Kitty? I wonder if they'd give you milk in there? Perhaps Looking-glass milk isn't good to drink' (containing L-lactose instead of D-lactose). It might be the right thing for a Looking-glass cat?!

Thus, the chemical synthesis of amino acids has an obvious major drawback – half the product would be the tasteless and therefore useless D-glutamate, whereas *Corynebacteria* produce nothing but tasty L-glutamate. The same applies to lysine and threonine – only the L-forms are biologically active, and make up 100 % of bacterial production.

Microbial synthesis is always far superior to chemical synthesis when it comes to stereochemically complicated syntheses and conversions, involving chemically identical groups with differing spatial arrangements in a molecule.

The ability of microbes to perform such targeted syntheses and conversions is also utilized in the production of L-ascorbic acid (vitamin C).

■ 4.10 L-Ascorbic Acid (Vitamin C)

Humans, primates, and – for some reason – guinea pigs are unable to synthesize vitamin C (Fig. 4.29) and must take it up with their food. Guinea pigs and chimpanzees could also contract scurvy because of a lack of the enzyme gulonolactone oxidase.

From today's perspective, it seems bizarre that in 1900 the German emperor ordered his sailors to eat citric acid by the spoonful to avert scurvy. It had been observed that the vitamin contained in citrus fruit had prevented scurvy, but the emperor's orders only gave his sailors diarrhea. It was not known then that it was the **ascorbic acid** (**vitamin C**), not the citric acid in lemons that protected from scurvy. Sauerkraut, however, proved to be healthy food – as long as it was not kept in cans with lead-soldered lids (see Box 4.8).

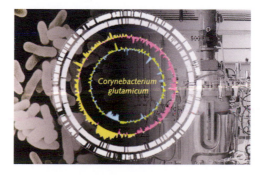

Fig. 4.22 *Corynebacterium glutamicum* carries the number 13032 in the American Type Culture Collection (ATCC). The circles representing the genome are coding regions. The bioreactors are also shown.

Box 4.4 Biotech History:
Kinoshita and the start of Japan's bioindustry

The era of microbial amino acid production with *Corynebacterium glutamicum* and the scientific study of this organism began almost 60 years ago with its discovery as a glutamate-secreting bacterium. This microorganism is today one of the most important organisms in biotechnology and is used to produce about two million tons of amino acids per year, of which more than 1 million tons are accounted for by sodium glutamate, used as a flavor enhancer in the food industry, and more than 0.55 million tons by L-lysine, employed as a feed additive. This market volume is constantly expanding.

Shukuo Kinoshita (born 1915)

Shukuo Kinoshita was the pioneer of amino acid fermentation in Japan. I knew he is in high age but asked him for a contribution. I received the following nice email (see box on the right):

How we started at Kyowa

In 1956, we started a research program at **Kyowa Hakko Kogyo Co., Ltd., Tokyo**, that was aimed at obtaining a microorganism that could accumulate glutamic acid extra-cellularly.

Among many isolates we found a colony that might be fit for the purpose. We named this isolate *Micrococcus glutamicus* No. 534. Further study revealed that this microorganism could accumulate glutamic acid at a limiting concentration of biotin present in the medium. This suggested that biotin must play a key role in the physiology of the cells and their glutamate-forming capability. By microscopic observation of cultures at various stages, we found that the cell form can change considerably. For this reason, and due to further taxonomical studies, we renamed the bacterium *Corynebacterium glutamicum*.

Dear Prof Dr Reinhard Renneberg

My name is Ms. Yuko K. Aoki, the former secretary of Dr Shukuo Kinoshita in Kyowa Hakko Kogyo Co.Ltd. I was asked by him to send his message to you as following.

First of all, I would like to express my thanks for your kind e-mail dated on 16th November. You kindly suggested me to write my story concerning the discovery of amino acid producing bacterium, Corynebacterium glutamicum, in a BOX in the Elsievier-textbook (English Edition). However, I have been suffering from glaucoma for long time and almost lost my eyesight for these days. Therefore, I am very sorry for that I can not accept your kind proposal extended to me.

As you know very well, I have endeavored to establish the industrial process of amino acid production by fermentation in Kyowa Hakko Kogyo Co. Ltd. since 1956. Owing to the big efforts of us and competitors, almost all amino acids can be produced by fermentative process at present. Our invention of industrial production of amino acids by fermentation process resulted in decreasing the manufacturing costs dramatically. Therefore, amino acids are widely used in various fields such as processed foods, pharmaceutical active ingredients, animal feed stuff and etc. in these days. The fundamental principles of our process are the excellent combination of metabolic regulation and genetic modification of microorganisms. I believe that the development of industrial production of amino acids by fermentation is the most distinguished and successful topic of modern biotechnology flourished in 20th century.

Furthermore, I think any other process to overwhelm the fermentative process will not appear in future. I am very proud of our accomplishment done by me and my colleagues. Two years ago, I contributed to the Handbook of Corynebacterium glutamicum, published by Taylor &Francis. In that book, I described how I found the amino acid producing microorganisms and why amino acids industry flourished in Japan. I would like to express to you my deepest gratitude if you could write my comments mentioned above in your textbook.

Thank you very much again and please give my best regards to Arny Demain.
Sincerely yours, Shukuo Kinoshita Ph.

D.1-6-1 Ohtemachi, Chiyoda-ku,
Tokyo, 100-8185, Japan

From mutational work on this organism, together with discoveries regarding key regulatory features, it was found that many amino acids, such as lysine, arginine, ornithine, threonine, etc., could be accumulated. Most of these amino acids are now produced commercially.

Amino acids produced by such a process are all in their natural L-form, and this gives **microbial production a big advantage over chemical synthesis**. Thus, a new industry called amino acid fermentation was born.

The commercial production of amino acids up to the discovery of C. *glutamicum* had relied on the decomposition of natural protein and the isolation of its constituent amino acids.

Our new process, on the contrary, was a biosynthetic process using carbohydrate and ammonium ions. Therefore our process can contribute to the amino acid supply, and also helps to increase the absolute amount of protein in the world.

Since the world population continues to increase year by year, so will the demand for amino acids and protein. After World War II, two new fermentation industries were born in Japan. These are the amino acid and nucleotide fermentation industries.

Kikunae Ikeda (1864-1936) and the first Japanese glutamate products.

The historical roots

I would like to explain briefly why amino acid production was born in Japan, and to do so we have to go back to the year 1908.

At that time Prof **Kikunae Ikeda** (1864-1936) at the University of Tokyo found that **monosodium glutamate** (MSG) had a potent taste-enhancing power [1]. He found this phenomenon through a careful examination of the decomposition products of *kombu*, a type of seaweed. During these studies he found a small crystal. This was glutamic acid, which he discovered had a sour taste. Then he added NaOH to a glutamic acid solution and tasted again. Surprisingly, it had changed into a beautiful taste. This was the aim of his studies, since he was searching for the potent essence of a flavor or taste enhancer. By the addition of only a few milligrams of MSG to various foods, their taste was noticeably improved. What a splendid achievement this was!

Amino acid production in Kyowa Hakko's Hofu plant.

Here, we have to consider the original ideas that led him to conduct such research. His real intention was to improve nutrition and increase the short life expectancy of the Japanese at that time. However, to **provide large amounts of microbial proteins** competitive with natural protein sources like soy or wheat protein was economically impossible. He thought it over, **searching for a good idea to relieve malnutrition in Japan**. He finally got the idea that even if the same food was eaten, its value might be increased if the taste is enhanced. In this sense, an improvement in taste might contribute to relieving malnutrition. Therefore, he began to search for the essence of good taste. *Kombu* had

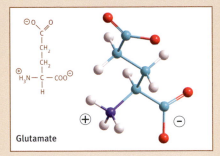

Glutamate

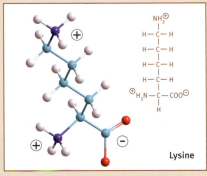

Lysine

been traditionally used in Japanese food as a taste enhancer, so he believed it should contain the essence of flavor. This led to the discovery of MSG, whose commercial production was essential to make use of its taste-enhancing properties for the daily food of the Japanese.

Mr. Saburosuke Suzuki was the man who supported Prof Ikeda's desire. Wheat gluten was chosen as the raw material to obtain MSG. But this task was very difficult.

Concentrated HCl must be used for decomposition of gluten, but no anticorrosive vessels were available in those days. So clay pots were used, but they were fragile and their use was very dangerous. Moreover, the gas from HCl caused serious damage to the health of the residents living near the factory. He had to face an onslaught of accusations and complaints. Consequently, he had to move his factory to a remote location. His struggle to produce MSG continued for ten years, before he finally became confident of commercial success. Once MSG appeared in the market, its miracle power overwhelmed the food market and it became an essential food additive. Mr. Suzuki's company is now known as **Ajinomoto Co., Inc**.

Our motivation: to feed the hungry after the terrible war...

After World War II, Dr **Benzaburo Kato** set up Kyowa Hakko Co., Inc. in 1945. Because of the shortage of food, the Japanese suffered great

hunger. Everywhere malnourished patients were seen. Dr Kato was deeply worried by this situation and thought of an idea for relieving the miserable situation by supplying plenty of protein as food. To implement his idea, he asked me to establish a commercial process that could supply food protein by a fermentation process.

"To produce food protein by a fermentative process?" I couldn't believe my ears. I was deeply impressed by his sincere desire to relieve malnutrition in Japan, but it was impossible to produce protein in a price range that was competitive with natural proteins. If protein production was not feasible, then how about amino acids? Their nutritive value is very similar, so my judgment was that an attempt to produce amino acids may not be the wrong choice. Thus, our challenging program started and was finally successful as described above.

Bioreactors for amino acid production.

This is the background of the birth of the amino acid industry in Japan. As was shown, Dr Ikeda's and Dr Kato's original motivation was the same: relieving the malnutrition of the Japanese. It is interesting to note that the answer to this problem comes out as two entirely opposite processes, one decomposition and the other biosynthesis.

Shukuo Kinoshita (in 2007)

Cited Literature:

1. Ikeda, K. New seasonings [translation]. *Chem. Senses* 27:847-849 (2002).
2. Kinoshita, S. Thom Award Address. Amino acid and nucleotide fermentations: From their genesis to the current state. *Developments in Industrial Microbiology* 28:1-12 (1987).

Fig.4.23 In "You Only Live Twice", James Bond (Sean Connery) posed as a businessman in order to find a super secret rocket developing plant. In order to distract the boss of a Japanese gang from his world rescue mission, he cunningly spelled out "*mo-no-so-di-um-glu-ta-ma-te*".

Fig. 4.24 In his book "Through the Looking Glass and What Alice Found There", published in 1871, Lewis Carroll describes how Alice passes through a mirror and enters a mirrored world.
"*Now, if you'll only attend, Kitty, and not talk so much, I'll tell you all my ideas about Looking-glass House. First, there's the room you can see through the glass – that's just the same as our drawing room, only the things go the other way... How would you like to live in Looking-glass House, Kitty? I wonder if they'd give you milk in there? Perhaps Looking-glass milk isn't good to drink.*"

Fig. 4.25 Configuration of the right-rotating (D, dextrorotatory) and left-rotating (L, levorotatory) form of a compound. They are like mirror images of each other or like the right and the left hand.
The four different substituents of an asymmetric carbon atom are assigned a priority according to atomic number. The configuration about the carbon atom is called S (from Latin "sinister", left) if the progression from the highest to lowest priority is counterclockwise. It is called R (from Latin "rectus" right) if it is clockwise. Only L-amino acids are constituents of proteins. For almost all amino acids, the L-isomer has an S absolute configuration (rather than R).

In 1933, sensational news came from the basement labs of ETH Zurich – vitamin C or L-ascorbic acid had been successfully synthesized (Box 4.3).

Polish-born **Tadeusz Reichstein** (1897 - 1996) chemically degraded glucose in more than ten intermediate steps. The resulting xylose was then converted into vitamin C, using hydrogen cyanide. Unfortunately, this method proved to be far too complex for mass production and only resulted in a small yield. After all, it is vitamin C humans need in larger quantities than other vitamins – about 100 milligrams per day.

Reichstein and his young colleague **Grüssner** therefore decided to explore an alternative route. They wanted to produce sorbose as an intermediate product by reducing glucose with hydrogen and a catalyst under pressure, resulting in a 100 % yield of sorbitol.

However, oxidizing sorbitol to produce the vitamin C precursor sorbose proved to be very complicated, but somebody else, the French chemist **Gabriel Bertrand** (1867 - 1962), had already described the next step in 1896. The acetic acid-producing bacterium *Acetobacter suboxydans* converts sorbitol into sorbose; it has been renamed in modern classification *Gluconobacter oxydans*.

Reichstein then did something very unusual for a chemist in his day – with biotechnology in mind, he bought pure cultures of *Acetobacter* from microbiologists. However, having no knowledge in microbiology, he could not get the bacteria that he had bought to work. He therefore had to resort to another crazy method also described by Bertrand – capturing fruit flies (*Drosophila*) that had wild sorbose bacteria living in their intestines.

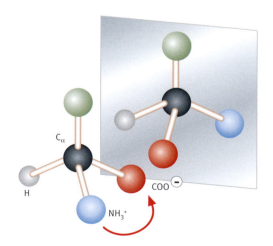

Reichstein used a **combination of chemical synthesis and biotechnology** to convert D-sorbitol into L-sorbose (Fig. 4.27).

In 1933, Reichstein offered his idea to the Swiss pharmaceutical firm Hoffmann-La Roche who had been very interested in adopting **Albert Szent-Györgyi**'s (1893 - 1986) method of isolating vitamin C from capsicum vitamin C had earned him the Nobel Prize in 1937.

The Reichstein method (Fig. 4.27) can be described in the following steps:

- D-Glucose is catalytically reduced (nickel catalyst, pressure 150 bar) to D-sorbitol.

- The sorbitol is then taken up by *Acetobacter suboxydans* and selectively converted into sorbose by the bacterial enzyme sorbitol dehydrogenase. This submerged fermentation process using a 20 - 30 % sorbitol solution achieves almost complete conversion within one or two days.

- Finally L-sorbose is oxidized chemically into 2-keto-L-gulonic acid (2-KLG), which is then treated with acid, and water is cleaved off to produce L-ascorbic acid.

After decades of success, this chemico-biotechnological method has come under threat by more recent **purely biotechnological methods** (Fig. 4.32, and Box 4.3).

Bacteria of the *Erwinia* genus (Fig. 4.33) convert glucose to 2.5 diketo-D-gluconic acid (2.5 DKG), in which process three enzymes (enzyme 1 to 3, Fig. 4.32) are involved. Like *E. coli*, *Erwinia* is a Gram-negative bacterium.

The three membrane-bound enzymes are contained in the **periplasm**, the space between the outer membrane and the cytoplasmic membrane in Gram-negative bacteria.

Other bacteria, such as the Gram-positive *Corynebacterium*, have a simpler wall structure (90 % murein, 10 % teichoic acid). It is therefore possible to distinguish Gram-positive from Gram-negative bacteria by staining them. *Corynebacterium* has no periplasm, but interestingly, its cytoplasm contains 2.5 diketo-D-gluconic acid reductase. This enables it to convert 2.5-DKG to 2-keto-L-gulonic acid, which, in turn, can be easily turned into a vitamin C ring molecule.

Erwinia thus supplies the starting product for *Corynebacterium*, and it would be worthwhile to bring them together in a co-fermentation process. To do this on an industrial scale, however, is very difficult, due to differences in pH and temperature at which optimal growth is achieved in the two species.

This is where genetic engineering comes in, as it has been possible to combine the genetic properties of the two microorganisms in a single one that takes up pure glucose and secretes the vitamin precursor directly into the medium.

To create this organism, the reductase gene from *Corynebacterium* was introduced into *Erwinia* cells – which makes sense. After all, *Erwinia* only lacks the enzyme, whereas in *Corynebacterium*, there would be no periplasmic membrane. The reductase in *Corynebacterium* is not membrane-bound, but is active freely in the cytoplasm.

This is what now happens in **recombinant Erwinia cells** (Fig. 4.32)

- D-Glucose is taken up from the medium into the periplasm.
- The three periplasmic enzymes synthesize 2.5-DKG step by step.
- 2.5-Diketogluconic acid reductase (from *Corynebacterium*) catalyzes 2.5-DKG to 2-KLG (2-ketogulonic acid) within the *Erwinia* cytoplasm, from where it is secreted into the medium.
- 2-KLG is then treated with acid, water is cleaved off, and L-ascorbic acid is produced.

Thus, genetic engineering elegantly combines the metabolic capacities of two different microbes. The recombinant *Erwinia* cells produce approximately 120 grams of 2-KLG within 120 hours, amounting to a glucose conversion rate of 60%.

The ingenious combination of chemical and biotechnological synthesis made Hoffmann-La Roche the world leader in vitamin C production for decades. Production has now been taken over by the Dutch firm DSM.

Nowadays, the market turnover of vitamin C is about $600 million. 65% of the production already comes from Chinese biotech firms that produce vitamin C considerably below the world market price. It can therefore be expected that vitamin C

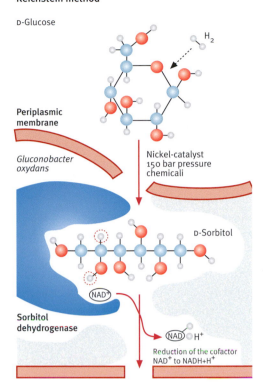

Reichstein method

D-Glucose

H_2

Periplasmic membrane

Gluconobacter oxydans

Nickel-catalyst 150 bar pressure chemicali

D-Sorbitol

NAD+

Sorbitol dehydrogenase

NADH+ H+

Reduction of the cofactor NAD+ to NADH+H+

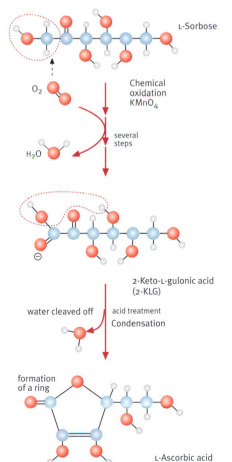

L-Sorbose

O_2

Chemical oxidation $KMnO_4$

several steps

H_2O

2-Keto-L-gulonic acid (2-KLG)

water cleaved off

acid treatment Condensation

formation of a ring

L-Ascorbic acid

Fig. 4.26 Left: Reichstein's vitamin C synthesis.

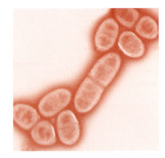

Fig. 4.27 *Gluconobacter oxydans* contains sorbitol dehydrogenase that converts L-sorbitol into L-sorbose.

Fig.4.28 Crystals of ascorbic acid (vitamin C) under a polarizing microscope. It shows interferences due to the dual refraction of light.

Fig. 4.29 The filamentous ascomycete *Ashbya gossipii* produces vitamin B2 (riboflavin).

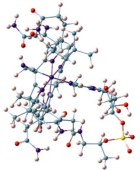

Fig. 4.30 Dorothy Crowfoot Hodgkin and vitamin B$_{12}$. Dorothy Crowfoot Hodgkin (1910 - 1994) was the third woman to be awarded the Nobel Prize in Chemistry, after Marie Curie (1867 - 1934) and her daughter, Irene Joliot-Curie (1897 - 1956). Aided by the first computers, she worked out the structure of penicillin by 1949. The structure of vitamin B$_{12}$, a very large molecule (below), was discovered by her in 1955. When she was awarded the Nobel Prize in 1964, the British press got the wrong end of the stick when announcing that the Nobel Prize had been given to a grandmother. In 1969, Dorothy Crowfoot Hodgkin publicized the final structure of insulin – the result of over 35 years hard work!

production will continue to rise while sales revenues will scarcely increase.

Tadeusz Reichstein was awarded the Nobel Prize for Medicine in 1950 for his work on cortisone, a hormone produced by the cortex of the adrenal gland (p. 135).

In 1936, Roche sold approximately 815 lbs (370 kilograms) of vitamin C at a price of 1140 Swiss Francs (CHF) per kilogram. By 1938, this price had dropped to 550 CHF, by 1940 to 390 CHF, by 1950 to 102 CHF, and by the early 1960s, a kilogram of vitamin C would cost 80 CHF. Currently, the price is about 20 CHF per kilogram – in other words, **vitamin C is 50 times cheaper** than it was 70 years ago!

Vitamin C is not only used for disease prevention and as a health supplement, but also as an innocuous **natural antioxidant** that is added to soft drinks as a preservative. The chemistry and Nobel Peace Prize winner **Linus Pauling** advocated vitamin C for capturing free radicals that could cause physical harm. He swallowed several grams of vitamin C per day, reached the wise old age of 92 and never had a cold. Although his theory has remained controversial, a diet rich in vitamins and various proteins rather than a diet based on fat and red meat will certainly increase life expectancy.

How much vitamin C should one take in? The German Society for Nutrition recommends a daily intake of 150 milligrams per day, while the "forever young" movement favors an intake of one to three grams per day. It is hardly possible to take an overdose, as vitamin C is water-soluble and does not accumulate in the body like the fat-soluble vitamins A, D, E, and K. Any surplus is secreted with the urine. However, an excess of vitamin C may cause acidity problems in the stomach and hit your wallet, too.

Most other vitamins are produced by a purely chemical process, e.g. the carrot pigment β-carotene, which is used in animal feed or the plant compound tocopherol (vitamin E). Only the vitamins C, B$_2$, and B$_{12}$ are mainly biotechnologically produced, i.e. using microbes.

The fungus *Ashbya* gossypii (Fig. 4.29) a filamentous ascomycete, produces **riboflavin (vitamin B$_2$)**. The industrially grown strain now produces about 20,000 times the amount of its wildtype relations. The production process goes back to 1947.

In the beginning, *Ashbya* only produced very small amounts of riboflavin. A related fungus, *Eremothecium ashbyii*, was also harnessed for the production, as well as the bacterium *Bacillus subtilis*. Some strains of that bacterium produce a surplus of the vitamin, which is secreted into the medium. In contrast to the fungi, these bacterial strains are the result of targeted genetic engineering.

Riboflavin occurs naturally in milk, liver, chicken, eggs, sea fish, nuts, and lettuce. It is supposed to be essential for muscle build-up and fitness and the production of the stress hormone adrenalin.

Pseudomonas denitrificans and *Propionibacterium shermanii* produce cobalamin, a precursor for **cyanocobalamin (vitamin B$_{12}$)** (Fig. 4.31) at a rate that lies 50,000 times above the wildtype production. Over a growth period of four days in sugar beet molasses, *Pseudomonas denitrificans* produces up to 60 milligrams per liter cobalamin in an anaerobic one-step process. The cobalamin is secreted into the medium. Molasses contains betaine, which increases the yield considerably. The culture is heated at the end of the process in the presence of cyanide, resulting in cyanocobalamin.

Vitamin B$_{12}$-deficiency results in pernicious anemia. In 1926, pernicious anemia was treated successfully for the first time on a diet of one to two pounds of bovine liver. In 1934, the Nobel Prize for medicine went to **George H. Whipple**, **George R. Minot,** and **William P. Murphy** for their insight into the disease.

Vitamin B$_{12}$ is given to help blood formation and to protect the liver. Half the annual 20 tons of production goes into animal feed to encourage growth and bone-building. In humans, it is also supposed to contribute to a calm disposition and mental alertness.

■ 4.11 Aspartame – Sweet Success of a Dipeptide Ester

In 1965, **James Schlatter**, a chemist working for the US pharmaceutical firm G. D. Searle Co., was testing peptides, short chains composed of amino acids, for the treatment of gastric ulcers. Rumor has it that by accident, he spilled a drop of one of his preparations on his hand. Later, when trying to pick up little pieces of paper, he absent-mindedly licked his finger tip and noticed a sugary-sweet taste (others claim it happened when he was lighting a cigarette in the lab, against all best practice rules!). Subsequent tests showed that the compound had 200 times the sweetening power of beet or cane sugar.

Aspartame, the new super sugar, is a peptide, a methyl ester of the two amino acids, **aspartate** and **phenylalanine**.

Phenylalanine and aspartate can both be produced by bacteria or enzymatically in bioreactors. They can be bonded either chemically or by the reverse reaction of proteases (such as trypsin) in a 2-phase system using organic solvents to form the peptide (see Ch. 2).

Although aspartame is broken down by digestive enzymes in the intestine, one gram of aspartame, the daily intake of an adult, only yields four kilocalories, not even a hundredth of the energy humans usually take up with sugar. It is not only low in calories, but tastes almost like sugar (except for a lack of body) and lacks the unpleasant aftertaste of its rivals saccharin and cyclamate.

The time was just right for aspartame, as the fitness movement swept the U.S. Pure aspartame or aspartame mixes (Fig. 4.35) are used in *diet* **products**. Its main drawback, however, is that it starts decomposing after six to nine months. The soft drink industry can live with that because 95% of their output does not remain in the shelves for longer than three months, as statistics tell us.

Currently, aspartame is more expensive than saccharin and also more expensive than enzyme-produced fructose (Ch. 2). If, however, it is possible to genetically engineer microbes to produce aspartame as a ready-to-use product or at least its two components, then aspartame could soon be ahead of its rival products. Due to its negligible nutritional value, there will be tough times ahead for **tooth-decaying bacteria**, such as *Streptococcus mutans* (Fig. 7.53). Whether this also applies to fat bulges in humans is questioned by many nutritionists. Diet drinks make the body believe that energy his being provided. As this remains an unfulfilled promise, the craving must be satisfied elsewhere. Some nutritionists even warn against the consumption of aspartame. Aspartame is anathema to some groups in the US because when it is digested, methanol develops, which can lead to blindness. However, in order to have an effect, absolutely enormous quantities of aspartame would have to be taken in.

Genetic engineering method

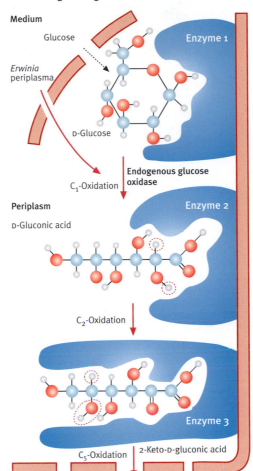
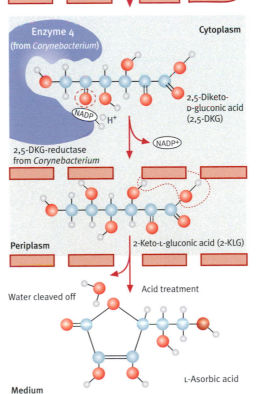

Fig. 4.31 Vitamin C synthesis in genetically modified *Erwinia* cells.

Fig. 4.32 *Erwinia* is a Gram-negative bacterium that causes dry and wet rot in plants by excreting pectinases into plant cells.

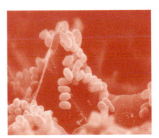

Fig. 4.33 *Corynebacterium* provides the enzyme 2,5-DKG reductase, which is expressed in the cytoplasm of *Erwinia* cells.

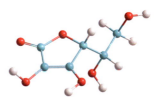

Box 4.5 Biotech History:
Vitamin C and Reichstein's Fly

Tadeusz Reichstein
(1897-1996)

Tadeusz Reichstein was an assistant of the ingenious synthetic chemist **Leopold Ruzicka** (1887-1976) at ETH Zürich. Ruzicka had been trying to sell his hormone-synthesizing method to Roche, but the rather unconventional and original Ruzicka and Roche manager **Barell**, who was a stickler for organization and discipline, "could not come to an agreement." (Roche Jubilee publication). So Ruzicka went to Ciba, Basel, instead. Reichstein had similar experiences with Roche in 1932.

French researcher Gabriel Bertrand (1867-1962) described an ingenious way of obtaining sorbose from *Drosophila* flies. It worked for Reichstein as well as for the author of this book in Hong Kong 120 years later.

50 years later, Reichstein recalls what happened:

"I do not remember the exact wording, but this is roughly what he wrote: Take wine, add a little sugar and vinegar and let it stand in a glass for a day. This mixture will attract swarms of little flies called *Drosophila* or fruit flies, which have bacteria in their intestines. Once the flies start sucking the liquid, they excrete these bacteria which then produce sorbose. When I was going to try this out, it was late in the year and I could not see any *Drosophila* around any more. However, the weather was fine and I could not wait for another year, so thought I'd give it a try. Instead of sugar, I added sorbitol to the wine, I added vinegar as instructed, but I also added some yeast bouillon because a bacteriologist had once said to me that yeast bouillon contains everything bacteria need. And after all, I wanted them to be well fed! I put out five beakers with this mixture on the window sill in front of my basement lab from where you could just catch a glimpse of the sun. That was on a Saturday, and

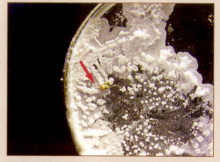

The sorbitol experiment, repeated by the author. The red arrow points to the fly.

I thought 'Well, if the flies come, it's fine, if they don't, nothing is lost.' When I came back on Monday, everything had dried up, but two of the beakers were full of crystals. We looked at these crystals, and they were pure sorbose! In one of the beakers, we also found a Drosophila that had drowned, and from it extended rays of sorbose crystals. Those wild bacteria managed to produce more sorbose in two days than the bought ones in six weeks. They had to be good! We then inoculated a second culture with them, and indeed, the rapidly proliferating bacteria managed to convert the substrate into sorbose within 24 to 48 hours. Thus, it took us only a few days to produce 50 grams of sorbose. Not surprisingly, our cultures contained a wild mix of bacteria, but that did not matter. They obviously liked an acidic environment, reaching their optimum productivity at pH 5. At this level of acidity, there is hardly any growth of other bacteria. Converting sorbose into vitamin C was very easy. We immediately obtained grams of it, and we could already predict that it would be possible to produce it by the ton.

I think we managed to get 30 to 40 grams of vitamin C out of 100 grams of glucose before even starting to fine-tune the synthetic process!

There was certainly room for improvement. Roche Zürich took over the license in spite of mutterings from the boss of the then fairly small firm. He was worried that the microbiological reaction might not be popular with his chemists, but Reichstein replied: 'Popular or not – this bacterium is the only lab assistant capable of achieving a 90% sorbose yield from sorbitol. No human can match that, while the bacterium does it in two days using nothing but air. The only thing you have to do is give it some yeast to eat.'

In its publication celebrating its centenary, the Roche company wrote: "After a long search, Hungarian researcher **A. Szent-Györgyi** had succeeded in isolating the anti-scurvy vitamin, and Roche was about to adopt his method of producing vitamin C from capsicum.

However, **Reichstein's** ingenious synthesis process turned out to be more promising for the future and started off the era of vitamin production at Roche. ..Barell still looked at vitamin C as a rarely used biochemical product of which a maximum of 10 kilograms per year could be sold and was taken by surprise when after a few months already, it turned out to be very much in demand."

Roche's vitamin C product Redoxon, launched in 1934, has undergone several improvements and is still sold today.

The worldwide production of aspartame was 14,000 tons, worth $850 million in 2004.

Two other, even sweeter, proteins have been discovered in African shrubs – **thaumatin and monellin** (Fig. 4.37). They consist of 208 and 94 amino acid components respectively and are supposed to be 2500 to 100,000 times as sweet as cane sugar. As the recovery of the compounds from plants is an expensive process, attempts are being made to produce them using genetically modified microbes.

Many cats love thaumatin, and there is a persistent rumor that thaumatin or certain amino acids may cause addiction in cats to certain food brands (Box 4.2).

4.12 Immobilized Cells Producing Amino Acids and Organic Acids

Japan is the most advanced country (Ch. 2) in the development of immobilized cells. **Ichiro Chibata** and **Tetsya Tosa** (Ch. 2, Fig. 2.23) of Tanabe Seiyaku, Osaka, who had already been at the forefront of enzyme immobilization, managed to utilize a process in 1973 to produce **aspartate**.

Deep frozen *Escherichia coli* cells, that were enclosed in a gel, synthesized 600 tons of aspartate per year (aspartate is one of the building blocks of aspartame) from fumaric acid. Only after 120 days did the activity of aspartase in *E. coli* drop to fifty percent, whereas free cells have a half-life of only ten days. Production with immobilized cells costs only 60% of the production cost with free cells. Thus, the catalyst cost goes down from 30% to 3%, and the cost for staff and energy is reduced by 15%. A 1000-liter column reactor yields approximately two tons (!) of L-aspartate per day.

In a similar process using immobilized *Brevibacterium cells*, 180 tons per year of **L-malate** were produced from fumaric acid.

Each of the processes uses just one of the microbial enzymes (aspartase or fumarase respectively), leaving the cells in their structural units for economic and practical reasons. Their cell membranes have been partially disrupted before use, so they are no longer living cells. The substrates are not held back by the cell membrane, but the enzymes remain stable.

Microorganisms are the obvious solution in **processes involving several steps**, for example, in alcohol production with immobilized yeasts.

There is enormous potential for immobilized cells, as their productivity surpasses that of free-living cells by far.

A process developed by Tanabe Seiaku in 1982 even had two different immobilized microorganisms switched in series. First, fumaric acid is converted into **L-aspartate** by immobilized *Escherichia coli* in a 1000-liter bioreactor, which is then converted by immobilized *Pseudomonas dacunhae* cells into **L-alanine** in a 2000-liter reactor. The pilot plant produced 100 tons L-aspartate and 10 tons of L-alanine per month.

4.13 Mutations as A Way of Targeting Microbial Programming

In order to obtain a tailor-made microorganism to be used in industrial production, undesirable traits of the wildtype must be eliminated, desirable properties enhanced, and perhaps even new properties added.

This can be achieved in several ways – by using **spontaneous mutations** or by artificially targeted mutations.

The simplest case is a **point mutation** where one base pair in the DNA is replaced by another, or a base pair or short DNA section may disappear or insert in a different place. Such changes happen naturally in DNA – probably faults in the copying process.

However, spontaneous mutations involving a specific base are very rare. In *E. coli*, the average mutation rate for a replicated base is 10^{-10}, i.e. one base out of ten billion! Most of these mutations remain silent or are repaired by enzymes (Box 4.6).

The mutation rate can be increased by a factor of 1000 in microorganisms that are **treated with mutagens**. These include ultraviolet (UV) radiation (e.g. causing two neighboring thymines in the DNA to form dimers, Box 4.4), ionizing radiation (X-ray, gamma or neutron rays), and a host of apparently unrelated chemical compounds that react with DNA bases (e.g. nitrous acid reacts with amino groups) or interfere with the copying process.

Fig. 4.34 How fashions can change! 100 years lie between these two advertising pictures. Top: The original drink, rich in calories, advertised in the US. Bottom: The lite product, as advertised in Asia.

Fig. 4.35 Next to the ingredients, the Diet Coke/Coca Cola Light bottle also carries a notice, warning phenylketonuria patients of aspartame.

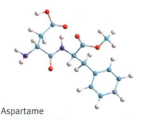

Aspartame

Fig. 4.36 Its sugar-like taste makes aspartame one of the most popular sweeteners.

Fig. 4.37 Thaumatin is obtained from the fruit of the African *ketemfe* plant.

Box 4.6 Why UV Light Kills Microbes

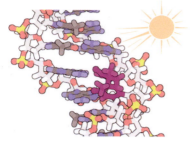

UV light is used to sterilize clean benches.

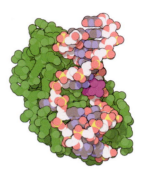

A DNA repair enzyme (green) finds the TT dimer (purple), corrects it and puts adenine in its pocket in the neighboring DNA.

UV radiation gives rise to a TT dimer (purple) in the DNA double helix, as a cyclobutane ring forms between the two thymine bases (top). The dimer causes a twist in the helix (shown at the bottom), thus interfering with the normal interaction between thymine and adenine and causing a slight kink in the DNA backbone.

All radiation of shorter wavelength (and higher energy) than visible light has a disruptive impact on organic molecules. The most serious damage to DNA is the formation of pyrimidine dimers. In the pyrimidine dimer, two adjacent pyrimidine bases – in most cases two thymine (T) residues – are linked in an abnormal structure which distorts the shape of the DNA double helix and blocks its replication. As little as one TT-dimer per cell can be lethal for the cell.

Luckily for us, these damages are usually repaired in seconds after they occur. Dozens of proteins cooperate in what is known as a nucleotide excision repair process. A 30-base pair-long segment is cut out and is replenished with the correct nucleotides. This is our only UV protection (apart from sun cream). An unsuccessful repair may lead to skin cancer. Microorganisms, on the other hand, use endonucleases that cut out the damaged base in a single step. Endonuclease V in the T4 bacteriophage handles the DNA it is "repairing" rather roughly.

Surprisingly, the enzyme does not recognize the dimer, but does note the weakening in the helix by the dimer. The enzyme puts a kink in the DNA on the side of the lesion and also excises the complementary adenine base and puts it in its pocket.

Fig. 4.38 Petri dishes with colonies of microorganisms. Screening for new antibiotics in microorganisms in Jena (Germany), the place where East German antibiotics production began after the war.

Some compounds are only turned into mutagens by liver enzymes such as cytochrome P-450, (e.g. aflatoxins in mold fungi, Fig. 4.10). They turn into highly reactive epoxides that react with guanine to form stable compounds.

Cancer in humans and mammals arises from mutations in genes that are involved in growth control and from faults in DNA repair (in many kinds of intestinal cancer).

It is generally impossible to produce a targeted mutation of just one gene. In order to enhance a microbial strain by mutation, very sensitive screening tests are required to spot desired accidental mutations.

How microorganisms can be **programmed by mutations** to function in a desired way has already been shown in Sect. 4.7, Production of Amino Acids. Through the combined use of natural mutations and targeted selection, a bacterial strain was found that produces large amounts of the vital amino acid lysine.

The situation is entirely different for **antibiotic-forming fungi and bacteria**. The amount of antibiotics produced depends on dozens of genes, and it is therefore impossible to find single mutations that could switch the poor yield of a few milligrams per liter to an industrial-scale yield. It took many mutation and selection cycles to obtain the highly developed industrial strains that now produce 20 or more grams of antibiotics per liter of nutrient medium (Box 4.6).

Cultures are treated with a **mutagen** in every cycle, and the resulting colonies are tested. When among the thousands of colonies a mutant has been found that shows significantly higher productivity, it becomes the starting point for the next cycle of mutation and selection. Thus, the laboratory evolution of the organism takes an unnatural direction until a strain has developed that is economically viable. It is a painfully slow and labor-intensive method with no guaranteed results.

Not only genes, but also **culturing conditions** have an impact on the yield of an antibiotic. At the beginning of the process, suitable mutants can be found by measuring the output of the various colonies growing on the nutrient plates. Later, however, the strains that have been enhanced in the lab must be tested under conditions similar to those in a huge bioreactor. In spite of all these difficulties, several antibiotics, such as penicillin are currently produced by highly productive strains that have

Box 4.7 Screening, Mutagenesis, and Selection – Prerequisites for the Creation of Powerful Antibiotic Production

In a first step, a promising habitat is chosen and samples taken.

Since a cubic centimeter of soil contains millions of microbes, the samples must be diluted with water. Diluted samples of the varying strengths are distributed to Petri dishes that contain nutrients. These are incubated at 25°C or 37°C for six days until the microorganisms have formed small, clearly distinguishable colonies. Those colonies that produce antibiotics will secrete inhibitors into their environment. When test bacteria (e.g. streptococci) are sprayed onto the capturing plates, they grow into dense lawns, except around antibiotic-producing colonies which are surrounded by a clear zone.

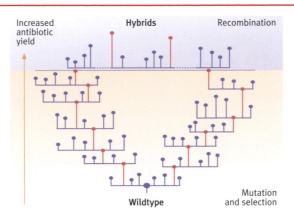

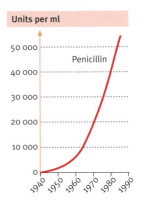

Selection of the best daughter colonies (shown in red). Right: The graph shows the increase in penicillin through mutation, selection, and improved fermentation from the beginnings into the 1990s.

Visual screening for dead zones around antibiotic-producing microbes

A thin platinum wire is disinfected by heating it red-hot and then used to take a sample from the antibiotic-producing colony. The microbes are deposited in a line on fresh nutrient medium. After a short incubation, test microbes (e.g. staphylococci, streptococci, *Escherichia coli*, or *Candida* yeasts) are put in lines perpendicular to the first line. Those test strains that are sensitive to the new antibiotic will not grow near the first inoculation line, while resistant strains will not be affected.

The newly found antibiotic-producing microbe is then cultured, and when the culture has grown sufficiently, the pure antibiotic must be isolated from it to test it on animals.

The antibiotic yield can be enhanced through mutagenesis and subsequent screening. The most productive daughter colonies are chosen and exposed to another mutagen, then tested for their productivity. After several cycles of mutation and screening, the best mutants are crossbred. The recombination of genes produces thousands of genetic variants, many of which will achieve an even higher yield of antibiotics. An excellent example of successful screening, mutagenesis, and selection or "evolution in a Petri dish" is penicillin. Current industrial strains of *Penicillium chrysogenum* produce 150,000 units per milliliter (about 90 g per liter), compared to the four units per milliliter yield of Fleming's original fungus.

Whereas in the early 1940s, the stock of filamentous fungi comprised about 1200 strains, the US Department of Agriculture (USDA) Culture Collection in Peoria, Illinois, now contains over 80,000 microbial strains. Towards the end of the 1990s, microbiologist **Stephen W. Peterson** discovered 39 new *Penicillium* species in addition to the 102 known species that include the famous melon strain.

Plated microorganisms in Petri dishes.

been developed in 20 or 30 selection steps over two or more decades.

■ 4.14 *Penicillium notatum* – Alexander Fleming's Miraculous Fungus

Most of us do not remember the times when doctors were powerless in the face of severe bacterial infections. Bacterial endocarditis was almost invariably fatal, and meningitis caused by meningococci left those few who survived it with severe mental disability. Pneumonia caused by pneumococci was known as "the old man's friend" because it let old people slip away gently.

Against this background, **penicillin** with its high effectiveness on a wide range of pathogenic bacteria and almost negligible toxicity must have seemed to be a wonder drug. In 1928, **Alexander Fleming** (1881–1955), whom we met earlier in the context of the **lysozyme** discovery (Ch. 2), found that *Penicillium notatum* inhibits growth of certain bacterial cultures (Fig. 4.41).

Penicillin ushered in a new era in the fight against disease. Box 4.6 tells the story of this discovery that Fleming himself had initially underestimated, and

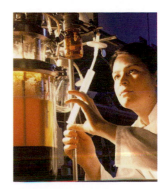

Fig. 4.39 Bioreactor culturing conditions are specifically optimized for the various microorganisms in the lab.

Box 4.8 Biotech History:
Alexander Fleming, Penicillin and the Beginnings of the Antibiotics Industry

In the autumn of 1928, microbiologist **Alexander Fleming** (1881-1955) was examining several cultures of pus bacteria (staphylococci) in his lab at St Mary's Hospital London. The lab was stuffed with Petri dishes that had bacteria growing on agar nutrient media – a bit untidy perhaps, but sometimes genius comes from chaos.

Before his summer holidays, Fleming had inoculated some dishes with bacteria. All of them were now covered in clearly visible colonies, and in some of them, some mold was growing. The summer had been cool, which had slowed down the growth of bacteria.

Fleming's colleague **Melvin Pryce** witnessed one of the finest hours in science. It is described in **André Maurois'** book "The Life of Sir Alexander Fleming".

While talking, Fleming lifted up a few dishes containing old cultures and took the lids off. "'As soon as you uncover a culture dish, something tiresome is sure to happen. Things fall out of the air.' Suddenly, he stopped talking, then, after a moment's observation, said, in his usual unconcerned tones: 'That's funny…'; On the culture at which he was looking there was a growth of mould, as on several of the others, but on this particular one, all round the mould, the colonies of staphylococci had been dissolved and, instead of forming opaque yellow masses, looked like drops of dew…" Apparently, the mold prevented the growth of the staphylococci.

"…Noticing the keen interest with which Fleming was examining the phenomenon, (Pryce) said: 'That's how you discovered lysozyme.' Fleming made no answer. He was busy taking a little piece of the mould with his scalpel, and putting it in a tube of broth."

"Fleming put the Petri dish aside. He was to keep it as a precious treasure for the rest of his life. He showed it to one of his colleagues: 'Take a look at that' he said, 'it's interesting – the kind of thing I like; it may well turn out to be important.' The colleague in question looked at the dish, then handed it back with a polite: Yes, very interesting.'"

Over the following days, Fleming grew the fungus on broths that had been made microbe-free by heat. He then placed various kinds of Gram-positive bacteria around the fungus, including chain-forming streptococci, clusters of staphylococci, and pneumococci. Sure enough, none of them could spread to the immediate neighborhood of the fungus. Gram-negative bacteria, however, such as *Escherichia coli* and *Salmonella* species, grew unperturbed. Fleming identified "his" mold as a member of the *Penicillium* family, *Penicillium notatum*, to be precise.

He began to grow the fungus in a bigger broth container. The top of it was soon covered by a greenish mold layer, similar to a lawn. The liquid turned golden yellow after a few days, and new experiments with bacteria showed that the broth alone could slow down their growth. The fungus was apparently secreting something bactericidal or bacteriostatic into its environment. Fleming called it penicillin after its origin.

Ernst Boris Chain (1906-1979)

Howard Florey (1898-1998), according to his Australian biographer, is the only Australian ever who did something useful for the whole of mankind.

Streptococci, staphylococci, anthrax, diphtheria, glandular fever, and tetanus could all be stopped in their tracks by penicillin.

Little did Fleming know what a breakthrough this meant and that it would save the lives of millions. Although further experiments showed that penicillin attacked only bacteria, but not rabbits, Fleming never tried to obtain pure penicillin to treat bacterial pathogens in lab animals.

Hardly anybody noticed his article in the *British Journal of Experimental Pathology*, and even in 1940, Fleming wrote that it would not be worthwhile to produce penicillin. It seems that his main interest was in the selective effect on various bacterial species, as it helped to classify the species. By that time, however, other scientists had become aware of its bacteria-inhibiting properties.

In 1938, **Ernst Boris Chain** (1906-1979) in Oxford became interested in the fungus. The outbreak of the Second World War led to a surge in demand for medication to treat bacterially infected wounds in soldiers. Under the supervision of Australian researcher **Howard Florey** (1898-1998), he and colleague Norman Heatley worked frantically, separating penicillin from other compounds in the nutrient broth and purifying it. The yellow powder was then tested on mice that had been infected with pathogenic bacteria.

The mice recovered within a very short time, which was sensational. The UK and US governments began to support the efforts to produce sufficient amounts of penicillin, but kept the project top secret because of its military relevance.

In the summer of 1941 when Germany was expected to attack Great Britain, rumor has it that Florey and his colleagues had decided to destroy their lab completely, should the enemy land on British shores. The only exception was the miraculous penicillin fungus which the researchers rubbed into their clothes in order to be able to start new cultures elsewhere.

1941 was the year that saw penicillin tried out on a patient for the first time. He had a life-threatening staphylococcal and streptococcal infection. The patient, who first seemed to recover, died a month later, as the amount of available penicillin was insufficient, although it had been recycled from the patient's urine, brought to the lab every day by Florey's wife. Florey and Chain had to produce larger quantities of penicillin before they could cure their first patients.

In July 1941, **Howard Florey** and **Norman Heatley** began a collaboration with the USDA in Peoria, Illinois, and soon, companies such as Merck, Squibb, Lilly, and Pfizer joined in, only a few months before the US entered the war.

When Heatley and Florey arrived in the US in 1941, the amount of penicillin they had been able to produce was nothing to shout about – four units per milliliter (1 unit = 0.6 micrograms). The National Academy of Science was consulted and recommended Charles Thom, an expert on *Penicillium*, who, in turn, pointed them to the fermentation unit of the newly founded Northern Regional Research Laboratory (NRRL) of the USDA in Peoria, Illinois. Charles Thom (1872-1956) had been the first scientist to describe *Penicillium roqueforti* and *P. camemberti* as the active fungi in cheese production (see Ch. 1).

Norman Heatley, the quiet, pragmatic hero in the penicillin success story. On March 25th 1940, Heatley injected eight mice with 110 million streptococci each. Half of them were injected with penicillin an hour later and survived. Florey, his boss, slightly revolted at the sight of those mice, commented: 'Looks quite promising'
In 1990, Heatley was not awarded the Nobel Prize, but something far rarer – the first honorary doctorate in Oxford University's 800-year history.

The problem with Fleming's *Penicillium* strain was that it would only grow on the surface of its nutrient (surface culture). So, the search, which also involved the US army, began for a more productive *P. chrysogenum* strain that could be grown in a submerged culture.

Reports cannot agree whether it was a humble housewife in Peoria or staff member **Mary Hunt**, known as Moldy Mary, who found the moldy melon in the market that carried the *Penicillium chrysogenum* strain. Grown in a submerged culture, it would produce 70/80 units per milliliter, and some mutants isolated from some conidia even produced 250 units per milliliter. *Penicillium* chrysogenum NRRL 1951, isolated from a Cantaloupe melon, was to become the ancestor of modern Penicillium strains.

The surface culture was soon replaced by an especially developed deep tank culture system that was suitable for mass production. The antibiotic industry was born. At the end of November 1941, **Andrew J. Moyer**, expert in the nutrition of molds, cooperated with Norman Heatley and succeeded in multiplying the output tenfold by using corn steep liquor as a nutrient. Corn was the dominant crop in US agriculture (see also Ch. 3 on fructose syrup).

One of the reasons for founding the lab in Peoria had been to find ways of disposing of liquid corn production waste. It comes in large quantities and contains a mixture of starch, sugars, and minerals.

It was later discovered that apart from sugars, it also contains a chemical precursor of penicillin, thus making life easier for the *Penicillium* fungus.

The US War Production Board initiated projects all over the US, e.g. at the University of Wisconsin Madison where **J.F. Stauffer** and **Myron Backus** tested thousands of UV-induced mutants (see Box 4.5).

The lab in Peoria.

Charles Thom (1872 – 1956) classified *Penicillium* strains.

Backus and Stauffer raised the production level from 250 to 900 units/mL, and through further mutagenesis, production reached 2500 units/mL. Other universities such as Stanford, Minnesota, and the Carnegie Institution in Cold Spring Harbor, N.Y. joined in, and by the end of 1942, 17 US companies were involved in the penicillin project.

On March 1st, 1944, the first industrial plant with submerged cultures was opened in Brooklyn, NY. The production went up from 210 million units in 1943 to 1663 billion units in 1944 and 6.8 trillion units in 1945.

In 1943, it was possible to treat 1500 military personnel, and only one year later, countless wounded in the D-day landings were saved by penicillin. The yield had been increased from 1% in 1 liter flasks to 80-90% in 10,000 gallon tanks.

Penicillin production in Oxford 1940.

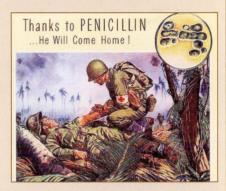

From May 15th 1945, penicillin was freely available in every drugstore in the US.

Fleming, Florey, and Chain were awarded the Nobel Prize in 1945. Until this day, the British regret that, for ethical reasons, they had asked Florey not to file for a patent on penicillin.

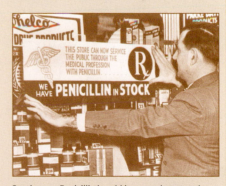

Good news: Penicillin is sold in every drugstore in the US in 1945.

The University of Oxford never got its share from the fabulous profits made from penicillin in the US, and, to add insult to injury, the UK had to pay licensing fees to US companies.

Fleming definitely had a nose for lysozyme!

Fig. 4.40 Alexander Fleming and his lab note about the Petri dish with the penicillin-lyzed bacteria.

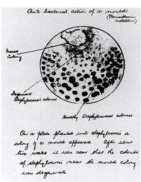

Fig. 4.41 The first really productive penicillin mold was found on a cantaloupe melon. Experiment repeated and photographed by the author.

of the contributions by **Florey**, **Heatley**, **Chain**, and others. The recoveries from bacterial infection as a consequence of penicillin treatment came close to a miracle. However, the production of penicillin was still too laborious and too expensive. For the treatment of a single patient, about 265 gallons (1000 liters) of "fungus soup" had to be prepared and processed (Box 4.7).

Three problems were waiting to be resolved:

- The *Penicillium* strain with the highest penicillin output had to be found (Sect. 4.15).
- Huge quantities of the fungus had to be grown.
- A method had to be developed to separate pure penicillin from the nutrient medium (Sect. 4.17).

■ 4.15 Screening – Biotechnologists In Search of Molds

All over the world, the search was on for fungi that would produce more penicillin than Fleming's *Penicillium notatum*. The fungi that were found were cultured on media and tested for their ability to produce penicillin. Screening programs were developed to pinpoint suitable microorganisms (Box 4.5).

Significant progress was only made after the US government supported this war-relevant research in the U.S. Department of Agriculture (USDA) labs in Peoria, Illinois. However, it was not until 1943 that a more productive strain than that discovered by Fleming was found.

The mold, later named *Penicillium chrysogenum*, was found nowhere else but on the doorstep of the Peoria lab, i.e., on a cantaloupe melon purchased in a Peoria fruit market (Fig. 4.42). Its productivity was so amazing that it became the major culture line, and all molds used in industrial penicillin production derive from the melon mold found in Peoria.

Just as in animal and plant breeding, the most productive mold varieties were propagated, and over many generations, the current high-performance strains were developed.

As early as the 1920s, researchers found that hereditary traits can be manipulated. X-rays and certain chemicals bring about a higher rate of **mutations** in cells. In order to obtain the most productive penicillin-producing fungi, researchers treated them with a combination of X-ray, ultraviolet radiation,

and chemicals (dichloroethyl sulfide, also known as mustard gas). The screening process ran through about 20 cycles (Box 4.5), and the best mutants yielded as much as seven grams per liter.

Modern high-performance fungi, however, differ considerably from their ancestors in the early 1950s, producing on average 100 grams per liter nutrient solution, which is 2,000 times more than the fungus on the melon produced or 20,000 times more than the output of Fleming's fungus. Groups of microbes of one species with widely differing properties are called strains. The majority of high-performance strains are usually so sensitive and pampered by their breeders that just like many of our pets, they would be unable to survive in the wild.

High-performing supermolds do not gain anything from their ability to produce a thousand times more penicillin than their ancestors. It is humans who force them into overproduction by manipulating their genetic makeup and living conditions. Just like pets these days, they live in an artificial environment, enjoying comfortable temperatures, sufficient and suitable food, and protection from competitors and predators. They exist to produce.

■ 4.16 What's On The Microbial Menu?

What does a microorganism need to be comfortable? First of all, nutrients. Their favorite "foods" are simple, easily digestible organic compounds such as glucose, fatty acids, and amino acids. Microorganisms are usually unable to break down more complex compounds like starch, cellulose, pectins, and proteins within the cell. Their only way to digest them is via **extracellular enzymes** (Ch. 2). To break down starch into glucose, for example, the cells secrete amylases into their environment (the medium) in what could be considered an external digestion process.

In the medium, **amylases** (Fig. 4.44) break down starch until the microbes are surrounded by glucose molecules which they can easily ingest. To digest protein, the cells secrete protein-cleaving proteases and to break down cellulose, they secrete cellulases into the medium (Ch. 2).

Not only glucose, but also starch and beet sugar are used in industry. The starch comes from grain, the sugar from corn or molasses. Molasses is a dark and

Box 4.9 How Penicillin Works – Enzyme Inhibitors as Molecular Spoilsports

Although the development of penicillin was a very important era in the history of the modern biotech industry, it was not understood for a long time why penicillin kills certain microbes.

Now we know that penicillin is taken up with other nutrients by growing bacteria. When they divide, bacteria must build new cell walls with the help of specific wall-building enzymes that interlace amino acid side chains, and it is this process that is disrupted by penicillin. It acts as a molecular spoil-sport, i.e. penicillin is structurally so similar to the enzyme's substrate that it binds to the enzyme, thus preventing the conversion of the actual substrate to a building block of the cell wall. This is called competitive inhibition.

The mechanism was only brought to light in 1957 by **Joshua Lederberg** (b. 1925) who was later to win the Nobel Prize. In a simple experiment, he demonstrated that penicillin prevents the formation of bacterial cell walls. In a highly saline solution, the bacteria continued growing in spite of the presence of penicillin, but they did not form cell walls. When Lederberg transferred those protoplasts, i.e. naked bacteria without a cell wall, into a normal nutrient solution, they burst because they did not have a protective cell wall.

In 1965, **James Park** and **Jack Strominger** gave a precise description of the action of penicillin. The bacterial cell wall consists of long sugar chains (murein) that are interlaced with peptide chains (peptidoglycans). This bonding process is catalyzed by the enzyme **glycopeptide transpeptidase**. Penicillin blocks this enzyme by posing as a perfect imitation of a protein bridge, thus putting an end to the bacterial cell wall production. This has a devastating effect on the bacteria, as the cell walls become leaky and cannot withstand osmotic pressure any longer. They inflate and burst.

Thus, penicillin only inhibits the proliferation of bacteria, but does not kill fully grown bacteria, and this is why it is not enough to keep taking it just as long as one is feeling ill. As penicillin only begins to act when the bacteria undergo division, it must be taken over several days. Stopping a course of antibiotic treatment in the middle, against medical advice, is extremely risky and foolhardy, as the surviving microbes will recover quickly and multiply, causing a dangerous relapse in the patient. Some of the surviving microbes may even have become resistant to penicillin, raising the stakes in the race between humans and microbes.

Action of antibiotics containing a lactam ring. Being structurally similar to the peptide chain that should interlace with murein chains, they bind to them instead and prevent interlacing.

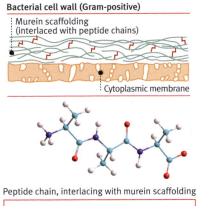

Bacterial cell wall (Gram-positive)
Murein scaffolding (interlaced with peptide chains)
Cytoplasmic membrane

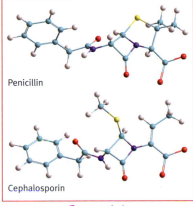

Peptide chain, interlacing with murein scaffolding

Penicillin

Cephalosporin

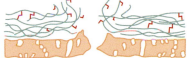

Cell wall bursts as a result of penicillin action

---

viscous byproduct of sugar production, containing 50 % beet sugar.

To make starch water-soluble, it is boiled and degraded into sugar with the help of microbially produced amylases. Other nutrients, such as **amino acids**, are derived from soy flour, corn steep liquor, or draff (malt residue) from breweries.

Finally, microorganisms also need minerals (ammonium salts, nitrates, phosphates) to thrive, and, depending on the species, more or less oxygen. Some microbes such as methane-producing species must even be protected from oxygen, as it is toxic to them.

When the first mold cultures were grown during the Second World War, nutrients that contained glucose as well as minerals were used.

The fungi used up the sugar and grew fast, but produced only little penicillin, until somebody in the Peoria lab hit on the ideal nutrient additive – corn steep liquor. It is cheap and encourages penicillin production. Ironically, a special department had been set up in Peoria to find environmentally compatible ways of disposing of this liquid.

As for *Penicillium*, a nutrient that caters to their specific needs had to be found for other microbes as well, and it had to be as cheap as possible. Basically all sugar-containing byproducts, e.g. from agriculture or from cellulose factories, can be fed to microbes.

Some microorganisms even have the ability to break down compounds that are hard to digest – such as crude oil residues and plastic waste. These can thus be put to a useful purpose instead of posing a threat to the environment. Not only are they cheap, but it could also be very expensive to dispose of them in any other way.

Biotechnology not only enables the use of otherwise wasted food resources, but also helps to keep the environment clean.

Fig. 4.42 Modern penicillin production in a submerged culture.

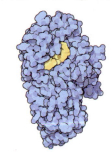

Fig. 4.43 Extracellular α-amylase degrades starch in the medium (here, 5 glucose units out of thousands have been bound to the active site).

Box 4.10 Biotech History: Preserving Techniques – Heat, Frost, and the Exclusion of Air

Bottling food to preserve it had been invented by French chef **Nicolas Appert** – without any awareness of the existence of microbes and fifty years before Pasteur discovered heat sterilization!

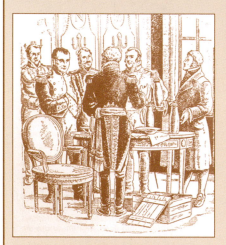

Chef Nicolas Appert (1749 – 1841), founder of the food preserves industry, meeting Napoleon.

In 1795, **Napoleon** offered a prize to the person who could come up with a practical method of long-term food preservation so it could be taken on army campaigns. Appert worked on it for 14 years, heating food in bottles and jars, making air-tight seals with cork. The working principle of vacuum sealing had been pioneered by **Otto von Guericke** (1602 - 1686) and **Denis Papin** (1647 - 1712) with their experiments.

Advertising for Weck bottling equipment (text: unequalled, unrivalled, unmatched in preserving food) and a Weck jar containing pineapple from 1897.

Otto von Guericke became famous by his experiments with the Magdeburg hemispheres, in which he demonstrated the enormous power of air pressure to a baffled audience at the Reichstag in Regensburg in 1654. He created a vacuum by pumping out the air in two hemispheres he had put together and then used up to sixteen horses to try and wrench them apart, and they failed! Little did he know that his discovery would be essential in vacuum-sealing food preserves.

Appert's preserves would keep for months, and they were also tested by the French navy. In 1809, Appert received the prize of 12,000 gold francs – a fortune worth about $315,000 in today's money, according to historians. He publicized his method and thus became the founder of the modern food preserving industry.

The only problem that remained was glass breakage. This was resolved by British industrialist

Peter Durand in 1810. He produced steel containers lined with tin. The meat preserved in those cans was called embalmed meat.

Sir John Franklin: did he and his men die from the lead of tin cans?

On May 19th 1845, an expedition set off from London with 134 of the best British officers and crew of the British navy, under the command of **Sir John Franklin**. They were hoping to explore the Northwest Passage. Never before had an expedition been so luxuriously equipped: a hot water boiler for heating, a steam-powered propeller, iron ice shields and cans and heating materials for three years. Who would have thought that none of those who took part in the expedition would be seen alive again? One of the major mysteries that remained was why such an experienced crew was unable to rescue themselves. In 1986, three of the sailors buried in the ice were carefully examined, and it turned out that what killed Franklin's crew was not, as had been surmised, hunger, cold, or scurvy, but massive **lead poisoning**. The lead had come from

Fig. 4.44 Autoclaves sterilize media by pressurized steam.

■ 4.17 A Modern Biofactory

In a modern biotech factory, there are no smoking chimneys, but bright, tiled halls with huge upright stainless steel containers surrounded by a maze of pipes, valves and monitors, and outside, there are more steel tanks, the size, of furnaces. These are the **bioreactors** or **fermenters**.

Modern bioreactors (Box 4.10) are technological miracles, the result of decades of research. Their development was spurred on by the hunt for penicillin. When Florey, Chain, and Heatley were looking for containers to grow molds in, they started off with small flat glass dishes.

With the fungi floating on the surface of the nutrient solution (emersed culture), it was not possible to produce enough penicillin for the treatment of

patients. The flat glass dishes also took up a lot of space. If it were possible to grow the fungus not just on the surface, but throughout the whole medium, i.e. submerged in the liquid, it would be far easier to grow them as well as save space.

While Fleming's *Penicillium notatum* wildtype strain could replicate only on the surface of a liquid, *Penicillium chrysogenum*, the new Peoria type, thrived under water, as long as there was a good supply of oxygen. Pumping in oxygen as one would into an aquarium, **Andrew Moyer** succeeded in growing *Penicillium chrysogenum* in **submerged culture** (see Box 4.6).

Nowadays, penicillin is mainly produced in 100,000 to 200,000 liter bioreactors. In laboratory and small pilot plant conditions, bioreactors of a few liters capacity are sufficient for the study of microor-

the canned preserves – a then still fairly recent invention.

The ships carried 8,000 cans which had been soldered with lead. This must have been done under enormous time pressure, so the solder was applied to the inside of the cans and not sealed. Large amounts of extremely toxic lead thus seeped into the food and were ingested with it. Heavy metals destroy the disulfide bridges in proteins and act as enzyme inhibitors (see Chapter 3).

Symptoms of lead poisoning are loss of weight, tiredness, irritability, paranoia, loss of concentration, inability to make decisions. The tissue samples taken from the three sailors showed a tenfold lead overdose. However, the lead hypothesis is only one of many attempts to explain the disastrous end of the expedition.

Johann Weck

Domestic users clearly could not go down the tin can route, as they needed re-usable, clean containers. Breakage was not the huge problem it would have been on an industrial scale, so glass containers were still the vessels of choice.

The original cork-and-seal method was replaced by glass or metal rings with rubber rings.

Georg van Eyck

In 1842, **John Kilner** opened a production plant for suitable jars in Yorkshire, England, but failed to file a patent on his jars. Other firms picked up the idea, and the firm went bankrupt in 1937. **John L. Mason** in New York filed a patent for his jars in 1858. Both types of jars are still in use, as are the Weck jars, popular in continental Europe and the US. They were based on a method invented by **Rudolf Rempel** (1859 - 1893) and patented in 1892.

Johann Weck and **Georg van Eyck** bought the patent from him and founded the Weck company in 1900.

About 100 years after Franklin's doomed expedition, US naturalist **Clarence Birdseye** (1886 - 1956) went on an expedition to Labrador for the U.S. Geographic Service and noticed that fish and caribou meat that had been exposed to the icy artic air still tasted fresh when cooked months later. Coming to the conclusion

that rapid freezing to extremely low temperatures must be the secret to this freshness, he went home to develop a multi-plate quick freeze machine which he patented.

This machine could freeze meat instantly to -40 °F/°C. In 1925, he produced the first frozen fish fillets, and soon, other food products followed. After some initial problems, frozen food became a staple American food.

When, during the Second World War, there was a shortage of tin, due to the initial control of the Pacific by the Japanese, all available tin cans in the US went into army provisions. These were ideal starting conditions for Birdseye's frozen food, as it did not require any tin.

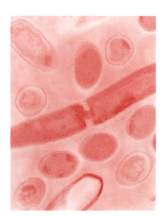

More Weck company advertising.

ganisms. In a next stage, scientists and engineers have to cooperate to ensure that a promising biotechnological process also works on a 10,000 times larger industrial scale. This is called the scaling-up process and can prove quite tricky (Box 4.10).

The right **temperature** of the nutrient medium is another crucial aspect of optimizing living conditions for microbes. The temperature at which most microbes will thrive lies between 68 and 122°F (20 - 50°C) – pretty much the temperate to tropical climate range. This is why biotech factories can do without chimneys, whereas the production of chemical compounds often requires temperatures of several hundred degrees Celsius.

By contrast, many biotechnological processes even require **cooling**, which sometimes is more expensive than heating. Just like people cooped up in a small room, molds and other cells in a bioreactor give off heat during their metabolic processes. In order to prevent fatal overheating, the outer walls of bioreactors must be cooled with water.

■ 4.18 Heat, Cold, and Dry Conditions Keep Microbes at Bay

Invisible foes threaten the existence of bioreactor cultures throughout their lifetime (Box 4.10).

What use is the best penicillin-producing strain if unwanted microbes in the nutrient medium gobble up the nutrients, inhibit the fungal growth, or secrete toxins into the medium? All nutrients and the air that is pumped in must therefore be briefly heated to become microbe-free.

Fig. 4.45 Microorganisms (here bacteria) require specific nutrient mixes.

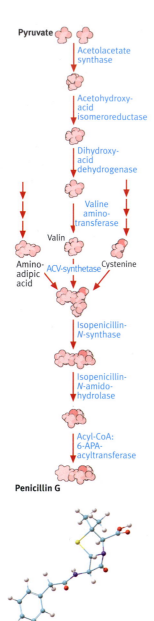

Pyruvate

Acetolacetate synthase

Acetohydroxy-acid isomeroreductase

Dihydroxy-acid dehydrogenase

Valine amino-transferase

Valin

Amino-adipic acid   ACV-synthetase   Cystenine

Isopenicillin-*N*-synthase

Isopenicillin-*N*-amido-hydrolase

Acyl-CoA: 6-APA-acyltransferase

Penicillin G

Fig. 4.46 The biosynthesis of penicillin from pyruvate in a cell.

Fig. 4.47 Chinese *Dim sum* are steamed in bamboo baskets, which is an effective way of killing microbes.

The bioreactors are also **sterilized using steam** before any cultures are grown in them. There is a second strategy in place to avoid contamination – as long as the **pressure** inside the vessel is slightly higher than the outside pressure, it is difficult for any germs to penetrate inside. The contamination risk is highest around inlets and outlets. In order to keep them safe, hot steam is blown into pipe openings next to the valves.

Heat is also used in the home (Fig. 4.47) and in the food industry to kill germs, i.e. noxious microbes. Think of boiling or pasteurizing milk, bottling fruit or preparing canned food – in all these processes, heat is used to kill bacteria and a large amount of fungal spores (Box 4.10). As most containers are tightly sealed, microbes and oxygen, which most microorganisms need to thrive, are kept out. The older generation remembers incidents with bottled fruit when the seal was not airtight and the contents became moldy.

Apart from extreme heat, **cold** (Fig. 4.49) can also be used to stop unwanted growth of microorganisms. As most microbes require warmth to grow, it makes sense to keep food in the refrigerator or freeze it. This is only a temporary measure and does not kill the microbes. Many microorganisms survive temperatures as low as -196 °C, as in liquid nitrogen. Food that has been thawed once must therefore be used up immediately, so as not to become a breeding ground for microbes brought back from the cold.

Microbes also need **water** to survive, and that is why drying can inhibit their growth (prunes, dried meat and sausages, stockfish). Water is also extracted from microbes by osmosis through pickling (herrings) or the use of high-sugar solutions (syrup). The microbes shrivel up and are unable to grow any further. **Freeze-drying** has become one of the most popular modern preservation techniques (Box 4.10). Concentrated alcohol works as a preservative not only by making bacterial cell membranes permeable (see Ch. 1), but also by dehydrating microbial cells.

Finally, a wide range of microbe-inhibiting disinfectants has been developed.

4.19 Downstream Processing

Once microbial control has been dealt with and the process in the bioreactor has been completed suc-

cessfully, it is harvest time. The mix of mold, left over nutrients, and penicillin is drained from the container. As the fungi have been secreting penicillin into the medium, obtaining the product is a straightforward process. The microbial cells are filtered off, and from the clear nutrient solution, the dissolved penicillin is precipitated. The resulting crystals can be easily separated.

In analogy to the substrate preparation process for the bioreactor, known as **upstream processing**, the steps leading from the fermenting solution to the end product are called **downstream processing**.

However, most products (e.g. many proteins) are not simply secreted into the nutrient solution, and in order to obtain the product, the microbial cells must be disrupted and the product separated from the rest of the cell content. This can be quite a laborious process, perhaps involving affinity chromatography (see Ch. 3) and adding to the cost of the procedure.

4.20 Streptomycin and Cephalosporins – The Next Generation of Antibiotics

Penicillin is very effective against a broad range of Gram-positive bacteria. When it was introduced into clinical practice, it led to a rapid and complete recovery from many bacterial infections, such as pharyngitis caused by streptococci, pneumonia caused by pneumococci, and most staphylococcal infections. It also cured the often fatal meningococcal meningitis and some forms of fatal bacterial endocarditis. Such spectacular clinical success triggered an intense **search for other natural antibiotics**.

There were two motives behind these efforts: Firstly, **penicillin is not effective on Gram-negative bacteria** such as *Escherichia coli*, *Salmonella*, *Pseudomonas*, and *Mycobacterium*. Secondly, it turned out that even some Gram-positive bacteria are or can become resistant to penicillin.

Selman Abraham Waksman (1888 - 1973) (Fig. 4.50) and his collaborators at Rutgers University developed a new technique that allowed the routine screening of the metabolic products of soil-inhabiting microorganisms to find antibiotic compounds. They isolated a new antibiotic produced by actinomycetes of the *Streptomyces* genus. Actino-

Box 4.11 Biotechnologically Produced Antibiotics – Sites and Modes of Action

Almost 12,000 antibiotics have been isolated from microorganisms and a further 7,500 from higher organisms.

Antibiotics can be roughly grouped into three main categories, according to their mode of action:

1. Cell Wall Inhibitors

The **β-lactam antibiotics** (**penicillins, cephalosporins** and their derivatives), featuring a four-membered lactam ring, are the most important antibiotics with a production of over 60,000 tons per year. They prevent peptide interlacing in bacterial cell walls. The main representatives of this group are penicillin G and cephalosporin C which form the bases of semi-synthetic antibiotics. The cell membranes of yeasts and other fungi – not bacteria – are mainly disrupted by **polyenes** produced by streptomycetes. Some of them, nystatin and amphotericin B are used to fight *Candida* infections, whereas pimaricin is used in cheese pro-

duction. The glycopeptide vancomycin, derived from *Amycolatopsis orientalis*, inhibits the synthesis of bacterial cell walls and is used as a last resort to combat resistant *Staphylococcus aureus* infections.

2. Protein Synthesis Inhibitors

Tetracyclines are broad-spectrum antibiotics and the second most used group of antibiotics. Their annual sales exceed one billion dollars. They bind to the 30 S-subunit of ribosomes (see Chapter 3) and are exclusively produced by streptomycetes. **Anthracyclines** such as doxorubicin (adriamycin) inhibit DNA replication by binding to the furrows in the helix and inhibiting topoisomerases (topoisomerases modify the spiralling of the DNA helix, thus affecting replication, transcription, and recombination).

Chloramphenicol from *Streptomyces venezuelae* is historically the oldest broad-spectrum antibiotic, blocking peptidyl transferase in ribosomes. **Griseofulvin** is a fungistatic antibiotic, inhibiting cell division and the spindle apparatus in fungi. It is used to treat fungal skin infections and to protect plants from mildew.

3. DNA Inhibitors

Macrolide antibiotics inhibit Gram-positive bacteria by binding to the 50 S- subunits in ribosomes, thus interrupting the growth of polypeptides. Erythromycin and spiramycin are used in respiratory tract infections. The related tylosin (effective against mycoplasma) is used by pig farmers in China, but banned in the EU. Its market value is $2.6 billion.

Ansamycins such as rifampicin, a semisynthetic derivative, are the most important antibiotics in

the fight against tuberculosis and leprosy. They inhibit bacterial RNA polymerase, but not the eukaryotic equivalent (see Ch. 3) by binding to its β-subunit. The compound (rifamycin) used for semi-synthesis of rifampicin is produced by *Nocardia mediterranei* (new genus name *Amycolatopsis*). Peptide antibiotics such as bleomycins produced by *Streptomyces verticillus* are major cancerostatic antibiotics. Cyclosporin is produced by *Tolypocladium inflatum* and used for immunosuppression. The worldwide tonnage produced every year is worth over a billion dollars.

Bacitracin is a mixture of related cyclic polypeptides produced by *Bacillus licheniformis*. Its unique name derives from the fact that the bacillus producing it was first isolated from a knee scraping of a little girl named Tracy. As a toxic and difficult-to-use antibiotic, bacitracin doesn't work well orally. However, it is very effective topically.

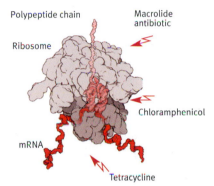

Tetracyclines, chloramphenicol, and macrolide antibiotics all act on the ribosome, inhibiting protein synthesis of the cell.

mycetes are Gram-positive, mottled and immobile bacteria that form a network of filaments.

Waksman took a novel and systematic approach, and unlike Fleming whose discovery of penicillin was more or less a chance find, he and his students grew a series of isolated soil microbes on agar plates growing under a variety of culture conditions. They screened them by looking for growth inhibition zones surrounding single microbial colonies.

They then proceeded to test the inhibition on specific pathogenic bacteria, a painstaking work: thousands of cultures of different microbes were isolated and then tested for antibacterial activity, but only a small percentage proved to be effective. These were

then further tested: would they yield microbe-fighting substances in sufficient quantities? And if so, were they not too toxic for therapeutic use?

The first agent isolated in 1940 under the initial screening program was **actinomycin** by **Harold Boyd Woodruff**, a Waksman graduate student (for details, see Box 4.2).

Streptomycin is an aminoglycoside, i.e. it consists of sugars and amino acids. The effectiveness range of aminoglycoside antibiotics is considerable, but on the downside, they are also slightly toxic. Incidentally, the typical smell of fresh soil is caused by streptomycetes. Waksman did not stop there, but discovered further new antibiotics, clavacin, grisein

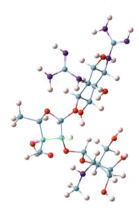

Fig. 4.48 Streptomycin

Box 4.12 Bioreactors – Creative Space for Microbes

At the heart of any biotechnological production plant is a bioreactor. It is, broadly speaking, a further development of the good old fermenting tank, but in contrast to its predecessor, it is surrounded by a wide range of elaborate technological paraphernalia. The development of bioreactors started with simple holes into which waste was dumped. Then, covered containers made of leather, wood or ceramic were used for the production of alcohol or acetic acid. Real progress was made only after Pasteur's work was accepted and concepts such as pure cultures, sterile environments and pure production became part of standard practice.

Bioreactor types of historic significance are the **surface reactor** for the production of citric acid and the **trickle film reactor**, the percolating filter in aerobic sewerage processing. Both are easy to handle, but their time-space-yield ratio is low.

In the 20th century, yeast production and the production of organic compounds (acetone and butanol) and eventually penicillin called for bioreactors that permit precise management and control of the entire process. The most popular

solution became the stirred tank reactor (STR). It is fitted with a thermostat, a stirring mechanism and gas supply, sterile inlets, and sampling valves.

Processes involving bioreactors fall into two categories – discontinuous and continuous processes. In **discontinuous** or **batch processing** methods, the tank is filled with the entire starting material and the microorganisms before biochemical conversion begins. This can take between a few hours and several days. Eventually, the tank is emptied, and the end product is purified. The tank can then be used for another cycle.

In a **continuous process**, the reactor is continuously supplied with starting material while equivalent amounts of the reaction mix containing the end product are taken out. Supply of new material and removal of the product must be finely balanced to achieve a dynamic equilibrium. The production flow resembles that of an oil refinery, while batch production could be compared to the production process in a bakery.

Both processing types can also be combined in a **semi-continuous production** situation where microbes remain in the reactor for 90 days and are supplied with fresh broth on a daily basis.

The choice of the process is determined by economic factors. In principle, continuous processes are more suited to large volumes (e.g. in sewage plants) than discontinuous procedures. However, batch production is still very popular where only small amounts are required.

A bioreactor can rapidly be switched to the production of other substances and is easy to keep sterile.

The various types of bioreactors can be classified in various ways. According to volume, we distinguish

- Laboratory reactors (<50 L)
- Experimental reactors (50 - 5,000 L)
- Industrial reactors (>50 L up to 1,500,000 L)

The height/diameter ratio leads to the distinction between tank reactors (H/D= 3) and reactor columns (H/D >3).

According to their energy input, reactors can be classified as
- Mechanically stirred reactors which are the most versatile (e.g. for the production of antibiotics)
- Hydrodynamic reactors (deep jet reactors) with an external liquid pump for the production of yeast for animal feed

Stirred Tank Reactor (STR)

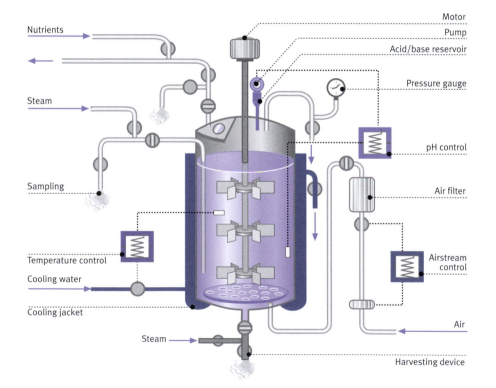

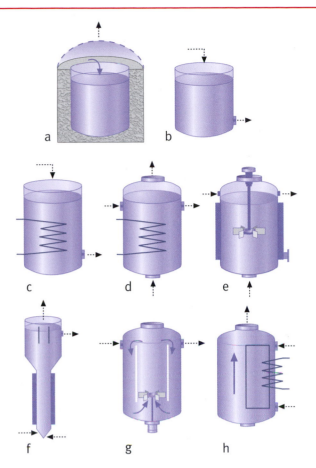

Historic Development of Bioreactors (fermenters)

a   covered hole in the ground (human waste, biogas).

b   simple container made of wood, leather, metal, or plastic (wine, beer, alcohol, vinegar, sour milk).

c   open reactor (brewery) with temperature control.

d   sterile reactor for controlled fermentation (yeast, specific chemicals).

e   stirred reactor (antibiotics).

f   tubular tower reactor (beer, wine, vinegar).

g   airlift reactor with internal recirculation (yeast from crude oil).

h   airlift reactor with external recirculation (bacteria from methanol).

Bacterial culture: *Escherichia coli* and *Klebsiella pneumoniae* on Levine agar.

Production plant for growing microorganisms.

Process development in a technical center for fermentation processes.

Molecules of different sizes are separated using membranes in an ultrafiltration plant.

Separation of cells and fermentation medium using centrifuges.

Low-contamination filling at the end of the production process.

● Pneumatic reactors (energy input through gas compression, airlift reactors with internal or external recirculation, e.g. for the production of single proteins (yeast from crude oil, bacteria from methanol) and sewage purification ("Tower Biology").

When growing free cells, **continuous recirculation** of the fermenting content is crucial, and aeration is also needed because most processes are aerobic. For these reasons, stirring bioreactors are the most widely used type. Recirculation is achieved with the help of a stirring mechanism or pumps. A **stirred tank reactor** can have mechanical stirring turbines or propellers that blow in and disperse air. It is often a combination of both. Especially where the height/diameter ratio is greater than 1.4, it is essential to have multiple stirrers.

More robust fermentation processes (submerged vinegar or yeast production processes or sewage works) use **bioreactors without active air distribution mechanisms**. The stirrer (propeller or turbine) creates turbulences to distribute the air evenly. Such reactors, however, are not suitable for more susceptible mammalian cells (Ch. 3). In beer and yeast production, gases develop during the actual fermentation processes (CO_2). In hydraulic stirring reactors, pumps are responsible for recirculation.

All bioreactors we have looked at so far use homogenous mixes of nutrient medium and cells.

In membrane reactors, by contrast, the catalyst (cell or enzyme) is separated from the product by a membrane that cannot be penetrated by the biocatalyst. Such a reactor was shown in Chapter 2, in the enzymatic production of amino acids. In a specific type of membrane reactor, the **hollow fiber reactor**, the enzymes or cells are immobilized on the outer surface of the fibers. The medium flows to the biocatalyst through the hollow fiber. Throughput and thus productivity in a small space can be raised by using bundles of hollow fiber.

Fixed bed reactors are filled with carrier material (e.g. porous glass, cellulose) with a large surface. They are used as enzyme reactors (Ch. 2) and for sensitive cells of higher forms of life (Ch. 3).

Box 4.13 Primary and Secondary Metabolites

In Chapter 1, we looked at classical metabolic end products such as ethanol and lactic acid. They are produced in fermentation processes to ensure the mere survival of the microbes.

Essential **primary metabolites** are synthesized by the cell because they are indispensable for their growth. Economically significant are amino acids (threonine, lysine, phenylalanine, tryptophan, and glutamate), vitamins (B_2 and B_{12}), and nucleotides (inosine-5-monophosphate [IMP], guanosine-5-monophosphate, GMP).

Natural microbial production of these substances is not excessive, as it requires extra energy and carbon. Producing a surplus would put microbes at a selective disadvantage compared to those that only produce what they need. In order to enhance microbial production, it is necessary to interfere with their regulatory mechanisms.

Secondary metabolites are produced only in a later stage of the life of a cell and are not needed for the actual growth process. Often, their natural function is not known. Many of these secondary products, e.g. antibiotics, give the microbes an advantage over competitors by inhibiting their growth, but there are other compounds as well, such as mycotoxins and pigments.

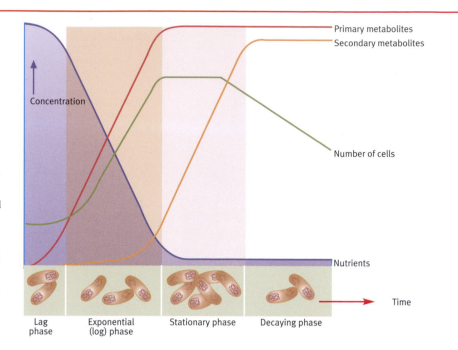

Primary metabolites (e.g. ethanol and amino acids) accumulate in the bioreactor during the growing phase, whereas secondary metabolites (e.g. penicillin) reach their maximum only after the growing phase.

This is significant for their production.

All that is required for the production of primary metabolites is to optimize the growing conditions for the bacteria or yeasts. As long as they proliferate, the amount of primary metabolites will increase. By contrast, if penicillin, the secondary metabolite of a mold fungus, is produced, its concentration will peak only after the fungus has stopped growing.

It is thus a matter of finding a balance between ideal growth conditions and high yields.

Fig. 4.49 In clean rooms, special care must be taken, e.g. protective clothing must be worn.

Fig. 4.50 Selman Abraham Waksman (1888 - 1973) coined the term "antibiotic". He began his Nobel Prize speech in 1952 with a citation from Ecclesiasticus: *"The Lord hath created medicines out of the earth; and he that is wise will not abhor them."*

(1946), neomycin (1948), fradicin, candicidin, candidin, and others.

It was Selman Waksman who coined the term *antibiotics*.

As early as 1945, Professor **Giuseppe Brotzu** (1895 - 1976) (Fig. 4.53) at the Institute of Hygiene, Cagliari, Italy, had been looking into a problem that has now grown into gigantic proportions – the pollution of the Mediterranean. He took water samples at a sewer outlet on Sardinia. He speculated that where bacteria causing intestinal diseases are present, their natural enemies cannot be far away. He duly discovered the mold *Cephalosporium acremonium* (now renamed *Acremonium chrysogenum*, Fig. 4.52) which produces compounds inhibiting a whole host of bacteria.

Brotzu's findings were published in a rather obscure Italian University review, and it was pure chance that it came to the attention of a group of antibiotic hunters in Oxford. They then found out that the *Acremonium* produces not just one, but several related antibiotics. One of them, **Cephalosporin C**, proved particularly effective against penicillin-resistant Gram-positive pathogens (see Box 4.9).

Penicillins, cephalosporins, and streptomycin may have been the most important discoveries in the early years of antibiotic research, but they were by no means the only ones.

The number of newly discovered antibiotics increased between the late 1940s to the early 1970s to approximately 200 a year. At the end of the 20th century, it hit a record mark of 300 compounds per year. Up to now, 8,000 antibiotics have been isolated from microorganisms and 4,000 from higher organisms (Box 4.7). Over 60,000 tons of penicillins and cephalosporins are produced per year worldwide.

While cephalosporins are exclusively used in humans, penicillins are also widely used (and often misused) in animals reared for meat production.

Soon after the first successes in the war of humans against microbes, the counter-attack began, showing that evolution is very much an ongoing process.

4.21 The Race Against Microbial Resistance

The great **Louis Pasteur** wrote: "Bacteria will have the last word, Messieurs!" How right he was!

After penicillin had been successfully used for several years, there was a dysentery outbreak in Mexico in 1972, and in 1975, promoted by widespread prostitution, an epidemic of gonorrhea in the Philippines.

Gonorrhea is a sexually transmitted disease that has been known for centuries, but only antibiotics provided a definite cure.

It turned out that in the Philippines, prostitutes had taken penicillin over a long time in order to protect themselves from gonorrhea. They were thus providing an environment that led to the selection of resistant *Neisseria gonorrhoeae* bacteria. Bacteria that become resistant produce enzymes that inactivate penicillin. These hydrolases, which are called β-**lactamases**, break up the β-**lactam ring** in penicillin or cephalosporin (Box 4.9) into inactive penicillanic and cephalosporanic acids. Hospitals have become increasingly dangerous places for patients, who run the risk of being infected with antibiotic-resistant bacteria. It is estimated that they are responsible for 90,000 deaths per year in the USA and 20,000 deaths per year in Germany.

Attempts are being made to outwit the microbes. In the pharmaceutical industry, the main product is penicillin G which carries a benzyl side chain. The side chain is cleaved off using specific immobilized enzymes (penicillin amidases) and a new group is joined on. The result are novel **semisynthetic antibiotics** to which the microbes have not (yet) developed resistance (see Ch.2)

Soon after the first successes in the war of humans against microbes, the counter-attack began. Some compare the ongoing struggle to that between Tom and Jerry. Whatever way you look at it, it is a clear sign that evolution is very much an ongoing process. The race against bacteria is compounded by the massive use of antibiotics in animal breeding where it is used to enhance growth and to prevent infection. As we saw before, bacteria exchange **plasmids carrying genes for antibiotic resistance**.

What we observe in the contest between bacteria and antibiotics is evolution, as Darwin and Wallace described it – mutation, selection, survival of the fittest, and of the best-adapted.

4.22 Cyclosporin – A Microbial Product Used in Transplants

From 1957, whenever employees of Sandoz, Basel, went for a holiday or on a business trip, they were asked to bring back a plastic bag with soil samples from all over the world. These were then put in a database and screened for organisms producing new antibiotics. It took them until 1970 to make a discovery – the fungus *Tolypocladium inflatum* (Fig. 4.54) was isolated from soil samples from Wisconsin, USA and Hardanger Vidda in Norway.

Much to the disappointment of the researchers, the substance they found was only effective against some other fungi, and it did not even work very well at that. However, the antibiotic, which was given the lab classification 24-556, had one remarkable property. Although it had some antibiotic effect on fungi, it was only slightly toxic to lab animals, and this was the reason why it underwent further testing.

In 1972, the tests had sensational results: the new substance, a ring-shaped peptide consisting of eleven amino acids, suppressed the human immune system. There had been earlier **immunosuppressants** that suppressed the rejection of transplant organs. Their major drawback was that they usually put the whole immune defense system out of action, and the slightest infection could kill the patient.

With 24-556 or **cyclosporin** A (Fig. 4.57), as it is known today, this does not happen. When it is taken up, it binds to specific receptors in T-lymphocytes. The T-cells are the most important cells in the specific defense of the body (see Ch. 5). The compound initiates a reaction cascade that stops the production of **interleukin**-2. Interleukin is a messenger substance (lymphokine) required for the activation of the immune system. It is involved in inflammatory reactions.

By its inhibition of T-lymphocytes and blockage of interleukin-2 production, cyclosporin suppresses the immune system.

The microbial product cyclosporin and its derivatives led to a dramatic increase in successful heart, kidney and lung transplants since the early 1980s. Cyclosporin is one of the few peptides that can be taken orally.

Fig. 4.51 Hans Christian Joachim Gram (1853–1938).

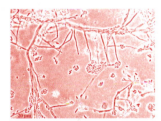

Fig. 4.52 *Acremonium* produces cephalosporins.

Fig. 4.53 Giuseppe Brotzu (1895–1976) discovered the cephalosporins.

Fig. 4.54 *Tolypocladium inflatum*

Box 4.14 Biotech History:
Mexico, the Father of the Pill and the Race for Cortisone

The son of Jewish physicians, Carl Djerassi (born 1923), grew up in pre-war Vienna and fled the Nazis in 1938.

When he was 16 years old, he arrived in New York penniless. A few years later, he graduated Phi Beta Kappa from Kenyon College in Ohio. His first job was as a junior chemist with the Swiss pharmaceutical company CIBA, where he was part of the team discovering the antihistamine pyribenzamine. He obtained his Ph.D. at the University of Wisconsin and then went to work at a little known Syntex lab in Mexico City. This Mexican lab became a center of steroid research and synthesis within a short time. They joined the worldwide race to synthesize cortisone. It was here, that Djerassi directed the synthesis of the first oral contraceptive for women and became known throughout the world as the 'Father of the birth-control pill'.

In the following, Carl Djerassi describes the exciting hunt for cortisone (extracted and shortened from two of his memoirs by me RR):

Carl Djerassi at Syntex with his assistant Arelina Gomez.

The telegram dated 8 June 1951, bore the name of **Tadeusz Reichstein**, who the year before had shared the Nobel Prize in medicine for his isolation and structural elucidation of cortisone. The telegram - sent from Basel to Mexico City, then the site of tiny Syntex, S. A.-said only, "*No depression.*"

Mexican yams

Carl Djerassi with Alejandro Zaffaroni pointing at the key chemical feature of all steroid contraceptives.

It signified that the melting point of the authentic steroid, isolated by Reichstein from adrenal glands, was not depressed when mixed with the synthetic specimen we had dispatched to Switzerland. (In 1951, a "mixed melting point" was one of the standard ways of establishing identity between two crystalline chemicals.)

Thus, our unknown research group – the oldest member of which, was, at age thirty-four, seven years my senior – won the race.

Those were the days of unrestrained optimism, when cortisone was believed to be a wonder drug for treating arthritis and other inflammatory diseases. **Philip Hench**, one of Reichstein's fellow Nobel Prize winners from the Mayo Clinic, had shown movies in 1949 of helpless arthritics receiving cortisone and then, in days, getting up to dance. The only problem was that cortisone cost nearly $200 per gram, and depended on not readily available starting material: slaughterhouse animals. In 1944, it had taken **Lewis H. Sarett** of Merck & Company in Rahway, New Jersey, 36 chemical steps to synthesize cortisone from cattle bile. Thus, a potentially unlimited source for cortisone was needed: if not through *total synthesis* – building from scratch from air, coal or petroleum, and water – then through **partial synthesis**, starting with a naturally occurring steroid and transforming it chemically into cortisone. Sarett had prepared the first few grams of cortisone by such a partial synthesis – somewhat akin to the conversion of a barn into a villa, the bile acid being his barn, cortisone his villa. Here I shall try to emulate Einstein's dictum: "We should make things as

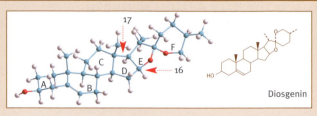

Diosgenin

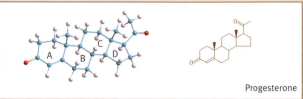

Progesterone

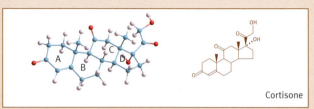

Cortisone

simple as possible, but not simpler." The starting material our Syntex team chose, as a more widely available alternative to Sarett's bile acid, was diosgenin.

At first glance, the choice seems to make neither chemical nor geographical sense. While its chemical structure does contain the four rings typical of steroids, diosgenin is burdened by extraneous chemical appendixes, encompassed by two more rings attached to positions 16 and 17 (see Figure, rings E and F).

Diosgenin, however, had been the raison d'être for Syntex's formation just a few years earlier. In the late 1930s and early 1940s, **Russell E. Marker**, a brilliant but unorthodox chemistry professor at Pennsylvania State University, conducted research on a group of steroids called "sapogenins," compounds of plant origin so-called because their chemical combination with sugars (termed "saponins") display soaplike qualities in aqueous suspension. Natives of Mexico and Central America, where saponin-containing plants occur wildly and in abundance, had long used them for doing laundry and to kill fish. Marker concentrated on the chemistry of the steroid sapogenin diosgenin, which was present in certain types of inedible yams growing wild in Mexico. He discovered an exceedingly simple process whereby the two complex rings (for us, molecular garbage) could be degraded to a substance then easily transformed chemically into the female sex hormone progesterone.

In 1944, unable to convince any American pharmaceutical firm of the commercial potential of diosgenin, Marker formed a small Mexican company named Syntex (from Synthesis and Mexico).

A few months later, Syntex started to sell to other pharmaceutical companies pure, crystalline progesterone prepared from diosgenin by *partial synthesis* in five steps. Within a year the partners had a disagreement, and Marker left the company... His partners **Emeric Somlo** and **Federico Lehmann**, looking for another chemist who could re-establish the manufacture of progesterone from diosgenin at Syntex, recruited Dr **George Rosenkranz** from Havana. A Hungarian like Somlo, Rosenkranz had received his doctorate under the Nobel laureate **Leopold Ruzicka** (one of the giants of early steroid chemistry) and read Marker's publications.

Within two years, Rosenkranz had not only reinstituted at Syntex the large-scale manufacture of progesterone from diosgenin but, even more important, had achieved the large-scale synthesis, from these same Mexican yams, of the commercially more valuable male sex hormone testosterone. Both syntheses were so much simpler than the methods used by the European pharmaceutical companies -such as CIBA, then dominating the steroid hormone field- that in a short while tiny Syntex broke the international hormone cartel. As a result, prices were lowered considerably, and these hormones became much more available.

In the late 1940s, Syntex served as bulk supplier to pharmaceutical companies throughout the world, but few people outside these firms even knew of the existence of this small chemical manufacturing operation in Mexico City, which was soon to revolutionize steroid chemistry and the steroid industry all over the world. By the spring of 1951, however, everyone was taking at least one Mexican coup seriously. By then our Syntex team had published a "Communication" in the JACS. Another "Communication" recorded our discovery of a novel path to the characteristic ring-A structure of cortisone or of other hormones, such as progesterone, from an intermediate that was particularly suitable to steroidal sapogenin precursors such as diosgenin.

At this point, we moved into the two-shift mode, which led to the synthetic crystals that were shipped to Reichstein in Switzerland. Within hours of the arrival of the "no depression" telegram, I wrote the first draft of our "Communication" entitled "Synthesis of Cortisone."

Our successful synthesis of cortisone from diosgenin had permanently placed Mexico on the

The press conference announcing the first synthesis of cortisone from a plant source at Syntex im Mexico City, 1951. Left to right: Gilbert Storck, Juan Pataki, George Rosenkranz, Enrique Batres, Juan Berlin, Carl Djerassi, and Rosa Yashin.

map of steroid research. It was, moreover, Upjohn's requirements for tons of progesterone – a quantity that at that time could be satisfied only from diosgenin – that started Syntex on the way to becoming a pharmaceutical giant. The process was accelerated by our synthesis – a few months later, again in Mexico City – of the first oral contraceptive. Reichstein's telegram, "*No depression*," of just a few months earlier applied to cortisone, but not to us. We were elated: *Que viva México*!

Life magazine featured our team, most in immaculate white lab coats, grouped around a gleaming glass table and apparently mesmerized by an enormous yam root, which overwhelmed the molecular model of cortisone lying next to it. Rosenkranz held a test tube, filled almost to the brim with white crystals – the chemist's equivalent of the climber's flag on top of Mount Everest. For the photographer's benefit, the tube had been filled with ordinary table salt, because at that time we had synthesized only milligram quantities of cortisone.

Ironically, none of the cortisone triumphs from our laboratory and those of our competitors at Harvard and Merck that appeared in the August 1951 issue of the *Journal of the American Chemistry Society* contributed to the treatment of a single arthritic patient because of the appearance of a newcomer, whose participation in the race was not even known. A few months after our publication, Syntex's management received an inquiry from the Upjohn Company of Kalamazoo (in Soth Africa), asking whether we could supply them with 10 tons of progesterone.

Since the world's entire annual production at that time was probably less than one-hundredth that amount, such a request seemed outlandish. No one in our group could conceive of a medical application of progesterone that could require tons of the stuff. We concluded that Upjohn was planning to use progesterone as a chemical intermediate rather than as a therapeutic hormone.

Our conclusion proved correct when, a few weeks later, we learned through a patent issued to Upjohn in South Africa (where patents are granted much more rapidly than in the United States) that two of its scientists, **Durey H. Peterson** and **Herbert C. Murray**, had made a sensational discovery: fermentation of progesterone with certain microorganisms resulting in an one-step key transformation on the way to cortisone.

What we chemists had accomplished laboriously through a series of complicated chemical conversions, Upjohn's microorganism with its own enzymes did in a single step!

Only a few months later, Djerassi found solace by synthesizing with Georg Rosenkranz and Luis E. Miramontes the progestin norethindrone which, unlike progesterone, remained effective when taken orally and was far stronger than the naturally occurring hormone. Djerassi remarked later that "not in our wildest dreams... did we imagine (it)", though he is now referred to as the Father of the pill.

Author of over 1200 scientific publications and seven monographs, Carl Djerassi is one of the few American scientists to have been awarded both the National Medal of Science (1973, for the discovery of oral contraceptives – 'the pill') and the National Medal of Technology (in 1991, for promoting the new approaches to insect control). He was inducted into the National Inventors Hall of Fame. Since 1959, Djerassi has been Professor of Chemistry at Stanford University.

In his second career, multi-talent Djerassi has turned from practicing chemistry to writing five "science-in-fiction" novels. Recently he has embarked on a trilogy of plays with emphasis on contemporary cutting-edge research in the biomedical sciences.

Cited Literature:
Djerassi C (1992) The Pill, Pygmy Cimps, and Degas Horse. BASICBOOKS, New York
Djerassi C (2001) This Man's Pill, Oxford University Press. Excerpts reprinted by permission of Oxford University Press
www.djerassi.com

Fig. 4.55 Microorganisms play a part in the production of therapeutically relevant steroids.

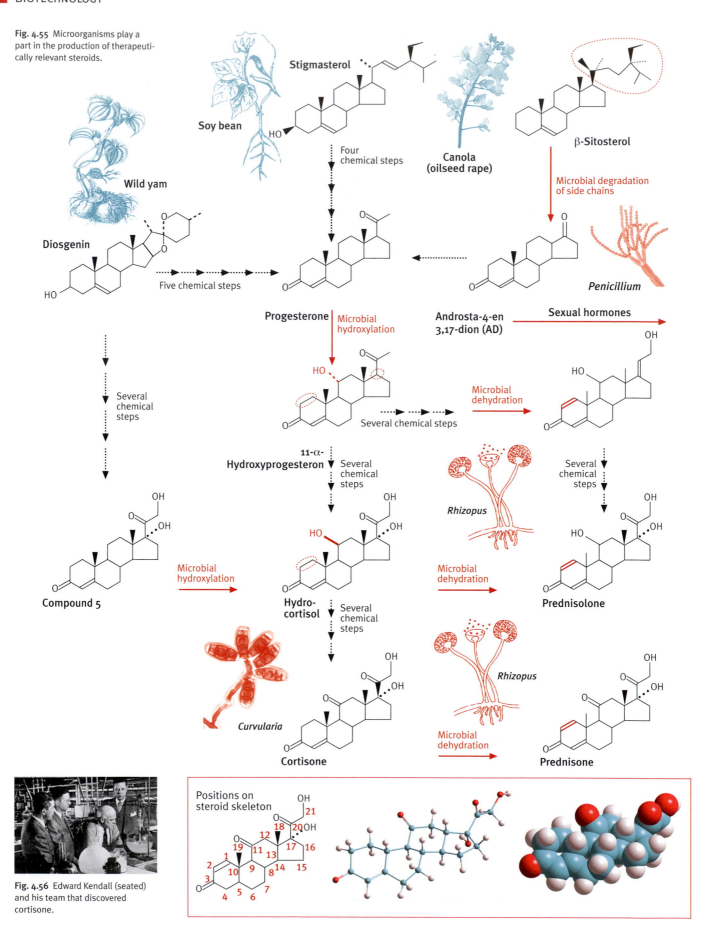

Stigmasterol

Soy bean

Canola (oilseed rape)

β-Sitosterol

Wild yam

Diosgenin

Four chemical steps

Microbial degradation of side chains

Penicillium

Five chemical steps

Progesterone

Androsta-4-en 3,17-dion (AD)

Sexual hormones

Microbial hydroxylation

Several chemical steps

Microbial dehydration

11-α-Hydroxyprogesteron

Several chemical steps

Rhizopus

Several chemical steps

Compound 5

Microbial hydroxylation

Hydro-cortisol

Several chemical steps

Microbial dehydration

Prednisolone

Curvularia

Rhizopus

Cortisone

Microbial dehydration

Prednisone

Fig. 4.56 Edward Kendall (seated) and his team that discovered cortisone.

Positions on steroid skeleton

4.23 Steroid Hormones – Cortisone and the Contraceptive Pill

While microorganisms are involved in all production steps of the industrial production of antibiotics, they may only be required for a few steps in the far longer production processes of other medications that partly involve non-biological synthesis. The production of steroid hormones is a case in point.

In the early 1930s, **Edward C. Kendall** (1866-1972, Fig. 4.56) of the Mayo Foundation and **Tadeusz Reichstein** (who had developed the synthesis of vitamin C, see above) of the University of Basel isolated **cortisone**, a steroid hormone produced by the adrenal cortex. This gained them the Nobel Prize, along with **Philip S. Hench**, in 1950. About a year later, it was discovered that cortisone provided pain relief to patients suffering from rheumatoid arthritis. Immediately, the demand for it went up considerably.

In view of the hugely increased sales, a chemical production process was developed, but it seemed to be very complicated. The procedure envisaged by Merck & Co involved **37 steps**, many of which requiring extreme conditions, to obtain cortisone from bovine bile acid. A gram of cortisone obtained in this way was worth $260. (Box 4.14).

Through **microbial hydroxylation** – in this case the fungus *Rhizopus arrhizus* – it was possible to cut down the synthesis process to just **eleven steps**, and bring down the price per gram of cortisone to about $9.70.

Microbial hydroxylation did not only cut synthesis time, it also did not require high pressures and temperatures or expensive non-hydrous solvents. Instead, the synthesis could be run at 37°C at the normal atmospheric pressure in ordinary water. This also contributed to the drop in production costs, and after several further improvements in the production method, the price per gram fell below $1.30.

The only source for raw material was the Wild Yam (*Dioscorea villosa*) root, which provided the starter substance diosgenin. Until 1975, more than 2000 tons per year were processed, which led the Mexican government to believe they had a monopoly on it. Inspired by the successful OPEC price negotiations on crude oil, the Mexican producer Proquivenex decided on a ten-fold price increase for dios-

genin. This did not go down well with the international pharmaceutical companies who usually dictate the price. They responded by looking into alternative methods, mainly the microbiological degradation of byproducts from soy bean production, the steroids sitosterol and stigmasterol (Fig. 60). Diosgenin was thus replaced within a very short time.

Two years after the Mexican firm had raised its prices so dramatically, it had to climb down in an equally dramatic way, but by then, nobody was interested in a product from such an unreliable source. The market for diosgenin had collapsed. There is a lesson to be learned – in biotechnology, there is no such thing as an irreplaceable source material!

The cortisone case also shows – as we have already seen when looking at vitamin C – that biotechnology and chemistry are not opposites, but can be combined. In the coming decades, we will see more combined biological and chemical processes. Microbes or their enzymes take over those production stages that would be very expensive or labor-intensive if carried out chemically.

Fig. 4.57 Structure of cyclosporin

Fig. 4.58 Plants and microorganisms used for steroid production.

Soy

Canola

Rhizopus

Curvularia

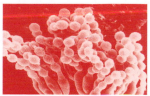

Penicillium

Organic Chemists' Prayer (unknown origin)

Dear God,
I pray on bended knees,
Make sure that all my syntheses
Will never be inferior,
To those conducted by bacteria

Prokaryotic and Eukaryotic Cells:
Microscopic Gigantic Synthesis Factories

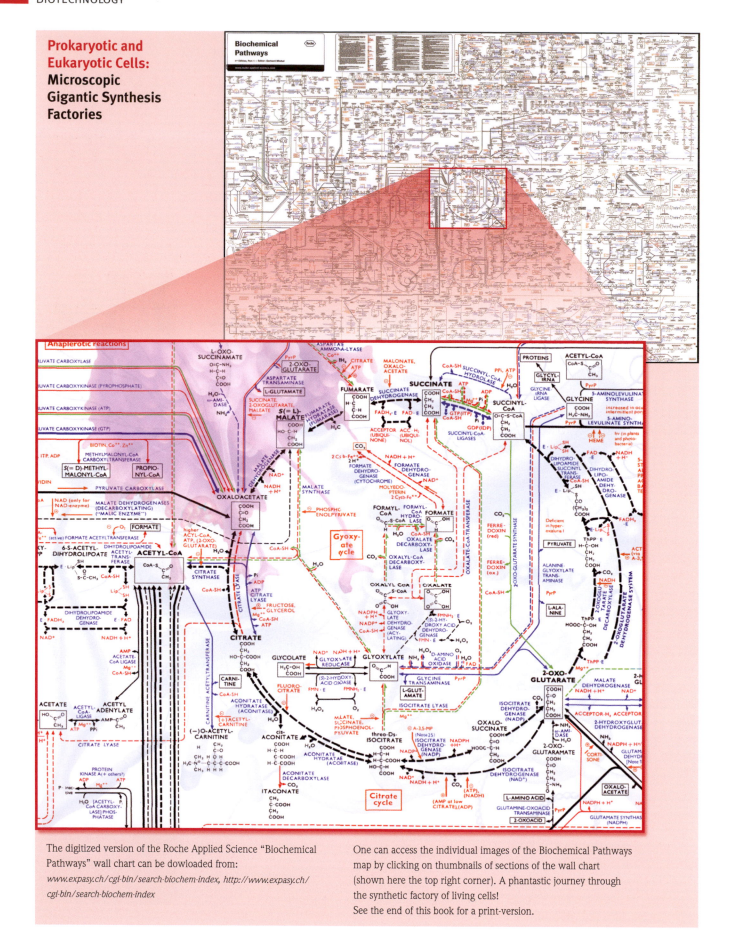

The digitized version of the Roche Applied Science "Biochemical Pathways" wall chart can be dowloaded from:

www.expasy.ch/cgi-bin/search-biochem-index, *http://www.expasy.ch/cgi-bin/search-biochem-index*

One can access the individual images of the Biochemical Pathways map by clicking on thumbnails of sections of the wall chart (shown here the top right corner). A phantastic journey through the synthetic factory of living cells!
See the end of this book for a print-version.

Cited and recommended Literature

- **Blanch H W** (1995) *Biochemical Engineering*. Dekker/CRC Press
  In my view, still unsurpassed as a textbook, unfortunately now 12 years old

- **Liese A, Seelbach K, Wandrey C** (edn.) (2006) *Industrial Biotransformations*.
  2nd edition. Wiley-VCH
  The most important reference book, application-oriented overview of one-step
  biotransformations (enzymes and cells) of industrial importance

- **Aehle W** (edn.) (2007) *Enzymes in Industry*, 3rd edn. Wiley-VCH, Weinheim
  A comprehensive and up-to-date book on industrial applications of enzymes

- **Schmid RD** (2003) *Pocket Guide to Biotechnology and Genetic Engineering*,
  Wiley-VCH, Weinheim. A very useful and concise overview of all important
  bioprocesses.

- **Ratledge C, Kristiansen B** (edn.) (2006) *Basic Biotechnology*, 3rd edn.
  Cambridge University Press New York
  The detail may be overwhelming for a beginner, but a good reference book

- **Buchholz K, Kasche V, Bornscheuer UT** (2005) *Biocatalysts and Enzyme Technology*.
  Wiley-VCH, Weinheim
  Very good overview of our current knowledge of biocatalysis and enzyme technology:
  Ullmann's Encyclopedia of Industrial Chemistry Wiley-VCH, now available not only in
  print (6th edn. 2002), but also in electronic form – either as a fully networkable DVD
  version or as an online database.

- **Michael Flickinger M, Drew S** (edn.) (1999) *The Encyclopedia of Bioprocess
  Technology Fermentation, Biocatalysis and Bioseparation (Wiley Biotechnology
  Encyclopaedia)* John Wiley & Sons, Inc. For specialists only: The quite expensive
  five-volume set of the Encyclopedia of Bioprocess Technology focuses on industrial
  applications of fermentation, biocatalysis and bioseparation.

Useful weblinks

- The Biochemical Engineering Research Network (BERN):
  http://www.icheme.org/enetwork/files/draft/59319/BERN/bernfrontpage.htm

- A starting point for all areas, the Wikipedia! For example Biochemical Engineering:
  http://en.wikipedia.org/wiki/Biochemical_engineering

- BRENDA database, created by Dietmar Schomburg (University of Braunschweig).
  A comprehensive compilation of information and literature references about all
  known enzymes: *http://www.brenda.uni-koeln.de/*

- Enzyme Structures Database links to the known 3-D structure data of enzymes in the
  Protein Data Bank: *http://www.ebi.ac.uk/thornton-srv/databases/enzymes/*

- Comprehensive site of biotechnology information maintained by the Biotechnology
  Industry Organization: *http://www.bio.org/*

- World Wide Web Virtual Library: Biotechnology: catalog of WWW resources related
  to biotechnology; contains over 3500 links to companies, research institutes,
  universities, sources of information and other directories specific to biotechnology.
  Emphasis on product development and the delivery of products and services:
  http://www.cato.com/biotech/

Eight Selftest Questions

1. What is *umami* and why is it not feasible/sensible to produce it through chemical synthesis?

2. How could a humble bread mold ruin the Italian citrus-growing industry within a very short time?

3. In what way did the discovery of lysozyme help in the later discovery of penicillin?

4. What phenomenon enables lactose to switch the production of lactose-transporting and degrading proteins in *Escherichia coli* on or off?

5. How did the fruit fly *Drosophila* turn the Roche company in Basel into the largest vitamin C producer in the world? Can you poison yourself with vitamin C?

6. How does penicillin work? How can a bacterium defend itself against penicillin? Why must penicillin be taken as prescribed, and why should treatment not be abandoned early?

7. What are the names of the eight essential amino acids (use mnemonic!), and what makes them essential?

8. Why does UV light kill microbes and possibly cause skin cancer in humans?

VIRUSES, ANTIBODIES AND VACCINES

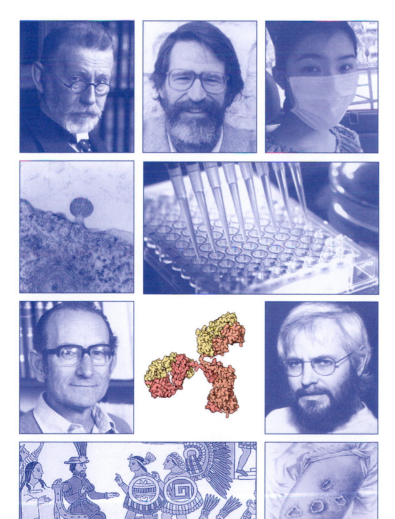

CHAPTER **5**

Fig. 5.1 New viruses pose a global threat to humanity. For two months, the city of Hong Kong was brought to a standstill by the SARS epidemic. The inhabitants of Hong Kong finally defeated the virus with discipline and charm.

Fig. 5.2 In spite of the SARS (Severe Acute Respiratory Syndrome) outbreak, my biotechnology lectures in Hong Kong continued. The topic, of course, was 'detection of viruses'.

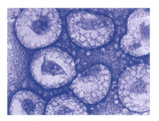

Fig. 5.3 The SARS virus is a coronavirus. It works in a different way from, say, the AIDS virus. It introduces its single-stranded RNA into the host cell and uses RNA-dependent RNA polymerase to produce mirror image copies.

Fig. 5.4 A new global threat could come from combined bird and human influenza viruses.

5.1 Viruses – Borrowed Life

Viruses fail to fulfill the classic criteria for a living organism, i.e., they have **no metabolism of their own** and they use the reproductive mechanism of their host animals, plant, or bacteria for their own reproduction. This **inability to reproduce outside** host cells makes it impossible for them to survive independently. If they are not living organisms, what are they?

They are basically programs that insert into the genome of their hosts, hijacking their reproductive mechanisms to produce more viruses – very much like computer viruses.

As viruses do not have their own metabolism, inhibitors such as **antibiotics have no effect on them** – there is no metabolism to be disrupted. The only point of possible intervention is viral interaction with the host.

There are two kinds of viruses – enveloped viruses and naked viruses. The genome of **naked viruses** is protected by a protein core or **capsid**, whereas **enveloped viruses** form their envelope by pinching off a bud from the host's cell membrane, into which they insert virus-coded proteins. This lipid-protein viral envelope contains a capsid, similar to that of naked viruses (Fig. 5.8).

The difference between microorganisms and viruses is that viruses only require their nucleic acid and a host cell for their reproduction. Some viruses, however, also contain the enzymes they need for replication. This applies to retroviruses that contain **reverse transcriptase** in their capsid.

Viruses are completely dependent on their **host cells**, as far as the reproduction of their nucleic acid and the synthesis of their proteins is concerned. Viruses go through two cycles – the **lytic cycle** (Greek *lysis*, dissolution – a word that also gave the name to Fleming's lysozyme, Ch. 2), during which the virus is released and destroys the host cell – and a **non-lytic cycle** in which the viruses form buds that are pinched off from the cell membrane (this is how it works in enveloped viruses such as influenza viruses or HIV).

All viruses contain one single type of nucleic acid (RNA or DNA). They are classified according to their nucleic acid, their protein covers, and their host specificity (Fig. 5.5).

RNA viruses include the AIDS virus HIV, influenza **viruses**, the measles virus, the rabies virus, and a plant virus known as tobacco mosaic virus (TMV, Ch. 3). The latter two are rod-shaped. Other viruses in this category include the picornavirus group, e.g., poliovirus and rhinovirus, responsible for colds (Figs. 5.3 to 5.5). The **SARS** (Severe Acute Respiratory Syndrome) **virus** is a major cause for concern, especially in Hong Kong, China and Canada (Figs. 5.1 to 5.3). This is another RNA virus, called a corona-virus, because its surface resembles a crown (*corona*, Latin for crown).

DNA viruses include, for example, papovaviruses that mostly cause warts, but some species may cause tumors. Others are *Variola* (smallpox) and *Vaccinia* (cowpox) viruses, Herpes viruses, adenoviruses (causing infections of the mucous membrane), bacteriophages that attack bacteria (Greek *phagein*, to eat), and baculoviruses that exclusively attack insects.

5.2 How Viruses Attack Cells

Viruses always bind first to the surface of cells (Fig. 5.6). DNA viruses such as **bacteriophages** inject their genetic material (double-stranded DNA) into the bacterial cell (Fig. 5.6, left). With the help of the bacterial cell, they produce enzymes (T4-DNA-polymerase) that are used to synthesize DNA and mRNA. The viral mRNA, produced from bacterial RNA, is read by the bacterial ribosome. Thus, the bacterial cell produces the viral protein envelope as well as its DNA from bacterial building blocks. The parts that make up a bacteriophage finally join to form a complete bacteriophage that lyses the host cell.

It is, however, possible that viral DNA is inserted into bacterial DNA **without lyzing** the cell. Such DNA is called **dormant viral DNA**, which will only be released and reproduced in later bacterial generations. In animal cells (Fig. 5.6, right), viruses bind to receptors on the cell surface, and the protein envelope merges with the cell membrane to let the virus enter.

In **RNA viruses** of the retrovirus group (e.g., HIV), single-stranded RNA enters the cell and is converted into double-stranded DNA with the help of an enzyme carried in the virus (**reverse transcriptase**, Ch. 3). The transcribed viral DNA is inserted into chromosomal DNA in the nucleus. The transcription mechanism of the host cell (RNA polymerase) first transcribes it into mRNA, which becomes a blueprint for the synthesis of viral proteins in the ribosomes. These include non-structural proteins which are the cause of the pathogenicity of many viruses. The newly produced viral RNA and the viral capsid combine to form new viruses which exit the cell.

Only very few viruses **integrate their genome** into that of the host. These include Herpes viruses and retroviruses. Genome integration enables a virus to remain stably in the genome of the host cell over many generations of cell division until it becomes active again. When some viruses, e.g., the hepatitis B virus or papillomaviruses, integrate their genome into the host genome, it results in an "abortive integration", which is partly responsible for the development of tumors. In these cases, the virus loses its ability to replicate during the integration process.

The search for **strategies against the attack of viruses** has intensified (Box 5.1). It is conceivable that specific **antibodies** (see later in the chapter) could interconnect and neutralize viruses before they dock onto the host cell and invade it. Antibodies could also prevent invasion by masking the relevant binding sites on the target cells so that the viruses would fail to recognize them (Fig. 5.9). Antibodies can also label viruses so that they can be recognized by **macrophages** and granulocytes and destroyed.

With RNA viruses, **inhibitors** of reverse transcriptase (Ch. 3) can be used to prevent the transcription of viral RNA into DNA. However, many of these inhibitors, which are used in the treatment of AIDS, are toxic. Viral RNA produced by the host cell could be blocked by **antisense RNA**, its exact mirror image (Ch.10), binding to it.

Another, fairly recent, strategy is to use short double-stranded RNA sections (**RNAi** – the i standing for interference. More details in Ch. 9). Artificially created RNAi, between 21 and 23 nucleotides long, was used by German scientist **Tom Tuschl** (Ch. 9) to silence mammalian genes without triggering a disruptive interferon response (which would have led to the degradation of all RNA present). Since then, it has been possible to silence specific genes – e.g., for HIV, the *nef, rev, gag,* and *pol* genes. There are also initial successes in the fight against influenza and hepatitis C viruses.

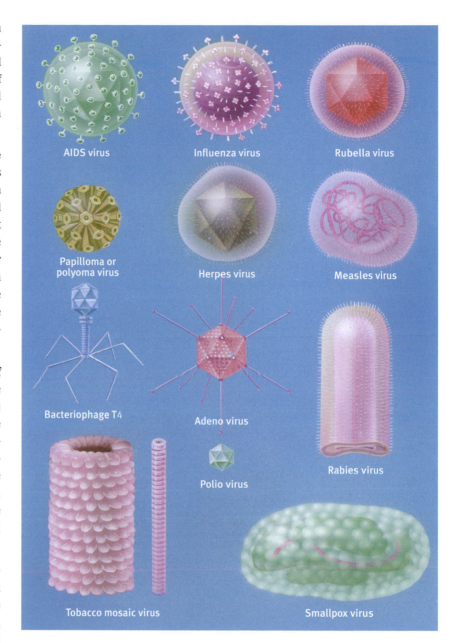

Fig. 5.5 DNA and RNA viruses (not to scale!) from top left to bottom right:

AIDS virus (HIV), a retrovirus with envelope and single-strand RNA, long latency period.

Influenza virus, an orthomyxovirus, several RNA strands, envelope, A, B and C type.

Rubella virus, single-strand RNA, envelope, a togavirus.

Papova (**papilloma and polyoma**) **viruses**, double-strand DNA, naked. Papillomaviruses can cause warts and some polyoma viruses can cause cancer in animals.

Measles virus, single-strand RNA, family of paramyxo-viruses, attacks mucous membranes and the cells of the immune and nervous system.

Bacteriophage T4, double-strand DNA, attacks bacteria such as *E. coli.*

Adenovirus, double-strand DNA, causes disease in the respiratory tract.

Rabies virus, single-strand RNA, a rhabdovirus.

Tobacco mosaic virus (TMV), single-strand RNA, rod-shaped. Whole crops of chili and capsicum have been devastated by TMV. Characteristic brown spots on leaves.

Polio virus, a picornavirus, single-strand RNA, naked, causes poliomyelitis, a serious disease of the nervous system.

Smallpox virus (*variola*), double-strand DNA, envelope, very large virus.

Box 5.1 **Antiviral Drugs**

There are many strategies to try and stop the spread of virus HIV. Attempts are being made to interfere with every single stage of the viral reproduction cycle.

1. **Docking on** to cells that have not yet been infected: The viral envelope protein gp120 binds to the CD4 receptor on the surface of helper T lymphocytes. Antibodies to CD4 could saturate the docking sites on the cell. An alternative would be to synthesize CD4 molecules, which could be injected into the bloodstream and bind to gp120 in the viral envelope, thus preventing infection. Both methods work in the lab, but could lead to immunological complications in vivo, as CD4 or anti-CD4 would interfere with the interaction of natural CD4 and its ligands.

2. The inhibition of **reverse transcriptase** is an effective method. As HIV is a retrovirus, its RNA

must first be transcribed into DNA. Compounds such as **azidothymidine** (AZT, also known as zidovudine), lamivudine, and dideoxyinosine (ddi) are structurally analogous to nucleotides, which is why the enzyme inserts them "by mistake" into the polynucleotide chain.

The first potent active agent against Herpes, i.e., **acyclovir**, is another drug that works as an inhibitor through structural analogy to nucleotides, blocking reverse transcriptase (revertase). It is very effective as an ointment to treat *Herpes simplex* and *Herpes zoster* (shingles). It is also comparatively non-toxic. Other inhibitors block the active site of HIV revertase (e.g., nevirapine and delavirdine).

3. **Antisense RNA** is an RNA copy that is precisely complementary to HIV. Antisense RNA does not code for proteins and has thus no function in the cell. The viral genome being a

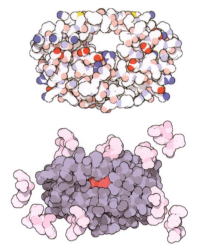

Computer-aided drug design: drugs can be designed and tested on computers. Automated docking methods are used to find the best docking sites on a biomolecule. If the predicted bond is strong enough, the molecule can be synthesized and its activity tested. The best site for saquinavir is shown in red.

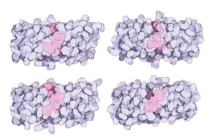

HIV protease (top) and AIDS drugs indinavir, saquinavir, ritnavir and nelfinavir (top left to bottom right).

single-stranded RNA, which is released during infection, it could immediately bind to the "waiting" antisense RNA to form a stable, "useless" RNA/RNA hybrid which would not be able to produce a provirus. This could be achieved through genetic therapy or stem cell treatment (Ch. 10).

The antisense drug fomivirsen has been successfully used for the treatment of a viral eye infection in AIDS patients and saved their eyesight.

4. Inhibition of **HIV protease**. Drugs that inhibit protease are a triumph of modern medicine and molecular design. The protease cleaves off the long polypeptide chains produced by the virus and cuts them into small fragments precisely when they are needed to pack the new viruses. The drug that firmly binds to the protease and blocks its action prevents the virus from maturing to its infectious stage.

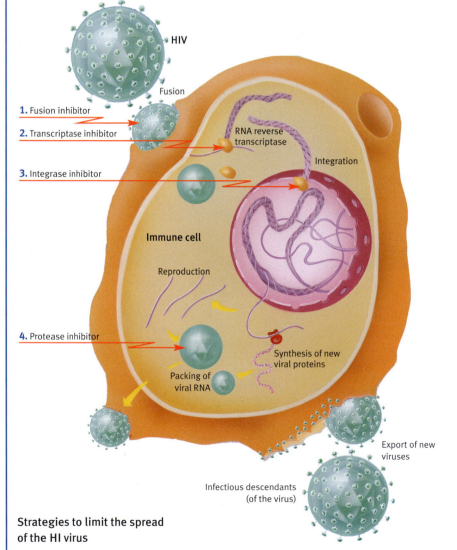

HIV

Fusion

1. Fusion inhibitor

2. Transcriptase inhibitor

RNA reverse transcriptase

3. Integrase inhibitor

Integration

Immune cell

Reproduction

4. Protease inhibitor

Synthesis of new viral proteins

Packing of viral RNA

Export of new viruses

Infectious descendants (of the virus)

Strategies to limit the spread of the HI virus

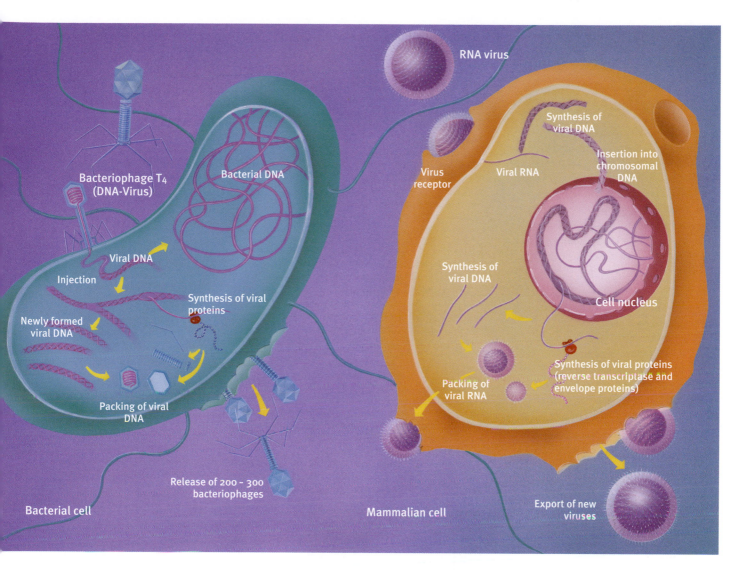

Fig. 5.6 How viruses attack cells.
Left: bacteriophages attack
Escherichia coli.
Right: HIV attacks a human cell.

Many HIV therapeutics are based on **inhibitors of virus-coded protease**, which plays an important part in the maturation process of viral proteins. All these different routes are followed in contemporary AIDS research (Box 5.1).

As the HI virus mainly attacks **T-helper cells** that organize the human immune response, it should be possible to strengthen the immune response by providing genetically engineered **cytokines** (e.g., interleukin-2). In a first step, the virus could be put out of action with "chemical weapons", and then immune cells could be stimulated by interleukin-2.

In other viral infections, **interferons** are used. Virus-infected cells naturally produce and secrete interferons. The secreted or artificially introduced interferon binds to specific receptor molecules on the surface of other cells and modifies their activities. Proteins are synthesized that make the cells resistant to viral infections.

Just like the lymphokine **interleukin-2 (IL-2)**, interferons were first enthusiastically hailed as the wonder drug of the future that would be able to cure a host of diseases from the common cold to cancer. However, they did not fulfill these unrealistic expectations. Their effectiveness as a cure is limited, whereas the side effects are often considerable. Like IL-2, they have their place in the treatment of human disease, but usually in combination with other medications (Ch. 9).

■ 5.3 How the Body Defends Itself Against Infections – Humoral Immune Response Through Antibodies

When Europeans conquered the Americas, they were assisted by biological weapons of which they were not aware at the time – bacteria and viruses. These killed a large proportion of the native inhabi-

Fig. 5.7 Body defenses against infections. An extensive description is given in Chapter 9. This picture shows macrophages, antibodies, and T-cells.

Box 5.2 The Expert's View: Testing for HIV infection

There are many reasons for wanting to establish an individual's HIV status, i.e., to test whether or not he or she is infected with HIV. For the individual, knowing one's status is key to benefiting from the enormous medical advances that have occurred over the past 25 years: with modern antiretroviral therapy, most infected people can enjoy a high quality of life while living with HIV rather than dying from AIDS. In addition, large numbers of HIV tests are done to ensure the safety of blood donated for transfusion and to screen pregnant women, so that measures can be taken to reduce the risk of mother-to-child transmission.

Before any HIV test is done, however, the **individual's informed consent** should be obtained. The term "AIDS test" should be avoided: AIDS is a clinically defined condition that develops in most HIV-infected individuals after a number of years of infection; testing is done to identify the presence of HIV.

HIV infection is normally diagnosed indirectly, through **detecting virus-specific antibodies**. Virtually all HIV-infected individuals will form such antibodies, but unfortunately these do not confer immunity, in contrast to most other viral infections.

There are various tests to detect antibodies. Most commonly, so-called **enzyme-linked immunosorbent assays** (**ELISA**) are used for screening purposes. Such screening tests have a very high sensitivity, i.e., they are able to identify positive samples; typically, far less than 1 in 1000 positive samples will yield a false-negative test result. This is achieved through the use of appropriate antigens (to which patients' antibodies react in the test) and careful optimization of the whole assay (designed to make the antigen-antibody reaction visible), in order to minimize the chance of obtaining a false-negative result and thus missing the diagnosis of an infection.

On the other hand, the **specificity**, i.e., the ability to correctly identify negative samples, of screening tests is usually not as high; this means that occasionally a sample will have a positive (or, rather, reactive) result although it does not actually contain antibodies against HIV. This non-specific reactivity can be caused by numerous factors, most of which are not associated with any pathological condition. A reactive ("positive") screening test on its own does not necessarily mean that the person tested is infected with HIV!

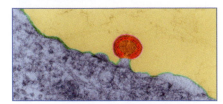

The Human immunodeficiency virus (HIV) is the cause of Acquired Immunodeficiency Syndrome (AIDS). It is a retrovirus – an enveloped virus that has an RNA genome.

To replicate, the RNA genome undergoes reverse transcription into DNA with the help of reverse transcriptase. Another enzyme, known as integrase, helps to integrate the viral DNA into the host genome. In AIDS, the immune system begins to fail, leading to life-threatening opportunistic infections. Infection with HIV can be transferred via blood, semen, vaginal fluid, preejaculate, or breast milk.

These bodily fluids may contain free virus particles or viruses within infected immune cells. The three major transmission routes are unprotected sexual intercourse, contaminated needles, and transmission from an infected mother to her baby at birth, or through breast milk. HIV primarily infects vital cells in the human immune system such as T-helper cells (specifically CD_4^+ T-cells), macrophages and dendritic cells. HIV infection leads to low levels of CD_4^+ T-cells through different mechanisms. When CD_4^+ T cell numbers fall below a critical level, cell-mediated immunity is lost, and the body becomes increasingly susceptible to opportunistic infections.

If untreated, eventually most HIV-infected individuals develop AIDS and die, while about one in ten remain healthy for many years, with no noticeable symptoms.

AIDS virus in the blood, surrounded by Y-shaped antibodies.

For this reason, each reactive screening test result must be confirmed by at least one confirmatory assay. This can be the so-called Western blot (obligatory in the USA and Germany) or a series of different tests applied in a defined sequence (algorithm). Only if this confirmatory testing confirms the sample's reactivity can HIV infection be diagnosed and the patient be told that he or she is HIV-positive. A second blood sample should then also be sent for testing to confirm the specimen's identity.

While the sensitivity and specificity of a specific HIV assay are usually known, another performance parameter is more relevant in practice. We do not actually know the "true" HIV status of the patient tested but we have to deduce it from the test results.

The **positive predictive value** (**PPV**) is the probability with which a positive test result indicates a truly infected patient; vice versa, the negative predictive value (NPV) is the probability with which a patient's negative test result reflects that he/she is truly uninfected. These predictive values depend not only on the sensitivity and specificity of the test used but also on the HIV prevalence of the population tested. Unfortunately, this statistical phenomenon is often misused to "prove" the alleged uselessness of HIV testing. In groups with a very low HIV prevalence (e.g., carefully selected blood donors), the majority of those with a reactive screening test result is indeed not infected. But this is exactly why all reactive screening tests must be confirmed before a diagnosis is made; it is not a reason to decry HIV tests as useless! In high prevalence populations, the vast majority of reactive test results sadly reflect true positivity; this is why the World Health Organization's guidelines stipulate simpler confirmatory algorithms in such settings.

In some settings, so-called **rapid/simple test devices** (also called point-of-care tests) are preferable to laboratory-based testing. These can often be done on capillary blood (obtained from the tip of a finger), are easy to perform, require minimal equipment, and test results are typically available within half an hour or less.

Rapid tests are valuable if the result is needed quickly: in emergency rooms, after needle stick injuries, etc. They can also help to reduce the rate of "unclaimed" test results (i.e., patients not returning to get their test results) and are the only feasible option in many resource-constrained settings. An algorithm consisting of different rapid tests may even be used for confirmatory purposes, obviating the need to send samples to a laboratory at all. However, quality control of rapid testing is a formidable challenge but extremely important.

All tests based on the detection of HIV-specific antibodies have one problem in common: **They do not recognize individuals in the very early stages of infection** while the body is still mounting an immune response. The length of time until antibodies become detectable is known as the "window period". Approaches are available to significantly shorten the "window": direct diagnosis, by isolating an infectious virus or by detecting viral antigens or viral genomic material (nucleic acid). Virus isolation requires cell culturing in a specialized laboratory and is

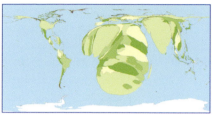

Prof Mark Newman (University of Michigan) and his team have created new types of maps: On the left, there is normal map of the world, showing the sizes of the countries of the world in proportion to their actual size on the surface of the planet and retaining their actual shapes. It is possible, however, to redraw the map with the sizes of countries made bigger or smaller in order to represent something of interest. Such maps are called cartograms and can be an effective and natural way of portraying geographic or social data. On the right, the cartogram shows the number of inhabitants with HIV/AIDS. (Maps courtesy of Mark Newman, UMICH)

HIV infection in humans is now a pandemic. As of January 2006, the Joint United Nations Programme on HIV/AIDS (UNAIDS) and the World Health Organization (WHO) estimate that AIDS has killed more than 25 million people since it was first recognized in December 1981. AIDS is one of the most destructive pandemics in recorded history. In 2005 alone, AIDS claimed an estimated 2.4 - 3.3 million lives, of which more than 570,000 were children. It is estimated that about 0.6% of the world's living population is infected with HIV.

A third of these deaths are occurring in sub-Saharan Africa, retarding economic growth and increasing poverty. According to current estimates, HIV is set to infect 90 million people in Africa, resulting in a minimum estimate of 18 million orphans. Antiretroviral treatment reduces both the mortality and the morbidity of HIV infection, but routine access to antiretroviral medication is not available in all countries.

therefore impractical and expensive. Testing for **viral p24 antigen** has become an integral component of testing using fourth-generation screening assays that detect viral antigen alongside specific antibodies and this method thus offers a much reduced "window period". Testing for viral genome (**nucleic acid testing** = NAT), e.g., using polymerase chain reaction (PCR), is done under certain circumstances: to exclude infectivity in blood donors and to diagnose HIV infection in patients with suspected primary infection and in babies born to HIV-infected mothers.

In such babies, passively (transplacentally) acquired maternal HIV antibodies are normally

detectable up to around 12 to 15 months of age. Therefore, children of HIV-positive mothers will initially have positive HIV antibody test results, until they have eliminated maternal antibodies. Fortunately, the majority of these babies will not themselves be infected; good prevention of mother-to-child transmission (PMTCT) programs can reduce the rate of vertical transmission to below 1%. NAT allows the early identification of infected babies and thus the initiation of appropriate clinical management.

When antibody tests are impractical, one can **test for proviral cDNA in leukocytes** using a qualitative ("yes or no?") assay to diagnose infection. The quantification of HIV RNA in blood plasma ("how much?"), the so-called **viral load**, may serve as a prognostic marker, to monitor the success of therapy and to estimate infectiousness. It has become a very important tool in the context of antiretroviral therapy. However, viral load tests are not intended to diagnose infection and may occasionally give false low-positive results in non-infected individuals.

In the right hands and done by skilled professionals, tests for HIV infection are nowadays **extremely reliable** and can give a definitive answer in almost all cases. If done in time, HIV testing allows avoidance of serious disease or even death through antiretroviral treatment and reduction of the risk of transmission to others through preventative measures.

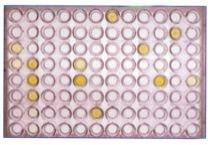

96-well microtitre ELISA plate

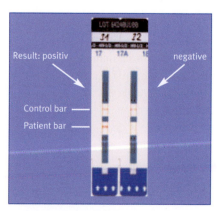

Example of rapid test (Capillus)

Born in Frankfurt am Main, Germany, **Wolfgang Preiser** studied medicine and later specialised in medical virology at J. W. Goethe University in his hometown and at University College London, U.K. During the outbreak of severe acute respiratory syndrome (SARS) in 2003 he was involved in the identification of the etiologic agent and was sent to China as a temporary advisor to the World Health Organisation (WHO). Since 2005, he has been Professor and Head of the Division of Medical Virology at Stellenbosch University in South Africa. His research focuses on developing and evaluating novel methods for laboratory diagnosis of viral infections and on tropical and emerging viruses.

Stephen Korsman is a medical virologist at the Walter Sisulu University/Mthatha National Health Laboratory Service Pathology Department in South Africa. He earned his MD in 1999 and became a Fellow of the College of Pathologists of South Africa as a clinical virologist in 2005. He was awarded an MMed in clinical virology in 2006. His fields of interest are mother-to-child transmission of HIV, emerging infectious diseases, molecular diagnostics, and medical education.

Cited Literature:

World Health Organization (WHO), HIV/AIDS Diagnostics: *http://www.who.int/hiv/amds/ diagnostics/en/index.html*

Center for Disease Control and Prevention (CDC), USA, Division of HIV/AIDS Prevention: *http://www.cdc.gov/hiv/testing.htm, http://www.cdc.gov/hiv/rapid_testing/*

U. S. Food and Drug Administration (FDA), Center for Biologics Evaluation and Research (CBER): Licensed/Approved HIV, HTLV and Hepatitis Tests: *http://www.fda.gov/cber/products/testkits.htm*

Preiser W, and **Korsman S** (2006): *HIV Testing*. Chapter 3 in HIV Medicine 2006. Full text available free online at *http://hivmedicine.com/textbook/ testing.htm* (This is the 14th edition of a regularly updated medical textbook that provides a comprehensive and up-to-date overview of the treatment of HIV Infection (825 pages, ISBN: 3-924774-50-1 – ISBN-13: 978-3-924774-50-9). Full text available free online at *http://hivmedicine.com/*)

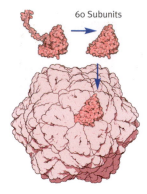

60 Subunits

Fig.5.8 Above: subunits of a virus combine to form a capsid. Above right: The polio virus, which has been fully synthesized in the test tube for the first time in 2002, consists of a long RNA strand in a hollow protein capsid (far right).

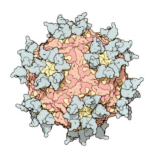

Fig.5.9 How antibodies neutralize a virus: Only the binding arms of the antibodies are shown here in light green, not their "feet". The diagram clearly shows how viral spikes are covered by the antibodies and thus neutralized.

Fig. 5.10 How antibodies bind to antigens: A deep cavity binds to a small fullerene molecule (top, Buckminster fullerene, shown in red); while the large flat surface of an antigen-binding site binds to a protein, lysozyme (bottom, also shown in red).

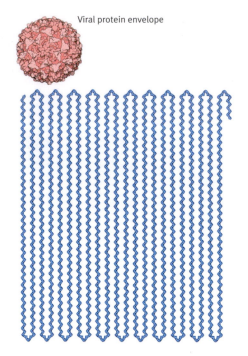

Viral protein envelope

tants. Between the 16th and the 19th century, the influx of conquerors and settlers into the Americas and Oceania brought measles, smallpox, influenza, typhoid, diphtheria, malaria, mumps, whooping cough, the plague, tuberculosis, and yellow fever, whereas the natives had just one fatal pathogen "on their side" – syphilis. It seems that epidemics were something unknown on the American continent. According to **Jared Diamond**, the inequality of weapons between Red Indians or Indios and Europeans developed through the large cattle herds of the settled Eurasian farmers. These herds became the breeding ground for acute and endemic diseases that later spread to similarly crowded human settlements (Box 5.4).

This explains why outbreaks of infectious diseases as we know them are a fairly recent phenomenon. Smallpox appeared for the first time in 1600 B.C, mumps and the plague at 400 B.C., and cholera and louse-borne epidemic typhoid only in the 16th century. Over several hundred years, the Eurasian population had been able to develop **immunity** to the diseases and reached a stalemate. The inhabitants of the New World, by contrast, did not have time to arm their totally unprepared immune system and were almost completely wiped out by the pathogens.

How does the immune system protect us? Let us just resort to a simplified answer at the moment, as the immune system is so complex that a full explanation would go beyond an introduction to

biotechnology. The immune system distinguishes between **"self" and "non-self"**. It has the ability to produce a hundred million (10^8) different antibody specificities and over a trillion (10^{12}) different T-cell receptors. The immune system consists of two closely interconnected systems with parallel action – humoral and cellular immune response.

The **humoral immune response** (Latin: *humor* = liquid) uses soluble proteins, antibodies (immunoglobulins, Box 5.2), as recognition elements. There are also humoral defense factors, lysozyme (Ch. 2) and interferons. Antibodies bind to alien molecules or cells, thus labeling them as intruders and encouraging phagocytosis by macrophages. The antibodies are produced by plasma cells, which, in turn, are derived from **B-cells**.

B-cells got their name B-lymphocytes from *Bursa fabricii* – a lymphatic organ that is unique to birds and lies in the end section of the cloaca (Fig. 5.11). Lymphocytes develop into B-lymphocytes in that bursa. If the bursa were removed from a chicken, it would become very susceptible to bacterial infections and would be unable to produce antibodies.

An alien macromolecule (or a cell or virus) is called an **antigen**. Antibodies do not target the whole antigen, but have an affinity to an exposed site in the molecule, which is known as **epitope** or **antigenic determinant**.

An **infection** mobilizes several co-operating immune cell populations. B-lymphocytes carry antibodies as recognition molecules (surface receptors) on their surface. However, they are usually not activated by circulating antigens. These are taken up by antigen-presenting cells – either **macrophages** or **dendritic cells**. They process the antigen, carry it on their surface and present it to T-helper cells. The antigen acts as a stimulus on the T-cells to produce interleukin-2, which, in turn, activates the B-cells that have also been in contact with the antigen. These begin to proliferate and form a cell clone (**clonal selection**) and differentiate. Some of their descendants become **memory cells** which ensure a rapid immune response in the case of re-infection, while others develop into antibody-producing plasma cells.

The freely circulating antibodies bind to the antigen, thus labeling it for destruction by other components of the immune system.

Box 5.3: Antibodies

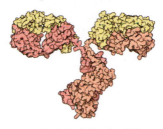

The two light chains (yellow) and the two heavy chains (red) combine to form a variable fragment that binds to the antigen and holds it in two "hands". The variable region contains up to 20 hypervariable amino acids that allow billions of combinations for the recognition of antigens.

Antibodies are components of the immune system of vertebrates that protect against intruders. Their task is to specifically recognize pathogens, bind to them and thereby label them for the immune system. They also neutralize the toxins released by many pathogens.

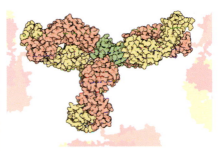

Three antibody fragments (Fab) bind simultaneously to the epitopes of an antigen (green). The enzyme shown is the familiar lysozyme (see Ch. 2). The contours of the second Fab and the "foot" of the antibody (Fc section) have been added.

Antibody molecules consist of four protein chains – two light chains or **L-chains** ("light"in this context means a molecular weight of 25 kDalton) and two heavy chains or **H-chains**

with a molecular weight of 55 kDalton. They are held together by disulfide bonds. The "foot" is the same for all antibodies and is called the constant region (Fc, fragment constant). The Fc part holds the antibody on the surface of the B-cell.

How is it possible for an organism to produce a specific antibody to virtually any antigen? How can it recognize the incredibly large number of possible antigens? As it would not be economical to have a specific gene for each of the approximately 100,000,000 different antibodies, the immune system resorts to an ingenious trick.

Immunotechnologists **Frank Breitling** and **Stefan Dübel** explain it as follows: Just as it is

possible to build any number of different houses with a few types of standardized bricks, cells use standardized polypeptide elements to build a modular antibody. Thus, only a few hundred of these polypeptide elements need to be coded for in the genome. Some large building blocks code for the constant regions in the antibody.

The antigen-binding specificity of an antibody, however, is mediated by a small proportion of the whole protein. These are the variable regions. They, too, consist of three to four different modules which are combined individually in each cell during the differentiation of B-lymphocytes.

Immunoglobulin G (IgG)

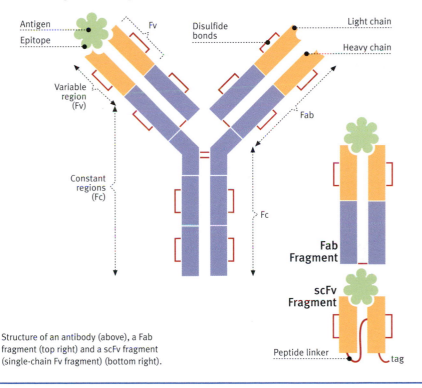

Structure of an antibody (above), a Fab fragment (top right) and a scFv fragment (single-chain Fv fragment) (bottom right).

These mechanisms are supported by the body's own **complement system**, a cascade involving about 30 proteins.

These are either dissolved in the blood plasma or are cell-bound and form a defense line against microorganisms (e.g., bacteria, fungi, or parasites). They have powerful cell-destroying properties which can cause tissue damage when not properly regulated, due to various diseases (heart attack, systemic *Lupus erythematodes*, or **rheumatoid arthritis**). Complement components damage cell membranes, thus destroying bacteria or damaged or degenerate cells (Ch. 9).

■ 5.4 Cellular Immune Response: Killer T-Cells

When the thymus gland is removed from young animals, they become susceptible to infection, similar to chickens after removal of the bursa. The number of lymphocytes (white blood cells) drops dramatically after a thymectomy. After their origin in the thymus gland, these lymphocytes were called **T-lymphocytes** or **T-cells**.

Soluble antibodies (Fig. 5.7), although very effective against pathogens outside cells, provide almost no protection against viruses and mycobacteria (such

Fig. 5.11 B-cells (B-lymphocytes) were named after the lymphatic organ Bursa fabricii that is found exclusively in birds, in the last section of the cloaca.

Box 5.4 Biotech History:
Jared Diamond: Lethal microbes

The importance of lethal microbes in human history. In his bestselling book "Guns, germs, and steel – a short history of everybody for the last 13,000 years", Professor Jared Diamond starts an inquiry into the reasons why Europe and the Middle East became the cradle of modern societies. He kindly permitted the reprint of a drastically shortened extract about the role of microbes in human history:

The importance of lethal microbes in human history is well illustrated by European's conquest and depopulation of the New World.

Far more Native Americans died in bed from Eurasian germs than on the battlefield from European guns and swords.

Those germs undermined Indian resistance by killing most Indians and their leaders and by sapping the survivor's morale. For instance, in 1519 **Cortés** landed on the coast of Mexico with 600 Spaniards, to conquer the fiercely militaristic **Aztec Empire** with a population of many millions. That Cortés reached the Aztec capital of Tenochtitlán, escaped with the loss of "only" two-thirds of his force, and managed to fight his way back to the coast demonstrates both Spanish military advantages and the initial naïveté of the Aztecs. But when Cortés's next onslaught came, the Aztecs were no longer naive and fought street by street with the utmost tenacity.

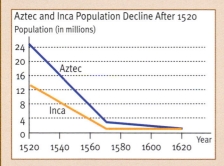

Aztec and Inca Population Decline After 1520
Population (in millions)

What gave the Spaniards a decisive advantage was smallpox, which reached Mexico in 1520 with one infected slave arriving from Spanish Cuba. The resulting epidemic proceeded to kill nearly half of the Aztecs, including **Emperor Cuitláhuac**. Aztec survivors were demoralized by the mysterious illness that killed Indians and spared Spaniards, as if advertising the Spaniard's invincibility. By 1618, Mexico's initial population of about 20 million had plummeted to about 1.6 million.

A woodcut of the Four Horsemen of the Apocalypse (1497 - 1498) by Albrecht Dürer (1471 - 1528, one of the greatest German artists).

Traditionally, the four horsemen stand for Pestilence, War, Famine, and Death. From the King James Version of the Bible, Revelation chapter 6, verses 2 to 8:

And I saw, and behold a white horse: and he that sat on him had a bow; and a crown was given unto him: and he went forth conquering, and to conquer. And when he had opened the second seal, I heard the second beast say, Come and see. And there went out another horse that was red: and power was given to him that sat thereon to take peace from the earth, and that they should kill one another: and there was given unto him a great sword. And when he had opened the third seal, I heard the third beast say, Come and see. And I beheld, and lo a black horse; and he that sat on him had a pair of balances in his hand. And I heard a voice in the midst of the four beasts say, A measure of wheat for a penny, and three measures of barley for a penny; and see thou hurt not the oil and the wine. And when he had opened the fourth seal, I heard the voice of the fourth beast say, Come and see. And I looked, and behold a pale horse: and his name that sat on him was Death, and Hell followed with him. And power was given unto them over the fourth part of the earth, to kill with sword, and with hunger, and with death, and with the beasts of the earth."

Pizarro had similarly grim luck when he landed on the coast of Peru in 1531 with 168 men to conquer the Inca Empire of millions. Fortunately for Pizarro and unfortunately for the Incas, smallpox had arrived overland around 1526, killing much of the Inca population, including both the emperor **Huayna Capac** and his designated successor. The result of the throne's being left vacant was that two other sons of Huayna Capac, **Atahualpa** and **Huascar**, became embroiled in a civil war that Pizarro exploited to conquer the divided Incas.

When we in the United States think of the most populous New World societies existing in 1492, only those of the **Aztecs and the Incas** tend to come to our minds. We forget that North Amer-

Francisco Pizarro asked the Inca leader Atahualpa to meet him and his body guards unarmed. He knew that if he had the Emperor he would have the entire Inca Empire, and all the gold which it held. With a blast of their cannons, Pizarro's men slaughtered all the Incas in the square of Cajamarca.

ica also supported populous Indian societies in the most logical place, the Mississippi Valley, which contains some of our best farmland today. In that case, however, conquistadores contributed nothing directly to the societies' destruction; Eurasian germs, spreading in advance, did everything.

The 11th and last Aztec emperor, Cuauhtémoc, surrenders in August 1521, and is presented to Cortés. The words in the legend "Ycpolinh q mexuca" are translated as "Now the Mexica [Aztecs] were finished."

When **Hernando de Soto** became the first European conquistador to march through the southeastern United States, in 1540, he came across Indian town sites abandoned two years earlier because the inhabitants had died in epidemics. These epidemics had been transmitted from coastal Indians infected by Spaniards visiting the coast. The Spaniard's microbes spread to the interior in advance of the Spaniards themselves.

De Soto was still able to see some of the densely populated Indian towns lining the lower Mississippi. After the end of his expedition, it was a long time before Europeans again reached the Mississippi Valley, but Eurasian microbes were now established in North America and kept spreading.

By the time of the next appearance of Europeans on the lower Mississippi, that of French settlers in the late 1600s, almost all of those big Indian towns had vanished. Their relics are the great mound sites of the Mississippi Valley. Only recently have we come to realize that many of the mound-building societies were still largely intact when Columbus reached the New World, and that they collapsed (probably as a result of disease) between 1492 and the systematic European exploration of the Mississippi.

When I was young, American schoolchildren were taught that North America had originally been occupied by only about one million Indians. That low number was useful in justifying the white conquest of what could be viewed as an almost empty continent. However, archaeological excavations, and scrutiny of descriptions left by the very first European explorers on our coasts, now suggest an initial number of around 20 million Indians. For the New World as a whole, the Indian population decline in the century or two following Columbus's arrival is estimated to have been as large as 95 percent.

The **main killers were Old World germs to which Indians had never be exposed**, and against which they therefore had neither immune nor genetic resistance. Smallpox, measles, influenza, and typhus competed for top rank among the killers. As if these had not been enough, diphtheria, malaria, mumps, pertussis, plague, tuberculosis, and yellow fever came up close behind. In countless cases, whites were actually there to witness the destruction occurring when the germs arrived.

For example, in 1837 the **Mandan Indian tribe**, with one of the most elaborate cultures in our Great Plains, contracted smallpox from a steamboat traveling up the Missouri River from St. Louis. The population of one Mandan village plummeted from 2,000 to fewer than 40 within a few weeks.

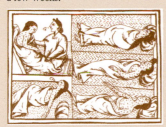

An Aztec drawing representing patients affected by smallpox at different stages.

While over a dozen major infectious diseases of Old World origins became established in the New World, perhaps **not a single major killer reached Europe from the Americas**. The sole possible exception is syphilis, whose area of origin remains controversial.

The one-sidedness of that exchange of germs becomes even more striking when we recall that large, dense human populations are a prerequisite for the evolution of our crowd infectious diseases. If recent reappraisals of the pre-Columbian New World population are correct, it was not far below the contemporary population of Eurasia. Some New World cities like Tenochtitlán have awful germs waiting for the Spaniards?

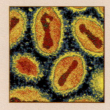

Smallpox (also known by the Latin name *Variola*) is a highly contagious disease unique to humans.

Instead, what must be the main reason for the failure of lethal crowd epidemics to arise in the Americas becomes clear when we pause to ask a simple question. **From what microbes could they conceivably have evolved?** We've seen that **Eurasian crowd diseases evolved out of diseases Eurasian herd animals** that became domesticated. Whereas many such animals existed in Eurasia, only five animals of any sort became domesticated in the Americas and the dog throughout the Americas.

In turn, this extreme paucity of domestic animals in the New World reflects the paucity of wild starting material. About 80 percent of the big wild mammals of the Americas became extinct at the end of the last Ice Age, around 13,000 years ago. The **few domesticates that remained to Native Americans** were not likely sources of crowd diseases, compared with cows and pigs. Muscovy ducks and turkeys don't live in enormous flocks, and they're not cuddly species (like young lambs) with which we have much physical contact.

The historical importance of animal-derived diseases extends far beyond the collision of the Old and the New Worlds. Eurasian germs played a key role in decimating native peoples in many other parts of the world, including **Pacific islanders, Aboriginal Australians**, and the Khoisan peoples (Hottentots and Bushmen) of southern Africa. Cumulative mortalities of these previously unexposed peoples from Eurasian germs ranged from 50 percent to 100 percent. For instance, the Indian population of Hispaniola declined from around 8 million, when Columbus arrived in A.D. 1492, to zero by 1535. Measles reached **Fiji** with a Fijian chief returning from a visit to Australia in 1875, and proceeded to kill about one-quarter of all Fijians then alive (after most Fijians had already been killed by epidemics beginning with the first European visit, in 1791).

Syphilis, gonorrhea, tuberculosis, and influenza arriving with **Captain Cook** in 1779, followed by a big typhoid epidemic in 1804 and numerous "minor" epidemics, reduced **Hawaii's** population from around half a million in 1779 to 84,000 in 1853, the year when smallpox finally reached Hawaii and killed around 10,000 of the survivors. The examples could be multiplied almost indefinitely.

However, germs did not act solely to Europeans' advantage. While the New World and Australia did not harbor native epidemic diseases awaiting Europeans, tropical Asia, Africa, Indonesia, and New Guinea certainly did. **Malaria** throughout the tropical Old World, **cholera** in tropical Southeast Asia, and **yellow fever** in tropical Africa were (all still are) the most notorious of the tropical killers. They posed the most serious obstacle to European colonization of the tropics, and they explain why the European colonial partitioning of New Guinea and most of Africa was not accomplished until nearly 400 years after European ship traffic, they emerged as the major impediment to colonization of the New World tropics as well. A familiar example is the role of those two diseases in aborting the French effort, and nearly aborting the ultimately successful American effort, to construct the **Panama Canal**.

There is no doubt that Europeans developed a big advantage in weaponry, technology, and political organization over most of the non-European peoples that they conquered. But that advantage alone doesn't fully explain **how initially so few European immigrants came to supplant so much of the native population** of the Americas and some other parts of the world. That might not have happened without Europe's sinister gift to other continents-the germs evolving from Eurasians' long intimacy with domestic animals.

For Jared Diamond's bios, see page 23 in Chapter 1.

Cited Literature:

Diamond J (1998) *Guns, germs, and steel. A short history of everybody for the last 13,000 years.* Vintage, Random House, London, Pp 210-213 and 214

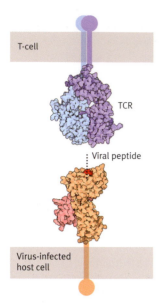

Fig. 5.12 Top: T-cell receptor (TCR), bottom: structure of a MHC class I protein with presented viral peptide (red).

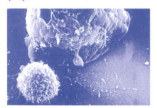

Fig. 5.13 above: T-killer cell (small, in the foreground) attacking a virus-infected cell.
Below left: T-cells with CD8 (green) and T-cell receptor (TCR, blue).
Below right: T-helper cell with CD4 (green) and T-cell receptor (TCR, blue). Both dock onto the virus-infected cell presenting a viral peptide (red).

as leprosy and tuberculosis). These are shielded from the antibodies by the host cell membrane. Evolution has, however, found a cunning defense strategy – **cell-mediated immune response**.

Cytotoxic T-lymphocytes are always on the lookout for alien components on the surface of all cells they encounter, and will destroy these cells where they find them (Figs. 5.12 and 5.13). This is not an easy task, especially because the intruders take great care not to leave any trace, but host cells have an ingenious "cut-and-present" mechanism (through proteasomes). They cut out a sample of small peptides found in the cytosol, which originate from the breakdown of the intruder's proteins. These peptides are transported to the surface of the cell membrane where they are presented by cell membrane proteins (Fig. 5.12) that have been coded by the **major histocompatibility complex, MHC**.

MHC proteins belong either to class I or class II (Fig. 5.13). **MHC proteins of class I** in an infected cell hold on very tightly to the peptides to be presented in order to enable the receptors of a T-killer cell to touch and examine them. When alien peptides are found, they act as a killer signal and trigger **apoptosis**, i.e., "cell suicide in the interest of the whole organism".

Cytotoxic cells carry an additional protein called **CD8** (CD stands for cluster of differentiation) that recognizes complexes of MHC class I protein and the presented alien peptide. When such a complex is recognized, the T-cell secretes the protein **perforin**

which opens pores of 10 nm in diameter in the membrane of the target cell. Proteases (granzymes) are secreted into the now permeable target cell. When they die, the target cells fragment their own as well as the viral proteins. The T-cell detaches itself from the target cell and proliferates, having proved to be an effective tool against the intruder.

Not all T-cells are cytotoxic or killer cells. **T-helper cells** are indispensable in the fight against extracellular as well as intracellular pathogens. They stimulate the multiplication of B-lymphocytes and cytotoxic T-cells.

T-helper cells are also activated through the recognition of antigens on the surface of antigen-presenting cells – usually **dendritic cells**. The antigen is presented to the T-helper cells as a peptide fragment that has been derived from the foreign protein in the antigen-presenting cell (what has been described as being "processed" earlier on). The recognition of the antigen by the T-helper cells depends on **MHC proteins class II** on the antigen-presenting cell.

When MHC proteins class II present a peptide, it is a call for help. "This cell has been in contact with a pathogen", whereas the message of class I is: "This cell has succumbed to the attacker. Trigger self-destruction."

T-helper cells use a T-cell receptor and a protein (CD4) on their surface that carries an extracellular immunoglobulin-like domain (structured like an antibody) (Fig. 5.13). The recognition of the complex triggers events that do not lead to the death of the cell, but stimulates the T-helper cells to secrete **lymphokines**, including inerleukin-2 and interferon-γ. IL-2 stimulates the proliferation of those B-cells that have also been in contact with the antigen. These turn into antibody-producing plasma cells in the blood. The HIV virus knocks out the immune system by destroying T-helper cells.

Neopterin is a newly discovered signaling substance that is activated in viral infections. This low – molecular – mass compound is secreted into the blood plasma by macrophages when stimulated by interferon-γ. As **Dietmar Fuchs** and his research group (Fig. 5.22) in Innsbruck, (Austria), showed, this can be used to provide a rapid test for viral infections. A high concentration of neopterin indicates that a viral attack is taking place, even if the culprit virus has not been identified yet. This can be

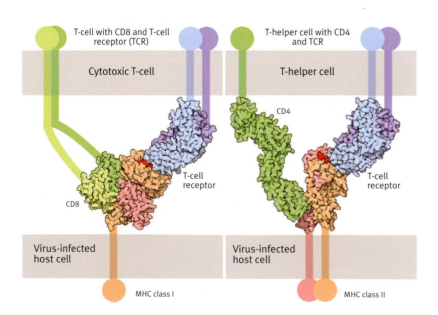

Box 5.5 Biotech History: Inoculation

"A non-medically qualified individual uses material of unknown composition and toxicity to treat patients, including a child, who may be suffering from a potentially fatal illness. The individual does not even try to obtain informed consent, but publishes patients' names and addresses to help publicise some astounding claims. Moreover, like fraudulent quacks the world over, the individual keeps details of the 'treatment' secret, so that its validity cannot be independently validated. Perhaps worst of all, this reckless person injects human beings with an extremely virulent microbe before conducting tests in animals. Some patients die, and a close collaborator who is a medical doctor dissociates himself from his colleague's work.

The person who took these risks, yet emerged with thunderous acclaim for his astonishing triumph in defeating rabies, was **Louis Pasteur**."

This is a quote from **Bernard Dixon**, *Power unseen. How microbes rule the world*. Pasteur himself said that he was extremely lucky: "chance favors the prepared mind". This also applies to his other pioneer achievements. Just like his predecessor **Edward Jenner**, who applied the first vaccines, he flouted several ethical principles. He postulated that the pathogen would be found in the spinal cord, although it was a completely unknown microbe then. The pathogen could be attenuated by removing the spinal cord from rabbits that had been inoculated with spinal cord material from a rabid dog and letting it age.

On July 6th, 1885, he injected little **Joseph Meister** with some of the aged rabbit spinal cord, although he could not be sure it contained the virus. Nowadays, the torpedo-shaped rabies viruses, members of the *Rhabdovirus* family, can be made visible under the electron microscope (Fig. 5.5). In Pasteur's day, they remained invisible. In contrast to Jenner's vaccine, Pasteur's had been produced in the lab.

Jenner, however, was on comparatively safe ground, having noticed that cowpox is innocuous. In the 11th century A.D., Chinese doctors had observed that people who had survived a smallpox infection were resistant to recurring infection. In ancient China, toddlers were therefore infected with smallpox on purpose. Given the high general infant mortality, the risks associated with the vaccination appeared acceptable. In China, there was a smallpox goddess called *Chuan Hsing Hua Chieh*, and the equivalent Hindu goddess was *Shitala mata*. Vaccine reactions were less pronounced when the vaccine came from patients with a mild form of smallpox. The use of such smallpox vaccines also spread to Europe and became quite popular in the second half of the 18th century under the name of "variolation".

In 1721, **Lady Mary Wortley Montagu** had her daughter "variolated" – the first person in England to have been "officially" vaccinated. Experiments had been carried out on prisoners and orphans, which gave British doctors the confidence to inoculate members of the Royal family.

However, it was **Edward Jenner** who demonstrated that there was a generally low-risk method to achieve the same purpose, and his vaccination method was one of the reasons for the decline in smallpox, without which the Industrial Revolution could not have taken place. The true significance of Jenner's discovery, however, was highlighted much later through Louis Pasteur's research. Pasteur had been experimenting with *Pasteurella multocidia*, a poultry pathogen, and one of his cultures had been forgotten in the lab for several weeks before Pasteur used it on his chickens. He found that not only did the chickens survive the infection, but also became immune to further infections with the pathogen.

Earlier on, in 1881, Pasteur had been able to demonstrate in public that sheep could be vaccinated to protect them from anthrax. He had also been the savior of silkworm breeding, as well as having expounded the scientific principles of wine fermentation. Pasteur was definitely successful, and his successful fight against rabies became the foundation of the Institut Pasteur in Paris, as Pasteur was able to draw conclusions about immunological reactions that led to the development of a whole range of vaccination methods.

Robert Koch (1843-1910) was the founder of medical bacteriology. In a controversy with the argumentative Louis Pasteur, he was the first to show that cholera, anthrax, tuberculosis, and the plague are caused by specific bacteria. His research on tuberculosis earned Koch the Nobel Prize for Medicine or Physiology in 1905.

Numerous research trips to India, Japan, and African countries yielded valuable results in tropical medicine and parasitology, such as information about the plague, malaria, sleeping sickness (African trypanosomiasis), and cholera pathogens.

Emil von Behring (1854-1917)

Emil von Behring (1854-1917) introduced a new vaccination procedure that became known as **passive vaccination**, which involves injecting antiserum (**serum therapy**). From 1880 to 1889, he was a medical officer (Stabsarzt) in the Prussian army and started teaching at the Institute of Hygiene and Infectious Diseases in Berlin where he became an assistant of Robert Koch. This is where he began his collaboration with the Japanese doctor and microbiologist **Shibasaburo Kitasato** (1856-1931).

Shibasaburo Kitasato (1856-1931): Behring's Japanese coworker.

In 1890, Behring had first successes in treating diphtheria in animals. Diphtheria was one of the most dreaded infections at the time, dubbed the "children's destroying angel". In cooperation with **Paul Ehrlich**, Behring succeeded in creating an antiserum to it in 1893 and saved the lives of many children.

In 1895, Emil Behring was nominated director of the Institute of Hygiene at the University of Marburg, Germany. In 1901 – four years before his much-revered teacher Robert Koch – he was awarded the Nobel Prize for Medicine and Physiology for his discovery of antibodies and the production of vaccines. He was given the noble title 'von'.

In 1904, he founded a factory called *Behring-Werke* in Marburg, where sera against diphtheria and tetanus were produced in large quantities.

Fig. 5.14 Right: One-step production of modern vaccines against both smallpox and hepatitis

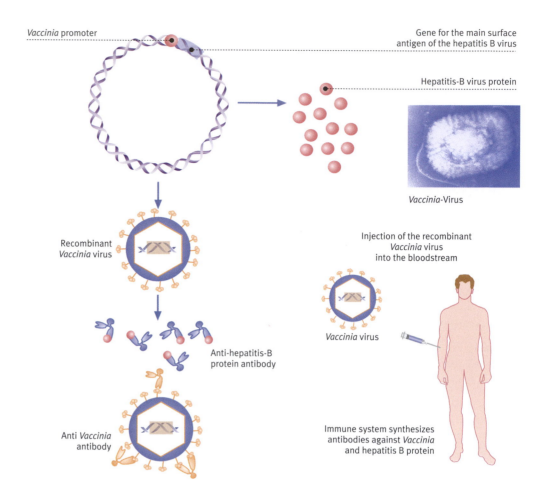

Vaccinia promoter

Gene for the main surface antigen of the hepatitis B virus

Hepatitis-B virus protein

Vaccinia-Virus

Recombinant *Vaccinia* virus

Injection of the recombinant *Vaccinia* virus into the bloodstream

Vaccinia virus

Anti-hepatitis-B protein antibody

Anti *Vaccinia* antibody

Immune system synthesizes antibodies against *Vaccinia* and hepatitis B protein

Fig. 5.15 Edward Jenner vaccinating a patient (top).
Jenner's drawing of the cowpox-infected arm of dairymaid Sarah Nelmes (below).

Fig. 5.16 The last patient in the world infected with smallpox, 23-year-old Somali Maow Maalin, in 1977. 200 to 300 million dollars have been spent on the eradication of smallpox in the world. Without antibiotics, Maow Maalin would not have survived the disease.

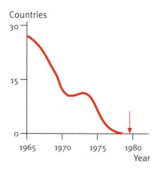

Countries

30 —

15 —

0 —

1965   1970   1975   1980
Year

Fig. 5.17 Number of countries with cases of smallpox.

useful in early diagnosis, and it is an invaluable indicator for blood banks who want to check their donors for viral infections. A neopterin test is routinely carried out by all Austrian blood banks.

■ 5.5 The First Vaccination: Cowpox against Smallpox

If the English country doctor **Edward Jenner** (1749 - 1823) were to repeat his famous experiment of 1796 (Fig. 5.15) today, he would soon find himself in prison. He injected eight-year-old James Phipps with a sample from a cowpox pustule of dairy girl **Sarah Nelmes** (Fig. 5.15). Then, two months later, he injected the boy with a potentially lethal dose of smallpox. This would nowadays be a violation of even the most lax safety guidelines, only exceeded by the experiments of **Louis Pasteur** (Box 5.3). However, the boy survived, and Jenner helped to revolutionize medicine. The first vaccine (Latin *vacca*, cow) in history had been found, although it has been said that there were active smallpox vaccinations at about 1000 B.C.

These days, biotechnology promises to revolutionize vaccination – making it possible to develop new **vaccines** within a very short time and dramatically reducing the risk involved. Jenner had correctly observed that overcoming cowpox conveyed lifelong immunity in humans, not only to cowpox, but also to smallpox. What Jenner did not know was that the **cowpox virus is closely related to the smallpox virus**. As explained above, there are lymphocytes in the blood that raise the alarm at the intrusion of antigens, passing on a command to other cells to produce antibodies. These label the pathogens for macrophages to destroy them (Fig. 5.7).

As mentioned above, B-cells proliferate after antigen contact and activation through T-helper cells. While a proportion of their offspring produce large quantities of antibodies to label the intruders, some of them become memory cells that enable a fast immune reaction, should the organism be reinfected.

These memory cells can remain in the system for life – conveying **lifelong immunity** to the relevant antigen.

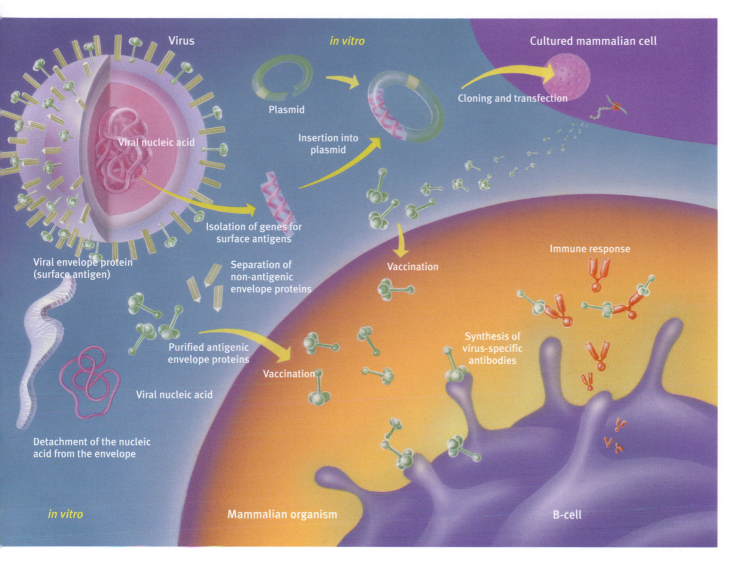

Virus

in vitro

Cultured mammalian cell

Plasmid

Cloning and transfection

Insertion into plasmid

Viral nucleic acid

Isolation of genes for surface antigens

Viral envelope protein (surface antigen)

Separation of non-antigenic envelope proteins

Vaccination

Immune response

Purified antigenic envelope proteins

Synthesis of virus-specific antibodies

Vaccination

Viral nucleic acid

Detachment of the nucleic acid from the envelope

in vitro

Mammalian organism

B-cell

Jenner was in luck because the similarity in the surface structure of cowpox and smallpox ensured that immunity was acquired not only for the innocuous cowpox, but also the far more dangerous smallpox. It is thus possible to prepare the immune system for the attack of life-threatening pathogens by exposing it to innocuous ones. The last smallpox patient in the world, the Somali **Ali Maow Maalin**, was discharged from hospital on October 26th, 1977. Over the following two years, the world population was scrutinized for smallpox and finally declared smallpox-free.

Only two laboratories in the world still hold smallpox viruses (one hopes!) – the international WHO reference labs in Atlanta (USA) and near Novosibirsk (Russia). There has been a debate about the wisdom of keeping or destroying the remaining stocks. In view of possible **bioterrorist** attacks, however, all industrial nations keep a stock of small-pox vaccines. Who would be confident enough to say that smallpox viruses will not get into the wrong hands? It may already have happened. For the first time in history, a disease has been eradicated due to vaccination.

At the beginning of the 19th century, half a million people per year contracted smallpox in Germany alone. One in ten died, and faces scarred by smallpox were a common sight.

Unfortunately, there are not many pathogens that have such close innocuous relations, and only with **Louis Pasteur**, born the year before Jenner died, did the systematic search for vaccines begin (Box 5.3).

■ 5.6 Contemporary Vaccination

Nowadays, we mainly use toxoids for vaccination, , killed or weakened live pathogens. After recent successes in genetic engineering, research is also being

Fig. 5.18 How vaccines are produced.

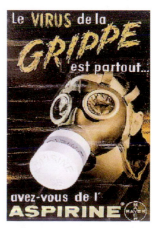

Fig. 5.19 Antibiotics are ineffective in viral infections, and the only useful medication is often an anti-inflammatory painkiller. (Poster text: The flu virus is everywhere... Have you got aspirin?)

Box 5.6 How Antibodies are Obtained

It has been known since the research by **Behring** and **Kitasato** around 1890 that specifically binding molecules can be isolated from blood.

The classical method is the **immunization** of laboratory animals, by injecting them with an antigen. The immunization process must be repeated several times successfully before antibodies can be isolated from the animals' sera. The Shanghai goat shown in Fig. 5.24 was inoculated with highly purified protein from a human heart muscle (h-FABP, which stands for heart Fatty-Acid-Binding Protein), and from its blood, h-FABP antibodies were isolated. The antibodies are a mixture of molecules that bind to various sites (**epitopes**) on the antigen. The strength of these bonds varies. Each of these specific antibodies is produced by its own B-lymphocyte clone in the blood, and the immune response relies on the reproduction of several different clones. These are called **polyclonal antibodies**.

Center: The method developed by Köhler and Milstein uses the **hybridoma** technique. In a first step, the lab animal (usually a mouse) is inoculated, and the antibodies to the antigen are first produced in the spleen, then circulate in the blood and lymph systems. These are a range of antibodies, derived from various cell clones, i.e., polyclonal antibodies. However, the aim was to produce large quantities of homogenous antibodies that only target a single epitope.

The antibodies are not taken from the blood this time, but from the spleen of the inoculated mouse. The large number of **B-lymphocytes** can be easily isolated. B-lymphocytes originate from bone marrow stem cells and reproduce in the spleen and lymph nodes, producing antigen-specific clones or differentiating into plasma cells or memory cells. A B-lymphocyte can only produce its very specific antibody.

The B-lymphocytes undergo *in vitro* fusion with **myeloma** cells, which are tumor cells that are easily cultured. The resulting hybridoma cells give rise to clones that produce uniform antibodies. These are called **monoclonal antibodies**.

The selected clones have the immortality of cancer cells, combined with the antibody production of lymphocytes and are therefore ready to be produced in unlimited quantities. However, the screening process is very labor- and material-intensive, as thousands of clones must be separately cultured and tested.

A third alternative is **recombinant antibodies** that are not produced in animals (*in vivo*), but in bacterial or cell cultures (*in vitro*).

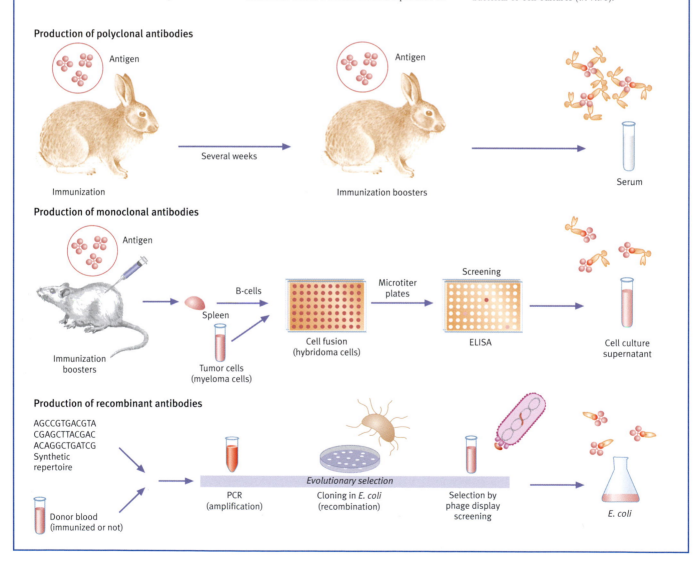

Production of polyclonal antibodies

Antigen — Antigen

Several weeks

Immunization — Immunization boosters — Serum

Production of monoclonal antibodies

Antigen

Immunization boosters — Spleen — B-cells — Tumor cells (myeloma cells) — Cell fusion (hybridoma cells) — Microtiter plates — Screening — ELISA — Cell culture supernatant

Production of recombinant antibodies

AGCCGTGACGTA
CGAGCTTACGAC
ACAGGCTGATCG
Synthetic repertoire

Donor blood (immunized or not)

Evolutionary selection

PCR (amplification) — Cloning in *E. coli* (recombination) — Selection by phage display screening — *E. coli*

done on modified live vaccines and peptide vaccines.

Toxoids are extracts from toxins released by pathogens. These are neutralized (sometimes using formalin) and stimulate the body's immune system, when injected. This applies, for example, to vaccines for tetanus and diphtheria.

The tetanus pathogen *Clostridium tetani* (the one who dwells in the ground), for example, infects a wound, injecting a neurotoxic protein into the bloodstream. This results in spastic paralysis – it used to be a shocking sight among soldiers wounded in battle.

The tetanus vaccination is one that requires a booster every now and then, to maintain a sufficient number of antibodies circulating in the system.

In cholera, polio, and typhoid vaccines, **chemically killed bacteria or viruses** are used as vaccines. Thus, the cholera vaccine cannot cause the disease, but it contains the neutralized toxin of the cholera bacterium.

The **active vaccine** is swallowed. It is called an active vaccine because the body produces its own antibodies to the killed bacteria and the toxin.

Cholera vaccination provides around 90 percent protection. Adults and children from six years on are given two vaccinations, between one and six weeks apart. Protection begins eight days after the vaccination and lasts about two years.

Attenuated (weakened) **pathogens** are used for Rubella and measles vaccinations. Unfortunately, there have been a number of incidents in which the pathogens had not been sufficiently attenuated.

Various **genetically engineered vaccines** for humans and animals have been in use since 1985, for example against foot-and-mouth disease in cattle. The first genetically engineered vaccine for humans was licensed in 1986 – at least in the US. It protects against hepatitis B.

The **hepatitis B virus** (HBV) is a DNA virus. It causes one of the most frequent infectious diseases worldwide, alongside tuberculosis and HIV. Up to 25 percent of people affected die from HBV sequels, such as cirrhosis or carcinoma of the liver. The virus is endemic in South East Asia and in sub-Saharan Africa. Due to vaccination programs, its presence in the Western European and US population has been reduced to 0.1 percent chronic virus carriers.

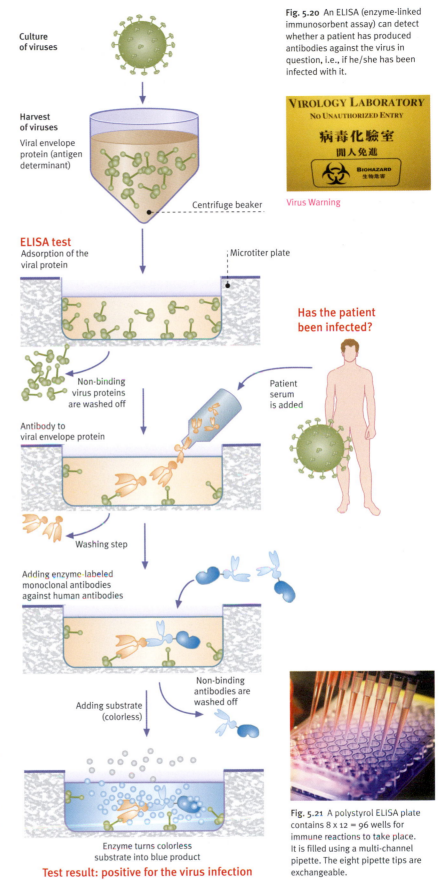

Culture of viruses

Harvest of viruses
Viral envelope protein (antigen determinant)

Centrifuge beaker

ELISA test
Adsorption of the viral protein

Microtiter plate

Non-binding virus proteins are washed off

Antibody to viral envelope protein

Washing step

Patient serum is added

Adding enzyme-labeled monoclonal antibodies against human antibodies

Non-binding antibodies are washed off

Adding substrate (colorless)

Enzyme turns colorless substrate into blue product

Test result: positive for the virus infection

Fig. 5.20 An ELISA (enzyme-linked immunosorbent assay) can detect whether a patient has produced antibodies against the virus in question, i.e., if he/she has been infected with it.

VIROLOGY LABORATORY
No Unauthorized Entry
病毒化驗室
閒人免進
BIOHAZARD
生物危害

Virus Warning

Has the patient been infected?

Fig. 5.21 A polystyrol ELISA plate contains 8 x 12 = 96 wells for immune reactions to take place. It is filled using a multi-channel pipette. The eight pipette tips are exchangeable.

Box 5.7 Biotech History: Monoclonal Antibodies

Georges Köhler who had studied in Freiburg, Germany, and written his PhD thesis in Basel, Switzerland, went to Cambridge in 1974 to work in **Cesar Milstein's** lab for two years.

Milstein had found a way around the problem that it is not possible to grow normal lymphocytes in cell colonies to provide a supply of antibodies. He cultured myeloma cells instead, which he had obtained from mice. These are cells descending from lymphocytes which have turned malignant. They have retained the ability of lymphocytes to produce antibodies, while also having the cancer cell characteristic of being immortal. If it were possible to interbreed them somehow, i.e., fuse them, with normal lymphocytes of known specificity, would the daughter cells produce those specific antibodies alongside the myeloma antibodies? In their experiment, they used erythrocytes (red blood cells) from sheep as an antigen and injected them into mice. Once the immune reaction had fully developed, theye took out the spleen where lymphocytes form in large quantities. They mashed the spleen tissue and mixed it with myeloma cell cultures, adding polyethylene glycol, a chemical compound that facilitates cell fusion. They hoped that this would result in hybrids with the desired properties.

And indeed, this cellular mix and match fulfilled expectations and produced spectacular results. Using the sheep erythrocytes as test antigen, Köhler and Milstein identified a considerable number of hybrid cells that recognized the erythrocytes as foreign and produced antibodies to them. Single hybrids could be grown in cultures, having inherited the immortality trait from their myeloma cell parent. They also produced the antibody of known specificity which they had inherited from their lymphocyte parent. These cells were called hybridoma cells.

After being awarded the Nobel Prize in 1984, Georges Köhler was asked in an interview why he and Milstein had not patented their method. After all, they could have earned millions, given the billions of dollars in sales of monoclonal antibodies. Köhler replied: "Prof Milstein told the relevant people on the Medical Research Council that we had found something that could be patented, but we did not get any reply. So we were not bothered and publicized our method. We are scientists, not businessmen.

I don't think scientists should patent anything. At the time, we did not think long and hard about it, but decided spontaneously, from the heart. It would have meant that I would have had to learn how to deal with money, how to negotiate licenses. It would have changed my whole personality, and I don't think it would have done me any good."

Georges Köhler (1946 – 1995)

Cesar Milstein (1927 – 2002)

Fig. 5.22 Top: Dietmar Fuchs (University of Innsbruck, Austria), the leading expert on neopterin, which is a signalling molecule produced by macrophages after a virus attack. Neopterin can be traced shortly after any kind of viral attack. Bottom: neopterin molecule.

The conventional production of hepatitis B vaccine was facing huge problems. In contrast to most other microorganisms, hepatitis B viruses cannot be bred in conventional nutrient media or animal embryos (such as fertilized chicken eggs). The vaccines had to be obtained from virus envelope proteins which cause an immune reaction, i.e., the surface antigens found in the blood of infected carriers. The viruses were mostly destroyed in the process, the envelope proteins were detached using detergents and then purified. These were used as vaccines, provoking an immune reaction.

Infectious blood is, of course, dangerous to work with. All members of the lab are therefore immunized, i.e., vaccinated, and the work is carried out in isolated secure labs. The matter was further complicated because each batch had to be tested on chimpanzees (of which, for ethical reasons, there were only a limited number available) in order to eliminate contamination with live viruses.

As the production of this vaccine takes nearly a year, there was only a limited amount of natural vaccine available. Only high-risk groups could be vaccinated.

The new genetically engineered vaccine against hepatitis is produced by **genetically modified yeasts**, which are eukaryotes, or by mammalian cells. It is based on a **viral surface protein**. Vaccine production in *E. coli* had not proved to be very effective, as the modified bacterial cells were not able to carry out the required protein modifications, glycosylations in particular (see end of Ch. 3).

DNA vaccines can be considered to be byproducts of genetic research. In DNA transfer experiments, it had been found that the resulting proteins often caused allergies, i.e., immune reactions.

Once the surface proteins that provoked the immune reaction had been identified, it was time to make a virtue out of necessity.

As antigens are proteins, it should be possible to isolate the responsible gene from the viral genome and insert it into that of harmless microorganisms such as bakers' yeast or mammalian organisms which could then produce large amounts of the protein. There is an added bonus to the method: viral contamination of the vaccine is impossible.

■ 5.7 Live Vaccines

Rabies has been an almost worldwide scourge, with the exception of North Western Europe, Japan, Australia and, a few Pacific islands.

The foxes in Europe's woodland are biotechnologically protected. Baits are laced with live vaccines. In live vaccines, innocuous viruses, such as the cowpox virus (*Vaccinia*) are used as vectors to transport alien genes. The double-stranded viral DNA comprises 180,000 base pairs.

In 1982, it was shown that at least two larger sections of the DNA are not needed for reproduction. These can therefore be augmented or replaced with foreign DNA. Genes that code for **envelope proteins with an antigenic effect** are thus stably inserted into the viral genetic material in a way that does not affect the ability of the vector virus to infect mammalian cells. Up to 20 foreign genes can be simultaneously introduced in this way, preferably coupled with a promoter gene that switches on their expression, i.e., their conversion into proteins (Fig. 5.14). The method has been successful in animal experiments, resulting in surface antigens of hepatitis B, rabies, *Herpes simplex*, and influenza viruses.

HIV vaccines entered clinical phase II tests at the beginning of 2005. These tests involve the trivalent vaccine "MRKAd5 HIV-1 gag/pol/nef", which is based on a modified common cold virus (adenovirus). Due to its modification, it has lost its ability to cause a cold, but serves as vector for three synthetically produced HIV genes. The HIV proteins that are expressed pose no risk whatsoever, but elicit an immune reaction in the organism.

Edible vaccines are highly controversial. In Chapter 7, we will look at the creation of transgenic plants. Specifically marked (perhaps through blue pigment, see Fig. 7.52) bananas or potatoes could be used like oral vaccines. The vaccine would be ingested via the gastrointestinal tract.

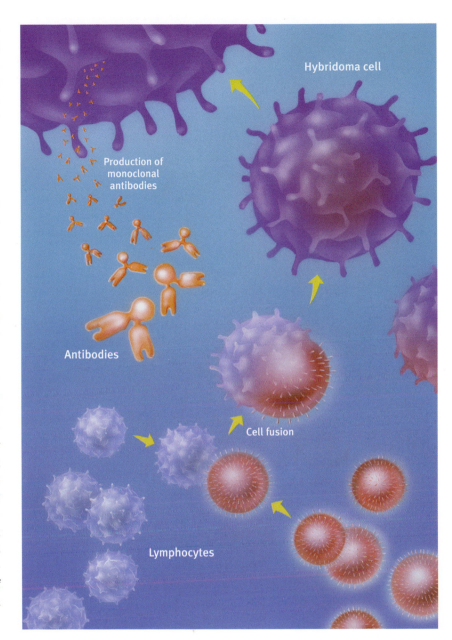

■ 5.8 Monoclonal Antibodies

August 7th 1975 saw a ground-breaking publication. **Cesar Milstein** (1927-2002, Box 5.6), an Argentinian who had fled the military dictatorship in his native country and worked in Cambridge, and the German **Georges Köhler** (1946-1995) described a method to produce **monoclonal antibodies**. These are antibodies with identical molecular structure and specificity. Monoclonal antibodies ushered in a new era in biochemical analysis and medical diagnostics. The two scientists were awarded the Nobel Prize together with **Niels Kai Jerne** (1911-1994) in 1984. The method involves immunizing an animal, retrieving **spleen** tissue from it,

Fig. 5.23 How hybridoma cells for monoclonal antibodies are made: Lymphocytes are isolated from the spleen of a mouse and fused (either using a chemical such as polyethylene glycol or electrofusion) with myeloma cells, i.e., immune cells that have turned malignant.

The resulting hybridoma cells combine the ability of lymphocytes to produce antibodies with the immortality of cancer cells.

Hybridoma cells cultured on medium grow into colonies of identical cells (clones) that synthesize the same (monoclonal) antibody and secrete it into the medium.

Fig. 5.24 Polyclonal antibodies are obtained from rabbits and – in larger quantities – from goats. Pictured here is "our" goat in Shanghai that was injected with antigens (in our case the heart attack marker FABP) to stimulate the production of antibodies. Some strains of mice have been exclusively bred to be inoculated and deliver spleen cells (bottom).

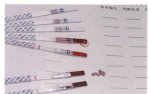

Fig. 5.25 A queue of volunteer donors at the reception desk of the Guangzhou bloodbank. Rapid "doorkeeper" tests for syphilis and hepatitis detect infected individuals upfront who will then be barred from giving blood.

growing a mixed culture of **myeloma** and spleen cells, **fusing** them and then selecting and breeding hybridoma cells (Box 5.5). What makes it so revolutionary is the potential of monoclonal antibodies to be converted into **large amounts of specific antibodies**.

The unique ability of the immune system to recognize certain structures at a molecular level can be harnessed by humans.

Opera buffs are familiar with the plot of **Carl Maria von Weber's** "Freischütz" where young Max tries to get hold of **magic bullets** that always find their target (Fig. 5.27).

Such magic bullets were also the dream of **Paul Ehrlich**, and monoclonal antibodies are just that in modern medicine, where they are widely used as diagnostic tools.

Take a viral infection, for example, when the body produces antibodies as a response to viral attack. Monoclonal antibodies against this particular virus can be used to identify the virus in the body fluids of the patient (Ch. 5, **ELISA test for viruses**). Such tests have already become routine for HIV and hepatitis (Figs. 5.20 and 5.21).

Often, however, it is not the virus itself that is identified. This would require very sensitive tests. What is picked up are the antibodies produced by the patient, and it can take weeks until these are produced in significant quantities. A very fresh HIV infection can thus go undetected.

When the SARS epidemic hit Hong Kong, it even proved tricky to detect viruses quickly and reliably through the **polymerase chain reaction** (PCR, see Ch. 10). The problem was compounded by the fact that the sensitivity of PCR is such that even contamination from the air in the lab was enhanced in the process. Healthy patients were identified as having SARS (false-positives) and became infected with SARS while in hospital.

What is true for viruses applies to other kinds of pathogens as well, and even for pathological surface structures of body cells. In certain kinds of cancer, for example, tumor cells display specific protein structures. These usually help the immune system to recognize them as tumor cells and are called tumor markers. It is possible to produce monoclonal antibodies to such **tumor markers**, which can help determine the size and position of the tumor in the body. Based on these results, a better decision can

be made on the treatment of the tumor, e.g., surgery or radiotherapy.

This is how the tumor is tracked down: A monoclonal antibody is radioactively labeled in a chemical *in vitro* process and then injected into the patient's bloodstream.

The antibodies disperse over the entire body, and those antibody molecules that encounter a tumor stably bind to them. Thus, after a while, a cluster of radioactivity emerges where the tumor is. In order to keep radiation to a minimum, only low-level radioactivity is used in the process, and the measuring tools are sensitive enough to precisely locate the concentration of this low-level radioactivity in the body. Treatment decisions can thus be taken on the basis of precise data about size and location of the tumor.

It is hoped that monoclonal antibodies also hold the key for an **improved chemotherapy**. The main problem of contemporary chemotherapy is that even the best drugs are extremely toxic, but cannot be targeted exclusively at cancer cells. Unwanted side effects are thus inevitable.

Monoclonal antibodies, coupled with the cytotoxic drugs, could be the magic bullets that deliver them to their targets.

The first major success in this area was the use of the monoclonal **rituximab** against non-Hodgkin-lymphoma (see below), based on a genetically modified antibody.

100 years have passed since **Paul Ehrlich** spoke of his vision of antibodies as magic bullets (Fig. 5.33), which finally seems to have come true. He talked of the effects of antibodies, which he named receptors and side chains and hoped they could become the basis for a whole new generation of highly selective drugs.

Further exciting developments are in store. **Bispecific antibodies** have been produced in which each of the two docking sites binds to a different antigen, and there are antibodies in the pipeline that contain components from several organisms (e.g., "humanized" antibodies, see below).

Antibodies with enzymatic activities open up further new avenues – **catalytic antibodies** or **abzymes**, which combine the sheer endless selection variability of antibodies with the catalytic power of enzymes.

■ 5.8 Catalytic Antibodies

Can antibodies act as enzymes? Given that both antibodies and enzymes bind to molecules in a similar way, i.e., through induced fit, **Richard Lerner** and his team at the Scripps Clinic in California had the idea of building antibodies with catalytic properties.

Enzymes drastically reduce the amount of activation energy required by binding not to the substrate itself, but to a **transition state** of the substrate (Ch. 2). Thus, the transition state is stabilized, while economizing on energy. The time needed to establish a reaction equilibrium can often be divided by a factor of 10^9.

The idea developed that if it was possible to develop an **antibody against transition states** of compounds, it should be possible for the antibody to catalyze the relevant reactions. In order to produce antibodies, antigens were required that would elicit an immune response in a lab animal. However, the transition state of a substrate is so unstable that it does not exist as a free molecule, and it is therefore impossible to make a lab animal produce antibodies. The problem was solved by chemically building models that perfectly resembled the chemical and spatial structure of the actual transition state of a substrate, but were stable (Fig. 5.26).

Here is an example: in ester hydrolysis (cleavage by inserting water into an ester), the reaction of the planar ester molecule (atoms lying in the same plane) involves a transition state in which the atoms are arranged as a tetrahedron (atoms sitting in the corners of the tetrahedron) before reaching the final planar shape (acid and alcohol). The ester is electrically neutral, while the transition state is polarized through positive and negative charges.

A **stable model compound** was found where phosphorus substitutes the central carbon atom of the ester (Fig. 5.26). The molecule imitates the geometry and charge distribution of the unstable transition state. The model compound was bound to a carrier molecule and successfully elicited an immune reaction in mice. Later, hybridoma cells were created, of which some clones with the ability to bind to the model compound were selected.

Ester was added to various monoclonal antibodies obtained in this way. Some of them showed no effect. These were probably specific to a part of the molecule that is not relevant to the transition state.

Desired chemical reaction

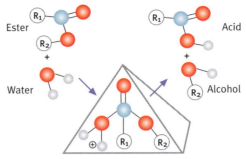

Ester + Water

Acid + Alcohol

Transition state (unstable)

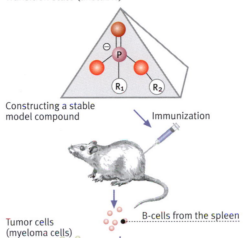

Constructing a stable model compound

Immunization

B-cells from the spleen

Tumor cells (myeloma cells)

Hybridoma cells

Catalytic antibody

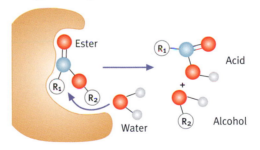

Ester

Water

R_1 Acid + R_2 Alcohol

Catalytically converts ester

Other antibodies, however, sped up the reaction by a thousand-fold.

The **Diels-Alder reaction** is a chemically very interesting phenomenon, as none of the enzymes involved occur naturally. Nevertheless, it has been possible to synthesize catalytic antibodies for it.

■ 5.10 Recombinant Antibodies

For many applications, the antigen-binding unit must be as small as possible. Before the emergence of gene technology, the traditional way of obtaining

Fig. 5.26 The production of catalytic antibodies, in this case catalytic antibodies for ester hydrolysis (diagram left).
Below: Despite its crucial significance in synthetic chemistry, the Diels-Alder reaction cannot be catalyzed by natural enzymes. The two starting compounds form a short-lived instable intermediate complex (shown in red) that decomposes into sulfur dioxide and the end product. Enzymes act by stabilizing the intermediate state. In order to turn an antibody into an enzyme, the intermediate state of the product must be chemically stabilized or mimicked.

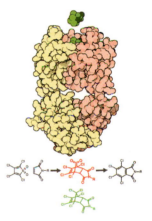

The green compound is such a stable imitation. It is then injected into mice and provokes an antibody reaction. The antibodies have catalytic action, i.e., they convert the starting products, although their effectiveness is not as dramatic as that of true enzymes. The diagram shows a top view of the two "hands" of the antibody.

My magic bullets are not diabolical, but monoclonal!

Fig. 5.27 If Samiel, the devil in the opera "Der Freischütz", had been around nowadays, he would have given away monoclonal antibodies rather than magic bullets.

Box 5.9 The Expert's View:
Frances S. Ligler: Sensing with Antibodies to Detect Something Lethal

In September and October of 2001, several cases of anthrax broke out in the US. It motivated efforts to define biodefense and biosecurity, where more limited definitions of biosafety had focused on unintentional or accidental impacts of agricultural and medical technologies. The Center for Disease Control (CDC) has defined and categorized **bioterrorism agents** according to priority. Category A agents are biological agents with both a high potential for adverse public health impact and that also have a serious potential for large-scale dissemination. The Category A agents are anthrax (by *Bacillus anthracis* bacterium), smallpox (by *Vaccinia* virus), plague (by *Yersinia pestis* bacterium), botulism (by *Clostridium botulinum*), tularemia (Rabbit Fever by *Francisella tularensis* bacterium) and viral hemorrhagic fevers (Ebola and Marburg viruses).

The human body has developed elegant methods for the recognition of hazardous molecules and pathogens. The most well known and best understood of these protective mechanisms involve antibodies. One of the most alluring characteristics of antibodies is that they can be generated to recognize most molecules with a unique shape and perform that recognition function in a way that is highly specific. However, in contrast to enzymes, antibodies do not convert the bound molecule (antigen) into a product.

How to generate a signal?

One needs a **label** to get a **signal** out of an antibody-antigen reaction and one needs a **sensor** to detect and quantify the signal.

ELISAs and immunodipsticks are already on the market: pregnancy tests are using colloidal gold to label a detector antibody to detect human choriogonadotropine (hcG). These tests are one-time tests, carried out when needed.

But, how can one monitor continuously the environment, for example to protect us from bioterrorist attacks? Early antibody-based sensor devices in the years 1975-1985 relied on two big leaps:

The first was that **antibodies could be integrated with optical and electronic hardware** to provide a direct readout of antigen binding that was more sensitive than the human eye.

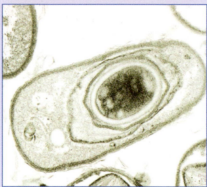

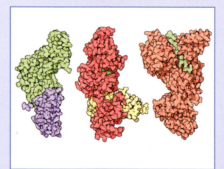

Recently, anthrax has gained the potential to be a major threat through bioterrorism. It is an effective weapon because it forms sturdy spores that may be stored for years and which rapidly lead to lethal infections when inhaled. Anthrax is caused by an unusually large bacterium, *Bacillus anthracis*. Once its spores lodge in the skin or in the lungs, it rapidly begins growth and produces a deadly three-part toxin. These toxins are frighteningly effective, designed for maximum lethality.

They combine two functions: part is a delivery mechanism that seeks out cells and part is a toxic enzyme that rapidly kills the cell. In anthrax toxin, there is one delivery molecule, termed "protective antigen" because of its use in anthrax vaccines (shown on the left). It delivers the other two parts, edema factor and lethal factor (center and right), which are the toxic components that attack cells.

The second concept was that antibodies could bind their target antigens after being **immobilized on a sensor surface** instead of binding the antigen in solution to generate a detection event.

The development of diode lasers, light emitting diodes (LEDs), photodiodes, CCD and CMOS cameras, and other small, inexpensive optical and electronic components sparked advances in the **sensor hardware** development. However, solving the problem of **maintaining antibody functionality after immobilization** had to be solved first in order to justify device development. This is where my lab made its first major contribution to the field in the late 1980s.

How to keep antibodies functional?

Intact IgG antibodies are Y-shaped molecules, with binding sites at the end of each arm of the Y. The prevalent scientific thinking was that the best approach to immobilization would be one that **bound the feet of the Y** to the sensing surface, leaving the arms to wave free into the sample.

The two primary methods of doing this relied on using either the carbohydrate side chains in the Fc (or "legs region") of the Y or to fragment the Y (by enzymes like papain, see Chapter 3) in order to generate a free thiol (–SH) group for attachment. While these approaches were moderately successful, they did not address the two most critical problems in preserving the antigen-binding capability of the immobilized antibodies: the **"fried egg" effect** and the **"Gulliver" effect**.

The "fried egg" effect

Antibodies, like many other proteins, like to adsorb to surfaces. The initial interaction of the hydrophilic outside of the protein with the more hydrophobic surface is weak; however, over time, the hydrophobic inside of the protein begins to interact with and bind tightly to the surface. Visualize a raw egg inside the shell with the hydrophobic "yolk" safely surrounded by the more hydrophilic egg white. When the shell is broken and the egg plopped onto a hot surface, the white spreads out leaving the yolk in close proximity to the frying pan. An antibody undergoing such a conformational change is definitely not in the optimum configuration for antigen-binding. In order to avoid the fried egg effect, the surface must be modified to be as hydrophilic as possible and direct interactions between the surface and the antibody discouraged.

In **Jonathan Swift's** *Gulliver's Travels*, Gulliver awakens on the shores of Lilliput to find himself tied down by the Little People using lots of tiny ropes. He cannot move his arms or legs. Initial methods for attaching antibodies to surfaces used crosslinking molecules that attached the antibodies to the sensor surface in lots of places, limiting the ability of the binding arms of Y to function away from the surface. An ideal crosslinker would bind the antibody to the sensor surface in only a few places, and preferably in the legs of the Y.

Gulliver tied down by the Lilliputians.

In the late 1980s, my lab first used a special class of crosslinkers to attach antibodies to a sensor surface modified with a hydrophilic film. These "heterobifunctional" crosslinkers had one end that bound only to the antibody and an opposite end that bound only to the modified surface so that antibodies could not be inadvertently attached to each other.

The use of the **silane films and heterobifunctional crosslinkers** was rapidly accepted and is now widely employed for antibody immobilization, avoiding the "fried egg effect".

The Gulliver effect

Our second breakthrough in antibody immobilization solved the Gulliver problem. It is well known that the B vitamin **biotin binds very tightly to the egg white protein avidin.** Though not the first investigators to attach antibodies through a biotin bridge to avidin on a surface, we demonstrated that antibody function could be optimized by attaching only two or three biotins to each antibody. The resulting

method of attaching the antibody to the surface had several advantages: The avidin provided a nice, hydrophilic cushion between the antibody and the sensor surface to prevent the "fried egg effect". The limitation on the number of crosslinks prevented the Gulliver effect. And finally, the limitation on the number of modifications to the antibody minimized the chance of directly damaging the active site of the antibody.

Biosensors using immobilized antibodies

The devices sensing the binding of antigens to antibodies have become incredibly sophisticated, but with that sophistication have become much simpler to operate and much more reliable in performance. For example, the first fielded fiber optic biosensor weighed 150 pounds, fluids were manually manipulated, and samples were tested for only one target at a time. Now you can purchase a fiber optic biosensor that is fully automated, tests for eight targets simultaneously, and fits in a backpack along with an air sampler. Another version of this system has been flown on a very small unmanned plane as part of a 10-pound payload and was able to identify bacteria while flying through the air.

Most of these **immunosensors use the following (simplified) principle** (see big Figure!): The capture antibodies are bound (via avidin and biotin) onto a glass surface. This can be a fiber optic or a simple glass slide. A laser beam is directed through the glass to the surface that carries the antibodies. If the glass is covered with a fluid, two different refraction indices apply. If the beam hits the surface at less than the critical angle, total internal reflection (TIR) of the light beam occurs, generating an evanescent wave on the glass surface which penetrates 100 nm into the liquid solution – precisely the operating region of detector antibodies! In fact, two detector antibodies are used. When the first or capture antibody binds to the antigen, a second antibody carrying a fluorescent tracer binds to the caught antigen to form a sandwich.

The evanescent wave excites the fluorescent tracers: they start to emit light, indicating that the antigen has been bound and detected. Unbound detector antibodies are not excited, as they are not within the reach of the evanescent wave. The fluorescent signal is detected, filtered, and amplified.

These immunosensors are **sensitive to parts per billion** or better – e.g., a tablespoon of

Frances S. Ligler

Antibody-based biosensor built for deployment in *Desert Storm* (150 pounds, manual operation, big laser and electronic components).

BioHawk: Man-portable air sampler and biothreat agent detector.

Swallow Unmanned Airborne Vehicle equipped with air sampler (tube extending from nose) and antibody-based biosensor for remote identification of biothreat agents.

See next page

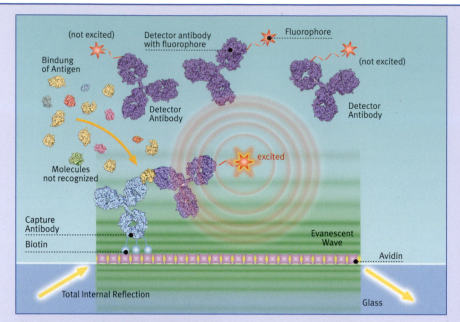

Immunosensor using the evanescent wave effect for exciting labels bound to detector antibodies (for details, see running text).

drugs or explosives in an Olympic-sized swimming pool.

Array biosensors have been developed in order to simultaneously monitor for large numbers of targets. These systems rely on an array of antibodies in discrete spots on a flat surface.

Sample and fluorescent tracer molecules cause fluorescent complexes to form on some of the antibody spots but not others. The identity of the target can be determined by the location of the spots that light up. The intensity of the signal provides quantitative information on the amount of the target present in the sample.

Users of antibody-based biosensors have become **less concerned with how they work than with how easily and cheaply they can be used**. To make them easier to use, small plumbing systems (microfluidics) are becoming increasingly used.

Advances in optics offer new opportunities for increased sensitivity and reductions in size and cost. Silicon technology is producing better and better integrated optical waveguides for highly multiplexed analyses. Arrays of single photon detectors may offer the opportunity to detect one target if the antibody binding is sufficiently long-lived and if the background can be sufficiently reduced.

The production of devices based on organic polymers, such as organic LEDs, transistors, and

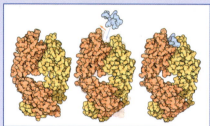

Principle of molecular recognition by antibodies

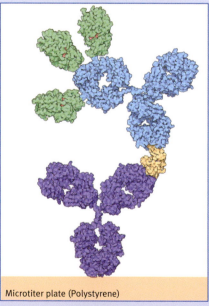

Microtiter plate (Polystyrene)

Principle of Enzyme-linked Immunosorbent Assay (ELISA): a capture antibody is immobilized to a surface and binds the antigen. The detector antibody is labelled with an enzyme and forms a "sandwich" structure. After washing steps and adding colorless substrates, a colored product is formed (proportional to antigen concentration).

photodiodes, is also very exciting because they should be relatively **simple to integrate with biological recognition elements** and polymer-based microfluidics, to form monolithic, inexpensive, or disposable sensors.

Novel antibody-based biosensors are highly sensitive and have proved their worth in the detection and monitoring of pesticides in agriculture, toxins and pathogens in homogenized foods, disease markers in clinical fluids, and biothreat agents in air and water have been demonstrated.

Thus, analytical biotechnology helps to make our lives safer.

The author, Dr Frances Ligler, competing in the hardest 100-mile horse race in the US, the Tevis Cup Race. This race is held in the Sierra Nevada mountains. She does like climbing mountains, in both scientific and equestrian endeavors.

Frances S. Ligler is currently the Navy's Senior Scientist for Biosensors and Biomaterials. She is a Member of the Bioengineering Section of the National Academy of Engineering. She earned a B. S. from Furman University and both a D. Phil. and a D. Sc. from Oxford University.

Currently she is working in the fields of biosensors and microfluidics. She has published over 290 full-length articles in scientific journals, which have been cited over 4000 times, and has 24 issued patents.

She is the winner of the Navy Superior Civilian Service Medal, and of awards like the National Drug Control Policy Technology Transfer Award, the Chemical Society Hillebrand Award, Navy Merit Award, NRL Technology Transfer Award, and others. In 2003, she was awarded the Homeland Security Award by the Christopher Columbus Foundation and the Presidential Rank of Distinguished Career Professional by President Bush.

Cited Literature:

Ligler FS, Taitt CR (2002) *Optical Biosensors: Present and Future.* Elsevier Science, NY

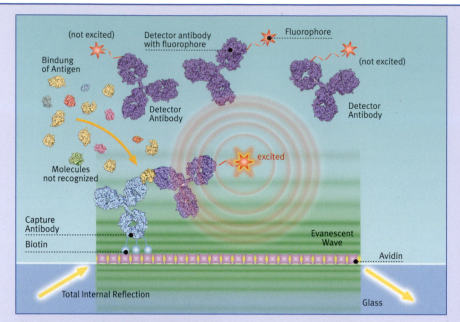 labels in figure: (not excited), Detector antibody with fluorophore, Fluorophore, Bindung of Antigen, (not excited), Detector Antibody, Detector Antibody, Molecules not recognized, excited, Capture Antibody, Biotin, Evanescent Wave, Avidin, Total Internal Reflection, Glass

antibodies was by immunization (**polyclonal antibodies** coming directly from animal serum) or by hybridoma technology (**monoclonal antibodies**) (Box 5.5)

In a recent development, antibodies have been produced *in vitro*, i.e., outside animal organisms, either in bacterial or other cell cultures. They mostly consist just of the antigen-binding fragment (**Fab fragment**). This is known as "fragment antigen binding". Heretofore, Fab fragments could only be obtained from complete antibodies by cleavage with proteases. The procedure has now been replaced by recombination.

The crucial sites for antibody binding are the variable regions (Fv). They lack the stabilizing disulfide bridges of constant chains and need extra stabilization. This is usually done by chaining peptides that turn them into single protein strands. They are also elongated by a tag that modifies their biochemical or surface-binding properties. This results in **single-chain Fv fragments** (scFv fragments, Box 5.3, bottom right).

Why is it necessary to produce **recombinant antibodies**, given that the immune system provides a virtually inexhaustible reservoir? Animal lovers will be pleased to hear that antibodies can thus be produced without animal or human involvement, which is certainly also an advantage in the presence of BSE, hepatitis, and AIDS!

It goes without saying that recombinant antibodies are also monoclonal, i.e., they are derived from a single clone and are highly specific, binding precisely to their epitope.

The production of antibodies with *E. coli* cells is vastly cheaper as well as faster than production in animal cultures via hybridoma cells, and, what is more, the entire range of *E coli* technology, with its comparatively simple methods of sequencing, analysis, and modification, is available.

■ 5.11 Recombinant Antibody Libraries

Even a very successful fusion of myeloma (cancer) and spleen cells into hybridoma cells can only be the basis for the production of a few dozen different antibodies – a hundred at the most. This is not terribly impressive, keeping in mind that our own immune system is capable of producing

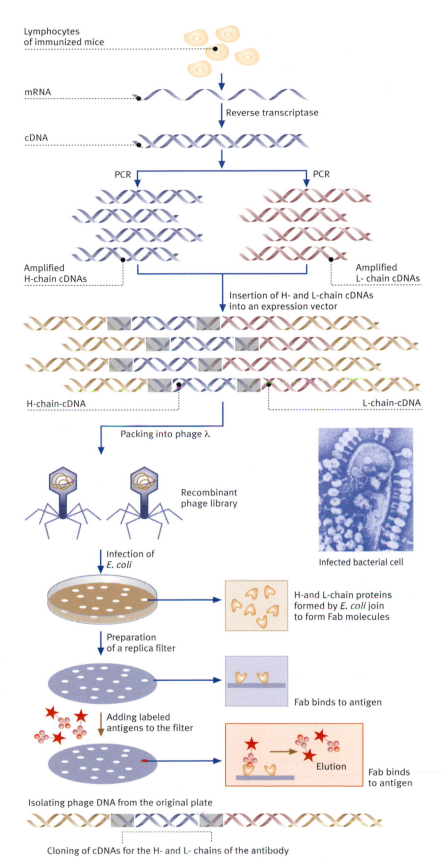

Lymphocytes of immunized mice

mRNA

Reverse transcriptase

cDNA

PCR

PCR

Amplified H-chain cDNAs

Amplified L-chain cDNAs

Insertion of H- and L-chain cDNAs into an expression vector

H-chain-cDNA

L-chain-cDNA

Packing into phage λ

Recombinant phage library

Infected bacterial cell

Infection of *E. coli*

H- and L-chain proteins formed by *E. coli* join to form Fab molecules

Preparation of a replica filter

Fab binds to antigen

Adding labeled antigens to the filter

Elution

Fab binds to antigen

Isolating phage DNA from the original plate

Cloning of cDNAs for the H- and L- chains of the antibody

Fig. 5.28 Building a recombinant antibody library.

163

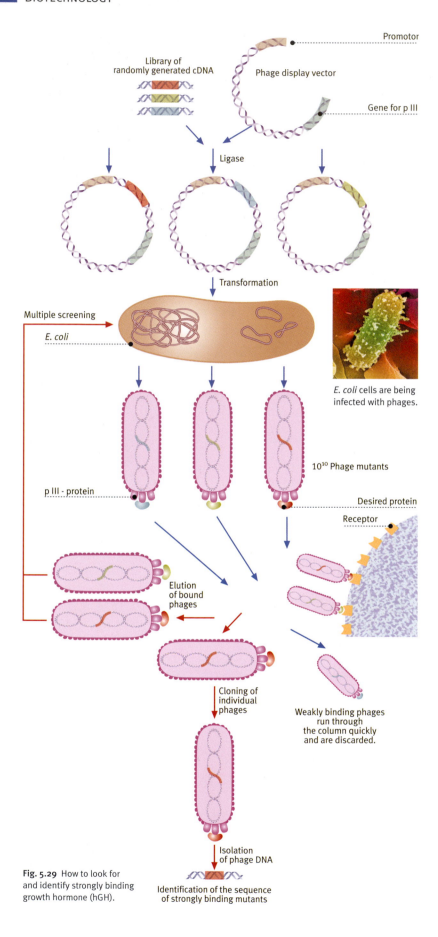

Library of randomly generated cDNA

Phage display vector

Promotor

Gene for p III

Ligase

Transformation

Multiple screening

E. coli

E. coli cells are being infected with phages.

10¹⁰ Phage mutants

p III - protein

Desired protein

Receptor

Elution of bound phages

Weakly binding phages run through the column quickly and are discarded.

Cloning of individual phages

Isolation of phage DNA

Identification of the sequence of strongly binding mutants

Fig. 5.29 How to look for and identify strongly binding growth hormone (hGH).

100,000,000 antibodies with different specificities. How could we tap into this vast potential?

The basically quite ineffective fusion step can be circumvented by injecting the antigen into a mouse. After immunization, spleen cells are collected, and mature mRNA is isolated (Ch. 3), which contains only the exons, i.e., the actual genetic material minus junk (introns). Using reverse transcriptase, the single-stranded mRNA is turned into double-stranded copy DNA (cDNA). PCR (Ch.10) is then used to produce millions of copies of cDNA for the light and heavy chains of the antibodies (Fig. 5.28). The cDNA copies are cut with restriction endonucleases, resulting in fragments with sticky ends, which, in turn, are pasted into vectors for bacteriophage λ. Thus, two different **libraries** are created, one containing the heavy chains (**H-chains**), the other the light chains (**L-chains**). The two phage libraries are then recombined into a third recombinant **phage library**. The phages contain accidental pairs of cDNA of an L- and an H-chain. They are then put on an *E. coli* lawn and infected with bacteriophages.

This increases the availability of possible antibodies at least by a thousand-fold, compared to hybridoma technology.

■ 5.12 Piggyback or Phage Display – The Next Revolution

The B-lymphocytes in the immune system use a selective trick, carrying a membrane-bound antibody on their surface (see above). An antigen (e.g., of a virus) that binds to it and IL-2, produced by T helper cells, stimulate the B-lymphocytes to divide (clonal selection). The antibody indicates the "street number" of the relevant gene in the lymphocyte. In other words, the antibody carries its gene piggyback, as if in a huge knapsack.

Thus, it is easy to identify the cell with the desired antibody from among large quantities of B-lymphocytes. A genetic engineer's dream would come true if it were possible to read the "street number" in bacteria to make sure that they do contain the desired gene.

This dream was turned into reality by **George P. Smith** (b. 1941, Fig. 5.30) at the University of Missouri in Columbia, Missouri, USA, in 1985. He did not find the solution for bacteria, but for the

bacteriophage M13. Like phage λ (Ch. 3), it is a filamentous phage, but with a much smaller genome. The genome consists of a tangle of ring-shaped single-stranded DNA, containing only a few genes for capsid proteins and a simple infection cycle, but no genes for insertion into the host genome.

The cunning virus enters *E. coli* via its sex pili, introducing its single-stranded DNA through an F-pilus. The DNA acts as a template to form double-stranded DNA in the cell, which is not inserted into the bacterial genome, but replicates to between 100 and 200 copies. With the division of the bacterium, each daughter cell receives copies of the phage DNA.

M13 is considerate enough not to kill its host and only to slow down its growth. After single-stranded DNA copies have been made and packed into filamentous protein envelopes, the phage particles are released. A tubule is formed by 2700 type pVIII proteins. Most remarkable, however, are the five proteins type pVII and pIX at one end and pIII and pIV at the other end of the phage (Fig. 5.29).

What makes these special is this: **pIII remains functional** even after integration of foreign sequences. In other words, if a foreign gene is inserted into the *pIII* gene of the phage, it should show up as a foreign protein on the phage envelope, e.g., on its surface. It worked!

This method, called **phage display** was first tried out to find the gene for a strongly binding growth hormone.

■ 5.13 Phage Display for High Affinity Growth Hormone

The search was on for variants of the **human growth hormone** (hGH) with enhanced bonding to the hGH receptor (Fig. 5.29). The fragments in the peptide sequence that were responsible for receptor bonding were already known. "Degenerate" oligonucleotides were created that coded for all possible amino acids in these positions. These **hGH gene variants** were inserted into a vector of M13 next to the *pIII* gene, to ensure that they would form fusion proteins *pIII*-hGH. This library was then introduced into *E. coli*.

This led to the emergence of infectious M13 phages carrying the normal pIII protein in their envelope, with an hGH variant attached to it. Each phage dis-

played one single hGH variant. How could the most suitable be identified?

The phages were put through a separation column filled with polymer beads to which the hGH receptor (an isolated protein) was firmly bonded (more about the method in Ch. 3, purification of insulin using bonded antibodies). Phages that do not carry hGH on the surface pass through the column straightaway, not bonding anywhere (Fig. 5.29). Phages with only weak bonds also land in the lab waste, whereas the **"good" phages bind to the receptors** and can be detached later, using a weak acid.

The phages that have been thus selected are reintroduced to *E. coli* and selected again in the column. The procedure was repeated six time, resulting in the selection of the phage with the highest affinity to the hGH receptor from the available variants. Some super phages were cloned and their DNA sequences identified. Thus, an improved version of growth hormone could be produced – an example of **protein design** (Ch. 2 and 3).

Can this revolutionary procedure also be used for antibodies? Very much so! Genes for the antibody fragment scFv were obtained from an antibody library and packed into M13 phages. They reproduced in *E. coli* and carried the relevant fragment on the surface like a street number! The selection process is called **panning**, in analogy with the sorting process in gold pans. A new gold rush has begun, and the targets are cancer cells.

■ 5.14 New Hope for Cancer Patients – Rituximab, a Recombinant Antibody

"Imagine a new cancer treatment, based on the **cruise missiles principle**. A submicroscopic rocket with an automatic search head is launched in the body through injection. It searches out cancer cells and destroys them without attacking normal healthy tissue. Such a wonder weapon does not exist yet, but there are indications that it could be available in the near future." Words to that effect were written in the *Wall Street Journal*, which is usually not prone to hype, in 1981.

Nowadays, some monoclonal antibodies specific to cancer cell surface antigens are available. As in tumor diagnostics, the antibodies are used to recognize the tumor-specific antigens, but instead of

Fig. 5.30 George P. Smith used the bacteriophage M13 in his phage display technique.

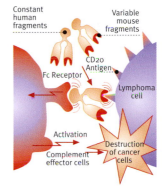

Fig. 5.31 Possible action of Rituximab.

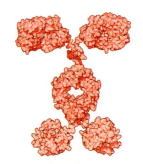

Fig. 5.32 Immunotoxins in action: An immunotoxins (red) consists of ricine and a Y-shaped antibody. It binds to a cell receptor of a leukemic B-cell (blue) and penetrates into the cell (using a cage structure of three-armed clathrins) and is secreted into the cytoplasm.

Recombinant antibodies that have been licensed by the FDA

400 clinical studies on antibody therapies are currently under way, 42 of which have already reached phase III and II/III and 174 are in phase I or I/II (according to Stefan Dübel, May 2005)

Avastin: intestinal cancer
Herceptin: breast cancer
Humira: rheumatoid arthritis
Leukosite: B-cell leukemia
Mylotarg: acute myeloid leukemia
Raptiva: psoriasis
Synagis: RSV (Respiratory Syncytial Virus) infection
Tysabri: Multiple Sclerosis
Xolair: allergic asthma
Zenapax: rejection of kidney transplants
Cotara: various types of cancer
Erbitux: intestinal cancer
Remicade: rheumatoid arthritis, Crohn's disease
Reopro: prevention against blood-clotting
Rituximab and *Alemtuzimab*: Non-Hodgkin lymphoma
Simulect: rejection of kidney transplants
Zevalin: lymphoma, rheumatoid arthritis

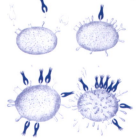

Fig. 5.33 Paul Ehrlich (top) with his Japanese disciple Hata (center) Bottom: Ehrlich's side chain theory.

labeling them with a radioactive isotope, they carry a **highly potent cytotoxin**, such as the protein ricin. No more than one ricin molecule is needed to kill a cell (Fig. 5.34).

The monoclonal antibodies provide the "ZIP" or "**postal code**" to which the cytotoxin is delivered, as well as the delivery vehicle. First results with some specific tumors look quite promising. However, the method is still a long way away from becoming hospital routine. A whole host of problems need to be resolved first. Taking a toxin molecule to a tumor cell does not mean that the toxin will arrive inside the cell and reach its deadly potential. The cellular uptake mechanism for antibody-toxin complexes (**immunotoxins**) is not yet fully understood, and for many variants of cancer, the surface antigen is not yet known or there are no monoclonal antibodies against them.

While ricin and other bacterial toxins can kill tumor cells very specifically, they are also toxic for liver and kidney and elicit violent immune responses. Meanwhile, enzymes produced by the human body can be used as toxins – **RNases**. One RNase molecule is sufficient to kill a cell, but only when it has been specifically introduced by an antibody. In the blood, such human immunotoxins are neither toxic nor immunogenic. This may change if further specific tumor markers can be found on the surface of cancer cells.

Bispecific antibodies offer another possibility for cancer treatment in the future. Here, no toxins or radionuclides are brought to the tumor cell, but the second antibody claw drags along a whole immune cell, e.g., a killer T-cell or a T-lymphocyte. Thus, the tumor cell is made visible for the body's immune system, which can now turn its weapons against them.

In 2004, a recombinant antibody had sales of 1.7 billion US dollars. 300,000 patients have been treated so far with **rituximab** (-mab refers to monoclonal antibody), the cancer treatment of choice. It targets **non-Hodgkin's lymphoma** (**NHL**).

NHL is a malignant disease of lymphatic tissue which can affect many parts of the body, for example, lymph nodes, spleen, thymus gland, adenoids, tonsils, or bone marrow.

Rituximab is an antibody produced in the lab that targets a specific surface antigen found in over 90 percent of a non-Hodgkin's B-cell lymphoma. It

is called antigen **CD20**. Rituximab interacts exclusively with cells that carry the CD20 antigen on their surface. It binds to them through various mechanisms that activate the body's own immune system and kills the cell (Fig. 5.31).

Rituximab acts in a way that makes it stand out from all other forms of cancer treatment, as it attacks only those cells that carry the CD20 antigen, while leaving all other cells unscathed.

Due to its targeted effect, rituximab is well tolerated by patients. It is a **humanized recombinant antibody**. Its malleable "hands" that bind to CD20 have been developed from immunized mice, while the rest of the molecule (90 percent) is of human origin. This is why rituximab is not recognized as foreign and does not elicit an immune reaction.

In 2006, the sales for rituximab (and its derivatives) alone climbed to 2 billion US$. It is now also tested successfully against some forms of multiple sclerosis (MS). Modern immune technology has now created a market for diagnostics and therapeutics worth billions of dollars. Paul Ehrlich's vision has come true (Fig. 5.33).

Paul Ehrlich called it the magic bullet (see p. 158), a name emphasizing its selective toxicity for certain pathogens. The first magic bullet was a chemotherapeutic drug called salvarsan and is known in the English-speaking world as arsphenamine. It was developed in 1909 by **Paul Ehrlich** and **Sahachiro Hata** (Fig. 5.33). The German name is derived from Latin *salvare* – to save, *sanus* – healthy, and arsenic and could be interpreted as "healing arsenic". It is a milestone in drug research, as for the first time, a targeted anti-microbial drug was available to tackle a dangerous infectious disease. The effectiveness of arsphenamine was such that some infections like syphilis could be cured with a single injection.

Modern immunotherapy makes Ehrlich's dream of magic bullets come true.

Where do we go from here? Bioreactors for cell cultures are reaching the limits of their capacity, and more and more entire antibodies or fragments of them are produced using transgenic animals (Ch. 7 and 8).

> He that will not apply new remedies must tolerate old evils
>
> Adapted from Francis Bacon (1561 - 1626)

Cited and recommended Literature

- **Cann A** (2005) *Principles of Molecular Virology* (4th edn.) Academic Press New York. In my opinion, the best textbook on this subject. Cann explains: that there is more biological diversity within viruses than in all the rest of the bacterial, plant, and animal kingdoms put together.

- **McKane L, Kandell J** (1996) *Microbiology* (2nd edn.) McGraw-Hill, New York. Colorful and easy-to read-textbook, highly recommended.

- **Goldsby RA, Kindt TK, Osborne BA and Kuby J** (2003) *Immunology*, 5th edn., W.H. Freeman, New York. Reliable introduction in immunology.

- **Sompayrac LM** (2002) *How Pathogenic Viruses Work*, Jones and Bartlett. A survey of 12 of the most common viral infections, good reading.

- **Sompayrac LM** (2002) *How the Immune System Works* (2nd edn.) Blackwell Cambridge. From same author. again., a good read!

- **Dübel S** (2006) *Recombinant therapeutic antibodies (Mini-review)* Appl Microbiol Biotechnol. DOI 10.1007/s00253-006-0810-y. Short and very informative review, comprehensible even for newcomers.

- **Dixon B** (1994) *Power Unseen: How Microbes Rule the World*. WH Freeman/ Spektrum Oxford NY Heidelberg, Very amusing and interesting reading!

Useful weblinks

- Great resources:
  www.microbe.info and
  www.microbiology-direct.com

- Microbes on the Move: Alan Cann (University Leicester) provides microbiology and immunology resources and video:
  http://www-micro.msb.le.ac.uk/

- The Center for Disease Control and Prevention website
  http://www.cdc.gov/

- Nature Reviews Immunology. With an impact factor of 32.7, it is the leading monthly review title for immunology:
  http://www.nature.com/nri/index.html

- The Merck Manual of Diagnosis and Therapy, Section 12: Immunology; Allergic Disorders
  www.merck.com/pubs/mmanual/

- HIV-test:
  http://en.wikipedia.org/wiki/HIV_test

Eight Self-Test Questions

1. Why can viral infections not be cured by antibiotics?

2. How can RNA viruses interact with host cell DNA?

3. What would happen to Jenner or Pasteur if they were to carry out their vaccination experiments today?

4. How can killer cells tell that a host cell has been invaded by viruses?

5. How are monoclonal and polyclonal antibodies obtained? What was Milstein and Köhler's ingenious idea?

6. Can the high specificity of antibodies be combined with the catalytic power of enzymes?

7. How can a genetic street number be placed on the surface of viruses and be used for the optimization of proteins?

8. How can antibodies be harnessed in the battle against cancer?

ENVIRONMENTAL BIOTECHNOLOGY
From One-Way Streets to Traffic Circles

CHAPTER **6**

Fig. 6.1 Robert Koch, discoverer of the cholera, tuberculosis, anthrax, and other pathogens, Nobel Prize in 1905.

Fig. 6.2 Cholera epidemic, the horror vision of dying masses

Fig. 6.3 Ciliates, as drawn by biologist and artist Ernst Haeckel in "*Art Forms in Nature*" in 1904. Ciliates like *Vorticella* are indicators of good active sludge. A milliliter of wastewater can contain up to 10,000 of these. If they are in the majority, the degradation process is on track.

■ 6.1 Clean Water – A Bioproduct

Harold Scott, author of A History of Tropical Medicine, 1939 described cholera as perhaps the most frightening of all epidemics. It spreads so fast that a man who got up hale and hearty in the morning could be dead and buried before the sun sets.

In 1892, the inhabitants of Hamburg still thought it was safe to drink water straight from the rivers Elbe and Alster – a fatal mistake that cost the lives of 8,605 citizens. The rivers had become an ideal breeding ground for microbes, including the cholera pathogen. The neighboring community of Altona, by contrast, let the river water run through simple sand filters and was unaffected. Sewage treatment became of paramount importance for large cities.

Although **Robert Koch** (1843-1910) (Figs. 6.1 and 6.2) had identified the cholera pathogen *Vibrio cholerae* as early as 1884, improvements in the water supply and sewage treatment were just beginning. Cholera and typhoid certainly had their share in speeding up the process. The diagram in Box 6.1 shows the correlation between the number of typhoid death and water hygiene.

Clean water! Every day, we use 200 to 300 liters (53 - 80 gal) – on hot days even up to 1000 liters (264 gal) per person. Food remains, fat, sugar, protein, excrement, everything ends up in the sewers, including washing machine detergent.

Biodegradability of waste water is measured as **5-day Biochemical Oxygen Demand** (**BOD**) in the lab. This tells you how many milligrams of dissolved oxygen would be metabolized by microbes within five days in order to completely degrade the organic pollutants in a water sample. The average organic pollution of domestic wastewater is 60 grams 5-day BOD. This has also been defined as **population equivalent** (**PE**).

In the case of **domestic wastewater**, 60 grams of oxygen would be needed to degrade 1 PE. Given an average solubility of oxygen of approximately ten milligrams per liter of water, oxygen dissolved in 6,000 liters of clean water would be required to remove the daily pollution of just one city-dweller.

Here is a simplified account of the chain of events triggered by untreated sewage ending up in lakes and rivers: **Aerobic microorganisms** degrade organic pollution while rapidly metabolizing the available oxygen. Some areas in the water (mostly near the ground) are so deprived of oxygen that only anaerobic bacteria can thrive.

Fish and other oxygen-dependent organisms begin to die, while **anaerobic microorganisms** produce poisonous ammonia (NH_3) and hydrogen sulfide (H_2S) with its typical rotten-egg smell. These two compounds kill the remaining water organisms that have survived in the low-oxygen environment, turning rivers into smelly sewers.

In a **farming environment**, one cow produces as much wastewater as 16 city-dwellers. 1.4 billion cattle worldwide produce about as much wastewater as 22 billion humans.

Even more worrisome are the huge quantities of wastewater produced by industry. They contain large amounts of inorganic compounds, not only salt or hydrogen sulfide, but poisons such as mercury and other heavy metals that cannot be easily broken down by microbes or may even kill them. Paper mills produce 200 to 900 PEs per ton of paper, breweries 150 to 300 PEs per 1,000 liters (264 gal) of beer.

As the natural cleaning power of microbes in rivers has long been insufficient, wastewater must be broken down by microbes in huge **sewage plants** before it can safely be released into rivers and lakes. Sewage treatment plants are the largest biofactories we have at present. They convert wastewater into large quantities of a **bioproduct called clean water**.

Microorganisms in sewage plants work extremely hard for their living. In a respiration process using oxygen from the air, they convert sugar, fat, and proteins in the wastewater into carbon dioxide and water, while growing and building new cells in the process. Sewage plants provide ideal conditions for the reproduction of microbes and the breakdown processes they are involved in.

The breakdown of one gram of sugar requires more than one gram of oxygen, but only ten milligrams of oxygen can be dissolved in a liter of water. Thus, the microorganisms use up the oxygen dissolved in water very quickly, and wastewater must be continuously stirred and aerated in order to provide enough oxygen – a very energy-intensive process.

Box 6.1 Biotech History:
Disposal of Wastewater And Sewage Farming in Berlin

Until 1878, rain, wastewater, and kitchen waste, to some extent even feces, were disposed of through the gutters, which were open ditches between walkway and road, ending in a runoff ditch or a stream or river. Where there was no immediate access to a waterway, the discharge went into large subterranean canals which were built ad hoc without a sufficient gradient, but had far too large diameters.

The decaying waste in those canals caused very unpleasant smells, not to mention the nuisance the gutters caused for moving traffic and pedestrians alike. Feces were collected in dung pits that were emptied manually at regular intervals and the contents transported in horse carts along the streets. Neither streets, nor gutters or horsecarts were properly sealed, which resulted in a considerable contamination of the soil and ground water. People drew their drinking water from wells in or near their houses, and the contaminated water caused the spread of diseases.

In 1816, the Prussian Government took a first decision to flush out gutters and streets with water in order to remove dirt and evil smells. However, this was never put into practice because there was neither enough water nor enough pressure. A government committee in the service of Friedrich Wilhelm IV of Prussia decided in 1846 that waterworks should be built. They also gave permission to the introduction of flush toilets. In 1856, the waterworks were built near the Stralau gate. In 1873, the waterworks were taken over by the city of Berlin. However, flushing out the gutters did not have the expected success. Through the easier access to water and the rapid expansion of flush toilets, the amount of wastewater increased dramatically, and although the gutters had been widened and deepened and measured

Deaths from typhoid per 10,000 inhabitants — connections to sewage

With the improvement of Berlin's water supply, typhoid death rates fell in inverse proportion

now one meter (over a yard) across, they simply could not cope. In spite of wastewater becoming a major problem, conservative Berliners still objected to the introduction of a sewage system. People thought it would just increase the foul smells, as did the existing canals.

Rudolf Virchow (1821–1902) (l.), medical historian and hygienist who founded the discipline of cell pathology, and James Hobrecht (1825–1903)

In 1860, Councillor **Wiebe**, who was in charge of public building in Berlin, was sent to study the sewage systems of Hamburg, Paris, and London and come up with a plan to solve Berlin's wastewater problems. He traveled the same year, accompanied by the head of rail and waterway construction, **James Hobrecht**. Wiebe's idea was to collect all wastewater in subterranean canals and discharge it untreated into the river Spree outside Berlin. This plan was highly controversial among engineers, medical doctors, and politicians. In February 1867, a committee was established, led by the well-known medical microbiologist **Rudolf Virchow** who, in turn, put Hobrecht in charge of further investigations. Virchow presented his report to the city council in November 1872. In May 1873, a decision was taken to build a sewage system. The city was divided into several sections, each of which was provided with an

independent canal system (radial system). The system had the advantage that if there was disruption somewhere, the whole system was not disabled, and the remaining canals could drain the wastewater from the affected canal. The water flowed through burnt and glazed clay pipes. Steam-powered pumps pumped it up to be distributed over sewage farms outside the city. Thus, instead of ending up in sewer riverbeds, the water was treated by the soil bacteria and provided fertilizer for agriculture. Hobrecht divided Berlin into twelve approximately equally sized districts, each of which was allocated a large acreage for sewage farming.

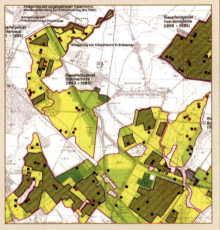

Sewage farms in the South of Berlin, some of which were in use for 100 years, from 1876 until 1976

Once the sewage system was in operation, gutters and pits with their health and hygiene hazards were a thing of the past, and the areas that had been acquired for sewage farms could be adapted to their new purpose. They were leveled, dams were built, and, depending on the gradient, flat beds or embankments were created on which the water was sprayed evenly. The countryside underwent a complete change, as the sewage farms not only dealt with wastewater, but were also used for agricultural production.

Schematic representation of a sewage farm where various crops (grain, vegetables, turnips) were grown. Regular and even flooding of the fields was essential.

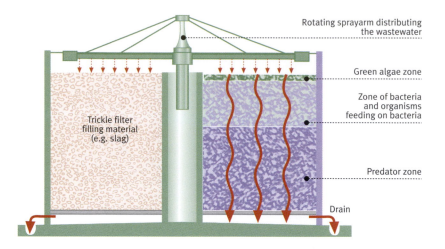

Rotating sprayarm distributing the wastewater

Green algae zone

Zone of bacteria and organisms feeding on bacteria

Predator zone

Trickle filter filling material (e.g. slag)

Drain

Fig. 6.4 Trickle filter technology.

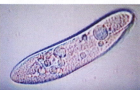

Fig. 6.5 Fresh water organisms in wastewater (*Vorticella, Paramecium, Daphnia*).

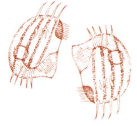

Fig. 6.6 *Aspidisca costata* is a nitrate indicator in the wastewater treatment process. Up to 25,000 can be found in a milliliter of wastewater. Not only the number of microorganisms, but also the protrusion of their ribs is an indicator. The more nitrates in the water, the more the ribs will stick out.

■ 6.2 Aerobic Water Purification – Sewage Farms, Trickling Filters, and Activated Sludge

Sewage farms are among the oldest sewage treatment methods that have been developed around European cities such as 19th century Berlin (see Box 6.1). The wastewater is mechanically pre-cleaned and then allowed to trickle into the soil where organic pollutants are microbially degraded.

Sewage farms need a lot of space, so the search began for space-saving alternatives. **Trickling filters** (Fig. 6.4) were developed in England in 1894. They work like fixed-bed bioreactors (see Ch. 4) and are filled with large-pored material such as lava, slag, sintered glass, or plastic. A revolving sprayer head mechanically sprays the pre-cleaned wastewater over the material.

The large surface of the filling material is a prerequisite for the activity of wastewater organisms. A lawn of bacteria, fungi, cyanobacteria (blue green algae), algae, protozoa, rotifera, mites, and nematodes develops in the trickling filter – an excellent aquatic community or **biocoenosis** (Figs. 6.3 and 6.5).

A third sewage-processing method used these days involves **activated sludge** (Figs. 6.7 to 6.9). It also was introduced about 100 years ago. After mechanical treatment, aerobic bacteria, fungi, and yeasts in the wastewater form large flakes full of nutrients, held together by bacterial slime. These symbiotic sludge flakes, called floc, provide support scaffolding for microbes. They float about in gigantic active sludge basins (Fig. 6.7 and 6.8) in which paddles or brushes rotate to literally beat air into the water. This ensures that oxygen-using microbes prolifer-

ate. The throughput is much higher than in a trickling tank because of the better supply of oxygen.

Part of the sludge settles in a settling pond or lagoon, where smells can become a problem due to the open construction.

Wastewater may contain high levels of the nutrients nitrogen and phosphorus. Excessive release of wastewater (or fertilizers) into the environment can lead to a **buildup of nutrients** (**eutrophication**) in lakes and rivers. This encourages the uncontrolled growth of algae, resulting in a rapid surge of the algae population or algal bloom. The decomposition of these algae by bacteria uses up so much of oxygen in the water that most or all animals in the water die, which creates more organic matter to be decomposed by bacteria. Anaerobic bacteria take over, producing toxic hydrogen sulfide and methane and killing the remaining life.

Nitrogen is removed from wastewater through the biological oxidation of nitrogen from ammonia (nitrification) to nitrate, followed by denitrification, the reduction of nitrate to nitrogen gas. Nitrogen gas is released to the atmosphere and thus removed from the water.

Nitrification itself is a two-step aerobic process, each step facilitated by a different type of bacteria. The oxidation of ammonia (NH_3) to nitrite (NO_2^-) is most often facillitated by *Nitrosomonas* species, nitrite oxidation to nitrate (NO_3^-) facilitated by *Nitrobacter* species.

Phosphorus can be removed biologically in a process called enhanced biological phosphorus removal by bacteria. When the biomass enriched in these bacteria is separated from the treated water, these biosolids have a high fertilizer value. Phosphorus removal can also be achieved by chemical precipitation, usually with salts of iron (e.g., ferric chloride) or aluminum. Drinking water is then treated with chlorine or ozone to kill germs. A small proportion of the active sludge is brought back into the active sludge basin to ensure that the microbial population is up to the job.

The efficiency of active sludge microbes is truly amazing. A cubic meter (264 gal) of sludge can purify 20 times its volume of polluted wastewater. Up to 20% of the basin content is taken up by active sludge.

The settled sludge undergoes special treatment in digestion towers in the absence of oxygen.

Methane bacteria (nowadays called methanogens, members of the Archaea kingdom, see Fig. 6.14) convert the remaining organic matter into methane, which can be used as a source of energy (biogas, see below).

Sewage treatment plants with all their basins take up a lot of space, and space is at a premium, particularly in industrial areas. Space-saving **tower bioreactors** (so called Tower biologies) that are 15 to 30 meters (45 to 90 feet) high have been developed by Bayer and Hoechst for their own biological sewage purification (Fig. 6.9). Industrial wastewater treatment in the Rhein-Main area (one of Germany's key industrial regions) deals with approximately 15 billion cubic meters per year.

Deep-shaft reactors that go deep into the ground are another possibility. Their depth ensures that gas bubbles remain in the liquid for a longer time and oxygen solubility is enhanced by the increased pressure within the reactor.

Both reactor types are supplied with a rich flow of oxygen from the bottom and are able to dissolve 80% of it (compared to just 15% in ordinary reactors). They are thus able to break down large quantities of organic matter, taking up very little space and keeping unpleasant smells at a minimum. However, at present, they are expensive and complicated devices.

■ 6.3 Biogas

Will o' the wisp in fens is an existing phenomenon. It has been referred to as "fire in the marshes" in the ancient Chinese book *I Ching* as early as 3000 years ago. In Europe, it was the Italian physicist **Alessandro Volta** who described the "burning air above the swamps." Those who do not believe in UFOs attribute "flying saucers" seen in marshy areas to the same phenomenon. Fermentation gas rising from the ground contains **methane**, which is flammable. A methane bubble that has caught fire is a perfectly rational explanation for will o' the wisp as well as UFOs. What remains unclear, however, is how exactly the gas spontaneously ignites.

The formation of methane can be observed when stirring up mud in a lake. The stomach (rumen) of ruminants could also be described as a miniature biogas factory. The microbial population of a **cow's rumen** turns eight to ten percent of food into 100 to 200 liters of the greenhouse gas methane. It is released by belching or passing of gas. The huge cat-

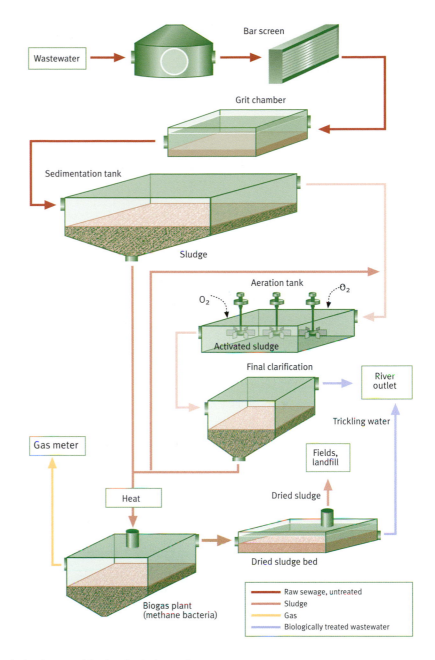

tle herds owned by fast food chains that graze on burnt-down jungle areas contribute considerably to the methane balance of the planet. The useful bacterial inhabitants of the human intestine produce not only vitamins, but also gases that are only partly reabsorbed by the colon. Depending on the food taken in, up to half a liter of gas per day is released – mainly odorless methane, and only minute quantities of hydrogen sulfide.

Every year, the microorganisms on the planet produce approximately 500 million to one billion tons of methane – which is roughly the equivalent of what is extracted from natural gas fields. Around five percent of the carbon assimilated by microbes

Fig. 6.7 Diagram of a sewage treatment plant with mechanical and biological treatment steps (active sludge basin) and recycling of the sludge in biogas plant.

Fig. 6.8 Active sludge basin at the modern treatment plant at Kai Tak airport in Hong Kong. In the plant, oils and detergents must first be separated from the water before it reaches the activated sludge basin.

Box 6.2 The Expert's View:
Rainforests, Volatile Antibiotics, and Endophytes Harnessed for Industrial Microbiology

The diversity of microbial life is enormous and the niches in which microbes live are truly amazing, ranging from deep ocean sediments to the earth's thermal pools. However, an additional relatively untapped source of microbial diversity is the world's rainforests.

Gary Strobel on a field trip collecting new samples

Endophytic microorganisms exist within the living tissues of most plant species and are most abundant in rainforest plants.

Microorganisms have long served mankind by virtue of the myriad of the enzymes and secondary compounds that they make. Furthermore, **only a relatively small number of microbes are used** directly in various industrial applications i.e., cheese/wine/beer making as well as in environmental clean-up operations and for the biological control of pests and pathogens. It seems that we have by no means exhausted the world for its hidden microbes. A much **more comprehensive search of the niches on our earth** may yet reveal novel microbes having direct usefulness to human societies. These uses can either be of the microbe itself or one or more of its natural products.

Rainforests: untapped source of microbial diversity

The rainforests occupy only 7% of the earth's land surface and yet represent 50-70% of all of the species that live here. This is nicely exemplified by the fact that over 30% of all of the bird species on earth are associated with Amazonia alone.

The question is **why is so much diversity focused in rainforests?**

Paul Richards writing in *The Tropical Rainforest* indicates that "The immense floristic riches of the tropical rainforest are no doubt largely due to its great antiquity, it has been the focus of plant evolution for an extremely long time."

In my opinion, the diversity of macro lifeforms is a template for what probably has also happened with microorganisms. That is, areas of the earth with enormous diversity of higher life-forms are probably also accompanied by a high diversity of microorganisms. **It seems that no one has looked!**

One specialized and unique biological niche that supports the growth of microbes are the spaces in between the living cells of higher plants. It turns out that each plant supports a suite of microorganisms known as endophytes. These organisms cause no overt symptoms on the plants in which they live. Furthermore, since so little work on these endophytes has been done, it is suspected that untold numbers of novel fungal and bacterial genera exist as plant-associated microbes and their diversity may parallel that of the higher plants.

Untapped resources for new microbial species: Epiphytes in the rainforest

Some endophytes may have coevolved with their respective higher plant, thus having an already existing compatibility with a higher life-form. Thus, we have begun a concerted search for novel endophytic microbes and the prospects that they may produce novel bioactive products as well as processes that may prove useful at any level. Any discovery of novel microbes will have implications in virtually all of the standard processes of industrial microbiology since scale-up fermentation of the microbe will be necessary.

In order to give the reader a sense of the excitement of discovery that comes with research on

rainforest microbes I have concentrated mostly on the discovery of only one novel endophytic fungal genus – *Muscodor*.

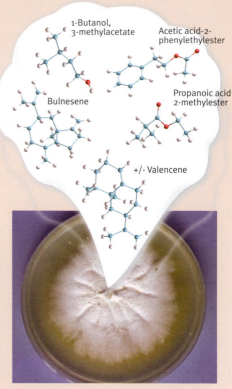

1-Butanol, 3-methylacetate

Acetic acid-2-phenylethylester

Bulnesene

Propanoic acid 2-methylester

+/- Valencene

Muscodor albus and some representative inhibitory volatile compounds produced by "Stinky"

The discovery of *Muscodor albus*

In the late 90s I was on a collecting trip in the jungles near to the Caribbean coast of Honduras. I had selected this area to visit since all of Central America is one of the world's "hot spots of biodiversity". One modestly sized tree, not native to the new world, was introduced to me as *Cinnamomum zeylanicum* and I obtained small limb samples. Unfortunately, most plant samples from the tropics are infested with microscopic phytophagous mites.

These nasty creatures infest the bench tops and parafilm-sealed Petri plates. Thus, in order to eliminate this nagging mite problem we decided to place the Petri plates, with plant tissues, in a large plastic box having a firmly fitting lid. This maneuver would make it difficult for the tiny animals to find their way from the bench surfaces to the inside of the box.

After a few days most plant specimens had sported endophytic fungal growth. Eventually the plates were removed and the individual hyphal tips were transferred to fresh plates of potato dextrose agar.

After two days of incubation we noted that **no transferred endophyte grew except one**. Had the placement of the endophytes in the large plastic box killed the endophytes by limiting oxygen availability? Quite the contrary, it became obvious that the one endophytic fungus (designated isolate 620) remaining alive was producing volatile antibiotics or volatile organic compounds (VOCs).

The hypothesis that **an endophyte can make volatile antibiotic substances** with a wide range of biological activity was born. It was quickly learned that although many wood-inhabiting fungi make volatile substances, none of these possessed the biological activity of isolate 620.

Isolate 620 is a whitish, sterile endophytic fungus possessing hyphal coiling, ropyness, and right-angle branching. Therefore, in order to taxonomically characterize this organism, the partial regions of the ITS- 5.8 rDNA were isolated, sequenced, and deposited in GenBank.

The Gas Chromatography/Mass Spectromety (GC/MS) analysis of the fungal VOCs showed the presence of at least 28 VOCs. These compounds represented at least five general classes of organic substances including lipids, esters, alcohols, ketones, and acids with some representative structures shown in the Figure. Final identification of the volatile compounds was done by GC/MS of authentic compounds obtained from commercial sources or synthesized by us or others and compared directly to the VOCs of the fungus.

With all of the data in hand we felt secure in proposing a binomial for this fungus derived from the Latin – *Muscodor* (stinky) *albus* (white).

Ultimately, artificial mixtures of the compounds were used in a biological assay system to demonstrate the relative activity of individual compounds. Although over 80% of the volatiles could be identified, this seemed to be adequate to achieve an excellent reproduction of the lethal- antibiotic effects of the VOCs that were being produced by the fungus.

Volatile antibiotics from "Stinky white fungus"

This mixture of gases consists primarily of various **alcohols, acids, esters, ketones, and lipids**. We then examined each of the five general classes of VOCs in the bioassay test and each possessed some inhibitory activity with the esters being the most active. Of these the most active individual compound was 1-butanol 3-methyl acetate.

However, no individual compound or class of compounds was lethal to any of the test microbes which consisted of representative plant pathogenic fungi, Gram-positive and Gram-negative bacteria, and others.

Obviously, the antibiotic effect of the VOCs of *M. albus* is strictly related to the **synergistic activity** of the compounds in the gas phase. We know very little about the mode of action of these compounds on the test microbes. Thus, this represents an interesting academic avenue to pursue in the future.

Using the original isolate of *M. albus* as a selection tool, other very closely related isolates of this fungus have been obtained from tropical plants in various parts of the world including Thailand, Australia, Peru, Venezuela, and Indonesia. They possess high sequence similarity to isolate 620 and produce many, but not all of the same VOCs as isolate 620.

"Mycofumigation" with *Muscodor*

The VOCs of *M. albus* kill many of the pathogens that affect plants, people and even buildings. The term "mycofumigation" has been applied to the practical aspects of this fungus. The first practical demonstration of its effects

against a pathogen was the mycofumigation of covered smut-infected barley seeds for a few days resulting in 100% disease control. This technology is currently being developed for the treatment of fruits in storage and transit. Soil treatments have also been effectively used in both field and green house situations. In these cases, soils are pretreated with a *M. albus* formulation in order to preclude the development of infected seedlings.

AgraQuest, of Davis, Calif., is in the full scale development of *M. albus* for numerous agricultural applications with an anticipated release of a product in 2006. The US-EPA has given provisional acceptance of M. albus for agricultural uses. Also, it appears that the concept of mycofumigation has the potential to replace the use of otherwise hazardous substances that are currently applied to crops, soils and buildings and the most notable of which is methyl bromide for soil sterilization. AgraQuest is expected to have a product on the market for fruit treatment by now.

Hopefully, the discovery and the development of information on *M. albus* will have broad implications for the discovery and development of other rainforest microbes.

It will at the same time add another impetus for the conservation of the world's precious rainforests being under tremendous human thread now.

Gary Strobel is Professor of the Department of Plant Sciences at Montana State University in Bozeman.

Cited Literature:

Demain A (1981) *Industrial microbiology.* Science, 214: 987-995.

Bull AT (ed.) (2004) *Microbial Diversity and Bioprospecting,* ASM Press.

Kayser O, Quax W (2007) *Medicinal Plant Biotechnology.* Vol 1 and Vol 2 Wiley-VCH.

Strobel GA, Daisy B (2003) *Bioprospecting for microbial endophytes and their natural products.* Microbiol Mol Biol Rev, 67: 491-502.

www. agraquest.com

Fig. 6.9 Bayer's "tower biology" plant achieves a high oxygen supply and rapid degradation without unpleasant smells.

Fig. 6.10 Biogas plant near Berlin.

Fig. 6.11 Top: Mao Tse Tung, the Great Helmsman himself, inspecting a cooker powered by a biogas plant in the 1950s. The Chinese writing praises the benefits of marsh gas. Bottom: 30 years later, biogas is higher on the agenda than ever. The Great Reformer Teng Hsiao Ping in front of a biogas plant.

Fig. 6.12 Biogas plant being built under the auspices of the United Nations (UNEP).

Fig. 6.13 Right: Development stages of methane.

during photosynthesis is converted into methane – a major factor in the carbon cycle. Huge **methane hydrate** stores lie under the seabed. They capture the imagination of ecologists as well as thriller writers. If the methane were to be released, it would to lead to a major disaster. In contrast to other greenhouse gases, the amount of methane in the Earth atmosphere has not increased lately, according to Nobel laureate **Paul Crutzen** (Fig. 6.15). For once, the largest producers of methane are not humans, but insects – termites, to be precise. One termite (Fig. 6.22), i.e., the microbes in its intestine (mainly one-celled flagellates), produces only half a milligram of methane per day, but given the incredible number of these insects, their yearly production amounts to 150 million tons.

How does **methanogenesis** happen? Various groups of bacteria are responsible for the anaerobic conversion of large biomolecules such as cellulose, protein, or fat into methane and carbon dioxide. First, anaerobic clostridia and facultative anaerobic (with the ability to live with or without oxygen) enterobacteria and streptococci secrete enzymes into their environment that degrade high-molecular material into their basic components, i.e., sugar,

amino acids, glycerol, and fatty acids (see Fig. 6.13) in what is called the **hydrolytic phase**. This is followed by the **acidogenic phase** in which the components are broken down mainly into hydrogen, carbon dioxide, acetic acid, and other organic acids as well as alcohol. Methane arises from **hydrogen**, **acetate**, and **carbon dioxide**.

Methanogens are the most oxygen-sensitive organism on Earth. As these archaic creatures lack cytochromes and the hydrogen-cleaving enzyme **catalase**, oxygen that enters the cell causes the cytotoxic hydrogen peroxide to accumulate and destroy cell structures.

Methanogens, together with salt-loving halobacteria and thermophilic sulfur-reducing microorganisms form the phylum *Archaea* (formerly *Archaebacteria*). Their structure and metabolism is very different from ordinary *Bacteria* (formerly *Eubacteria*), and they are mostly found in extreme environments (Fig. 6.14) where conditions resemble those of earlier periods of the history of our planet when oxygen was scarce.

■ 6.4 Biogas Could Save Forests!

Approximately two million people worldwide still depend on burning biomass (wood, agricultural waste, and dried dung) for their energy supply – a direct and inefficient way of recovering energy from biomass with disastrous consequences for agriculture as well as the environment. In developing countries, wood has become as scarce as food.

Biogas, on the other hand, would be a viable option in agricultural regions where it could be produced by **small reactors fed with animal and human excrement** as well as plant waste. They would also provide natural fertilizer in the shape of sludge, which contains nitrogen, phosphorus, and potash, and thus would reduce the need for artificial fertilizers. Closed, airtight bioreactors (Fig. 6.12) would also kill pathogens. Biogas could help save the forests in many developing countries.

In **India**, a breakthrough was expected from what became known as the **Gobar project** (*gobar*, Hindi, meaning cow dung). It is estimated that the project runs 2.5 million biogas plants, bringing the country closer to Mahatma Ghandi's dream of self-supporting village units.

However, the project is still hampered by the semifeudal social structure of Indian villages where only

Biogas
CH₄
Acetic acid
CO₂
H₂
H₂O
Alcohol
Amino acids, sugars, glycerol, fatty acids
Bacteria **Hydrolysis**
Biomass

rich farmers can afford to buy a stainless steel biore-actor and to provide it with enough waste products. Poor peasants, by contrast, often do not even own a cow, and, even if provided with a free reactor, could not even provide the material to feed them.

The rich farmers would buy up dried dung very cheaply, thus robbing the poor of the fuel they used to collect from the streets. In addition, for many Indians, touching excrement is taboo for religious reasons. As a result, all the wood that anybody can get hold of is used for heating.

The United Nations look at **China** as a positive example. Several hundreds of thousands of biogas plants (Fig. 6.11) are said to be run there, often as communal projects. The reactors were built in cheap concrete reactor technology. Waste, energy, and fertilizer are economically used and wisely distributed.

◼ 6.5 Biogas in Industrial Countries – Using Liquid Manure

For obvious reasons, it is more difficult to produce biogas in non-tropical countries. However, a wide range of promising projects are under way in Europe and North America. Biogas reactors can help **solve the waste problem** of industrial farming. Liquid manure from those farming units is produced in such large quantities that it would heavily pollute the soil if used as conventional liquid manure, and carting it away would also be very costly. A dairy cow produces 75 liters (nearly 20 gal) per day. In Switzerland, farmers used old oil tanks to build a biogas plant which now provides them with the equivalent of 300 liters (nearly 80 gal) of heating oil per cow per year. Biogas can also be produced very efficiently from sewage treatment sludge.

Another source of biogas could be household waste. A lot of methane develops in landfill sites and can be quite a hazard. If these sites were properly covered up, methane could be extracted. According to US calculations, one percent of US energy consumption could be supplied by biogas. This may not seem much, but bear in mind that one percent of the US consumption is equivalent of the total energy consumption of several developing countries.

In the industrialized world, biogas will perhaps cover one to five percent of the total energy

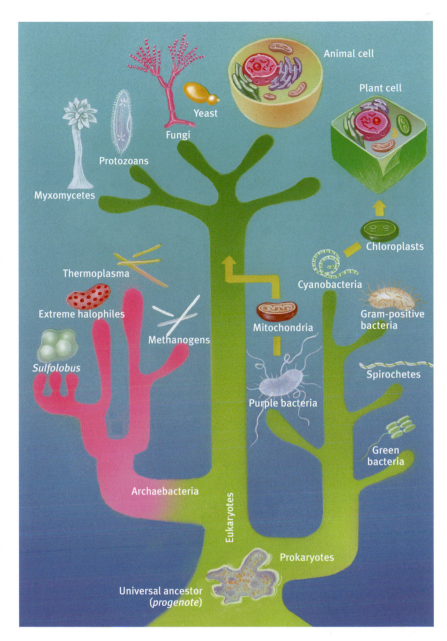

Fig. 6.14 The Tree of Life

Before the discovery of **Archaebacteria**, the tree of life consisted of two major lines of evolution – the prokaryotic line, and the eukaryotic line which was derived from the former.

At the beginning, there must have been anaerobic bacteria that produced energy through fermentation. Once the atmosphere contained enough oxygen, certain anaerobic cells that had lost their cell walls (mycoplasma) ingested smaller bacteria and developed an endosymbiotic relationship with them.

Among these "swallowed up" bacteria, one aerobic bacterium (with oxygen respiration) became a **mitochondrion**, a

photosynthetic cyanobacterium developed into a **chloroplast**, and a spirochete may have turned into a flagellum. The ancestor of eukaryotic cells thus developed.

The **dendrogram** of real bacteria, as far as it is known today, branches off into five major lines: Gram-positive bacteria, which include bacilli, streptomycetes and clostridia, have thick cell walls. The photosynthetic Purple bacteria (e.g., *Alcaligenes*) and some of their close relatives that are not photosynthetic, such as *E. coli*, form another group.

Spirochetes are long, spiral-shaped bacteria.

Cyanobacteria, which are photosynthetic and produce oxygen, probably gave rise to the chloro-

plasts in plant cells. Some spherical bacteria with an atypical cell wall (micrococci) are extremely sensitive to radiation. The green photosynthetic bacteria (*Chlorobium*) are anaerobic organisms.

In the 1970s, it was discovered that the prokaryotic archaebacteria differ considerably from all other organisms in their cell structure.

They are a "third form of life", a group apart, which played a dominant role in the primeval biosphere. Due to their oxygen sensitivity, they now only survive in ecological niche habitats.

Box 6.3 Biochemical Oxygen Demand (BOD5) – Measuring Unit for Biodegradable Substances in Wastewater

Oxygen has low solubility in water. At 15 °C (59 °F), approximately 10 milligrams of oxygen per liter, at 20 °C (68 °F) only 9 milligrams per liter are dissolved. When wastewater is discharged into lakes, rivers, and the sea, the amount of dissolved oxygen in the water drops dramatically, as it is used up by aerobic bacteria and fungi to break down the organic matter.

Sewage treatment plants therefore require an extra supply of oxygen in order to produce clean water.

Biochemical Oxygen Demand (BOD) is a test developed in England in 1896 to measure the organic contamination of water. BOD_5 gives an idea of the easily biodegradable proportion of the total of organic compounds in the water. It is derived from the oxygen demand of heterotrophic microorganisms.

The amount of oxygen removed from water at 20 °C during the breakdown is calculated for a certain number of days – in this case, five days. Water samples are diluted and air bubbled into shaking flasks (in order to fully saturate the water with oxygen). It is seeded with a mixed

culture of wastewater microbes. The oxygen content is then measured using a Clark oxygen electrode. The flasks are then sealed and kept at 20 °C in the dark where they are shaken. After the incubation period, the oxygen content is measured again. The difference between the first and fifth day (multiplied by the dilution of the sample) results in the BOD_5 value.

Measuring the BOD value using a Clark oxygen electrode in a culture flask

If the water did not contain any biodegradable substances, then there was nothing for the microbes to break down, and they will not have been respiring and multiplying. No oxygen will have been used up. The difference between day one and five will be zero. BOD_5 would then be 0 milligrams of oxygen per liter. The water is free of easily biodegradable material.

If, by contrast, the water were high in biodegradable substances, then the added microorganisms would have multiplied and used up the oxygen.

If, for example, the difference is 9 milligrams/liter, O_2 has been completely metabolized at a 100x dilution rate, the BOD value is 900 milligrams/liter.

900 milligrams of oxygen would be needed to completely break down one liter of wastewater. In other words, the oxygen dissolved in 900 liters of clean water would be needed to degrade one liter of wastewater.

The BOD_5 contamination of wastewater by one person per day is quoted in **population equivalents** (**PE**). One PE is approximately 60 grams BOD_5/day. The BOD_5 value can be affected by nitrification, algal respiration or toxic compounds that inhibit the growth of microorganisms.

The BOD_5 value provides a comparability framework for wastewater, and wastewater charges are based on it. However, the BOD_5 does not give any information about non-biodegradable compounds.

BOD testing has a major drawback, i.e., the long testing time. Five days of measuring time make it impossible to use the test for the flexible management of treatment plants, whereas microbial biosensors measure the BOD of wastewater within five minutes. However, they can only flag up low-molecular compounds that can penetrate a protective membrane.

Fig. 6.15 Nobel laureate Paul Crutzen (center), co-discoverer of the ozone hole, after a lecture in Hong Kong in April 2005. "*1.4 billion cattle on Earth produce massive amounts of methane.*"

Fig. 6.16 Strip of land near the former Berlin wall, sealed off because of herbicide contamination.

demand, due to the limited amount of available biomass. It is, however, highly significant as an energy source for agricultural regions all over the world. Not only the supply of energy, but also the disposal of waste is becoming a pressing issue, and the protection of soil and woodland is a global ecological problem.

■ 6.6 Fuel Growing in the Fields

When you turn the ignition key, you expect to smell the usual unpleasant exhaust gases, but instead, there is just a whiff of alcohol. This is an everyday experience in **Brazil** (see Box 6.5 for more detail) where ethanol-powered and flexible-fuel vehicles are manufactured. Their engines burn hydrated ethanol, an azeotrope of ethanol (around 93% v/v) and water (7%).

In 2004, around 42 billion liters of ethanol were produced in the world, most of it for use in cars. Brazil produced around 16.4 billion liters and used

2.7 million hectares of land area for this production. Of this, around 12.4 billion liters were produced as fuel for ethanol-powered vehicles in the domestic Brazil market.

It was the first oil crisis in 1973/4 that sparked the gigantic ethanol-for-fuel production in Brazil. A second boost for the industry came during the second oil crisis in 1979.

There was already a sufficient supply of raw material, i.e., sugarcane. As a reaction to the high world market prices for oil, the state of Brazil launched an ethanol-for-fuel program called *Proalcool* in 1975. The project was politically as well as economically motivated and has been controversial. There is, on the one hand, the question whether fuel production should take priority over food production, and on the other, the **environmental impact** through water pollution and soil erosion is significant. Each liter of pure ethanol produces 12 to 15 liters (3.2 to 3.9 gal) of sugar residuum and 100 liters (26.4 gal) of washing water, both of which, for economic reasons, are mostly discharged into the rivers untreated. Sugar residuum that could have been used as fertilizer turns rivers into sewers instead. The organic waste created to produce one liter of alcohol is the equivalent of the wastewater of four city-dwellers. One single distillery producing 150,000 liters (around 40,000 gal) per day produces the same amount of wastewater as a city of 600,000 inhabitants.

The conclusion? **Biotechnology does not automatically deliver eco-friendly solutions!**

During the alcohol-producing season, the drinking water supply of whole cities was temporarily disrupted. Now, the Brazilian government is pulling the plug. The alcohol industry must use its washing water in a closed cycle, sugar residuum must be turned into animal feed and fertilizer, and the sewage sludge into biogas. The biofuel problem highlights the economic as well as the political aspects that need to be considered in the future development of new technologies (Box. 6.5).

Lester Brown, director of the Worldwatch Institute, emphasizes that growing crops for biofuel would increase the already excessive pressure on fertile arable land in many parts of the world and finally lead to soil erosion and destruction. According to Brown, 1,000 square meters (0.25 acres) are needed to feed a human over a year, whereas to produce enough crops to fuel just one US car for a year would require an area of 30,000 square meters

(7.4 acres). In other words, **one car "eats" the food supply for 30 people!**

The United States are now increasing their own production. According to the Renewable Fuels Association, there were 107 grain ethanol biorefineries in November 2006, with a capacity of 5.1 billion gallons of ethanol per year. 56 additional refineries are under construction in the U.S., increasing the existing capacity by a further 3.8 billion gallons.

■ 6.7 Ananda Chakrabarty's Oil-Guzzlers

Could supermicrobes play a part in saving the environment? The Indian-born biotechnologist **Ananda Mohan Chakrabarty** (b. 1938, Fig. 6.18 and Box 6.4) who lives in the US, grew bacteria that could break down the **herbicide** 2,4,5-T, when working for General Electric. The herbicide was used in large quantities during the Vietnam war as a component of **Agent Orange** (which also contained mutagenic dioxin) to defoliate vast areas of jungle (Fig. 6.17). The catastrophic consequences include congenial deformities and cancer in the Vietnamese population and among the offspring of US war veterans.

After successfully producing herbicide eaters, Chakrabarty went on to grow veritable oil-guzzlers (Fig. 6.19). He took plasmids from four *Pseudomonas* strains, each of which degrades respectively **octane, camphor, xylene,** and **naphthalene.**

Fig. 6.17 Vietnamese jungle, intact (top) and after defoliation with herbicide Agent Orange (2,4,5-T) (bottom). Ananda Chakrabarty bred Agent Orange-degrading bacteria.

Fig. 6.18 Ananda Chakrabarty obtained his patent in Washington in 1981. Today, Chakrabarty, graduate of the University of Calcutta, has the title of Distinguished Professor at the University of Illinois in Chicago (see also special box).

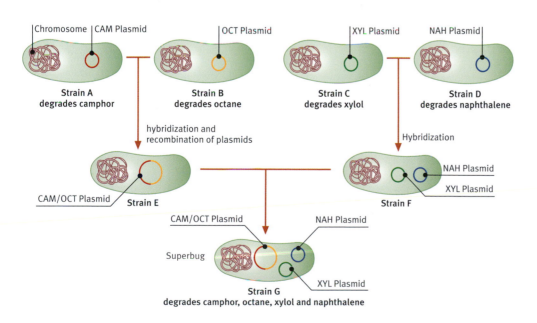

Fig. 6.19 The construction of a superbacterium or superbug that can break down the higher hydrocarbons of petroleum.
First, a camphor-degrading plasmid (CAM) was introduced into a plasmid that already contained an octane-degrading plasmid (OCT).
Both plasmids underwent fusion. Enzymes for both degradation pathways were thus encoded for in one plasmid.
xylene- and naphthalene-degrading plasmids, by contrast, co-existed already in one cell.
Finally, Chakrabarty put all these plasmids together in a single bacterial strain that thrives on petroleum, using camphor, xylene, octane, and naphthalene as carbon sources.

Chromosome | CAM Plasmid | OCT Plasmid | XYL Plasmid | NAH Plasmid

Strain A degrades camphor
Strain B degrades octane
Strain C degrades xylol
Strain D degrades naphthalene

hybridization and recombination of plasmids
Hybridization

CAM/OCT Plasmid — **Strain E**

NAH Plasmid
XYL Plasmid
Strain F

CAM/OCT Plasmid — NAH Plasmid
Superbug
XYL Plasmid
Strain G degrades camphor, octane, xylol and naphthalene

Fig. 6.20 The *Exxon Valdez* disaster. Day 1–5 The tanker *Exxon Valdez* at Bligh Reef, March 26th 1989.

Beaches heavily polluted with oil (Smith Island, April 1989). Attempts to clean up the beaches. Contamination was found even 350 miles away on the Alaskan peninsula (August 1989).

The Aftermath of the *Exxon Valdez* Disaster in Figures

38,000 tons of crude oil (the contents of 125 Olympic swimming pools) ended up in the sea.
1,300 miles of coast were polluted. The estimated number of animals killed is
250,000 sea birds,
2,800 otters,
300 seals,
250 bald-headed sea eagles,
22 killer whales and vast numbers of fish.

These plasmids were used to create super-plasmids which were then re-introduced into the bacteria and conveyed the ability to digest all four compounds. These transformed bacteria attacked poisonous crude oil residues with a vengeance and were intended to be used to quickly clean up large oil slicks. The massive amounts of microorganisms would later be polished off by organisms higher up in the food chain. However, Chakrabarty's oil guzzlers have never been used, due to the restrictions on release of genetically modified bacteria into the environment.

When in 1989 the oil tanker *Exxon Valdez* was stranded on the Alaskan coast, most of the thick oil layer was pumped off and filtered (Fig. 6.20), while the oil coat on rocks and pebbles was degraded by conventionally bred bacteria. Their growth was enhanced by the addition of phosphate and nitrate "fertilizers". Until this day, this is all that can be legally done.

What does the future hold for such improvers of the environment? In his novel *Sexy Sons*, German biologist and writer Bernhard Kegel develops a frightening scenario: Not only does the cloning of the boss of a global environmental protection company go horribly wrong, but newly engineered oil-digesting microbes that have been doing their job very effectively are maliciously introduced into oil wells, with disastrous consequences. Thankfully, this is just science fiction!

Spectacular oil tanker disasters like that of the *Exxon Valdez*, however, are only responsible for a few percent of overall oil pollution. Millions of tons of crude oil still end up in the sea every year, 25% due to illegal cleaning of empty tankers in the open sea, one third through effluents into rivers.

The cleaning-up of **petroleum-contaminated soil** (under gasoline stations, for example) is more complicated. After German reunification, soil remediation became a major issue in the East. Seven-foot-high heaps of soil were inoculated with specialist microbes, aerated and stirred. It often took no more than two weeks until 90% of the pollutants were broken down.

In Berlin, the strip of land that ran along **the Wall** that once divided the city had been heavily sprayed with **herbicides** for decades to keep the view clear for the border guards. The chemicals used were mainly chlorinated hydrocarbons such as 2,4-D (Fig. 6.16). Sixteen years later, the contaminants

have been degraded by naturally occurring soil microorganisms.

In 1980, the US Supreme Court made a decision in the case of *Diamond versus Chakrabarty* [447 U.S. 303 (1980)]. **Chakrabarty** had filed a patent on an organism in 1971 which had been fighting its way through the courts. His oil-digesting bacterial strain became the first engineered organism in history to be patented in the U.S. (Fig. 6.19), creating a precedent for the biotech industry allowing microbes to be patented (see Box 6.4 for the whole story!)

■ 6.8 Sugar And Alcohol From Wood

The ideal raw material for the production of sugar, alcohol, and other industrial chemicals is **starch**. However, as starch is a major nutrient, its use for industrial purposes is highly controversial (see Box 6.5).

The most significant sustainable raw material for alcohol production, however, is **lignocellulose**, which cannot be used as a food resource.

Lignocellulose consists of three components – **cellulose**, a linear polymer made up of glucose molecules, **hemicellulose**, also a polysaccharide, containing sugars of five carbon atoms such as xylose, and **lignin**, an aromatic molecule complex. In wheat straw and wood, these components occur in a ratio of 4:3:2.

Although the price per ton of dry cellulose biomass is considerably lower than that of grain starch, when it comes to conversion into sugar, it simply cannot compete. The **firm structure of lignocellulose**, vital for plants, becomes a major obstacle in the process, as cellulose is only available in crystalline form, enclosed by hemicellulose and lignin. Anybody who still owns genuine old wooden furniture will confirm that this type of cellulose is not water-soluble, whereas starch is.

Most microbes are unable to break down wood without some enzymatic pre-treatment, and this is precisely the reason why timber has become such a **popular building material**, although it needs to be protected from woodworm, termites, and white rot and blue stain. All these organisms produce **cellulases** that break down cellulose into sugar and thus destroy the wood. In methane-producing termites, it is the protozoa in the intestine that produce cellulases (Fig. 6.22).

Box 6.4 The Expert's View:
Ananda Chakrabarty – Patents on Life?

The first patent on new lifeforms in history was granted to Ananda Chakrabarty. He earned his PhD at the University of Calcutta in India in 1965. In his days as a young scientist at General Electric in the United States, he developed a *Pseudomonas* strain that could break down crude oil components into less complex substances on which aquatic life can feed.

The strain was the subject of the landmark 1980 U.S. Supreme Court decision that forms of life created in the laboratory can be patented. Shortly after the decision came through, a spokesperson of pioneer biotech company Genentech (founded by R.A. Swanson and H.W. Boyer, see Ch. 3) said that the Supreme Court's action "assured this country's technological future."

Chakrabarty's battle for patent protection is now widely thought to have paved the way for future patenting of biotechnological discoveries.

Here is his story:

"Who owns life?" This is a rhetorical question for which simple answers do not exist. The definition of what life is or when it begins has eluded a satisfactory answer since the days of the abortion debate or even earlier, when philosophers and scientists tried to distinguish between living and nonliving objects.

To complicate matters: What is meant by ownership?

In a sense, we all own life, if ownership means having the ability to breed new lives (as we do with cattle, chickens, fish, or plants) and to terminate such lives at will.

My own case of patenting a life form

My involvement with the two issues, viz., patenting life forms or judicial lawmaking, goes back to the 1970s when I was a research scientist at General Electric (GE) R&D center in Schenectady, N.Y.

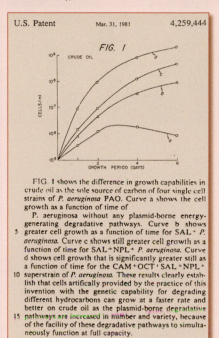

FIG. 1 shows the difference in growth capabilities in crude oil as the sole source of carbon of four single cell strains of *P. aeruginosa* PAO. Curve a shows the cell growth as a function of time of
P. aeruginosa without any plasmid-borne energy-generating degradative pathways. Curve b shows
5 greater cell growth as a function of time for SAL⁺ *P. aeruginosa*. Curve c shows still greater cell growth as a function of time for SAL⁺NPL⁺ *P. aeruginosa*. Curve d shows cell growth that is significantly greater still as a function of time for the CAM⁺OCT⁺SAL⁺NPL⁺
10 superstrain of *P. aeruginosa*. These results clearly establish that cells artifically provided by the practice of this invention with the genetic capability for degrading different hydrocarbons can grow at a faster rate and better on crude oil as the plasmid-borne degradative
15 pathways are increased in number and variety, because of the facility of these degradative pathways to simultaneously function at full capacity.

There I developed a genetically improved microorganism (see this book on page 179) that was designed to break down crude oil rapidly and was therefore deemed suitable for release into oil spills for their cleanup. Because such an environmental release would make them available to anyone who wanted to have them, GE filed for a patent for both the process of constructing the genetically manipulated microorganism and the microorganism itself.

The **Patent and Trademark Office** (**PTO**) granted the process patent but rejected the claim on the microorganism based on the fact that it was a product of nature. GE appealed to the PTO Board of Appeals, pointing out that the genetically engineered microorganism was genetically very different from its natural counterparts. The Board conceded that it was not a product of nature but still rejected the claim on the patentability of the microorganism because of the fact that the microorganism was alive.

Convinced that such a rejection had no legal basis, GE appealed to the U. S. Court of Custom and Patent Appeals (CCPA). The CCPA ruled three to two in GE's favor. Judge **Giles S. Rich**, speaking on behalf of the majority, contended that microorganisms were "much more akin to inanimate chemical compositions such as reactants, reagents, and catalysts than they are to horses and honeybees or raspberries and roses."

Another judge, **Howard T. Markey**, concurred, saying, "No congressional intent to limit patents to dead inventions lurks in the lacuna of the statute, and there is no grave or compelling circumstance requiring us to find it there."

The PTO then appealed to the Supreme Court. Initially, the Supreme Court sent it back to the CCPA for reconsideration in light of another patent case, but the CCPA again ruled in favor of GE, this time emphasizing that the court could see "no legally significant difference between active chemicals which are classified as dead and organisms used for their chemical reactions which take place because they are live."

The CCPA decision prompted the Solicitor General to petition the Supreme Court for a ruling, which was granted. Several amicus briefs were filed by many individuals and organizations to support or oppose the case, which came to be known as **Diamond v. Chakrabarty** because **Sidney Diamond** was the new commissioner of patents. One amicus brief filed on behalf of the government came from People's Business Commission (PBC), which argued that patenting a life form was not in the public interest and that granting patents on a microorganism would inevitably lead to the patenting of higher life forms, including mammals and perhaps humans. The "essence of the matter," PBC argued, was that the "issuance of a patent on a life form was to imply that life has no vital or sacred property and it was simply an arrangement of chemicals or composition of matter."

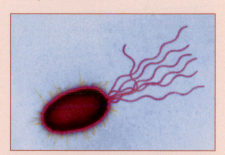

This phantastic picture of *Pseudomonas* was taken by Dennis Kunkel (see www.denniskunkel.com). He explained it: "This is a Transmission Electron Microscope (TEM) view of a negatively stained bacterium. It was just luck that I retained most of the polar flagella on this Pseudomonas – this species has a bundle of polar flagella only. The negative stain allows the flagella to stand out nicely when viewed with a TEM."

In 1980, **8 years after the initial filing**, the Supreme Court, by a vote of 5 to 4, agreed with Judge **Giles Rich** of the CCPA that the microorganisms in question were a new composition of matter, the product of human ingenuity and not

See next page

of nature's handiwork, and thus a patentable subject matter. Justice **William Brennan**, arguing on behalf of the minority, pointed out that the case involved a subject that "uniquely implicates matters of public concern" and therefore belonged to congressional review and a new law. Ownership of life forms became a reality with interesting consequences!

What happened next?

The decision of the Supreme Court was narrowly focused on what was described as a "soulless, mindless, lowly form of life," even though the court emphasized that anything under the sun made by man could be patented as long as it met the criteria. The PTO took the decision as a board interpretation of the patenting of life forms and has issued hundreds of U.S. patents covering microorganisms, plants, animals, fish, bird and human genes, mutations, and cells.

The issuance of **patents on animals and human body cells** conferred a degree of ownership that was an uncharted territory with unprecedented legal ramifications. Many disputes involving patent infringement cases emerged because of questions related to obviousness, enablement, or the priority of invention that had to be decided by the courts.

More difficult were the questions about ownership rights and privileges. For example, in the patent "Unique T-lymphocyte line and products derived therefrom," the inventors used the spleen of a patient, Mr. **John Moore**, who suffered from hairy cell leukemia and came for treatment to Dr. **David W. Golde** at the University of California at Los Angeles (UCLA). As part of the treatment, Mr. Moore's spleen was removed and Dr. Golde developed a cell line with enriched T-lymphocytes that produced large amounts of lymphokines useful for cancer or AIDS treatment. Without Mr. Moore's initial knowledge or consent, but requiring Mr. Moore's repeated visits to the hospital, Dr. Golde and UCLA applied for a patent on the cell line derived from Mr. Moore's spleen, which was granted in 1984. Mr. Moore subsequently sued Dr. Golde and UCLA, claiming theft of his body part. The trial court ruled against Mr. Moore, but the ruling was reversed by the Court of Appeals, and the case finally ended up at the California Supreme Court. Both the Appeals Court and the Supreme Court recognized the novelty of Mr. Moore's claim, which was without any precedent in law. Nevertheless, the California Supreme Court ruled against Mr. Moore on the issue of conversion (unauthorized use of his

Ananda Chakrabarty in his lab when working with "superbugs"

body part) but recognized his right to be informed of what the physician was doing involving his health and well being.

Who should own intellectual property when a given invention not only requires human ingenuity but also human tissues?

On October 30, 2000, a lawsuit was filed in Chicago federal court by the Canavan Foundation of New York City, the National Tay-Sachs and Allied Diseases Association, and a group of individuals. These individuals and organizations raised money, set up a registry of families, and recruited tissue donors to help develop a genetic test for the **Canavan disorder** caused by a mutation on chromosome 17 in a gene encoding the enzyme aspartoacylase. The disease predominantly affects Ashkenazi Jewish children and a genetic test was deemed to be useful in screening potential parents harboring such mutations. The foundation provided diseased tissues and money to Dr. **Reuben Matalon**, who developed a genetic test while employed at the Miami Children's Hospital (MCH). Once the test was developed, a patent on it was obtained by MCH. However, MCH was alleged to charge a high fee for the test, restrict access to the test, and put a limit on the number of tests that a licensee can perform.

This was contrary to the intention of the individuals and foundations, which was to help develop a cheap test that could be made widely available to prospective parents to prevent Canavan disease.

The question of how important the contribution of the issue donors is, without which tests cannot be developed and patents cannot be obtained, is going to resonate as more and more such cases end up in courts of law.

Another important consideration is the cost of the tests. Diseases are often caused by mutations, deletions, or genetic rearrangements in human genes. Therefore, genetic screening for

some diseases may involve several genetic tests to ensure detection of all possible genetic alterations. For a disease such as cystic fibrosis, where 70% of the patients harbor a single mutation (a trinucleotide codon deletion) in the cftr (cystic fibrosis transmembrane conductance regulator) gene, a single genetic test may not be enough and multiple mutations may have to be screened, thus greatly increasing the cost of genetic tests.

Because detection of genetic alterations and development of such tests are time consuming and costly, the developers seek patents on such genetic tests and try to recoup the expense by assessing license fees for conducting the tests. **Who should pay** for these costs? **Who decides** the cost of such genetic tests? Should market forces determine who should live disease-free and who should suffer? The recent incidence in South Africa where **drug companies allowed low-cost generic production of their patented AIDS drugs** to be made available to HIV-infected patients is an interesting example of the societal responsibility of large drug manufacturers.

Ananda Chakrabarty after winning his case

Then there is the question of **patentability of genetic tests** that do not detect all the mutations or genetic variations in a gene. **Myriad Genetics** of Salt Lake City obtained a European patent in 2000 on genetic testing of BRCA1 and BRCA2, called BRAC Analysis Technique, which used automated sequencing to scan for BRCA mutations and deletions. Mutations in BRCA1 and BRCA2 account for almost 10% of all breast cancers, and their early detection is important to treat breast cancer. However, certain deletions and genetic rearrangements in BRCA1, involving about 11.6 kb of DNA, cannot be detected by the Myriad tests and can only be detected by a patented process called combed DNA color bar coding. Such genetic deletions may comprise about 36% of all BRCA1 mutations. The claims of the Myriad Genetic European patent allegedly make it difficult for European clinicians to use the combed DNA color bar coding technique patented by Institute Pasteur, raising difficult legal questions on who should control or establish ownership of human genetic mutations and their detection and on human genetic makeup in general.

There are also **privacy and civil right issues** in patented genetic tests. For example, the U.S. Equal Employment Opportunity Commission (EEOC) sued Burlington Northern Santa Fe Railroad for requiring genetic tests of its employees who filed claims for certain work-related hand injuries (carpal tunnel syndrome).

The railroad wanted to determine which employees may be predisposed to this syndrome, which was believed to be due to a specific genetic deletion on chromosome 17. The railroad was alleged by the EEOC to have threatened to fire employees who refused the test, thus violating the employees' civil rights. This case has since been resolved to the satisfaction of the EEOC.

What is human?

In 1998, a patent application was filed to the United States PTO for a **human-animal hybrid** which was not actually made, but the concept was based on the fact that hybrids can be made out of sheep and goats. It was pointed out that the DNA sequence identity between human and chimpanzees is of the same order as that between sheep and goats. Consequently, the applicants argued that it should be possible to make hybrids between humans and chimpanzees, and such hybrids could be useful for organ harvesting or other medical purposes. In reality, these applicants did not want the patents but were simply raising the issue to prevent future patents on engineering of human genes and human reproduction. The patent office rejected the application based on the 13th Amendment of the U.S. Constitution, which is an antislavery amendment that rejects the ownership of human beings.

Although the patent application has been rejected, it has raised some interesting questions. If the hybrid human-animal is a human and therefore cannot be owned, how much human genetic material or human phenotypic trait must be present in an organism or an animal to confer on it the characteristics of a human?

Several human genes have been inserted in non-human organisms without raising problems of patentability. Is there an upper or lower limit of the presence of such human genes or traits to deter patentability based on the 13th Amendment? A Massachusetts company has claimed to have cloned a human embryo even though the

embryo did not undergo enough cell division to give rise to a blastocyst. Given the large number of animals that have been cloned, it is likely that human embryos will be cloned in the future. There are many uncertainties regarding the health and well-being of cloned animals, and certainly there would be enormous resistance to the cloning of human beings. Given scientists' curiosity and relative ease of cloning, it is likely that somebody, somewhere, would conduct nuclear transfer to enucleated human oocytes. What would happen if somebody transfers a **chimpanzee nucleus into an enucleated human oocyte** and implants it into a human uterus? Alternatively, one could take a chimpanzee egg and replace the nucleus with that of a human and then implant it in the womb of a chimpanzee. Even with the contribution of the cytoplasmic material, which controls gene expression, or the mitochondrial DNA of the egg, it is likely that the transferred nucleus will determine the primary genotype of the embryo.

Genome **Comparative Genetics**

Is a baby chimpanzee delivered of a human mother a **chimpanzee or a human**? Alternatively, is a mostly human baby delivered of a chimpanzee mother a human or a chimpanzee? Can such babies be patented if they are not products of nature?

Epilogue

This is both an exciting and a difficult time for a biologist.

The technology of animal and human reproduction, as well as the techniques of genetic manipulation, are progressing so rapidly it creates situations that transcend our legal structure and directly affect our social and moral fabrics.

It is high time that the United States Congress takes a serious look at where the science is going, where it needs to go to make a positive contribution, and perhaps define the boundaries of our venture into the unknown biological mys-

teries of nature. Of course, no Congressional mandate will ever cover all future scientific directions or orient human ingenuity, and the judiciary will play increasing roles in resolving conflicts involving human genetic reproduction, genetically manipulated plants and foods, and environmental restoration. There is thus a great need to continue dialogues between the judiciary, the legal community, the legislature, the interested public, and the scientific community, to provide guidance in scientific developments that may have major impacts on society.

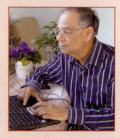

Ananda Chakrabarty
today

Ananda Mohan Chakrabarty was born at Sainthia (India) and is at present a Distinguished University Professor at the University of Illinois College of Medicine at Chicago. Chakrabarty's career illustrates a talent for turning research always into practical means. He is currently dealing with a new exciting finding: The ability of certain infecting pathogenic bacteria to allow tumor regression in human patients has been known for more than hundred years. The reason for regression was, however, thought to be due to the production of cytokines and chemokines by an activated immune system.

Chakrabarty has now shown that bacteria such as Pseudomonas aeruginosa produce the protein azurin that is secreted when the bacteria are exposed to cancer cells. Azurin, and a modified form of azurin called Laz produced by Neisseria species, are highly effective in forming complexes with various proteins involved in cancer growth and surface proteins of the malarial parasite Plasmodium falciparum and the AIDS virus HIV-1, thereby significantly inhibiting their growth. Thus single bacterial proteins might have potential therapeutic application against such unrelated diseases as cancer, malaria or AIDS, including coinfection of AIDS patients by the malarial parasite.

Cited Literature:

Chakrabarty A M (2003) *Patenting life forms: yesterday, today, and tomorrow*, in: Pespectives on properties of the Human Genome Project (Kieff, FS, Olin, JM, eds.), pp 3-11, Elsevier Academic Press, Amsterdam, Boston.

Fig. 6.21 Hypothetical enzymatic degradation of lignocellulose.

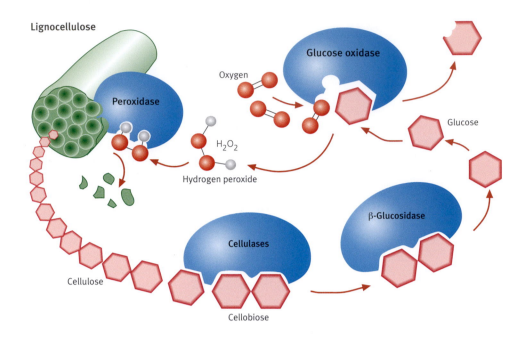

Lignocellulose

Glucose oxidase

Peroxidase

Oxygen

H_2O_2

Glucose

Hydrogen peroxide

β-Glucosidase

Cellulases

Cellulose

Cellobiose

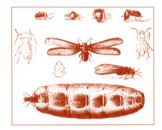

Fig. 6.22 Top: Termites produce biogas through cellulases produced by the intestinal flagellates they carry in their guts.
Center and bottom: A drawer in the author's cupboard in Hong Kong that had not been used for a while revealed a termite colony that turns the wood of the cupboard into methane.

We have just begun to fight!

PEARL HARBOR
BATAAN
CORAL SEA
MIDWAY
GUADALCANAL
NEW GUINEA
BISMARCK SEA
CASABLANCA
ALGIERS

Fig. 6.23 In World War II in the Pacific against the Japanese, the cotton clothes and cartridge belts of U.S soldiers disintegrated at an alarming rate. Cellulase-producing fungi such as *Trichoderma* were to blame.

The current favorite among cellulase-producing organisms is a fungus named after the eminent American cellulase researcher **E.T. Reese**, *Trichoderma reesei* (formerly *Trichoderma viride*), which secretes cellulase extracellularly. It was discovered on a rotting cotton cartridge belt from World War II in New Guinea (Fig. 6.23). At the time, microbiologists were following up alarming reports about cellulose-containing equipment of the US forces disintegrating in tropical regions with frightening speed.

Meanwhile, mutants have been found that are ten times more productive than the wildtype strain, in turning cellulose into sugar. The process, however, has remained uneconomical. After 30 days, even the most efficient lignin-degrading fungus, *Chrysosporium pruinosum*, leaves 40% of lignin intact.

The degradation products of lignin are toxic to many microbes – perhaps a natural tree defense mechanism – and no sensible use for lignin has been found so far. **Pre-treating lignocellulose** with acid is therefore necessary to ensure an efficient enzymatic breakdown of cellulose, and the removal of the acids is an expensive business. The cost could be reduced by using steam explosion or freeze explosion technologies involving liquid ammonia to remove the acid.

The long search for lignin-degrading microbes, however, yielded some unexpected results several years ago. Experts had expected that high-molecular-weight compounds would always be cleaved first by hydrolytic enzymes (hydrolases) produced by microorganisms and secreted into a medium. This is the case for amylases and cellulases. In lignin degradation, however, a fungus was found that causes what is known as **white rot** in wood. These fungi break down 60 to 70% of lignin into fragments, carbon dioxide, and water, freeing white cellulose (hence the name white rot).

The lignin degradation mechanism is sensational. The *Phanerochaete chrysosporium* and *Coriolus versicolor* fungi secrete not hydrolases, but extracellular peroxidases into the medium. These break the bonds between the phenols in lignin (Fig. 6.21).

The origin of the hydrogen peroxide, without which peroxidases do not function, remains unexplained. It is thought that **extracellular oxidase**s such as glucose oxidase provide H_2O_2 when oxidizing glucose. The unexpected involvement of peroxidases in the degradation shows that enzyme research has so far uncovered no more than the tip of the iceberg, and further amazing and useful applications may lie ahead.

The second problem concerning the degradation of wood is the inhibition of cellulases by their own product (**product inhibition**) glucose and its dimer cellobiose, and the comparatively low activity of cellulases. Even the most effective cellulases only exhibit one thousandth of the activity of commercial amylases. **Protein engineering** might be a way to make cellulases relinquish their voluntary self-control. Genetic engineers have been cloning cellulase genes from several fungi into bacteria in order to produce large quantities of cellulase economically. Another possibility of improving the effectiveness of lignocellulose conversion is to convey to bacteria the ability to use the five-carbon sugars (e.g., xylose) of hemicellulose.

As a last resort, microbes that metabolize lignocellulose directly, such as *Clostridium thermocellum*, can be used. The wild strains offer a wide range of lignocellulose products, particularly ethanol and organic acids. The production of any of these products can be genetically enhanced, i.e., production can become economically viable.

A company in North America uses the white rot fungus *Ophiostoma piliferum* to pre-treat woodchips (biopulping), and within a few weeks, the cellulose yield grows considerably. *Ophiostoma*-inoculated chips break down lignin while at the same time crowding out bluestain competitors.

■ 6.9 Basic Chemicals From Biomass?

Approximately **100 chemical compounds** employed in industry make up **99% of all chemicals** used. Three quarters of these are made out of five essential ingredients: ethylene, propylene, benzene, toluene, and xylene. Currently, all of these compounds are derived from petroleum and natural gas, and all fluctuations and turbulences of the oil markets are passed on to the chemical sector. In addition, the cracking of petroleum and precautions against environmental damage require a lot of costly energy. Experts believe that in principle, about half of the top 100 chemicals could be produced from renewable sources. What is the situation today? Ethanol, citric acid, and acetic acid can be produced cost-effectively through biotechnology (Fig. 6.24).

Ethanol is an important industrial chemical (Ch. 1) which is utilized as solvent, extractant, or antifreeze, as a starter substance for the synthesis of other organic compounds used as pigments, glue, lubricants, medicinal products, detergents, explosive, resins, and cosmetics.

When fossil fuels prices were low and sugar and starch for fermentation were expensive, chemical ethanol synthesis involving ethylene hydration at high temperatures and catalysts widely replaced fermentation. Due to rising fuel prices, we now see a renaissance of the oldest biotechnological method

in the world. Through continuous processing, new highly effective thermophilic and ethanol-tolerant microorganisms, and energy-efficient distilling techniques, the distillation process can be made competitive.

Currently, **acetic acid** production by oxidation of ethanol by *Acetobacter* is exclusively used for human consumption, whereas highly concentrated industrial acetic acid is obtained through the chemical carbonylation of methanol, which is more cost-effective. Approximately 200,000 tons of acetic acid worldwide are produced from ethanol by fermentation.

An environmentally friendly application of acetic acid is being tested in the US. Acetate, an acetic acid salt, is produced from acetic acid by mixing it with limestone. The acetic acid is obtained from biomass. **Calcium magnesium acetate** (CMA) has a melting point of minus eight degrees Celsius (17.6° F) and is used as an environmentally friendly salt to keep roads free from ice. Car drivers are happy with it because it also protects cars from corrosion. The sodium chloride widely used in Europe for roads, by contrast, is a tree killer because it competes with essential plant nutrients.

n-Butanol (1-butanol) is an important organic solvent. The n stands for 'normal' and implies that the compound consists of a straight chain with no side chains. n-Butanol is used in the production of plasticizers, brake fluid, fuel additives, synthetic resins, extractants, and paints. 1.2 million tons of butanol are produced from petroleum every year. The production of butanol began due to shortages in England during World War I and is directly connected to the foundation of the State of Israel (Box 6.6).

Clostridium acetobutylicum produces butanol, with the solvent acetone as a byproduct. **Acetone** was very much in demand in World War I. In Britain, it was needed to produce the explosive cordite. After the war, butyl acetate was used for nitrocellulose varnishes, and there was a shift in interest towards n-butanol. Only in the 1940s and 50s, when prices for petrochemical products fell below those of starch and molasses, did the microbial production of n-butanol drastically decline. Only South Africa, where petroleum was scarce due to an international embargo, kept running 90-cubic-meter reactors that yielded 30% solvents, of which six parts were butanol, three parts acetone, and one part ethanol.

Fig. 6.24 Basic chemical compounds that can be produced from biomass.

$CH_3—CH_2—OH$

Ethanol

Acetic acid

n-Butanol

$CH_3—CH_2—CH_2—CH_2—OH$

Acetone

Citric acid

Lactic acid

D-Gluconic acid

Box 6.5 The Expert's View:
Nobel Prize Laureate Alan MacDiarmid About Agri-Energy

You start your car, and off it goes, leaving a wonderful smell of booze behind…

The old Esso tagline 'Put a Tiger in Your Tank' should now read 'Put BIO in Your Tank'. Many countries, China, the EU, and even the US under George Bush, have decided to move towards an increased use of bioethanol. Against a background of sharply rising oil and fuel prices, it made sense to fall back on the oldest biotechnology in the world, the yeast fermentation of sugars into alcohol. Many see biofuel as the savior of our future.

Prof Alan MacDiarmid (University of Texas at Dallas and University of Pennsylvania) has just given a lecture about agri-energy at my university in Hong Kong. In 2000, he was awarded the Nobel Prize for the revolutionary discovery that modified plastic polymers work as electric conductors. He recently set up a research center in China which looks into China's supply of bioenergy. China has vast desert-like areas where hardy energy-supplying plants that do not compete with food crops could be grown and provide poor farmers with an income.

Ethanol-fuelled car in amongst the sugarcane that provides sustainable fuel.

Nobel Laureate Alan McDiarmid explained to the students:

As we deplete oil, coal, and natural gas reserves globally, the recovery of the remaining fossil fuel may be technologically feasible, but it will not be economically recoverable. It will simply cost too much. We see that clearly today with the price of oil at $70 a barrel and rising.

This reality enhances the financial viability of any alternate form of renewable energy because that viability is hitched to the price of oil per barrel on the international market. If the price of oil falls, which is unlikely, the economic feasibility of alternate forms of energy decreases. If the price of oil continues to rise, which seems likely, then so does the economic viability of alternative energy.

The most promising source of alternate energy comes from nature's own solar cells – the leaves of trees, bushes, and grass. They absorb sunlight and convert it into various organic materials - stored energy from the sun.

Sugarcane (*Saccharum officinarum*)

In past eras, people hunted for fish, animals, berries, and roots. Then we learned to grow them through farming. But we are still hunting for energy from forests that grew millions of years ago – that is, coal, petroleum, and natural gas.

Now that we are more enlightened, we will surely do the same for energy as we have done for food and begin to farm our energy needs.

Like "farms for food" long ago invented by humans, now the time has come to create **"farms for fuel."**

In the future, we will get most of our energy from the growth of plants instead of waiting for them to decay over millennia and then using expensive and complex technology to get that energy out of the ground.

The most prevalent type of fuel that can be grown is bio-ethyl alcohol, **or ethanol** as it is commonly called, which we are already putting into gasoline in significant quantities. It is made primarily from the fermentation of sugar and certain parts of corn.

Bio-diesel, which is oil obtained from soybeans, sunflower seeds, jatropha, or the like, is another form of bio-fuel. In this case, the oil is extracted and no fermentation is involved.

Since a given country may be limited by its climate or soil conditions in the growing of sugar or corn for fermentation into bio-alcohol, of great interest at the moment are advances in creating fuels out of cellulosic materials such as are found in wood-chip waste or dry waste from farm products.

If this cellulose can be broken down by enzymes into sugars that can be easily fermented into ethanol, then there will be few limits to widespread use of biofuels. Pilot projects on this have been implemented, and the results are promising. Every day, the costs are being reduced because the cost of the cellulose enzyme is becoming cheaper and cheaper with active research.

In time, the fuel of the world will be derived from trees, shrubs, and grasses that can grow in essentially any climate in the world, as well as from corn and sugarcane in certain climates. Instead of a petroleum economy, we will live in a bio-alcohol economy.

The beauty of using fuel obtained from living plants is that any carbon dioxide released into the atmosphere when a biofuel is used is then reabsorbed by the leaves of plants. So we get a cycle in which the amount of carbon dioxide released into the air is neither increased nor decreased. Therefore, if we use biofuels of any type, we **do not add to global warming** by increasing the amount of carbon dioxide into the air, as the burning of fossil fuels does.

Of course, no one form of renewable energy, whether biofuels, wind, or solar, can cover all the energy needs of a given country. Necessarily, there will have to be a **mix of fuels according to the local conditions** of climate, soil, and terrain. We will no doubt see wind and hydroelectric powers in one part of the country and biofuels in another. And still some "economically recoverable" fossil fuels in another.

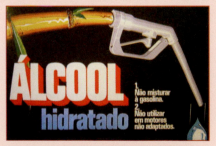

What I'm describing has already happened in **Brazil**, which today is essentially independent of any imported petroleum. It has already produced **6 million automobiles** that run either on pure ethyl alcohol or some combination with gasoline. These are known as flex-fuel cars. You can drive into any gas station and you will see two types of

Jatropha (pictured above), also called physic nut, is currently grown on around 2 million hectares across China for its non-edible oil used in candle and soap production. It is now expected to become the main crop for the production of biodiesel. The 13- million-hectare forest, mostly in southern China, should yield nearly 6 million tons of biodiesel every year. The jatropha trees also provide wood to fuel a power plant with an installed capacity of 12 million kilowatts – about two- thirds the capacity of the Three Gorges Dam project, the world's biggest.

pumps – one labeled alcohol and the other labeled gasoline, although the gasoline already has about 22 percent ethanol mixed in. Sensors in the gas tanks sense the relative amounts of ethanol and gasoline, and the engine adjusts accordingly. At the moment, a gallon of ethanol is cheaper than a gallon of gasoline.

The number of cars in Europe with a flexible fuel mix are increasing. Ford Motor Company has said that it will produce 250,000 flex-fuel cars in the United States. One wonders: If Brazil can convert so thoroughly to biofuels, why can't the sole superpower now trapped in its dependence on Middle East oil with all the attendant conflicts that involves and the contribution to global warming that entails?"

It seems so straightforward, but is it? We had a controversial friendly discussion with Alan MacDiarmid afterwards:

Recently, the Chinese government has pulled the plug on biofuel. China had been subsidizing biofuels since 2002. In **China**, this meant producing ethanol mainly from corn, sorghum, cassava, sweet potato, and beet. Just two months before the change of policy, it had been announced that bio-diesel produced from animal and vegetable fat would be introduced to Chinese gas station. One of the key projects in the 10th five-year-plan had been to substitute unleaded gasoline with ethanol. Five provinces and 27 cities, including Heilongjiang, Jilin, and Liaoning, have already switched. According to the official *Xinhua News Agency*, one fifth of all car fuel used in China was bioethanol. It is thought that bio diesel from oilseed releases fewer carcinogenic substances.

Biofuels, mixed with gasoline or diesel could help reduce the increasingly intolerable air pollution in China and reduce the country's dependency on oil. As recently as five years ago, China's annual corn and wheat production increased steadily, and turning the surplus into ethanol made economic sense.

So what caused this **change in policy**? There is simply not enough arable land available, and grain prices rose steeply over the past few months. The increase between October and November 2006 was a whopping 5%. According to the China National Grain and Oils Information Centre, corn prices per ton in the port of Dalian where most of the corn for export passes through, had risen to 1530 Yuan (CNY) in November, an increase of 200 Yuan since October 2006.

Continuing to produce alcohol from food could set prices spiraling out of control and lead to social instability and unrest. Therefore, anyone wanting to produce bioethanol now needs a licence to do so, and acquiring one can be a complicated process.

China could be setting a trend that many countries will follow. The **EU** plans to cover 20 % of all its energy needs from agricultural production by 2020. In 2004, 34 million tons of grain were turned into ethanol in the US alone. This figure is now set to double to 69 million, mainly from corn conversion. Corn-importing countries such as Japan, Mexico, and Egypt are worried about a possible decrease in US exports, which make up 70% of the world trade. Corn is the staple diet of the poor in Mexico.

A poor grain harvest worldwide in 2006 exacerbates the problem. At 1,967 million tons, the harvest was 73 million tons below estimated consumption – the largest deficit for years.

What is the solution? Grain could be replaced by **a second generation of energy-providing organic raw material** – straw, wood, manure, and waste material.

Another point critics often make is the **lack of efficiency** of biofuel programs. Brazil seems to be the only country that gets more energy out of sugarcane than it puts in because it also ferments byproducts like crushed pulp. India and China, by contrast, cannot produce sufficient amounts of corn or sugarcane for ethanol production unless they subsidize their agriculture to do so. This is not only due to small farming units and a lack of fertilizers, but also to climatic factors (soil fertility, rainfall) China and India will have to find their own solutions.

Meanwhile, biotechnologists are busy developing energy-rich crops such as **corn with a higher starch content**. They work on optimizing enzymes and microorganisms to accelerate the ethanol production process and on methods that allow the use of plant substrates or agricultural waste, including the use of cellulose.

Alan MacDiarmid was born in 1927. He was educated at Hutt Valley High School and Victoria University in New Zealand. After completing an MSc in Chemistry he was awarded a Fullbright Fellowship for a PhD at the University of Wisconsin. He then won a Shell Scholarship, which enabled him to complete a second PhD Cambridge University. In 1955, after a year at St Andrews University in Scotland, he took a position at the University of Pennsylvania. He became a full Professor in 1964 and the distinguished Blanchard Professor in Chemistry from 1988. In 2000, he was awarded the Nobel Prize for Chemistry shared with physicist Alan Heeger (USA) and chemist Hideki Shirakawa (Japan) for the discovery and development of conductive organic polymers. Prof MacDiarmid is now organizing an international network to solve the energy problems partly by "agri-energy"!

Box 6.6 Biotech History: How a Bacterium Founded a Country

As a young man, **Chaim Weizmann** (1874–1952) was forced to leave his Byelorussian homeland due to anti-Semitism. He studied in Switzerland and Germany and then began to work with the famous chemistry professor William Perkin in Manchester in 1904.

Early in 1915, he came to the attention of the then Minister of Munitions **David Lloyd George**. There was a severe shortage of highly explosive cordite, a mix of nitroglycerin and cellulose. The acetone that was required for its production was particularly scarce, as it was distilled from wood.

Lloyd George and Weizmann met and immediately took to each other. Was there a way to produce acetone through fermentation? Weizmann remembered **Pasteur**'s findings on the fermentation of sugar into alcohol and began to search for suitable bacteria or yeasts in the soil, on corn and other grain.

The state-founding bacterium *Clostridium acetobutylicum*

Within a few weeks after the encounter, Weizmann isolated *Clostridium acetobutylicum* with the wonderful ability not only to produce acetone, but also the even more valuable butanol.

Butanol is needed for the production of synthetic rubber, i.e., for strategically important rubber tires.

Lloyd George was absolutely thrilled and offered to recommend Weizmann to the Prime Minister to be specially honored. Weizmann, however, refused categorically and talked instead of the need for a permanent homeland for all Jews. When Lloyd George himself became Prime Minister, he discussed Weizmann's wish with his Foreign Minister **Earl Balfour**. This led to the historic Balfour declaration on November 2nd 1917 and finally to the foundation of the state of Israel in 1948 whose first president was Weizmann. Weizmann did more than just find an ingenious method of producing two chemicals. With him, the growth of the fermentation industry and thus modern biotechnology began, long before the production of penicillin.

Lloyd George (1863–1945), British statesman, became Minister of Munitions 1915, secretary of war and Prime Minister from 1916 to 1922

Chaim Weizmann, to this day is the only biotechnologist who was ever pictured on a bank note.

Chaim Weizmann and Albert Einstein

Fig. 6.25 Glycerol

Fig. 6.26 Diatoms provide the kieselguhr for the production of dynamite. Etching by Ernst Haeckel from *"Art Forms in Nature"*.

With rising crude oil prices and advances in biotechnology in the 1980s, the butanol biotechnological process became more attractive again. Its effectiveness could only be increased if a way could be found to solve a major problem – the toxicity of butanol to bacteria. Using immobilized microorganisms in a continuous process increased butanol yields by a factor of 200.

Glycerol is a versatile solvent and lubricant (Fig. 6.25) that was already produced by microbes (yeast) during the First World War for dynamite production in Germany, just as Britain used bacterial acetone to make cordite. In order to produce dynamite, glycerol is constantly cooled and dripped into nitrous acid (sulfuric acid plus nitric acid). Small amounts burn off easily with no major risk involved, whereas large amounts, when suddenly heated or hit, explode violently. **Alfred Nobel**, who later created the Nobel Prize Foundation, stabilized **nitroglycerin** through kieselguhr absorption. Kieselguhr consists of natural deposits of the silica shells of diatoms (Fig. 6.26), which are highly absorbent due to their large pores.

Sodium sulfite was added to ethanol-producing yeast cultures to bind an important intermediate product of ethanol synthesis. Thus, glycerol was produced alongside ethanol. 1000 tons of glycerol per month were thus produced. After the war, however, this method was replaced by the chemical saponification of fats or by glycerol production from propylene and propane.

Citric acid (Fig. 6.24) is produced by the *Aspergillus* fungus. It is used as a completely innocuous flavoring, as a preservative or cleaning agent. 700,000 tons are produced annually worldwide, worth $ 700 million.

In 1780, **lactic acid** (lactate) was discovered in sour milk by the Swedish chemist **Carl Wilhelm Scheele** (1742 - 1786).

Carl Wehmer started its production from glucose by *Lactobacillus delbrueckii* in a small firm called A. Boehringer in 1897. The company later went on to become a world name in biochemistry. Lactic acid is used as an acidifier in the food industry as well as a canned food preservative, in the dyeing of textiles and in plastic production. Approximately 50,000 tons are produced worldwide. Almost half of all lactic acid produced in Europe is produced by microbes, while in the U.S., chemical methods are used exclusively. Isolating lactic acid from the culture medium remains an expensive process.

Two molecules of lactic acid can be dehydrated to lactide, a cyclic lactone. A variety of catalysts can polymerize lactide to **polylactate** (**PLA**), which as biodegradable polyesters with valuable properties are currently attracting much attention (see the end of this Chapter).

Another bioprocess used in the early days that is currently being rediscovered is the production of **gluconic acid** (Fig. 6.24) from glucose with *Aspergillus niger*. The fungus contains a glucose oxidase that is vital for the process and is also used as biosensor to measure glucose content (Ch. 10). It converts glucose into gluconolactone and hydrogen peroxide, metabolizing oxygen. Hydrogen peroxide, which is toxic to cells, is rapidly broken down by catalase, while gluconolactone spontaneously hydrolyzes into gluconic acid. Gluconic acid is very versatile and is mainly used as a detergent additive because it binds metal ions and prevents calcium stains on glasses. It also gently dissolves existing deposits without corroding metal vessels. Approximately 60,000 tons of gluconic acid were produced worldwide.

There are other acids that can also be produced biotechnologically, e.g., fumaric acid (salt **fumarate**) and malic acid (salt **malate**). Fumaric acid can be produced by the fungus *Rhizopus nigricans* from sugar or by *Candida* yeasts from alkanes (paraffins). However, chemical synthesis remains the cheaper alternative. For malate, by contrast, a highly efficient biotechnological process using immobilized killed microorganisms was developed by the Japanese company Tanabe Seiyaku in 1974 (see Ch. 2). As the only enzyme obtained from the killed *E. coli* cells, fumarase is used for the production of malic acid from fumaric acid.

Is the **production of industrial chemicals from renewable sources financially viable**? For large volumes of products with little value added, strict cost control is required and mostly favors fossil resources.

Many industries will not change their ways in the near future, and the progress of "biotechnological takeover" in the chemical industry very much

depends on the oil price and the development of economically viable biotechnological processes. Such processes only stand a chance if they can lower present production cost by between 20 and 40%, and, after all, chemical methods and catalysts are under constant review, and improvements are being made.

In the end, **economic considerations** determine whether biotechnological or chemical processes or a combination of both will be adopted. With sufficient interest, as well as investment and economic pressure, most industrial chemicals could be produced biotechnologically from renewable resources. Biotechnology comes into its own with novel products that cannot be produced chemically, and low-volume, high-value fine chemicals, such as amino acids.

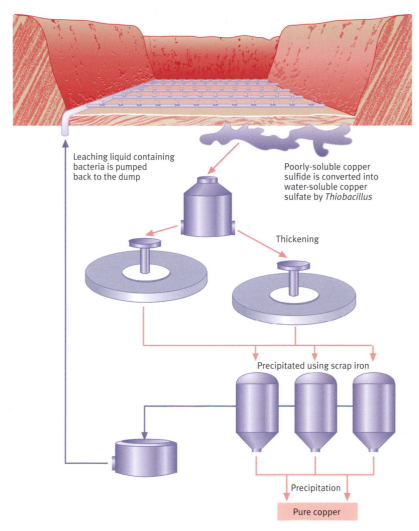

Leaching liquid containing bacteria is pumped back to the dump

Poorly-soluble copper sulfide is converted into water-soluble copper sulfate by *Thiobacillus*

Thickening

Precipitated using scrap iron

Precipitation

Pure copper

Fig. 6.27 Schematic diagram of microbial copper leaching.

Fig. 6.28 Historic copper factory in the US.

Fig. 6.29 Copper mine with low-copper rocks, Kennecott Utah Copper.

Fig. 6.30 Microbial copper leaching: Liquid containing thiobacteria is sprayed over the ore and then collected (blue liquid).

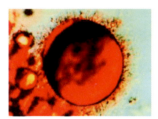

Fig. 6.31 Drop of oil with microbes sitting on it.

"Go find the oil, Xanthy!"

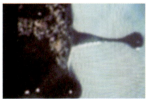

Fig. 6.32 Xanthan extracts oil from pores in the rock.

Fig. 6.33 These bacteria (*Alcaligenes* or *Ralstonia eutropha*) consist almost entirely of bioplastic (polyhydroxybutyrate).

■ 6.10 Silent Mining

Copper, sometimes nicknamed "red gold", has been mined so extensively in recent years that copper ore with a high copper content has become rare. It has to be retrieved from further below the surface, which makes energy and mine development costs rise. The answer to this problem was found in the Mediterranean region as early as 3000 years ago. Copper was obtained from mine water. It has been documented that the Spanish obtained copper through solution mining in the Rio Tinto. Until thirty years ago, nobody was aware that bacteria played an active part in the extraction process, helping to turn **poorly-soluble copper sulfide** into **water-soluble copper sulfate** (Fig. 6.27). Nowadays, microbes turn billions of tons of low-value copper ore into pure copper. In the US, Canada, Chile, Australia, and South Africa, they produce one quarter of the total copper production through bioleaching. More than 10% of gold and 3% of cobalt and nickel are biotechnologically produced.

The most important microbes involved in the bioleaching of copper are the sulfur bacteria *Thiobacillus ferrooxidans* and *Thiobacillus thiooxidans*. *T. ferrooxidans*, is an acidophilic bacterium that oxidizes bivalent iron into trivalent iron and attacks soluble sulfur as well as insoluble sulfides to convert them into sulfates. *T. thiooxidans* by contrast, only grows on elemental sulfur and soluble sulfur compounds.

The two bacterial strains work closely together via two different mechanisms. In **direct leaching**, the bacteria obtain energy (ATP) through a transfer of electrons from iron or sulfur to oxygen on the cell membrane. The oxidized products are more soluble. In **indirect leaching**, the bacteria oxidize bivalent iron to trivalent iron, which, in turn, is a strong oxidant. In sulfuric acid solution, it oxidizes other metals into easily soluble forms. Again, bivalent iron is produced, which is rapidly oxidized by bacteria into its aggressive trivalent form. In practice, there will be overlap between the two leaching processes. Ultimately, copper leaching results in sulfuric acid and the conversion of insoluble copper sulfide into readily soluble blue copper sulfate.

Millions of tons of mining waste containing small, but valuable amounts of copper are brought to collection points for bacterial leaching. Such spoil piles can be up to 400 meters (1300 feet) high and contain four billion tons of rock material (Fig. 6.29). The rock is sprayed with acidified water. While the water is seeping through, Thiobacteria, millions of which are present in each gram of rock, proliferate. A metal-containing liquid runs off the bottom of the pile into large basins. The copper is now easily recoverable, and the copper-free leaching liquid is sprayed over the waste heap again.

When **uranium is leached** (tetravalent uranium ions), bacteria turn pyrite (iron disulfide) or soluble bivalent iron into aggressive trivalent iron, which, in turn, forms hexavalent uranium ions that are readily soluble in dilute sulfuric acid.

Biosorption is very attractive from an environmental point of view: reeds filter toxins from wastewater. Algae have been discovered that bind large amounts of toxic heavy metals, such as cadmium and might offer a solution for cleaning up wastewater, while at the same time accumulating valuable metals. Various types of cabbage also accumulate heavy metals. Concentrations 30 to 1000 times higher than in the surrounding soil have been observed.

■ 6.11 A New Life For Tired Oil Wells?

A helicopter lands on an oil platform, and out steps a biotechnologist carrying a small case. The content of the case is supposed to revitalize an exhausted oil well – microbial cultures that are pumped into the oil reservoir where they multiply. Their products make the oil flow again.

With current primary oil production techniques, **two thirds of the oil remains in the ground**. Secondary methods involving water and gas are used to increase the pressure again (Fig. 6.32). In the North Sea oil fields alone, oil worth £300 billion ($590 billion) cannot be recovered. Even a few percent of additional output would pay back research and development investment. **Tertiary oil recovery** is the buzz word, also called **MEOR** (microbial enhanced oil recovery).

Several methods are being tried. The seemingly empty oil wells are inoculated with bacterial mixes that produce gases such as carbon dioxide, hydrogen, and methane and thus increase the pressure on the oil deposits. Other microbial strains are meant to produce **biosurfactants** that **shatter** oil into little droplets. These can then be squeezed out of minute pores in the rock (Fig. 6.32).

The supply of microorganisms with oxygen and nutrients is currently still a problem, unless they can feed on oil components. In addition, conditions in an oil-field are not for the faint-hearted. There is the high salt content of the sea, a lack of oxygen, the pressure is between 200 and 400 atmospheres and the temperature between 90 and 120°C (194-248 °F). Only extremophiles can survive here!

Oil companies are carrying out experiments with microbes that form and secrete long-chained **biopolymers** such as xanthan, which is produced from glucose by *Xanthomonas campestris*, a plant pathogen. Xanthan acts as a thickener, making water viscous. After soap-like biosurfactants have been pumped into an exhausted oil well to separate the oil from the rock, xanthan water is added. Like the plunger in a syringe, it exerts pressure on the oil and forces it out of the drill hole.

Xanthan is still too expensive to speed up oil production effectively, but it is widely used in the food industry for the production of soft ice cream, pudding, and low-calorie drinks, to give them body. Although xanthan acts as thickener, it is not a fattener, as humans lack the enzymes to break it down. In other words, people who want to be slim feel temporarily full, but have not ingested additional calories.

30 years ago, xanthan was one of the first modern biotech products. Other products followed, most of which are easily biodegradable.

■ 6.12 Bioplastic – From Dead End to Merry-Go-Round

In Japanese supermarkets, you will find vegetables and ready meals wrapped in a cellophane-like foil made of **pullulan**, a new bioproduct (Fig. 6.35). It is a polysaccharide, consisting of glucose components which are interlinked via carbon atoms one and six instead of one and four as in starch molecules. Thus, they cannot be broken down by starch-degrading amylases and cannot be digested by humans, i.e., they are low in calories. Like xanthan, pullulan enhances the viscosity of food.

The Hayashibara company produces pullulan from simple sugars, using *Pullularia pullulans*, a fungus. The viscous pullulan syrup is then poured out in

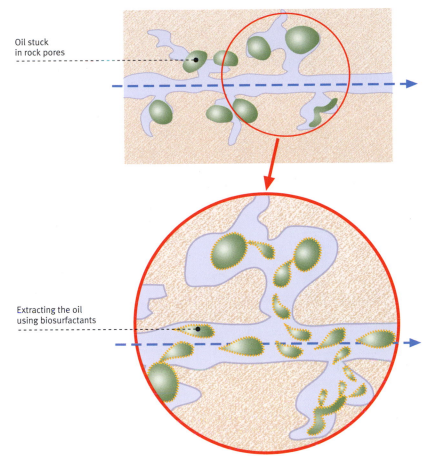

Oil stuck in rock pores

Extracting the oil using biosurfactants

Fig. 6.34 How microorganisms can enhance oil recovery (MEOR).

thin layers on flat surfaces to form a film when drying. These films provide excellent wrapping material, providing an airtight seal around the wrapped produce and later dissolving in hot water. These foils are also environmentally friendly and can be microbially degraded when wet. As a sales gimmick, they are now available in various flavors, e.g., fruit or garlic, which are supposed to retain the flavor of the wrapped food.

Other easily biodegradable products are made of **polyhydroxybutyrate** (**PHB**), which has the same properties as polypropylene, used in so many everyday plastic products. Whereas polypropylene is a petroleum product, PHB is produced by the bacterium *Alcaligenes eutrophus* (Fig. 6.33) from sugar. PHB is bacterial energy storage material – the bacteria mainly consist of plastic.

The new bioplastic material was first developed by biotechnologists of the British branch of ICI and produced by ICI daughter company Marlborough Biopolymers Ltd. in Cleveland, Britain (Fig. 6.38) under the name of **Biopol**®. PHB from *Alcaligenes* turned out to be a highly crystalline thermoplast with a melting point near 180°C (356°F).

Fig. 6.35 Pullulan foil for clear breath melts on the tongue.

Fig. 6.36 Poly-3-hydroxybutyrate (PHB, section).

Box 6.7 The Expert's View:
From Biomass Conversion to Sustainable Bioproduction: Fuel, Bulk, Fine, And Special Chemicals

A major motive for the development of industrial biotechnology was the responsible use of resources, long before this way of thinking became mainstream and shaped sustainable development strategies worldwide.

Medieval ethanol distillation

However, in the currently discussed green scenarios for the industrial use of renewable resources and waste biomass, it is often forgotten that by definition, there are three essential aspects of sustainable development:

- Ecological compatibility
- Social compatibility
- Economic viability

Often, the suggested policies and strategies are too one-dimensional and simply extrapolate from the current policy that relies on heavily subsidized agricultural production.

Canola/rapeseed for biodiesel production

What is required are truly sustainable approaches that also offer satisfactory economic solutions. Economic optimization should clearly aim at an improvement of the material and

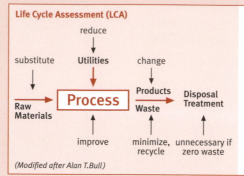

Life Cycle Assessment (LCA)

reduce → Utilities
substitute → | change →
Raw Materials → **Process** → Products / Waste → Disposal Treatment
improve ↑ | minimize, recycle ↑ | unnecessary if zero waste ↑

(Modified after Alan T. Bull)

Life Cycle Assessment (LCA)

is defined as "a process used to evaluate the environmental burdens associated with a product, process, or activity by identifying and quantifying energy conversion, materials used, and waste released to the environment, to assess the impacts of those energy and materials uses and releases to the environment, and to identify and evaluate opportunities to obtain environmental developments."

(Society of Environmental Toxicology and Chemistry (SETAC), 1993)

energy balances (protection of resources) as well as enhanced efficiency of the processes. The guiding principle in chemistry and biotechnology, starting from raw material, should be "minimal structural changes in raw material". In other words, the biomass used to extract industrial material should be used as intelligently as possible, i.e., maintain the high degree of organization and the C/H/O ratio as far as possible when converting it. To quote a positive example: The established oil and fat industry largely follows these principles, as it practically retains the starting weight throughout the process. When using biomass, the aim should therefore not be to produce chemical equivalents of petrochemically based products (which would be possible with considerable transformation losses), but functional equivalents that are structurally closer to the starting material.

In a fairly recent OECD report, The application of Biotechnology to Industrial Sustainability . a method called Life Cycle Assessment (LCA) has been put forward, which includes energy and raw material use, waste products and byproducts, ecological compatibility of processes and safety aspects in the assessment. This is documented by 21 detailed case studies.

Here are the ten conclusions of the report

- Global awareness of environmental problems will induce major efforts to achieve cleaner industrial processes.
- Biotechnology provides strong enabling technology that can ensure clean products and processes on a sustainable basis.
- The assessment of industrial production processes is essential as well as complex. LCA is currently the best available method for such assessments.
- The major driving forces behind industrial biotechnological processes are the economy

(market forces), government policy, and science and technology.

- In order to achieve more (market) penetration for the biotech sector, R & D efforts by government and industry must be united.
- In order to use the full potential of biotechnology for clean products and processes beyond the current applications, more R & D is needed.
- In view of the increasing significance of biotechnology, including recombinant DNA technology and its applications for the production of high-value products and the development of biocatalysts, harmonized and supportive rules and regulations (for the industry) are needed.
- Market forces can provide strong incentives to reach targets in environmental protection.
- Government policies to encourage clean industrial products may be the single decisive factor for the development and industrial application of clean biotechnological processes.

Searching for new enzymes in a screening test

- Communication and education will be needed to achieve market penetration for clean biotechnological products and processes in various industrial sectors.

Biotechnology and Biocatalysis: R & D Analysis From an Industrial Perspective

The chemical industry still has a key role to play in developing sustainable production strategies. An essential structural change must take place within the industry in order to establish more and more biological processes and biology-based thinking in R&D. The current situation (slightly caricatured) could be described as follows:

- Innovation in the catalytic sector is viewed and evaluated very differently in university and industry. Invention must not be confused with innovation!

- The chasm between purely academic and real industrial needs is becoming wider. Accumulation of knowledge does not necessarily lead to new applications and problem solutions.

- In the academic sector, there is a one-side emphasis on publishing and too little effort goes into innovative applications of research results (e.g., through cooperation with partners in industry or via spin-offs). Incentives for the development of applications must be put in place.

- Screening efforts are insufficient on both sides. Additions to the biocatalytic toolbox are essential.

In order to improve chances of success of the biocatalytic and thus the white biotech industry, we can define the following strategic objectives:

- The industry urgently needs new types of biocatalysts to catalyze new (so far unknown or not biologically accessible) types of reaction. The opportunities offered by the vast, as yet unresearched, biodiversity of living organisms must be used as much as possible. Intelligent and rapid targeted screening methods must be developed. These are challenges that can only be met by the cooperation of industry and universities.

- In some, but by no means all cases, the genetic, chemical, or physical improvement of known biocatalysts is necessary and desirable. In the interest of an optimal use of resources, universities and institutes should take more care when choosing the "right" systems. This can be done in close dialog with industry.

In 2005, the Saab 9-5 Biopower automobile, which can be fueled either with ethanol or gas/petrol, became a runaway success in Sweden. 5,000 cars have been sold, 2,000 more have been commissioned. Sweden imports three quarters of its ethanol from Brazil.

- New and well – characterized biocatalysts with large application potential should be made commercially available as fast as possible in order to have an impact on academic as well as industrial Research and Development. To quote a negative example: Nitrilases have been known for over 20 years, but only recently have they been available via some enzyme production companies.

Prof Oreste Ghisalba *has been Head of Bioreactions at CIBA-GEIGY (Central Research Laboratories and Pharma Division) since 1982 and later at Novartis Institutes for Biomedical Research, Discovery Technologies.*
From 1991 - 2002, he was Programme Director of the Swiss Priority Programme Biotechnology of the Swiss National Science Foundation and since 1999 Professor for Biotechnology at the University of Basel.
Since 2003, he is Head of CTI Biotech (Swiss Innovation Promotion Agency CTI) and since 2005 Member of the Board/Vice President of the Swiss Biotech Association SBA.

Cited Literature:

World Commission on Environment and Development. *Our Common Future* (The Brundtland Report), Oxford University Press, 1987

The Application of Biotechnology to Industrial Sustainability, OECD 2001

Oreste Ghisalba (2000) *Biocatalysed Reactions*. In: Gualtieri F (ed.) New Trends in Synthetic Medicinal Chemistry, Wiley-VCH, Weinheim pp. 175-219

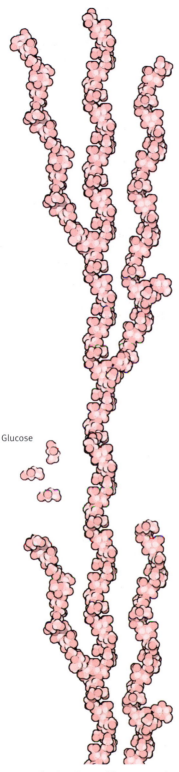

Glucose

Fig. 6.37 Renewable resources that support sustainable industries. Here is the multi-branched molecule of starch. Amylases cleave it into simple sugars which are then fermented to alcohol.

Fig. 6.38 Biopol® is biodegradable material.

PLA

Fig. 6.39 Biodegradable products in Japan: Self-degrading food containers and backpack.

Fig. 6.40 A biodegradable corn CD.

Goethe – an Advocate of Recycling?

My dream dog will not bite
Nor growl or snarl at you
It thrives on broken glass
And leaves diamonds in its ...

J.W.Goethe in *Tame Xenien*, vol. 8

In the first test runs, molded parts, foil, and fibers were produced. It proved to be good wrapping material, but not superior to polypropylene – a useful, but not particularly attractive alternative from a technical perspective. Things began to change when alongside 3-hydroxybutyrate components, 3-hyydroxypentanate was successfully produced by bacteria.

Its biodegradability makes Biopol® a very attractive candidate for medicinal use. In the future, stitches need not be taken out any more after surgery. Biopol® could also be used for capsules that slowly release long-term medication into the body. In horticulture and agriculture, Biopol®-coated nutrients and growth regulators could be put in the soil and be slowly released during the microbial degradation process.

More biotech products are on the horizon. The silk produced by the *Nephila* spider is so strong that it is used for fishing in the South Pacific. It can stretch by up to a third without tearing. Attempts are now being made to produce these **spidroins** as recombinant proteins in *E. coli* or even in goats (by the US company Nexia, under the name of Biosteel®).

The Japanese telephone company NTT DoCoMo sends bills to customers in eco-friendly envelopes. The plastic window in the envelope began its life not in an oil well, but in a corn field. It consists of polylactate (polylactic acid, PLA). In polylactate, lactate (lactic acid salt, see Ch. 2) is polymerized into a chain. It is obtained from glucose by microbial fermentation of corn starch (Fig. 6.39).

In 2002, Cargill built a plant in Nebraska (USA) with the ability to produce 140,000 tons of PLA per year. It is sold under the name of NatureWorks® PLA. Toyota recently announced that spare tire covers and floor mats will be made out of PLA, and Sanyo is promoting a CD called MildDisc made from PLA.

If it were possible to make all sorts of waste dissolve into thin air instead of polluting the environment, that would be a biotechnological success story, turning a **one-way-road** (raw-material-product-waste) into a succession of **natural cycles**.

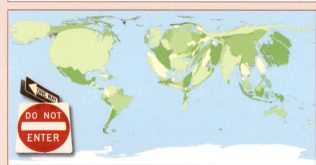

Cartograms made by Mark Newman (University of Michigan) show the territory (top) and the fuel consumption (below) of the world's countries.

North America and Western Europe import the highest values of fuel.

The region that imports the least fuel is Central Africa – where six of the ten territories reported no fuel imports. Imports per person are highest in Singapore and Bahrain. These are small island territories, where enough of the people living there are relatively rich. Singapore is a long-established trading port. Bahrain is a group of islands in the Arabian Gulf. Due to declining oil reserves Bahrain´s industry now imports crude oil, refines it, the exports it.

The Western European region records the highest total fuel imports partly due to trade within this region. Territory size shows the proportion of worldwide fuel imports arriving there.

Cited and Recommended Literature

- **Thiemann WJ, Palladino MA** (2004) *Introduction to Biotechnology.* Chapter Bioremediation. Benjamin Cummings, San Francisco Easy-to-read and interesting introduction.

- **Barnum S R** (2006) *Biotechnology*: An Introduction, updated 2nd edn., with InfoTrac® Brooks Cole. Great introduction.

- **Alexander M** (1999) *Biodegradation and Bioremediation*. Academic Press, NY

- **El-Mansi EMT, Bryce CFA, Demain AL, Allman AR** (2007) *Fermentation Microbiology and Biotechnology* (2nd edn.) CRC Press, Taylor and Francis, Boca Raton. Comprehansive handbook for the fermentation biotechnologist. Highly recommended!

- **Wackett LP, Hersberger CD** (2001) *Biocatalysis and Biodegradation*. ASM Press Washington DC Both are academic books written for specialists.

- **McKane L, Kandell J** (1996) *Microbiology* (2nd edn.) McGraw-Hill, New York Good chapter about Environmental Microbiology

- **Tchobanoglous G, Burton FL Stensel, HD** (edn.) (2002) *Wastewater Engineering: Treatment and Reuse* (4th edition) McGraw-Hill Everything about wastewater for specialists

- *Ullmann's Encyclopedia of Industrial Chemistry* Wiley-VCH Ullmann's Encyclopedia is not only available in print (6th edn. 2002), but also in electronic form – as a fully networkable DVD or an online database.

- **Flickinger M, Drew S** (edn.) (1999) *The Encyclopedia of Bioprocess Technology Fermentation, Biocatalysis and Bioseparation (Wiley Biotechnology Encyclopaedia)* For specialists: five-volume set of the Encyclopedia of Bioprocess Technology focuses on industrial applications of fermentation, biocatalysis, and bioseparation.

- **Gore A** (2006) *An Inconvenient Truth:* The Planetary Emergency of Global Warming and What We Can Do About It. Rodale Books. The accompanying book to the great documentary.

Useful weblinks

- Try Wikpedia, for example about biogas: *http://en.wikipedia.org/wiki/Biogas*

- All about microorganisms: *http://www.microbes.info/*

- Oilspills info from the Exxon Valdez Oil Spill Trustee Council: *www.oilspill.state.ak.us*

- Everything about Polylactic acid: *http://www.natureworksllc.com/*

Eight Self-Test Questions

1. Why is it necessary to pump extra oxygen into biological sewage plants?

2. What does a BOD5 of 900 milligrams/liter found in a wastewater sample mean? How many liters of clean water would be needed to treat just one liter of the wastewater?

3. Where can biogas be used very effectively?

4. What was the US patent granted to Ananda Chakrabarty for? What did he achieve in his experiments, and why was his patent a breakthrough?

5. Why is timber such a popular building material? Against what must it be protected?

6. What are the pros and cons of biofuel?

7. Which organism is the largest producer on Earth of the greenhouse gas methane?

8. What biodegradable product from sustainable sources is currently the most promising in the textile and disposable plastic industry?

Green Biotechnology

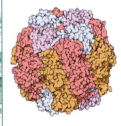

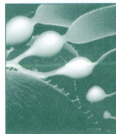

CHAPTER **7**

Fig. 7.1 Aztecs fishing for *Spirulina*.

Fig. 7.2 Brown algae kelp (*Macrocystis*).

Fig. 7.3 *Spirulina* farm in India.

Fig. 7.4 *Spirulina* tablets.

7.1 Microbes are Edible

Five to ten kilograms (ten to twenty pounds) of plant protein are needed to produce one kilogram of animal protein. A massive amount of protein is lost in this process, in addition to the huge losses resulting from pests, harvesting, transport, and storage. Microorganisms could help effectively: not only are they involved in the production of medicinal products, and wine and cheese – microbes themselves are edible. They contain valuable proteins, fats, sugars and, vitamins.

After the conquest of Mexico in 1521, the Spaniard **Bernal Diaz del Castillo** described how the **Aztecs** ate peculiar, small, cheese-like cakes. These cakes consisted of microscopic algae from the seas around Mexico and were called *Techuilatl* by the Aztecs. This was actually *Spirulina. Spirulina* is not a true alga but a cyanobacterium ("blue bacterium").

In **Montezuma's** Empire, servants had to provide the ruler with fresh fish daily; this was only possible by jogging over very great distances, so legend has it that the servants used *Spirulina* as an energy food. Even in the time of **Cortés**, it was supposedly available in the local markets and was eaten with bread and cornmeal products. Fishermen used fine nets to scoop up *Spirulina* from the saltwater lakes (Fig. 7.1) that had not yet dried out in those days. Today, *Spirulina* is probably only to be found in Lake Texcoco in Mexico. The other lakes, including the swimming gardens of the Aztecs, are infertile deserts today.

Thousands of kilometers away, the natives on Lake Chad in Africa (Nigeria) have also been consuming *Spirulina* since primeval times. The thin, hard, blue-green algae cakes can be bought at local markets on Lake Chad. The people of **Kanembu** call this product *Dihé*, and it is an important ingredient in 70 % of all dishes. It is added to sauces made from tomatoes, chili peppers, and various spices, which are eaten together with the basic diet of millet. The *Spirulina* cakes are produced by drying in the sun. The swimming algae are first scooped out of the water in protected areas of the lake. They are then drained and spread out on the hot sand where they dry quickly.

In Western countries and in Japan, *Spirulina* from algae farms is eaten as part of a diet to reduce cholesterol and cleanse the blood (Fig. 7.4). 100 grams of *Spirulina* contain approximately 70 grams of protein, 20 grams of sugar, two grams of fiber, and just two grams of fat, as well as important vitamins (A, B_1, B_2, B_6, B_{12}, E) and mineral salts.

7.2 Algae and Cyanobacteria

With the exception of prokaryotic *Spirulina*, algae are photosynthetic eukaryotes (Box 7.1).

Macroalgae are economically more significant than **microalgae**: green (chlorophytes), red (rhodophytes), and brown macroalgae (fucophytes or phaeophytes) are used at present.

Brown algae such as the jungle-like Californian kelp beloved by divers (*Macrocystis*) (Fig. 7.2) have been harvested since 1900. The gelatin-like alginate has been produced from it since 1921 in San Diego, where there are huge seaweed forests. **Alginate** is used today as a thickener and stabilizer in foodstuffs and ice cream, in the textile industry and as an encapsulator for medicinal products (and for enzymes and yeasts, Ch. 2).

Other algal products are the jelly-forming **agar** (previously called agar-agar, important for the cultivation of microbes and gel electrophoresis, see Ch. 10) and **carrageenan**.

The flavor-enhancing amino acid **L-glutamate** was first found in algae in Japan (Ch. 4). Other brown algae such as *Undaria* (Jap. *wakame*) and *Laminaria* (Jap. *kombu*) grow on the coasts of Japan and China and are used for salads, soups, noodles, or with meat (Figs 7.5 and 7.6). The annual market value of both algae is 600 million US dollars and 20,000 metric tons of *Wakame* alone are harvested annually.

The **red alga** *Porphyra* (Jap. *nori*) has been cultivated in Japan since the Middle Ages. These days, it is collected in huge quantities in bamboo clumps or horizontal nets on ocean farms and later air-dried.

The most important **microalgae** come from two different classifications: the already mentioned prokaryotic blue bacteria and the eukaryotic green algae.

The prokaryotic **blue bacteria** (cyanobacteria) were previously known as blue algae and include the economically significant varieties of the genus *Spirulina*. The water fern *Azolla* is cultivated in Asia and contains the symbiotic blue bacterium *Anabaena azollae* (Fig. 7.9), also known as symbiotic filaments, which supply the fern with nitrogen.

Box 7.1 Photosynthesis

Almost all free enthalpy used by biological systems comes from the energy of the sun. This is a huge quantity of approximately 4×10^{17} kJoule or 10^{10} metric tons per year of energy turned into sugar. The sun's energy is changed by **photosynthesis** into chemical energy. Water and carbon dioxide combine in a highly complex process to form carbohydrates (first glucose, then sucrose and starch) and molecular oxygen.

In the **chloroplasts** of green plants, pigment molecules (chlorophyll) produce high-energy electrons from captured light energy in the thylakoid membrane. They are used in light reaction to create NADPH$^+$H$^+$ and ATP (Ch. 1).

The photosynthesis of green plants is carried out by means of **two photosystems** joined together. It may be understood in simplified form as follows:

In photosystem II, the light excitation of P680 (one of the pair of chlorophyll molecules) leads to an electron transfer via several pigmented molecules to plastochinon A and then to plastochinon B. These energy-rich electrons are replaced by the withdrawal of low-energy electrons from water molecules, i.e. the oxygen-developing center takes an electron from water, transfers it to a tyrosine group and returns it to the chlorophyll, which is then able to take up another photon.

One molecule of oxygen is produced for every four transferred electrons. The electrons run from the plastochinon via a cytochrome bf complex to the plastocyanin and from there to photosystem I.

Photosystem I shown here is a trimer complex which "swims" in the membrane. Each of the three subunits has hundreds of cofactors (green chlorophyll, orange carotenoid). The colors are significant: chlorophyll absorbs blue and red light – which is why we see plants in a wonderful green color.

Photosystem I contains electron transfer chains as a center for the three subunits. Each is surrounded by a thick ring of chlorophyll and carotenoid molecules, which act as "antennae". These antennae absorb light and transfer energy to their neighbor. Everything is then transferred into the three reaction centers which generate electrons (a reducing agent).

To sum up, the light reaction in the thylakoid membranes of the chloroplasts leads first of all to a reducing agent (to create NADPH+H$^+$), secondly to the formation of a proton (H$^+$) gradient (a slope between the two sides of the thylakoid membrane

so that ATP is created) and thirdly to the production of oxygen.

Another step is **CO$_2$ fixation**. The enzyme ribulose-1,5-biphosphate carboxylase/oxygenase (**RUBISCO** for short) is a lyase (Ch. 2) and forms a bridge between life and non-life.

It binds the inorganic CO$_2$ with ribulose-1,5-bisphosphate, a short chain sugar with five carbon

atoms. From this, RUBISCO forms two 3-phosphoglycerates (with three carbon atoms in each case). Most phosphoglycerate molecules are recycled in order to form even more ribulose biphosphate, but every sixth molecule is used to form sucrose or starch (as a storage material).

16% of chloroplast protein is RUBISCO and with the enormous quantity in plants, RUBISCO is obviously the most common protein on Earth.

Ribulose 1,5-bisphosphate

CO$_2$

H$_2$O

ATP

starch, sugar

RUBISCO

2 NADPH +2H$^+$

3-phosphoglycerate

2 ATP

Photosynthetic reaction center

Light

Light-collecting complex

Light

PS I

Light

PS II

Light

Above right: structure of photosystems I and II
Left: process of CO$_2$ fixation by means of RUBISCO.

Fig. 7.5 Dried macroalgae form a cheap basis for making soup in China and Japan.

Fig. 7.6 Macroalgae have always been a staple food in Japan.

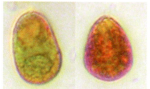

Fig. 7.7 The alga *Dunaliella* is cultivated in farms (top: in Western Australia); bottom: carotenoids accumulate in the cells shown on the right.

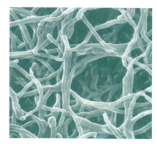

Fig. 7.8 *Fusarium graminareum* produces *Quorn*.

Like other bacteria, the *Spirulina* cell wall consists of mucoproteins and is therefore easily broken down by human digestive enzymes, an advantage in its use for dietary purposes. *Spirulina* is a thread-like, corkscrew-shaped organism that may consist of 150 to 300 individual cells and can reach a length of up to half a millimeter. The freshwater cyanobacterium *Nostoc*, which forms jelly-like colonies popularly called "fallen star", "witches' butter", or "star jelly" (Fig. 7.9), is also a blue bacterium.

The eukaryotic **green algae** (*Chlorophyceae*) include species from the genera *Chlorella*, the four-tailed *Scenedesmus*, the whip-like *Dunaliella* and *Chlamydomonas* (Fig. 7.9), and the algal colony *Volvox* which consists of up to 20,000 cells. *Volvox* is highly organized, with differentiated cells and is an important alga in sewage and plankton in lakes and oceans.

The cells of **Chlorella** and **Scenedesmus** are surrounded by a firm cellulosic wall with sporopollenin deposits (the substance in the walls of the pollen grains of higher plants). The cell walls impregnated with this are chemically difficult to attack, so the approximately ten micrometer large cells must be broken down before further use in foodstuffs.

Algae such as *Chlorella* are cultivated these days at a cost of approximately eight dollars per kilogram (four dollars per pound) and sold at about ten times that price predominantly as a diet food. The raw protein of these algae is about 50% of the total mass (compared to 35% for soya beans) and they are rich in unsaturated fatty acids and vitamins.

On the other hand, *Dunaniella* species are single-celled flexible flagellates which do not have a firm cell wall. Their most striking property is an extraordinary tolerance of salt. These halophilic algae occur in massive quantities in drying ponds of brine. Because of their enormous β-carotene content, they color alkaline solutions deep red (Fig. 7.7). To compensate the external osmotic value, they accumulate considerable amounts of glycerine, which is industrially extracted like the carotene.

Algae double their weight in just six hours, while grasses need two weeks to do this; chicks need four weeks, piglets six weeks and calves two months.

There are therefore **algae farms** in many countries. Large, flat pools of water are needed for these, so that the algae can get sufficient sunlight (Figs. 7.3

and 7.7), which is essential for forming sugars from carbon dioxide, water, and nutrient salts and then protein from the sugars. Light and air cost nothing, and just a few cheap nutrient salts are needed to get the algae to grow luxuriantly. Per acre of surface, *Spirulina* produces ten times more biomass and has a much higher protein content than wheat. At harvest, the algae are strained and air-dried, then mixed with flavoring agents and sold.

Why are there no huge algal farms in areas of famine? Unfortunately, even the simplest technology is lacking in these areas and in many regions water is also scarce and expensive. The crucial factor is that even under favorable production conditions, algal proteins cost ten US dollars per kilogram and soya protein only 20 cents per kilogram. Nevertheless, experts are convinced that algae have a great future.

Bacteria, yeast, and filamentous fungi (molds) grow even faster than algae. **Bacteria double their weight within 20 minutes to two hours**, and can consist of 70% protein. On average, yeasts can form protein per weight unit 100,000 times faster than a cow. A cow only passes back to us in the form of meat one eleventh of the plant feed it uses. Ten elevenths of the feed that produced the beef is therefore lost to the human diet. Bacteria, yeasts, and molds are different in that almost the total quantity of foodstuff is changed into usable protein, sugars, and fats for humans and animals. It makes sense to make use of this.

■ 7.3 Single Cell Protein: the Hope for Cheap Sources of Protein

The modern story of protein production by microbes began in Imperial Germany during the First World War with the breeding of **yeasts**. Because of the shortage of food, bakers' yeast was cultured on a large scale to "stretch" mainly sausages and soups. Yeasts have the great advantage of feeding on cheap, otherwise unusable sugary solutions and are able to change the sugar directly into valuable protein. During the 900 day siege of Leningrad, it was yeasts that kept thousands of people from starving to death during the Second World War and shortly after the war, yeast flakes stayed the hunger of many people.

It was not until the 1960s that anyone began to set up facilities for the production of protein using microbes in Europe again (Box 7.2), when the

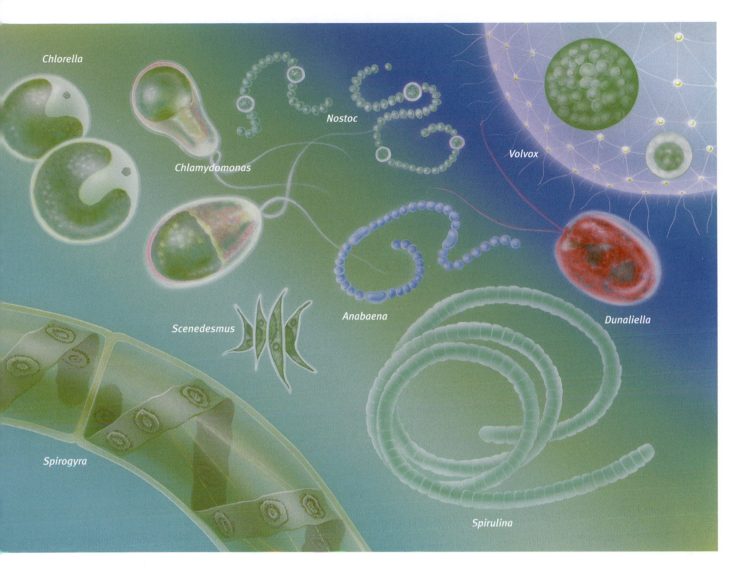

Fig 7.9 Algae and cyanobacteria.

human requirement for protein was increasing. It was assumed that there would be more famine in the future and in the meantime microorganisms were discovered that could feed not only on sugary food solutions but also on the hydrocarbon-containing components of crude oil, **alkanes** (**paraffins**), and **methanol**. The wax-like alkanes are not usable by humans or animals, and only microbes can change them into usable protein.

In Eastern Europe, on the assumption of permanently cheap crude oil, they concentrated on alkane-digesting yeasts (*Candida*), while in the West the emphasis was on yeasts and bacteria that use methanol, particularly the British company ICI (see Box 7.2 and Fig. 7.10). Both huge **Single Cell Protein** (**SCP**) projects were very hopeful but finally ended without success.

The alkane yeasts were only released in limited quantities as animal food, because of the fear of can-

cer-causing residues. The projects in both East and West eventually failed on economic grounds as a result of two oil crises. In the West, another cause for the failure of the methanol animal feed was EU subsidies which made skimmed milk powder unbelievably cheap as an animal feed additive.

However, biotechnologists collected inestimable experience in building and operating huge bioreactors.

■ 7.4 Mycoprotein is a Success With Consumers as a Plant Protein

On the other hand, **mycoprotein** (Greek *mykes*, fungus/mushroom) from Rank Hovis McDougall (RHM) and today Marlowe Foods, a subsidiary of ICI, is a very successful product. RHM is the fourth largest foodstuffs producer in Western Europe. It

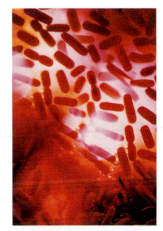

Fig 7.10 The bacterium *Methylophilus methylotrophus* produces high-quality protein from methanol.

Box 7.2 Biotech History:
Single Cell Protein (SCP)

In the 1960s, a lavish program was launched in the Soviet Union under **Nikita Khrushchev** to search in the best alkane feeders; this was inspired on the one hand by success in space but plagued on the other by constant crop failures. The idea was to **produce valuable protein from cheap crude oil**.

At the same time, they were searching for ways of breaking down cellulose in the Siberian forests to sugar and then to "seed" this with yeast to produce protein. The first trial facilities began working in 1963. Yeast strains of the genus *Candida* were grown on pre-cleaned crude oil samples seeded with "ravenous" alkanes. At the start of yeast production from crude oil alkanes, doctors and veterinary surgeons had reservations because the alkanes were difficult for humans and animals to digest, and the alkane yeast protein could be problematic for higher life forms or even cause cancer. In fact, many years of Russian experiments showed that yeast protein can probably be safely included in the human food chain. In the West, however, this met with great skepticism, which was probably justified. The first large alkane yeast factory was set up in the Soviet Union in 1973 with the production of 70,000 metric tons of yeast per year. The facility was in the petrochemical combine at Schwedt in the former GDR (German Democratic Republic or East Germany), the end point of the Soviet crude oil line *Drushba* (Friendship). It began continuous operation at the start of 1986. With deep jet reactors, it supplied 40,000 metric tons annually of the yeast animal feed product **Fermosin®**. The bioreactors were without doubt a master performance by East German engineers and biotechnologists. After the reunification of Germany, the Fermosin® process was stopped.

Alkane yeast bioreactor (42 m high) in Schwedt in the time of the GDR.

However, all *Single Cell Protein* projects also failed in the West. In Sardinia in 1971, British Petroleum (BP) participated in the manufacture of **Toprina®**, a yeast product grown on crude oil residues by the Italian company ANIC. Supposedly, the project failed as a result of the crude oil crisis, the soy lobby, the reduction in soy prices, discussions on the safety of Toprina® with its high nucleic acid content, which causes gout, and environmental considerations.

The gigantic ICI factory in Billingham, natural bioreactors in the foreground.

At the same time, there was research into the exploitation of methanol in Western Europe. Biotechnologists from the **British Imperial Chemical Industries** (**ICI**) made a discovery on a rugby field in County Durham in Great Britain. They discovered the bacterium *Methylophilus methylotrophus* (Fig. 7.10).

Approximately 10,000 microorganisms were tested over 13 years in the search for a microorganism that could grow quickly on petrochemical raw materials and supply a protein concentrate for pets. The result was **Pruteen®**. At first, ICI had concentrated on methane as the source of carbon, because the company had access to abundant North Sea gas. It seemed an elegant path from the simplest organic molecule to the complicated protein. Arguments against methane, however, were the danger of explosion of the gas as well as its low solubility and the problem of distributing it evenly in the medium. **Methanol**, oxidized (i.e., containing oxygen) methane, could, on the other hand, more easily satisfy the oxygen requirements of the microbes, is totally water soluble and would not lead to the development of high heat in the bioreactor.

The ICI researchers also decided in favor of the *Methylophilus* strain (the Latin means methanol-loving) because it was stable and free from toxic side effects. The decision to grow a pure culture of the bacterium in a continuous process still had a drawback in that the process had to be carried out under exceptionally sterile conditions. In contrast to this, the alkane yeast process was non-sterile, i.e., the yeast strain itself suppressed all competitors – which is the case with most biotechnological processes in food production.

The British company, **John Brown Engineers and Constructors**, built the gigantic ICI factory at Billingham for the largest sterile bioprocess in the world. The biofactory covered an area of eight hectares (20 acres). The bioreactor, its center-piece, was 60 meters (65 yards) high (with eight draft tube fermenters) and contained 150,000 liters (34,000 gallons) of perfectly germ-free nutrient solution in which lived the methanol bacteria. It was kept free of foreign microbes for three to four months at a time using a cleverly devised system of 20,000 valves and filters. At 35°C (95°F), *Methylophilus* lives only on methanol, ammonia, and oxygen from the air. Microbes were continuously removed from the bioreactor, killed with steam, gathered together in large clumps and dried, producing the granular, caramel-colored product Pruteen®. Everything seemed to be going perfectly.

One of the ICI bioreactors.

When the factory started in 1976, however, the company was confronted with **increasing energy prices** and an **outstanding soy harvest**, so it was no longer economical to produce single cell protein. *Methylophilus* was therefore improved using both genetic engineering and methods of standard genetics. The gene encoding for enzyme glutamate dehydrogenase (for more effective use of nitrogen from ammonia) was successfully introduced and the protein yield was improved by to 7%. The ICI factory was working, but demand remained less than expectation.

Nevertheless, ICI regarded the 100 million pounds sterling invested by 1982 as an "entry ticket to biotechnology". However, this did lead to the significant flow of experience into mass production of the fungal protein *Quorn* (see main text).

discovered its microbes in the 1960s and spent over 50 million dollars on the fungus, which can be transformed into passable imitations of fish, poultry, and meat (Fig. 7.13).

RHM researchers had collected more then 3000 soil samples from all over the world, but, as is so often the case, the main contender was close by. *Fusarium graminearum* (now known as *Fusarium venenatum*, Fig. 7.8) was found near Marlowe in Buckinghamshire, England. The name had previously only been familiar to plant pathologists, as the fungus caused root rot in wheat.

At that time, RHM produced 15% of the range of edible British mushrooms. Because of their unfortunate experience of psychological consumer prejudice against protein from bacteria, RHM stressed from the beginning that *Fusarium* **was a fungus like the mushrooms** and truffles that we eat without thinking twice about it.

Apart from the fact that *Fusarium* is virtually odorless and tasteless (Fig. 7.11), and **ideal for meat imitation**, its dry weight contains approximately 50% protein, a composition similar to grilled steak. However, the fungus has a lower fat content than steak, only 13%, and this is vegetable fat, with no cholesterol (ergosterin) and a fiber content of 25% - all of which appealed increasingly to the health-conscious public (Fig. 7.12).

The main advantage of producing fungi compared to bacterial cells is that they are **typically much larger** and are easily separated from the fermentation medium. On the other hand, **fungi grow much more slowly than bacteria**, doubling their quantity in four to six hours compared to 20 minutes for bacteria.

Even that can be turned into an advantage: slower growth also means that **less nucleic acid** is contained in the end product. Taken in by mammals and humans over a long period in high concentrations, nucleic acid can lead to gout.

While some bacteria contain up to 25% nucleic acid and yeasts up to 15%, RHM succeeded in lowering the content in the new mycoprotein food to below the acceptable top limit for humans of 1%. The fungus also has an amino acid composition recommended by the UN Food and Agricultural Organization (FAO) as "ideal".

Probably the most extraordinary property of the fungus is the manner in which it can be converted into **a full range of imitation food products**, from soups and biscuit to convincing imitations of poultry, ham, or veal (Figs. 7.11 to 7.13).

The key to this adaptability is that the length of the fibers can be controlled; the longer the fungus "is allowed" to grow in the bioreactor, the longer the fibers and the coarser the texture of the product. The medium consists of glucose syrup as the source of carbon with ammonia as the source of nitrogen. The syrup can be extracted from any available starch products (such as potatoes, corn, or cassava) and the process is much more efficient than the conversion of starch into protein by domestic animals.

ICI and RHM cooperated in the production of the fungus and the mycoprotein was released in Great Britain by MAFF (Ministry of Agriculture, Fisheries and Food) in 1985, the first product being a Savory Pie. Marlowe Foods was set up in Marlowe in the 1990s and in the meantime, the fungus protein was marketed in England as *Quorn*.

Fig. 7.11 Mycoprotein before processing. Left "beef", right "chicken".

Fig. 7.12 First trial meals with "refined" mycoprotein.

Fig. 7.13 *Quorn* products made of *Fusarium*: meatballs and sausages without meat.

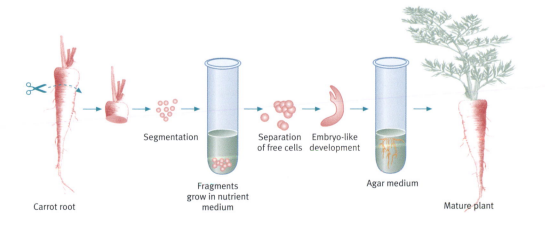

Carrot root — Segmentation — Fragments grow in nutrient medium — Separation of free cells — Embryo-like development — Agar medium — Mature plant

Fig. 7.14 Left: how plants can be reproduced *in vitro*.

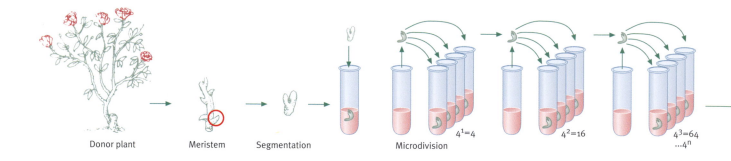

Donor plant   Meristem   Segmentation   Microdivision $4^1=4$ $4^2=16$ $4^3=64$ $...4^n$

Fig. 7.15 *In vitro* propagation of roses.

Fig. 7.16 A test-tube rose.

Fig. 7.17 Roses: from the wild rose to modern varieties.

Sales before 1993 were still less than three million US dollars per year, but by 2001 they had reached 150 million, thanks to powerful advertising. Since its US launch in 2002, Quorn products have become a best-selling frozen meat-free brand in natural food stores. In the year to 31 December 2004, total Quorn sales were £79m (representing 100% of the sales of Marlow Foods). The target groups are interesting: women from 25 to 65 in the USA and 25 to 45 in Great Britain.

The older generation is somewhat conservative, in spite of the fact that the company does not speak directly of microbes but of "vegetable protein".

Almost any country could in principle convert its **carbohydrate reserves into fungus protein**. In Europe, cereals and potatoes could be used and, in tropical countries, cassava, rice, or raw sugar. Many tropical countries traditionally use fungi as foodstuffs, such as *Tempeh*, a mixture of soya beans and mushrooms (Ch.1). This would mean there would be less psychological reservations to overcome against food from a bioreactor.

■ 7.5 "Green" Biotechnology at the Doorstep

Weeding, watering, fertilizing, composting, battling attacks of insects and fungi in as harmless a manner as possible are just a leisure activity for the hobby gardener, but for farmers, these processes are real economic factors which they would like to minimize drastically.

Will the soil be able to feed us at all in the future? In 1950, half as many people lived in Africa as in Europe but these days there are twice as many. Today we need to feed well over six billion people around the world and in 20 years we will be dealing with eight billion. These days, the Chinese eat four times more meat than 20 years ago: 32 kilograms (16 pounds) per person. The US American *Worldwatch Institute* predicts that in 2030, China will have to import 200 million metric tons of grain, the same total amount as is available on the world market today.

There have been even further increases in world hunger, although it is macabre that according to *Time Magazine*, the numbers of those undernourished and those overnourished were equal for the first time in 2004 (thanks to obesity in prosperous countries and those sections of other populations that have become "rich overnight").

None of this is the fault of the farmer. The **plant breeding revolution** of the 1960s and 1970s has changed the world economy significantly. It was not so much the very sophisticated new fertilizers and pesticides that led to progress; it was rather new high-performance types of important economically useful plants such as rice and wheat.

The **increase in productivity** came at great cost, however, as highly cultivated new varieties need constant **use of fertilizers and pesticides**. Nevertheless, the plant breeders now have a set of new tools which are more powerful than ever before, awakening new hope for overcoming the problem of feeding the world. Biotechnology promises the plant breeders the introduction of foreign genes into plants with the aim of improving their quality: e.g. achieving higher protein, vitamin, and energy content, developing **resistance** to "pests", diseases and frosts and becoming resistant to dry or salt-laden ground and to herbicides used to control weeds. They also create the possibility of producing drugs, cosmetics, and food additives using gene-manipulated (transgenic) plant cultures and/or laboratory cultures independently of farming.

■ 7.6 Fields in a Test Tube: *in vitro* Plant Breeding

Roses are the preferred subject for breeders, from simple forms to those that appear artificial. The his-

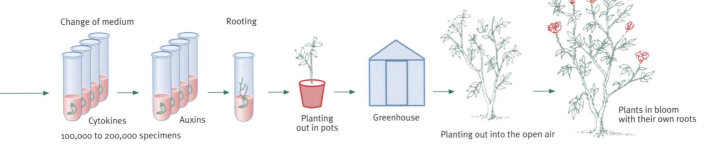

Change of medium

Rooting

Cytokines
Auxins
100,000 to 200,000 specimens

Planting out in pots

Greenhouse

Planting out into the open air

Plants in bloom with their own roots

tory of **rose breeding** was rather a late starter, although the cultivation of roses was known among the Medes and Persians before our time in the 12th century. It was *Rosa gallica* that was cultivated at the time, which makes it the first and oldest garden rose in the world (Fig. 7.17).

Only with the theoretical background of knowledge about genetics and its possible application could roses be bred as required. The first breeding by artificial pollination was carried out by the Frenchman **Jaques-Louis Descemet** (1761 - 1839).

Nowadays, **500,000 new plants** can be created in just one year from a single mother plant without grafting. Plant breeding in the test tube (*in vitro* reproduction) allows the breeder to reproduce new worthwhile varieties quickly (Fig. 7.15) and specific properties such as resistance to disease and herbicides can be selected during cell culture.

If a **hybrid plant** is involved (a cross between different varieties or breeds), sexual reproduction via seeds must be avoided, as the offspring will separate in accordance with **Mendel's** hereditary principles (Mendel's Law). Until now, the only alternative has been **vegetative propagation**, using such familiar and traditional processes as rooting strawberry runners or taking cuttings. However, not all plants lend themselves so easily to vegetative propagation.

Micropropagation is much faster and more effective than propagation by taking cuttings or using runners, but how is it possible that a whole plant may be regenerated from individual cells? All plant cells are **totipotent**, especially cells from roots and shoots. Each cell contains the full chromosome set, i.e., all the genetic information necessary for the development of a complete individual plant from one cell. Depending on the function taken on by the individual cells and the cell group during development and in the finished organism, only a specific part of the genetic information is reproduced and the remaining part is not used.

In contrast to animal cells, some cells of many dicotyledonous plants can be dedifferentiated. Their biological development clock can therefore be reset at the beginning and set in motion again under certain conditions, at which time the cells run through their development program from the beginning again, as dictated by their genes.

For successful regeneration of a plant from a single cell, it is necessary to add **growth hormones (phytohormones)** to the culture substrate. These control growth and differentiation of the plant cells and include **auxins** to regulate root growth and make sure that only the uppermost bud of a plant is produced and **cytokines** to induce the growth of shoots and inhibit root growth. The ratio of cytokines to auxins is crucial.

100,000 new little plants can be raised in a square meter of the laboratory (Fig. 7.15). It is therefore possible to propagate particularly lucrative plants, and all the plantlets will come from cells from one single super-plant. These offspring from a single plant are called **clones** – from the Greek *klon* which means much the same as shoot or branch. All clones, like identical twins, have the same hereditary characteristics. We will come back to transgenic animals later (Ch.8).

Meristem and haploid cultures are differentiated from the cell type of the original material, and callus or suspension cultures are differentiated by the type of cultivation.

■ 7.7 Meristem Culture

The most important modern method of plant cloning is meristem culture, using the **actively dividing tissue (meristems)** of plants (shoot, root, or axillary buds) (Fig. 7.21).

Meristems are actively dividing tissue found at different sites on the plant. The most important meristems are **shoot apical meristems** which are hid-

Fig. 7.18 The majority of our ornamental plants have come from micropropagation.

Fig. 7.19 Even orchids are mass-propagated *in vitro*.

Fig. 7.20 The oil palm supplies palm oil, which contains saturated fatty acids. Huge monoculture plantations of palms in Asia were grown in cell culture and "improved". They were planted at the cost of tropical jungle regions and are the subject of heavy criticism.

Box 7.3 The Expert's View: Marine Biotechnology

The oceans are an ancient ecosystem where life, in the form of bacteria, originated about four billion years ago. Marine organisms provide us with a valuable gene pool that is beginning to be tapped. To ensure a healthy and stable marine food web, pollution must be controlled, threatened and endangered species must be protected (and perhaps organisms cultured that are in high commercial demand), and the link between the ocean's primary producers (that is, plankton) and other organisms must be understood.

The oceans offer a wealth of untapped resources for research and the development of products (including medical). The vast majority of marine organisms have not been identified (the majority are microorganisms) and little is known about many of the organisms already identified.

Marine organisms are of great interest to scientists and industry for two important reasons:

- They represent a major proportion of the Earth's organisms. Most major groups of organisms on Earth are either mostly or completely marine. There is much genetic and metabolic information to be obtained from these organisms.

- Marine organisms have unique metabolic pathways and other adaptive functions, such as sensory and defense mechanisms, reproductive systems, and physiological processes, that allow them to live in extreme environments, from cold Arctic seas to very hot hydrothermal vents to pressures at great depths.

Aquaculture

Aquaculture has been practiced throughout the world for thousands of years. Ancient aquaculture, which was primarily freshwater, began in the Far East.

Chinese aquaculture dates back probably at least 3000 years. The earliest known record, in 473 BC, reports monoculture of the common carp.

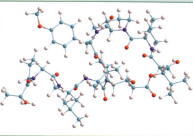

Didemnin B a cyclic peptide, is produced by the Caribbean tunicate *Trididemnum solidum*.

As methods were refined, polyculture was practiced, combining species with different food preferences; ponds contained varieties of finfish, shellfish, or crustaceans.

Today, high density aquaculture is practiced in many countries. In the Philippines, for example, the giant tiger prawn, *Penaeus monodon*, is cultured in ponds at densities of 100,000 to 300,000 prawns per hectare. Significant advances have been made in the culturing of oysters, clams, abalone, scallops, brine shrimp, blue crab, and a variety of finfish.

Traditional aquaculture encompasses low input technologies to propagate, harvest, and market fish, algae, crustaceans, and mollusks.

Unfortunately, aquaculture ponds have the potential of destroying delicate habitats through pollution by wastes or by decimating, for example, mangroves to make room for shrimp ponds. Biotechnology may be able to help solve some of the problems created by intensive aquaculture practices.

Marine and freshwater biotechnologies are used today to increase the yield and quality of finfish, crustaceans, algae, and various **bivalves** such as clams and oysters. Recently, **transgenic organisms** with desired characteristics have been generated for future commercial production.

Modern Japanese mariculture is more productive than freshwater aquaculture and has recently accounted for more than 92% of Japan's total aquaculture yield. The major food products have

been the Japanese oyster, *Crassostrea gigas*, and a red alga called nori (*Porphyra*). Nori provides the largest seaweed harvest. Other products include the fish yellowtail jack, the red sea bream, as well as a scallop and two types of brown algae (*wakame* and *kombu*).

The high worldwide demand for clams, oysters, mussels, abalone, crabs, shrimp, and lobsters favors the development of efficient culturing methods to increase their production. A variety of methods exists, from rearing in tanks to using floating platforms and substrates for the growth of organisms.

Genetic manipulation in culture promotes faster growth and maturation, increased disease resistance, and triploidy. Normal **diploid oysters** spawn in the summer and lose their flavor because they form a massive amount of reproductive tissue. Instead of the two sets of chromosomes that diploids have, triploid Pacific oysters have three (two from the female and one from the male). **Triploid oysters** are obtained in culture by treating eggs with cytochalasin B. This inhibitor of normal cell division doubles the number of chromosomes so that, when the eggs are fertilized with normal sperm, zygotes have three sets of chromosomes. Because triploid oysters are sterile and do not form reproductive organs, they are more flavorful and meatier year round. They also grow larger and faster than diploid oysters and can be harvested sooner. Triploid oysters represent a significant portion of the total oyster production in the United States (making up almost 50% of the Pacific Northwest ouster hatchery). Concerns about the safety of using cytochalasin B may soon lead to the pro-

Aquafarming in Hong Kong: clams and mussels in a seafood restaurant (top) and swimming farms in the Clear Water Bay region (bottom).

duction of triploid oysters by mating tetraploids (having four sets of chromosomes) and normal diploid oysters.

Bivalves such as oysters and **gastropods** such as abalone (which has a commercial value of $20 to $30 per pound) are cultured by the manipulation of the reproductive cycle. Hydrogen peroxide added to the seawater induces synthesis of prostaglandin, a type of hormone that triggers spawning. Larvae are then induced to settle by the addition of the amino acid γ-aminobutyric acid (GABA), an important neurotransmitter in animals. Larvae exposed to GABA settle onto the substrate and begin developmental metamorphosis and cellular differentiation.

Growth-accelerating genes are being identified and cloned for the eventual production of compounds from these genes, which are important in the efficient, controlled culturing of abalone and other shellfish. Productivity is also being increased by crossing specific genetic lines to increase, for example, the growth rate of oysters (by up to 40%). The use of recombinant fish growth hormone may also produce faster-growing shellfish.

Medical applications

For the past 100 years, much energy has been devoted to screening the world's organisms for useful chemicals. Approximately 20,000 chemicals have been characterized, and many of these come from marine organisms such as bacteria, algae, **sponges**, corals, jellyfish, bryozoans, mollusks, **tunicates**, and **echinoderms**. Most of the new natural products identified each year have come from terrestrial organisms: annual sales of pharmaceuticals from plants exceed $10 billion in the United States. However, plants were virtually untapped until recently, so natural products from marine organisms may hold similar promise.

Indeed, marine organisms have been shown to be an excellent source of new compounds that exhibit a variety of biological activities.

Most of these compounds came, in descending order of abundance, from sponges, ascidians, algae, mollusks, and soft and gorgonian corals. Sponges have recently become a primary object of investigation because of their wide range of biosynthetic capabilities. Metabolites from echinoderms (such as starfish and sea stars) also have been isolated for natural products.

A variety of compounds isolated from marine organisms show promise as anticancer drugs. Didemnin B, a cyclic peptide from a Caribbean tunicate (*Trididemnum solidum*), was shown

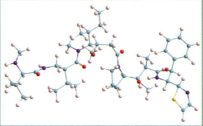

Dolastatin 10, a linear peptide, produced by the Sea Hare *Dolabella auriculata*.

experimentally to be an effective immunosuppressive agent. It is approximately 1000 times more effective than the standard cyclosporin A. Further study may prove it useful in repressing rejections of transplanted organs.

The Romans knew of the toxic effects of crude extracts from sea hares (a type of mollusk) containing dolastatins and by the year 150 were using such extracts in the treatment of various diseases. However, not until the 1970s were dolastatins known to be effective against lymphocytic leukemia and melanoma. Different structural forms of dolastatins that show antitumor properties have since been isolated.

Dolastatin 10 is a linear peptide from the sea hare *Dolabella auricularia*, found in the Indian Ocean. This compound is the most active of the isolates and is a potent antimitotic (inhibitor of cell division) because it inhibits the polymerization of tubulins into microtubules, which are involved in cell division. This compound is undergoing clinical trials and is being compared with other antitubulin polymerization drugs, such as vinblastine, which is isolated from the Madagascar periwinkle (*Catharanthus roseus*).

Many marine organisms-including cyanobacteria, green, brown and red algae, sponges, dinoflagellates, jellyfish and sea anemones-produce secondary metabolites with **antibiotic properties**. One promising class of broad-spectrum antibiotics, squalamine, has recently been isolated from the stomach of the dogfish (*Squalus acanthias*); it has activity against a wide variety of bacteria, fungi, and protozoa. Unfortunately, antibiotic isolation and characterization has not been extensive, and few compounds from

marine organisms are being studied, most likely because of the costs of identifying and testing commercially important antibiotics.

Toxins are thought to be produced by marine organisms for a variety of purposes, including predation, defense from predators and pathogenic organisms, and signal transduction in the autonomic and central nervous system. Many of these marine natural products are extremely toxic; their effects can range from dermatitis to paralysis, kidney failure, convulsions, and hemolysis. Livestock and humans have become ill and died when specific toxin-producing algae or cyanobacteria have contaminated water. Dinoflagellates produce saxitoxins that are 50 times more toxic than curare. When bivalves that have incorporated saxitoxins are ingested by humans, paralytic shellfish poisoning can lead to severe illness or death. Dinoflagellates also produce the potent ciguatoxin. When small grazing fish that have ingested dinoflagellates while feeding on algae are then eaten by larger fish, ciguatoxin is transferred and can poison humans who consume the fish.

Marine toxins are being studied for their antitumor, anticancer, and antiviral compounds, tumor promoters, and antiinflammatory properties, analgesics, and muscle relaxants.

The sea is a chamber filled with yet not explored treasures. One of the highest priorities of mankind is therefore to protect this precious ecosystem.

Susan Barnum is a Professor in the Department of Botany at Miami University Oxford.
She wrote an introductory biotechnology book that she uses in her introductory class. Her research involves the molecular evolution of nitrogen fixation genes and the evolution and excision of DNA elements found within nitrogen fixation genes of some types of cyanobacteria. A new initiative involves identifying molecular processes in heterocyst (the microaerobic terminally differentiated cell where nitrogen fixation occurs in some types of filamentous cyanobacteria) development and spacing.

Shoot meristem

Regeneration to
complete plants

Anthers

Ovaries

Callus culture for
product synthesis

Anther and ovary
cultures for
haploid plants

Fig. 7.21 Meristem culture (left) and anther/ovary culture (right).

Fig. 7.22 An example of desired virus attack: an *Abutilon* leaf with characteristic mosaic pattern.

Fig. 7.23 The symbol of Hong Kong, *Bauhinia*. The anthers can be seen clearly here.

den deep in the shoots. They are isolated and separated and the removed plant material is placed in a solid or liquid culture medium. Once it has grown, it can be divided again and with the addition of appropriate plant hormones, one or more plants can be regenerated from each of these pieces.

These days, thousands of types of plants are propagated in this way. It started with rare orchids, lilies (Figs. 7.18 and 7.19), chrysanthemums, and carnations and ended with such economically significant plants as potatoes, corn, cassava, grape vines, bananas, sugarcane, and soy beans.

The highly lucrative oil palms (Fig. 7.20) in Malaysia have been propagated by cell culture since the 1960s. They not only produce 30% more palm oil than normal palms, but are also decidedly smaller, making harvesting easier, although the excessive palm monoculture is a threat to the last tropical forests of Asia.

Standard vegetative propagation of strawberries using runners produces a maximum of ten plants from runners. On the other hand, modern meristem culture could produce up to 500,000 plants in theory per year from one mother plant.

There is another crucial advantage: even if the original plant has been affected by a **virus**, vegetative offspring can be produced that are virus-free (Fig. 7.21). Only virus-free plant parts are selected for propagation. The meristem tissue divides rapidly and seems to "outgrow" the spread of the virus. It therefore usually contains less viruses. Sometimes, however, plant viruses are welcome. The *Abutilon* mosaic virus creates interesting leaves on this houseplant (Fig. 7.22) and beautifully flamed tulips were caused by the *Poty* virus.

The meristems are cut, cultivated, and rooted (Fig. 7.21). The whole cycle is repeated until virus-free meristems are found, then these are used to pro-

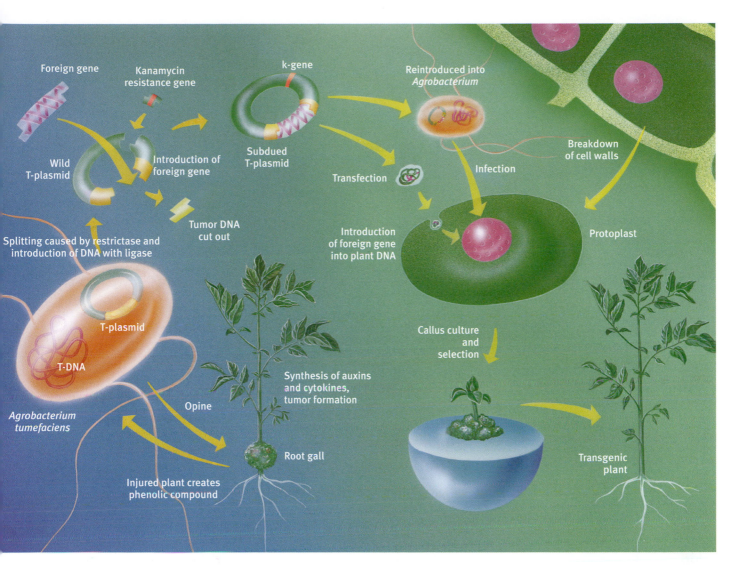

Foreign gene

Kanamycin resistance gene

k-gene

Reintroduced into *Agrobacterium*

Wild T-plasmid

Introduction of foreign gene

Subdued T-plasmid

Breakdown of cell walls

Transfection

Infection

Splitting caused by restrictase and introduction of DNA with ligase

Tumor DNA cut out

Introduction of foreign gene into plant DNA

Protoplast

T-plasmid

T-DNA

Agrobacterium tumefaciens

Opine

Synthesis of auxins and cytokines, tumor formation

Callus culture and selection

Injured plant creates phenolic compound

Root gall

Transgenic plant

Fig. 7.24 How transgenic plants are obtained using *Agrobacterium*.

duce virus-free plants. However, these are not immune to new virus attack if they are then planted out in the natural world. In the meantime, meristem culture has revolutionized horticulture and the estimated world market is more than three billion US dollars.

■ 7.8 Haploid Cultures: Anthers and Ovaries

Anthers, (Fig. 7.23) consist of a few male germ cells that are haploid, i.e., each cell nucleus contains only one set of chromosomes.

If young anthers are taken from the flowers under sterile conditions and placed on a culture medium (agar), they grow into complete, sterile, haploid plants after the addition of hormone supplements.

These plants are coveted in breeding, because you can save time in easily getting from them homozy-

gous diploid lines by experimental doubling of their chromosome set with the aid of the "mitosis poison", colchicine. Recessive mutations are not covered by the second chromosome available in diploids and are immediately visible, which is advantageous to the breeder.

In the flowering plants, an ovary is a part of the female reproductive organ of the flower. Specifically, it is the part of the carpel which holds the ovule(s) and is located above or below or at the point of connection with the base of the petals and sepals.

Whole haploid plants can also be produced from **ovaries** (Fig. 7.21) as well as anther cultures.

Haploid cultures are used in the breeding of tobacco, rape seed, potatoes, barley, and medicinal plants.

Fig. 7.25 Tobacco in callus culture.

Fig. 7.26 Protoplasts.

Box 7.4 Biotech History: Pomatoes and Biolipstick

A pipe dream: The **pomatoes** experimentally "created" by the author in just ten minutes!

The real **pomato** (centre) created at the Max Planck Institute for plant breeding research in Cologne. Left: tomato, right: wild potato.

Protoplast fusion is a possible biotechnical means of changing plant cells. "Naked" protoplasted cells are fused together by means of chemicals such as polyethylene glycol or by electrical impulses. Crosses between taxonomically distantly related varieties (somatic hybrids) had already proved successful, but it was not possible to cross them by natural means.

The first object was the **pomato**, a cross between the potato and the tomato: tomatoes should have grown on the top part of the plant and potato tubers on the underground runners. **Georg Melchers** at the Max Planck Institute for Biology in Tübingen carried out the first successful experiment in 1977. However, the pomato did not form either real potatoes or real tomatoes. The possible cause could have been

that cell division in such "unnatural" hybrids was abnormal. A real **pomato** was developed at the Max Planck Institute for Plant Breeding Research in Cologne in 1994 by **Inca Lewen-Dörr** and the company Green Tec. The plants were "vigorous". Apart from the fact that the flowers of the **pomato** were yellow, the plants were otherwise similar to the potato phenotype. Of course, the flowers of the potato are not normally yellow (Fig. 7.50).

Somatic hybridization is perfectly suitable for breeding purposes and is also used in practice for *Citrus* fruit or chicory. It is always the method of choice when complex, polygenetically determined characteristics such as frost tolerance (already done with potatoes) or flower color are to be transferred from a wild form into a cultivated form.

It is highly probable that the various characteristic components of the potato or the tomato cause them to be inedible when they are mingled. Although both potatoes and tomatoes belong to the same Solanaceae family, there are still great problems.

Shikonin

"A convincing lipstick without chemistry, with purely biological colorants already familiar to Traditional Chinese Medicine – for modern women – a high-tech Japanese product!"

The biolipstick was a big hit on the cosmetics market. Within a few days, two million items were sold in Japan in 1985, in spite of the hefty price of 3,500 yen (30 USD). The advertisement for it skillfully used the antipathy of the Japanese to chemical products and their awareness of tradition and pride in their high technology performance. The red colorant in the lipstick is a truly biological product. Shikonin (a naphthoquinone compound), a drug used in Traditional Chinese Medicine for centuries, used to be laboriously derived from the roots of the shikonin plant (*Lithospermum erythrorhizon*). It took three to seven years before the plant had collected a maximum of two percent of the active substance in its roots. In traditional medicine,

The Kanebo biolipstick, a short-term sales hit.

shikonin is used against bacterial disease and inflammation. The Japanese imported ten metric tons of shikonin raw material annually from China and South Korea for approximately 4500 dollars per kilogram. For such an expensive product, it was worth trying to get the shikonin-producing plant cells to grow in a nutrient solution. The Tokyo company **Mitsui Petrochemical Industries Ltd**. was very successful with this.

The manager of the cosmetics company **Kanebo** then had a brilliant sales idea. He suggested the splendid color was eminently suitable for lipsticks and powder, especially rouge. The gimmick was that the colorant is not just biologically created, it also protects against bacteria and has anti-inflammatory properties!

Lithospermum cells are today grown on an industrial scale in a 200-liter bioreactor with a growth medium, then transferred to a smaller reactor with a medium that promotes shikonin production. These cells are used to inoculate a 750-liter bioreactor where the shikonin is produced. The root cells in the bioreactor create 23% shikonin in only 23 days. Compare it once again with what nature can do: only two percent in three to seven years! Approximately five kilograms of pure shikonin are produced in each complete bioreactor run. Initially, Mitsui produced approximately 65 kg shikonin per year out of a Japanese requirement of 150 kg.

Cell aggregate of *L. erythrorhizon* from cultures in M9 medium, colored red by its shikonin content.

7.9 Callus and Suspension Cultures

Wound tissue growing in an unorganized manner is described in plants as a **callus**. It arises in plants from the cut surface where plant material has been removed. Calluses can either be bred on agar as surface cultures and then grown on as undifferentiated cell clumps, or phytohormones can be added to get complete plants again (Fig. 7.25).

Protoplasts offer a special opportunity for plant propagation; these are cells from which the cell walls have been removed (Fig. 7.26). Strips are cut from broad leaves and placed in a solution of enzymes that break down cell walls (i.e., pectinase, cellulase). The resulting "naked" cells, the protoplasts, are transferred to a culture medium where they regenerate their cell walls. The reconstituted walled cells allow the development of small calluses resulting from continual division and the addition of hormones induces small shoots in the calluses, which can be rooted with other hormones. Thousands of complete new plants can be created in this manner from a single leaf using protoplast culture (Fig. 7.21).

As isolated protoplasts are lacking a cell wall, they are relatively easily fused together by **cell fusion** (chemically) or by electric fields (electrofusion) (as with hybridoma cells using polyethylene glycol, Ch. 5).

The fusion cells are then grown to form somatic hybrid plants. This **somatic hybridization** led to tomato + potato hybrids (Box 7.4) in 1977, which were unfortunately inedible. Even if the tomato + potato hybrid had been successful, however, it would have been difficult to supply two nutrient stores, both tubers and fruits, at the same time.

On the other hand, the fusion of cells from two varieties of **thorn apple** (*Datura*) was successful and the new variety of thorn apple produced more alkaloids (scopolamine) than either of the two source varieties and grew better.

7.10 Plant Cells in a Bioreactor Produce Active Substances

To produce plant material in a reactor, you do not need differentiated plant parts, as a "cell clump" or callus is sufficient to synthesize the plant components.

Rooting

Enzymatic breakdown of cell walls

Formation of shoots after the addition of hormones

Isolated protoplasts

Formation of tissue complexes (calluses)

Growth in nutrient medium, new formation of cell walls

If you can isolate tissue fragments from a plant, you can grow an unlimited amount of plants on fully chemically-defined nutrient media. A plant part is first sterilized, then some of it is taken from the inside and cultivated on a culture medium hardened with gelatin-like agar.

The culture medium must contain inorganic nutrient salts and sugar as a source of energy, as well as some vitamins and hormones. If the correct composition for a particular cell culture has been found, a callus forms and this callus tissue can now be transferred into a liquid nutrient medium.

If the culture is shaken to provide the cells with sufficient CO_2, the plant cells continue to reproduce (Fig. 7.21).

Fig. 7.27 Principle of protoplast culture.

Fig. 7.28 The purple foxglove (*Digitalis purpurea*) supplies medicinal products that affect the heart.

Fig. 7.29 The antipyretic, salicylic acid, was produced from willow bark (*Salix*) (Chapter 9).

Fig. 7.30 *Catharanthus roseus* provides drugs against cancer.

Fig. 7.31 Eckard Wellmann (University of Freiburg) was the first to be successful with the *in vitro* cultivation of *Huperzia* (top). Chinese varieties of *Huperzia* hanging on a cupboard for Traditional Chinese Medicine (TCM).

Such cells can also be cultivated in large bioreactors with several cubic meters of contents. The substances available in the plant accumulate under suitable cultivation conditions and can then be isolated, while all the cell material can be freeze dried, for example, and extracted using a suitable solvent. Not every plant, however, produces its highly complex contents "voluntarily" in cell culture.

Lutz Heide's group at the University of Tübingen used two ubiquinone biosynthetic genes of *E. coli* as well as plant metabolism genes (encoding HMG CoA reductase) to transform the shikonin plant *Lithospermum erythrorhizon* (Box 7.4). In cell cultures of *Lithospermum*, certain biosynthetic steps are specifically influenced by the expression of this enzyme in the biotechnological production of **Shikonin**, a medicinal product and colorant.

Shikonin has become famous (Box 7.4) because of the biolipstick, and there are other plant products from the bioreactor, e.g. Paclitaxel, an excellent antitumor agent (see Box 9.3).

Even primitive man, a hunter and nomad, used **medicinal herbs**, such as willow bark (*Salix*) containing salicylic acid (Fig. 7.29) which reduces fever, or plants that speed up wound healing. Even today in the age of synthetic medicinal products, plant products are indispensable to medical treatment.

At the top of the list are steroids (produced from **diosgenin** in the roots of yams, Ch. 4), **codeine** (sedative and cough medicine), **atropine** (to enlarge the pupils in eye examinations and in poisoning), **reserpine** (to reduce blood pressure), **digoxin** and **digitoxin** (heart pills) from the foxglove (*Digitalis*, Fig. 7.28), and **quinine** (malaria drug and aromatic substance).

The disadvantages of **traditional extraction processes** are limited availability, variations in quality, danger of rare plant species dying out, demands of space for plantations leading to deforestation of rainforests and too little land for agriculture, occurrence of plant diseases and pests in monocultures, contamination by pesticides, heavy metals from polluted air, dependence of the active substance content of plants on climate, weather, time of year, age and site, and also the dependence on political crises or price cartels in the crop-growing countries. The last disadvantage mostly affects only the multinational pharmaceutical concerns, although these are well able to defend themselves -

as the example of steroid hormones has shown (Ch. 4). The fears of the developing countries, whose exports are very often dependent on plant products, doubtless have more foundation: for them, physical survival is often literally more important.

Plants surpass by far everything that will be synthesized by chemists for the foreseeable future. It is estimated that there are **tens of thousands of compounds with extremely complicated structures** and all of these substances have obviously arisen under the pressure of selection of animals and microbes and many are directed at warm-blooded creatures. We must therefore assume that there is a huge treasure trove of pharmacologically interesting substances waiting to be discovered. Only a fraction of the plant world has been investigated so far, some of it using obsolete, very outdated and inadequate methods to find evidence of active substances. It is to be feared that with the deforestation of tropical forests, plant species may even die out before their uses have been tested at all. As chemical synthesis of the active substances is either not yet possible or is very costly, the natural products in medical treatment produced from plants cannot be replaced.

In order to safeguard this supply for the future, we have to consider the production of active substances from plants with the aid of cell culture technology.

■ 7.11 What are the Active Substances from Plants that Will Follow Shikonin?

For the cell culture researcher, the Madagascar periwinkle (*Catharanthus roseus*) (Fig. 7.30) is a rewarding subject for investigation, as it contains a series of active substances, including the bisindol alkaloids **vinblastin** and **vincristin** which are of complicated construction and difficult to obtain synthetically. Both active substances cost several thousand dollars per pound and are imported from developing tropical countries.

The steroidal cardiac glycosides **digoxin** and **digitoxin** can be isolated in cell culture from the foxglove (*Digitalis*, Fig. 7.28) and it is predominantly digoxin and digoxin derivatives that are therapeutically interesting. Digoxin differs from digitoxin only in the absence of an additional hydroxyl group.

Eckard **Wellmann** at the University of Freiburg (Fig. 7.31) succeeded, in cooperation with our group in Hong Kong, in producing a highly interesting substance in cell culture, which has been used since primeval times by the Chinese against loss of memory.

We know the disease today under the name **Alzheimer's disease**. The lycopod *Huperzia* only grows slowly in the Chinese mountains and needs eight to ten years to produce the coveted **Huperzine A**. It has almost died out in the meantime as a result of collectors. Huperzine A can now be made for the first time in cell culture. If cultivation succeeds on a large scale, authentic huperzine A can be economically produced in large quantities.

Plant cell cultures have proved to be the method of choice for high-quality substances if synthetic production is too expensive or microbial manufacture is not possible (see Box 7.4).

Not least, they make a relevant contribution to the protection of rare plants and save valuable acreage urgently required for food crops.

■ 7.12 *Agrobacterium* – a Pest as Genetic Engineer

If you can cultivate plant cells in nutrient solutions in a manner similar to the growth of microorganisms, why should you not be able to manipulate them using genetic engineering (Ch. 3)?

The aim of the plant genetic engineer is to create plants that are resistant to drought, insects, and pesticides and able to grow on salt-laden ground, while containing more protein and nutrients.

"Natural" genetic engineers are bacteria of the species *Agrobacterium tumefaciens*, which lives in the ground and causes cancer-like growths or tumors on the root collar of damaged dicotyledonous plants. This discovery was made back in 1907.

The bacteria are attracted by elicitors, which are molecules released as a result of damage. They have a large plasmid (ring-shaped double-stranded DNA [see Chapter 3], 200 to 800 kilobase long), which triggers the tumor and is therefore called **Ti-plasmid** (tumor-inducing plasmid). The pest is therefore suitable as a "Trojan horse" or vector for foreign genes (Fig. 7.24).

The T-plasmids carry genes for opine detection and phytohormone formation. **Opines** are special amino acids formed in affected cells and can only be used by *Agrobacterium*. The bacterium therefore reprograms the plant cells for its own purposes. **Auxine** and **cytokine**, the phytohormones, stimulate growth and cell division of the transformed cells in the root collar tumor. Ti-plasmids also carry genes that recognize wounded cells and mobilize and transfer the **T-DNA** into the plant. After transmission, the transfer DNA is introduced into the plant DNA in the cell nucleus. That is a rare example of a gene transfer from a prokaryote into a eukaryote.

In 1983, **Marc von Montagu** and **Jeff Schell** (1935-2003) (Fig. 7.32 and Fig. 7.33) in Ghent (Belgium) and **Robert Fraley** at the Monsanto Co (St. Louis, USA) suggested a revolutionary method independently of each other.

They "deactivated" the "wild" Ti-plasmids by introducing a foreign gene into the 15-30 kb sized T-DNA. The phytohormones and the opine genes of the T-DNA were removed to create space for the introduction of the foreign DNA but this did not lead to any further growths occurring at the same time. Indeed, it is difficult to regenerate whole plants from cancerous tissue at all.

Foreign genes can be integrated at will into the thus "**tamed**" **Ti-plasmids**. In addition to the foreign gene, a gene resistant to antibiotics, usually kanamycin or ampicillin, is incorporated so that the "successful" cells can later be more easily selected (Fig. 7.24).

The recombinant plasmids are either transferred directly into "naked" plant cells, i.e., into the protoplasts, or incorporated in *Agrobacterium* which then attacks intact plant cells. The latter is the standard method.

Pieces of leaf are incubated with a suspension of recombinant *Agrobacterium* cells. The addition of hormones then induces the formation of shoots. The bacteria are killed by adding antibiotics and the successfully transformed plants are selected on a kanamycin medium. The kanamycin-resistant cells

Fig. 7.32 Marc van Montagu, one of the fathers of plant gene engineering.

Fig. 7.33 Jeff Schell (1935-2003), a pioneer of "green" biotechnology.

Fig. 7.34 Film star Hans Albers in the UFA film "Münchhausen": Ride on a Cannonball. Just like Münchhausen on the cannonball, DNA is shot into cells during biolistic gene transfer.

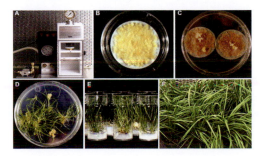

Fig. 7.35 (bottom left) Biolistic gene transfer in turf grass. Zeng-yu Wang of the Samuel Roberts Noble Foundation in Oklahoma transferred genes using biolistic gene transfer on a type of sheep's fescue (*Festuca*).
A: The PDS/1000 device for microprojectiles.
B: Cell suspension before bombardment.
C: Calluses after selection on hygromycin.
D,E: Transgenic plants are regenerated.
F: Transgenic *Festuca* plants in the greenhouse.

Box 7.5 The Expert's View:
Ingo Potrykus, "Golden Rice" and the split between good intentions and the public distrust in GMOs

"Golden Rice" is the first case of genetically modified bio-fortification and opens a new area of sustained intervention to reduce micro-nutrient malnutrition in poor populations in developing countries. The 'Father of Golden Rice', the Swiss scientist Ingo Potrykus, reports in the following (drastically shortened) article about his ambition to save the lives of the poorest and his disappointment at the years of delay caused by unjustified regulation:

Polished Golden Rice in comparison to untransformed control. The yellow color indicates the presence of carotenoids; the intensity is a measure of the concentration (in the example given approx. 4mg/g endosperm).

Poor members of rice-dependent societies are vitamin A-deficient because rice is their major source of calories, but does not contribute any pro-vitamin A to the diet. Hundreds of millions in Southeast Asia, Africa and Latin America, therefore do not reach the 50% level of the recommended nutrient intake (RNI) for vitamin A required to live a healthy live. Even with rice lines containing modest concentration of pro-vitamin A, a shift from ordinary rice to Golden Rice in the diet could save people from vitamin A malnutrition. A recent independent study (1) has established that Golden Rice, if supported by the government, could save up to 40,000 lives per year in India alone at minimal cost.

The underlying science

As a scientist, I was determined to contribute to **food security** and have been investing in genetic engineering technology for cereals since 1974. By the end of the 80s, we had transformation protocols ready for rice and had worked on projects to rescue harvests with the help of insect-, pest-, and disease-resistant plants. A wish-list from **The Rockefeller Foundation** alerted me to the problem of micronutrient malnutrition, specifically vitamin A-malnutrition.

In 1991, I appointed a PhD student, **Peter Burkhardt**, to take up the task to work towards pro-vitamin A biosynthesis in rice endosperm.

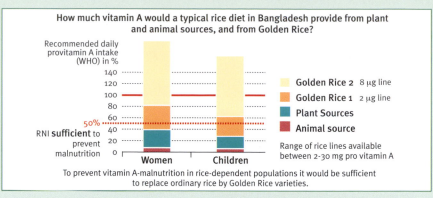

How much vitamin A would a typical rice diet in Bangladesh provide from plant and animal sources, and from Golden Rice?

Recommended daily provitamin A intake (WHO) in %

140
120
100
80
60
50% RNI **sufficient** to prevent malnutrition
40
20
0

Women   Children

Golden Rice 2 8 µg line
Golden Rice 1 2 µg line
Plant Sources
Animal source

Range of rice lines available between 2-30 mg pro vitamin A

To prevent vitamin A-malnutrition in rice-dependent populations it would be sufficient to replace ordinary rice by Golden Rice varieties.

The situation for Bangladesh: The diet provides only 40% (women) or 30% (children) of the recommended amount from plant (vegetables and fruit) and animal (fish and poultry) sources. Rice, though the major source of calories, does not provide any provitamin A. A shift to Golden Rice would raise both women and children above the critical 50% line; with 8-µ lines even above 100%. Men are not at risk. This is the only sustained, and most cost-effective, intervention.

As Full Professor of plant sciences at the ETH Zürich, I had some financial independence to start a project considered totally unfeasible. I soon realized that my PhD student needed a co-supervisor with solid knowledge in the regulation of the terpenoid biosynthetic pathway, and he found Dr **Peter Beyer** from the neighboring University of Freiburg, Germany. When approaching The Rockefeller Foundation for support, Dr **Gary Toenniessen** organized a brainstorming session in New York. The opinion was that the feasibility for success was extremely low. But as the outcome in case of success would be so overwhelming, the Foundation decided to finance one PostDoc both in my and Peter Beyer's lab. Our expertise was perfectly complementary and we decided to take up the challenge. We had to introduce four genes (see main text in this book) whose successful function could be tested only when all four would be working together.

The breakthrough came, after 8 years, from an experiment by a Chinese PostDoc in my lab, **Xudong Ye**. When Peter polished the harvest from a co-transformation experiment with five genes, the offspring from a transgenic line harboring all genes was segregating for white and yellow endosperm. HPLC analysis showed that the yellow color was based on the presence of carotenoids.

This was in February 1999, two months before my retirement. Fortunately, Peter was younger and could continue.

A scientific breakthrough alone is not enough for success

We presented our success to the public at my Farewell Symposium on 31 March 1999 and it was finally published in Science. Nature refused publication because of lack of interest, but the scientific community, the media, and the public

were quite interested, even became quite excited about this vitamin A rice.

TIME Magazine devoted a cover story to it on July 31, 2000 and there were hundreds of articles and airings in the media. Quite flatteringly, the readers of *Nature Biotechnology* voted Peter Beyer and myself as the most influential personalities in agronomic, industrial, and environmental biotechnology for the decade 1995 to 2005.

The **first disappointment** followed soon afterwards: nowhere in the public domain could we find a granting agency which would be willing to invest in the tedious process of "product development" of a humanitarian GMO.

TIME magazine devoted a cover story to Golden Rice and there were many hundreds of articles and TV airings from 2000 to 2003 reporting very positively about this first GMO case in the interests of the consumer, and from the public domain.

We had the choice to give up, or to find support from the private sector. Fortunately, Peter had taken the trouble to patent our invention. Therefore, we had something to offer. Having ourselves no commercial plans, we offered the agroindustry the patent rights for commercial exploitation in exchange for support of our humanitarian project.

How to make Golden Rice freely available to the poor and gain financial return from a commercial product? Dr Adrian Dubock (then at Zeneca, later Syngenta) helped us to find a solution.

"Humanitarian" was defined as including any action which did not lead to a higher income from rice than $ 10,000 per year, the territory to be developing countries, and export not being allowed. On this basis, our rights were transferred to Zeneca (later Syngenta) and licensed back for humanitarian use to the inventors.

This "public-private-partnership" also helped to solve the next big problem: all intellectual property rights involved in the technology, i.e., other peoples' patents! We did not know which patents we had been infringing in the course of our work. Two patent lawyers, commissioned by the Rockefeller Foundation, were studying the case and found alarming numbers: we had been using 70 patents belonging to 32 patent holders! So, we had no choice but to get free licenses. Fortunately, 58 patents were not valid in our target countries. Of the remaining twelve, six belonged to our partner Zeneca/Syngenta and for the rest, it was not a big problem to get free licenses.

These four personalities (Adrian Dubock; Ingo Potrykus, Peter Beyer, Gary Toenniessen, from left to right) were (and are) instrumental to the Golden Rice case: Peter Beyer was the "scientist" and Ingo Potrykus the "engineer" in establishing proof-of-concept. Gary Toenniessen, Rockefeller Foundation, created the idea, organized the brainstorming and financially supported proof-of-concept and beyond. Adrian Dubock (Zeneca/Syngenta) organized the public-private-partnership, the free licenses, established the Humanitarian Board and is the driving force behind product development and deregulation.

Golden Rice soon became the "poster boy" of agbiotech industry. For the same reason, "the darling of the agtech industry" immediately became the prime target of the antiGMO-lobby. The involvement of a commercial company caused the suspicion of the media and concerned environmentalists.

The Wonder of Gene Technology

The next deadly hurdle: how to pass through the regulatory procedures? As university scientists we had, of course, not the slightest idea what that meant. It meant in reality, the commitment of four years in repeating the same experiments many hundreds of times, with modifications of course, until we had transgenic events which were streamlined according to the ideas of the regulatory authorities. We improved the content of pro-vitamin A to 23-fold above the proof-of-

concept experiment because we had no precise data to calculate how much pro-vitamin A we had to provide with a routine daily diet. Much of this was due to the success of the Syngenta team working towards a commercial product. (3).

When the Syngenta management decided, much to our regret, not to continue on the lines of a product, all results were donated to the Humanitarian project.

For the necessary far-reaching strategic decisions, we scientists simply did not have the necessary expertise. Thus, we established a "Humanitarian Golden Rice Board" with expertise from all the different areas from which we needed advice.

Since 2001, the Board has been guiding the project. We needed GMO-competent public partner institutions in our target countries to develop locally adapted and agronomically successful varieties. Now, we have a Humanitarian Golden Rice Network in India, the Philippines, China, Bangladesh, Vietnam, Indonesia, Nepal and the Network is still expanding.

We needed managerial capacity beyond that of one overworked University Professor (Peter Beyer) and one retired one. We now have a

"Network Coordinator" with an office at the International Rice Research Institute, Philippines, and a "Project Coordinator" at the University of Freiburg. For more detailed information, visit our homepage under www.goldenrice.org

Lost years because of over-regulation

Because of the extreme requirements of the regulatory procedures in costs and data, we have to base all Golden Rice breeding work and variety development on one single selected transgenic event.

This event selection will be completed in 2007 and from then on, our partner institutions will be integrating this event into a carefully chosen selection of around 30 Southeast Asian popular rice varieties. Deregulation will be based later on the single event, not on 30 different varieties.

We have lost with Golden Rice 4-6 years in the preparatory adoption of regulatory requirements which do not make any sense scientifically, as well as from the perspective of guaranteed food safety. The most drastic case demonstrating how irrational regulatory authorities operate, is our experience with permission for a small-scale field testing of Golden Rice. No ecologist around the world has been able to propose any serious risk for any environment from a rice plant containing a few micrograms of carotenoids in the endosperm (the natural rice plant is loaded, exclusive of the endosperm, with carotenoids), and that is not surprising because this trait does

The Holy Father has an open mind for the challenges of new technologies: Ingo Potrykus with Prof Nicola Cabibbo (President of the 400-year-old Pontifical Academy of Sciences) and Pope Benedict XVI at the inauguration of the Academy. (Courtesy of Monsignore Marcelo Sánchez Sorondo, Chancellor of the Pontifical Academy of Sciences, see also www.vatican.va/roman_curia/pontifical_academies).

not provide for any selective advantage in any environment.

Thus, regulation is responsible for the death and misery of millions of poor people. And did regulation prevent any harm? Judging from all the regulatory reviews and all the data from all the "bio-safety research", the answer is: "most probably not".

Ingo Potrykus was born in 1933 in Hirschberg/Silesia, Germany. He studied biology at the University of Cologne and earned his doctorate with a thesis at the Max-Planck Institute for Plant Breeding Research. After several years at the Institute of Plant Physiology, University of Hohenheim, he became research group leader at the Max-Planck Institute for Plant Genetics. In 1976, he transferred to Basel to establish the area of plant genetic engineering at the Friedrich Miescher Institute. He was Full Professor of Plant Sciences, specifically of Biotechnology of Plants, at the Institute of Plant Sciences of ETH Zurich from 1987.

Since his retirement, Ingo Potrykus - as President of the international Humanitarian Golden Rice Board is devoting his energy to guiding Golden Rice towards subsistence farmers across the many hurdles of a GMO-crop.

Cited Literature:

Stein AJ et al. (2006) *Potential impact and cost-effectiveness of Golden Rice in India.* Nature Biotechnology 24(10).

Jajaraman KS et al. (2006) *Who's who in biotechnology.* Nature Biotechnology 24, 291-300.

Ye X et al. (2000) *Engineering the provitamin A (β-carotene) biosynthetic pathway into (carotenoid-free) rice endosperm.* Science 287, 303-305.

Paine JA et al. (2005) *Improving the nutritional value of Golden Rice through increased pro-vitamin A content.* Nature Biotechnology 23, 482-487.

Fig. 7.36 This is the devastating effect of Basta (phosphinothricin) on the author's lawn in Hong Kong.

Fig. 7.37 Logo of *Round Up*®, a glyphosate-based herbicide.

Fig. 7.38 Ecological farming allows wild flowers on the boundary ridge (top). A bunch of wild flowers from a wheat field.

(Chapter 3) must also contain the foreign gene. Root formation is then induced and the plants regenerated. No more root collar tumors are formed, because the T-DNA gene for auxin, cytokine, and opine production is missing.

In reality, the process is even more refined. The recombinant plasmids with intact Ti-plasmids interact and recombine in *Agrobacterium* cells; this is called a "binary or two vector system". Binary vectors remain in *Agrobacterium* cells as independently replicating vectors (see more below in the *antisense* experiments of "Flavr Savr tomatoes").

7.13 Biolistic Gene Transfer: DNA Shot from a Gun

However, not all plants can be transformed in this way. **Monocotyledons** are hardly ever infected, but among those that are, the monocotyledons corn and rice and other grasses are the most important food plants.

The solution did not come from protoplasts but from the USA, typically from the barrel of a "gene gun", hence the name **biolistic gene transfer**. The guns were developed at Cornell University in Ithaca (New York). Plasmid DNA is applied to gold or tungsten particles and then shot into tissue without damaging the cells. The adsorbed DNA remains in the cell as the shot comes through and is integrated into the DNA genome in admittedly rare but still a sufficient number of cases. Foreign DNA can be inserted into the nuclear DNA as well as into the DNA of chloroplasts. Other methods allow the introduction of foreign genes only into the chromosomes of the cell nucleus. The big advantage of

biolistic transformation is that it can also be used for any other organisms you like, including vertebrates.

7.14 Transgenic Plants: Herbicide Resistance

Plants that have been changed as a result of genetic engineering are known as transgenic plants and farmers all over the world are incorporating more of these genetically modified plants every year. Compared to 2003, areas cultivated with transgenic plants increased in 2004 by 20% to 200,000,000 acres. Seventeen countries have installed commercial green genetic engineering in their farming and areas are increasing almost everywhere (see Box 7.10).

More than 30 transgenic plants have already been registered in the USA, including transgenic cotton, potatoes, corn, rape seed, soy beans, and tomatoes, which are resistant to herbicides, insect pests, and viruses. They have already been introduced on over 86,000,000 acres of land.

Approximately 10% of crops on average are lost to "weeds" (Fig.s 7.36 to 7.38). The ideal **herbicide** should be active in low quantities, should not inhibit the growth of economically useful plants, should decay quickly, and should not reach ground water. Chemists work hard at the substance while bioengineers concentrate on the gene manipulation of economically useful plants. The herbicide market extends to six billion US dollars.

The American genetics company Calgene was the first to succeed in integrating bacterial genes into tobacco plants and petunias, which made the plants resistant to the substance **Glyphosate**. This means that these plants will tolerate quantities of the herbicide *Round Up ®*, the most popular herbicide in the USA, that would have long since destroyed normal plants. The gene company's client was Monsanto, the main producer of *Round Up®* (Fig. 7.37).

Glyphosate damages the plant by inhibiting an important enzyme for amino-acid metabolism (enolpyruvylshikimate phosphate synthase, **EPSP synthase**). Glyphosate-resistant *E. coli* strains were isolated by incorporating the EPSP synthase gene, then it was cloned. The gene was then transferred with *Agrobacterium* into petunia, tobacco, and soy bean cells, which were then grown again into full-grown plants.

The plants obtained now possessed a much higher concentration of the enzyme and were not damaged until they had been exposed to a much higher concentration of the herbicide. "Weeds" are therefore destroyed while the genetically modified plants survive.

However, this development has been criticized by ecologists. They regard with skepticism measures directed at full, scale destruction of a pest factor. To be fair, however, it should be noted that glyphosate is an excellent systemic herbicide that permits minimal application rates, keeping ecological pollution extremely low and leaving behind hardly any residues in plants or soil.

Ecologically more interesting than the protection of plants by increasing their enzyme content are transgenic plants that actively break down the herbicide and provide decontamination. This is the only way that accumulation of herbicide can be avoided.

This has been successful for the herbicide **Basta®**. The active substance is phosphinothricin, PPT. Basta® inhibits the synthesis of the amino acid glutamine in plants by means of glutamine synthase (GS) (Fig. 7.39). The poisonous ammonia which kills off cells is not processed and accumulates. The gene encoding the enzyme that modifies PPT by acetylation (PPT acetyl transferase) was isolated from a streptomycete and transferred to tobacco, potatoes, rape seed, and other plants. This neutralizes the inhibition of glutamine formation in transgenic plants, but not in the surrounding weeds.

Rape seed plants resistant to Basta® were used to find out their **ecological fitness** compared to conventional rape seed plants over three years in three different climactic regions of England. The results were published in *Nature* in 1993 and aroused

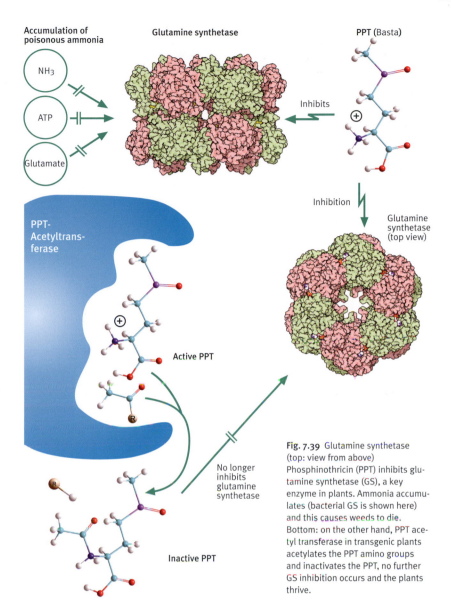

Fig. 7.39 Glutamine synthetase (top: view from above) Phosphinothricin (PPT) inhibits glutamine synthetase (GS), a key enzyme in plants. Ammonia accumulates (bacterial GS is shown here) and this causes weeds to die. Bottom: on the other hand, PPT acetyl transferase in transgenic plants acetylates the PPT amino groups and inactivates the PPT, no further GS inhibition occurs and the plants thrive.

widespread discussion. The transgenic rape seed plants had "on average less invasion potential" than non-transgenic ones. Opponents to release still criticize the model trials as faulty and seen overall, the transgenic rape seed was indeed generally inferior, although it could still be superior at specific individual sites in England.

■ 7.15 Biological Insecticides

In some developing countries, approximately 80% of the harvest is destroyed by insects or rodents, while a loss of only 25% to 40% occurs in Europe. Insects are also feared in tropical countries as carriers of malaria (*Anopheles* mosquito) or sleeping sickness (tsetse fly). Judging from the number of infected people, malaria is the most prevalent dis-

Fig. 7.40 Toxic protein crystals from *Bacillus thuringiensis.*

Box 7.7 The Expert's View:
Susan Greenfield about 21st century technology changing our lives: food and ageing

The acclaimed British neuroscientist Baroness Susan Greenfield shows in her new bestselling book that everything we take for granted about ourselves - imagination, individuality, memory, love, free will - could soon become lost forever. Only by harnessing technology in a humane way, she argues, can we preserve our unique sense of self and hold on to what makes us who we are.

The following short extract from her exciting book deals with our future lifestyle, the food we will eat and ageing:

…You settle down to your meal – this is one of the few remaining experiences that are 'real', involving an interaction with the atomic, physical world and a direct sensory experience solely within your physical body, and no one else's. As you indulge in chewing, sniffing, tasting and swallowing – you reflect how attitudes to **GM** (**genetically modified**) foods have changed in a relatively short period of time.

The first GM foods contained the actual substance of the source organism, as tomato purée does; more frequently now, however, they consist of purified derivatives that are actually indistinguishable from the non-GM organism, such as lecithin and certain oils and proteins from soya. Since the GM lecithin is chemically identical to its non-GM counterpart, it is hard to see how it presents any additional health risk.

The problem at the turn of the century was the impossibility of guaranteeing the **purity of each**

substance. Nowadays, focused tests are still needed to ensure that alien additional sequences are not absorbed by human tissues or into gut micro-organisms.

GM foods have now, for several generations, been part of the culture with no disastrous consequences, so the public is more accepting. Opinion first began to change when people realized that there was **simply no alternative way to feed** what was then known as 'the developing world', that vast majority of technologically disenfranchised humanity that cast such a shadow over the achievements of the early part of this century; according to UN estimates of the time, 800 million around the globe were undernourished. As well as combating mass starvation, the use of GM foods has reduced the number of **children suffering from blindness due to Vitamin A deficiency** from 100 million to zero. 400 million women of childbearing age no longer suffer from iron deficiency, which had increased the risk of birth defects. Rice is now routinely engineered to contain beta-carotene,

ready for conversion by the body into iron and Vitamin A.

And **pests had been a problem** for crop-growers: for example, 7 per cent of all corn had had to be destroyed. The use of GM crops, engineered to be pest-resistant, eliminated such wastage, as well as providing an attractive alternative to highly toxic pesticides.

However, the biggest incentive for wholesale acceptance of GM foods ended up being, quite simply, **personal gain.** Interestingly enough, the gain now has nothing to do with health or nutri-

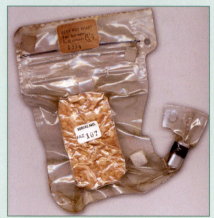

Astronauts' food: food of the future? Here is a typical entree shown aboard moon missions. A freezedried meal block would be reconstituted with a hot or cold water gun through the valve at lower right, then after a prescribed time, voilà: the astronaut opens the double ziplocks and spoons the food out.

Drink valve

Food Stick

Top: Cube-shaped water melons are a big-selling hit in Asia.
Middle: 'Kifli' is the name of a fingerling heritage potato received some years ago by the Canadian Potato Gene Resources Repository. Kifli is an attractive, smooth-skinned, curved small potato. Fingerling potatoes, of which the Repository has several, are particularly useful as fresh whole potatoes, or as salad potatoes. Bottom: A new fruit-creation: Strawbernana (Banana with strawberry taste), my two favorite fruits combined in one!

Will today's barbecue still be popular in the future?

tion. You look out into a world that is no longer composed of the muted mists and shades of previous centuries. Instead, because much of your time is spent processing artificial, heightened and bright colors, unmodified carrots, spinach or tomatoes would appear dull and unappetizing compared with their iridescent modern counterparts. Now that the food has become the main vehicle for that rare and prized phenomenon, direct stimulation of the senses, manufacturers have realized that by genetically modifying the non-essential features that sledge-hammer the senses they can make their consumers happy.

For a long time, the taste of each ingredient has been genetically engineered to be much more intense. Any edible substance can take on the **flavor** of any other. This flexibility is just as well since the production costs of 'real' chocolate, for example, preclude it as a viable product, as do the prohibitive amounts of sugar it contains.

The **colors of foods**, too, are far more vivid and standardized. And not just color – the physical form is more attractive too. Food also comes in a vastly more varied range of shapes and sizes. Thanks to genetic engineering, as well as precision manipulation of the atoms that constitute matter (nanotechnology), you can now choose to have **cuboid vegetables or meat in geometric shapes**. So bulk stacking in the fridge is easier, and the fridge can monitor consumption rates more easily, as the scanning tags are all in a uniform position.

Another advantage over 'natural' food is that everything is now bio-engineered to be **easier to prepare**; for example, grains now come with encapsulated liquid that is released when microwaved. Dripping, messy sauces are also a thing of the past and are now integral, not freeflowing but magnetized so that you can rub them off with a fork or a finger. Moreover, it is almost impossible to obtain food that has not been enriched with vitamins, minerals and other agents such as fish oils that ensure optimum body maintenance.

The **neutroceuticals industry is flourishing** as never before. Potatoes are engineered to combat constipation; cucumbers come in a purple variety, with as much Vitamin A as cantaloupe melons; carrots are maroon in color because they contain beta-carotene in massive amounts to improve night vision; and the salad dressing will lower cholesterol with regular use. Particularly popular is the bespoke neutroceutical option: food is custom-engineered for you, not just to deliver your particular taste preferences but also to cater for your particular health requirements.

Your personal medical profile is monitored and fed into a continually updated program that engineers foods with supplements according to your particular needs. Yet, like everyone else nowadays, you cannot easily draw a line between the genetic engineering employed to produce interesting food that hyper-stimulates your now cyber-jaded senses and that designed to optimize your health by direct intervention at the genetic level, be it for prevention or cure, of disease that is very much the dominant medical strategy.....

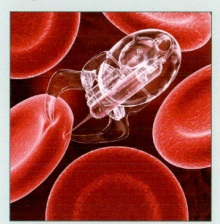

Nanoinjector with red blood cells. This image by Coneyl Jay, originally titled "Nanotechnology" as the winner of the 2002 Visions of Science Award by The Daily Telegraph of London and Novartis, was created to show one of the possible applications of nanotechnology in medicine in the future - microscopic machines roaming through the bloodstream, injecting or taking samples for tests.

Nanomedicine

… Another cornerstone of late-21st-century healthcare is nanomedicine: miniature devices patrol your body, giving early warning of possible problems or delivering just the right amount of medication to just the right place. These new therapies, combined with advances in traditional treatments and knowledge of disease as well as healthier lifestyles, all add up to longer life expectancy.

A thousand years ago life expectancy was just 25 years, but even by 2002, in Britain, men and women could expect to celebrate their 75th birthdays, and a female born at that time already had a 40 per cent chance of living 150 years. By 2050 there were 2 billion people over 60 years old worldwide, with such 'seniors' making up a third or more of many populations. Much had happened over the previous few decades for such long lifespans to be truly the norm.

Everyone now recognizes that **ageing is not a specific disease** but a general deterioration of the many processes that sustain body function; of these, the most feared is still a decrease in vitality and often degradation of intellect. Now science holds the promise not just of protracting mere existence but also of **actually extending active life**. For a long time now people have been aware of the need to view the mental abilities of older people in their own terms, rather than in comparison with immature, growing brains.…

Susan Adele Greenfield, Baroness Greenfield (born 1950) is a British scientist, writer, broadcaster, and member of the House of Lords. Greenfield is Professor of Synaptic Pharmacology at Lincoln College, Oxford University and Director of the Royal Institution. Susan Greenfield's research is focused on brain physiology, particularly the etiology of Parkinson's and Alzheimer's diseases, but she is best known as a popularizer of science.

Baroness
Susan Adele
Greenfield

Susan Greenfield has written several popular-science books about the brain and consciousness, and regularly gives public lectures and appears on radio and television. In 1994 she gave the Royal Institution Christmas Lectures, then sponsored by the BBC, entitled "The Brain". From 1995 to 1999 she gave public lectures as Gresham Professor of Physics. Greenfield created three research and biotechnology companies, Synaptica, BrainBoost, and Neurodiagnostics, which research neuronal diseases such as Alzheimer's disease.

Cited Literature:

Greenfield S (2004) *Tomorrow 's people. How 21st century technology is changing the way we think and feel*. Penguin Books, London

Fig. 7.41 Product to control the Colorado beetle, made from *B. thuringiensis*.

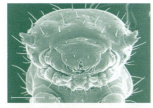

Fig. 7.42 Colorado beetle larvae.

Fig. 7.43 Pollen trap to ascertain the flight of pollen from transgenic plants. There is a fear that there may be interactions with other cruciferous plants such as charlock mustard or jointed charlock, especially with canola/rapeseed.

Fig. 7.44 Bee-friendly *Bacillus thuringiensis* products.

ease in the world, with 300 million new cases annually.

The fight against certain insects is therefore essential. In past, this has been carried out using chemicals but the **insecticides** do not just destroy pests, they also destroy many other insects with which they come into contact, disturbing the sensitive **ecological balance** and gradually poisoning other living creatures, such as insect-eating birds or the fish-eating bald eagle (the national bird of the USA), which is at the top of the food chain. In the end, insecticide residues also reach human food via the food chain and insect pests all over the world are developing **resistance**. To fight them, poison quantities are being increased or new types of insecticide are being used, creating a vicious circle.

Discovered in Thuringia (German county), *Bacillus thuringiensis* has proved its worth over the years in killing caterpillars. It is sprayed onto the fields and eaten by the caterpillars as they feed. At first, the microbes only form mildly poisonous crystalline protein (Fig. 7.40), which changes into poison in the caterpillar's bowel and perforates the bowel. The trophi are paralyzed at the same time, killing the caterpillar. *Bacillus thuringiensis* is bred in bioreactors on a large scale and a metric ton of the microbe product **Bt-Toxin** (Fig. 7.42) is sufficient to clear 740 acres of forest, fields of beet, cotton, or fruit plantations of insect pests.

New strains of bacteria are effective against the larvae of the **Colorado beetle** (Fig. 7.42). Forest pests such as the green oak moth and various types of spider can also be targeted and destroyed, as can house flies and the caterpillars of the brown tail moth, without killing other insects and bees.

The subspecies *israelensis* of *Bacillus thuringiensis* has proved its worth against mosquitoes and gnats. In the past, such use of insecticides often caused severe harm to the whole animal kingdom, but in German garden centers, it is already possible to buy a product for the garden pond derived from *B. thuringiensis* var. *israelensis*. It works specifically against mosquito larvae and not against beneficial insects such as bees (Fig. 7.44). It is therefore completely harmless to humans, fish and warm-blooded creatures. In spite of all the successes, however, the use of Bt toxin is low compared to other insecticides and is limited to specific cultures. It is rapidly degraded by UV rays and does not usually reach the roots of plants or insects in the plant stems.

The next logical step is to **isolate the Bt gene**, reproduce it and insert it into plant cultures with *Agrobacterium*. This would mean that only insects are affected that are parasites on the transgenic plant. Transgenic Bt corn (called "genetically engineered corn" by the critics, Box 7.10), resistant to the European corn borer, has been approved in the USA and tested in Germany under protest. Cotton, tobacco, tomatoes, potatoes, and poplars have been provided with Bt genes (Box 7.10).

Do insects not become resistant to Bt corn? This can occur if there is massive pressure of selection. However, no resistant creatures have yet been discovered.

It is not true that the creatures then migrate from the Bt fields to the fields without transgenic plants. Insects brought up on non-transgenic plants are not subject to the pressure of selection. Recessive (suppressed) insect resistance genes therefore do not have a selection advantage.

In the USA, farmers must cultivate fields with non-Bt plants next to Bt fields. Establishing such **refuge areas** is the most important measure in place there for avoiding the development of Bt-resistant pests and delaying their spread. These areas are intended to ensure that "normal", non-resistant pests survive to mate with resistant pests – if available – in the neighboring Bt corn field. With recessive resistance genes, however, mixed race offspring (with only one resistance allele) cannot survive in Bt corn. In order to ensure this, Bt plants should contain the Bt toxin in far higher quantities than are required to destroy non-resistant (and not homozygously resistant) insects.

It is easy for large farms in the USA to cope with losing so much land to refuges but it seems that enforcement of the regulations is somewhat indifferent. In Europe, however, this affects small alternative farmers who do not have any additional areas as refuges.

Another question under examination is whether herbicide-resistant genes can be passed on from crops to weeds? These wild plants would then be even more rampant (e.g. from transgenic rape seed to the related charlock mustard [wild mustard] and jointed charlock [wild radish]).

As a rule, genetically engineered plants cannot survive in nature, or at least, there has not been any evidence to refute that so far. They are grown where non-modified plants are also grown, and no new species have appeared with devastating consequences to the ecosystem like the giant hogweed in Europe or the rabbit in Australia.

And what about **pollen dispersal**? (Fig. 7.43) That has been intensively examined with regard to Bt corn. The results of experiments in Germany available so far clearly show that the share in the trial harvests rapidly declines the further it is from the Bt corn parcels.

The shares of GMO (genetically modified organisms) above the 0.9 % threshold value for definitive labeling are only found within a ten-meter strip immediately next to a Bt corn field. At greater distances, the GMO values are usually less than the 0.9 % threshold. Planting of early and late blooming corn species at the same time is unsuitable for preventing GMO introduction.

Conventional corn harvested in a ten-meter-wide strip around the outside of a Bt corn field should fall below the labeling threshold, as the GMO part must be greater than 0.9 %. Corn further away is not labeled because of its lower GMO component and can be used without limitation or restriction (Box 7.10).

Farmers who grow Bt corn should therefore position a separating strip of twenty meters around the areas with GMO corn, thus ruling out economic damage to neighboring operations as a result of cross breeding.

■ 7.16 Blue Carnations and Anti-Mush Tomatoes

The gene for the "glow-worm enzyme" **luciferase** was successfully introduced in cells of the tobacco plant (Fig. 7.45 shows the structure with the active center). If the luciferin substrate reached the plants in run-off water and was altered by luciferase with energy-rich ATP, the genetically modified plants actually began to glow greenish yellow! The luciferase gene is for scientists an easily recognizable marker that indicates which gene is "switched on" in which parts of the plant.

There are now almost no limits to what can be imagined. **Transgenic blue carnations** have been licensed by the Australian company Florigen to the Japanese whiskey company Suntory. To create them, a gene from petunias was transferred to the carnations (Fig. 7.46) and the petunia enzyme created a wonderful blue pigment in the carnation.

The main pigments in plants come from anthocyanins and yellow carotenoids, the three primary plant pigments being cyanidin, pelargonidin, and delphinidin. The enzyme dihydroflavinol reductase (DFR) modifies all colorless pigment precursors into colored products and any mutation in the DFR gene results in white flowers.

The delphinidin synthesis gene is missing in carnations. Florigen transferred the delphinidin gene from blue petunias to carnations and combined it with the mutated carnation DFR gene. The carnations were therefore colorless and were now slightly combined with the blue of the petunias, resulting in the blue carnations *Moondust* and *Moonglow, which* are a hit on the Japanese and American markets. However, the dream target is the blue rose (Fig. 7.47), which we should be able to create with RNA interference (**RNAi**, Ch. 9).

Genetically engineered foods are at the moment still rejected by the majority of Europeans, but when it comes to health, there is another group that approves of biotechnology, and there are hardly any European diabetics who refuse insulin that has been created through genetic engineering.

Tomatoes are the favorite object of plant breeders and users!

Transgenic tomatoes ripen more slowly, have a thicker skin and contain 20 % more starch, which pleases the ketchup producers. These "*Flavr Savr*" tomatoes are not ripe until the user requires them and they rot less quickly (Fig.s 7.48 and 7.49).

The idea came from the Californian company Calgene, Inc. Until now, tomatoes have been harvested while they are still green, to avoid them ripening and going soft before they reach the market. They are then treated with ethylene gas to turn them red but they do not really develop any flavor, but with genetically engineered tomatoes the fruit stays on the plant until it is ripe.

The gene for the enzyme **polygalacturonidase** (PG) is responsible for softening the cell wall of the tomato and rendering it defenseless against bacteria, breaking down pectin (Chapter 2). After all, the point of fruit is to distribute the seeds it contains as quickly as possible.

The gene has now been "turned off" and we have *Flavr Savr* ™ tomatoes. How do you switch off a gene? We use **antisense technology** which is a

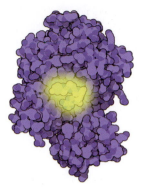

Fig. 7.45 The enzyme luciferase obtained from glow worms was the first biomarker to make the expression of cell genes visible.

Fig. 7.46 Transgenic blue carnations as a result of gene transfer from petunias (top right).
Below: the result of the gene transfer.

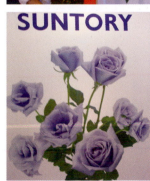

Fig. 7.47 Suntory's first green biotechnology result (top) and a vision of blue roses (below).

Fig. 7.48 A favorite object of both Green Biotechnology and the consumer: the tomato.
Top: Calgene poster trying to get people to understand the *Flavr Savr*™ tomato technology.
Bottom: no one in China wants to know whether these tomatoes have been genetically modified or not.

Fig. 7.49 Original protests against genetically modified food (by Greenpeace and Attac) with giant eight meter "tomato".

very important new technique, hence the explanation in the polygalacturonidase (PG) example. The gene (in the DNA double helix) for PG is localized in the cell nucleus by polymerase in single-strand messenger RNA for PG (PG mRNA), then reaches the ribosome through the cytoplasm and is expressed there. The relevant amino acids are tied to the PG protein, which is converted to the active PG enzyme.

How can that be prevented? A synthetic anti-PG gene is inserted into the DNA in the cell nucleus and this is changed into anti-PG mRNA during copying. This is complementary (or a mirror image) to the PG mRNA.

What happens now if both meet on the way to the ribosome? Both are single strands and bind to form a double strand. However, a double strand will no longer be accepted by the ribosome and will be destroyed by the cell as "foreign DNA". The production of PG does not therefore occur.

The same antisense technique is increasingly used for medicinal products as well: This has given rise to a new class of biotech products. Isis Pharmaceuticals (Carlsbad, California) received approval from the Food and Drug Administration (FDA) in 1998 for the first antisense drug Vitravene® with the active substance formivirsen. As a result, cytomegaly retinitis, a disease of the retina in AIDS patients, can be treated for the first time, whereas, in the past, it inevitably led to blindness. Drugs against cancer and inflammation are also now being developed on the antisense basis (Chapter 9).

■ 7.17 Danger from Genetically Modified Food?

The concepts of "genetically modified tomatoes", "genetically modified corn", or "genetically modified food" are all misleading.

All plants eaten by humans or animals naturally contain genes or DNA. Everyone absorbs about a gram of DNA every day with their "normal" groceries, no matter whether this DNA comes from animals, plants, or bacteria: These do not change people. Genes or DNA are just biological information. As a food component they are harmless chemicals that are quickly broken down in the stomach or bowel.

The **Flavr Savr tomato** was the first approved genetically modified food product in history. The

properties, not the production methods, were tested for user safety in accordance with the regulations of 1992 by the Food and Drug Administration (FDA).

The fact that a **gene for antibiotic resistance** was used was therefore not a safety risk for the FDA, which thought that the Calgene tomato did not differ significantly from a non-manipulated tomato and therefore had the essential characteristics of a normal tomato and were therefore as safe as a normal tomato.

In the end, the genetically modified tomato was not successful. It achieved "little market penetration". The source species was not of particularly high quality and tomatoes ripened on the vine were damaged by the harvesting machinery, which was designed for unripe fruit.

What happens though if a transgenic product now actually tastes better than the former products, contains less pesticide and more active substances such as vitamins and oils, and also costs less?

Transgenic potatoes (Fig. 7.50) with 25% higher amylopectin content (branched starch) have already been created. Amylopectin is more suited to processing than unbranched amylose.

A new type of potato has been developed at the Max Planck Institute for Molecular Plant Physiology in Golm (Brandenburg). Thanks to the introduction of two genes, it forms in its tubers not just starch but the special carbohydrate **fructane** (also known as inulin). This type of roughage is abundant in artichokes and chicory and has been attributed with a health-enhancing effect.

The fructane potato is indeed still a long way from being introduced onto the market, but it is an interesting model organism for safety research.

The basic element is fructose, which is associated with long-chain molecules. These bonds between fructose units cannot be broken by human digestive enzymes.

The consequence is that unlike plant starches, fructane passes through the stomach and bowel unchanged. In the large bowel, it stimulates the growth of certain useful bacteria. Fructane is therefore roughage with a probiotic effect; i.e., it improves the bowel flora and promotes "good" bacteria at the cost of "bad" ones. That means not just good digestion and an increased uptake of certain minerals: some studies have even found evidence of improved serum lipids (less cholesterol) and a lower risk of bowel cancer.

In the meantime, various food products enriched with fructane or other oligosaccharides are having great market success. In the main, things like yoghurts and milky desserts are combined with fructane additives (*prebiotics*) with certain "healthy" lactic acid bacteria so that their positive influence is enhanced.

After the Golm Max Planck Institute succeeded in finding the gene for the two crucial biosynthetic enzymes for forming fructane in the artichoke genome and transferred it into the potato genome, the potato tubers were actually producing fructane in the long-chain, particularly digestion-stable variation, which is typical of artichokes. The Golm potatoes may well consist of up to 5% fructane (in relation to dry weight), but it will still take several years until consumers can buy the new type of potato to balance the low supply of inulin in their daily diet.

Transgenic rapeseed and soya varieties with a higher content of the essential amino acid lysine have a higher nutritional value. Transgenic rapeseed with a changed spectrum of fatty acids is becoming a competitor for oil palms because lauric acid (C12), which is an important raw material for the production of cold water-soluble tensides, only occurs in coconut and palm kernel oils but now the transgenic rapeseed oil also contains lauric acid. Malaysia is the main producer of palm oil, and sees this as a real threat. Should rapeseed be grown instead of palms?

Virus-resistant sugar beets resist BNYVV (*beet necrotic yellow vein virus*) (Fig. 7.51), which attacks leaves and reduces the sugar yield. Resistance is introduced by inserting into the plant the genetic information for a viral coating protein so that the plant simulates infection with the virus.

Work is being carried out on plants resistant to *Phytophtora infestans*, the pathogen for **potato blight**, which led to the famines in Ireland in the 19th century. *Phytophtora* is the oomycete which forced the Irish ancestors of John F. Kennedy to emigrate. The irony of the story is that without "potato blight" Kennedy would never have become US President. Transgenic fungus-resistant plants produce cellulase which attacks and dissolves the cell walls of the fungus.

Genetically modified **grapevines** were released in 1999 in Franconia and the Palatinate in Germany. The aim of the genetic engineer is to eventually find an effective strategy against harmful fungal pathogens. These are a big problem in vine growing, particularly for the traditional varieties such as Riesling, Merlot, or Chardonnay.

■ 7.18 Should Genetically Modified Food be Labeled?

Must genetically modified food be labeled? **Oliver Kayser** of the University of Groningen (Netherlands) urges differentiation in the discussion: "No-one is afraid of getting a red head if they bite into a tomato, although of course they are consuming the full tomato genome at the same time. [...] In principle, human cells understand the information for the red color as well, but it cannot be produced because the pre-wired control elements do not function in humans."

So even if the "mush" gene of the tomato has been manipulated, humans do not have any tomato cell walls. However, there should of course be clear labeling of complex mixtures of recombinant products. Tomato ketchup contains both foreign DNA and the proteins coded by this DNA, so it should be labeled. This was unclear, for example, in highly refined sugar from root beard-resistant sugar beet. People thought sugar is sugar. New EU regulations have therefore come into force since 2005 (Box 7.4 the latest information at www.transgen.de). They are not all logical, but they are steps in the right direction.

■ 7.19 Gene Pharming

Genetically engineered hormones such as insulin and growth factors show that biotechnology is indispensable (Chapters 3 and 9). The need for such hormones is increasing tremendously fast, leading to a new dilemma, as we do not have sufficient production capacity.

The "poor" *E. coli* bacteria and mammalian cells will be totally overtaxed in the future. Experts such as **Jörg Knäblein** (Schering AG, Berlin) are therefore firmly convinced that **transgenic plants will be the answer** (Box 7.8).

Their main advantage is their cost. Producing plants is 10 to 50 times cheaper than using *E. coli* and up to 100 times cheaper than mammalian cells. A greenhouse costs ten US dollars per m² compared to

Fig. 7.50 Potatoes (top: with white flowers) produce fructosan. Center: preparation of harvested transgenic potatoes for animal feed trials.

Fig. 7.51 Grapes can be protected from fungi and beet from leaf viruses.

Box 7.8 The Expert's View: Plant expression system – a "mature" technology platform

One out of four new drugs today is a biopharmaceutical product whose active substances have been produced in a bioreactor, whether it be bacteria, beer yeasts, insects, or hamster cells, animals or plants. The market sector for these active substances is growing apace, but microbial fermentation is limited.

Profits of eight billion US dollars are expected in 2008 for antibodies alone. Ten monoclonal antibodies (see Chapter 5) cover more than 75% of industrial production capacity. 60 new monoclonal antibodies should reach the market in the next few years. There is a total of 1200 products "in the pipeline" based on proteins. A market of 100 billion US dollars is predicted for 2010. At present, transgenic plants are more frequently under discussion than bioreactors.

In a SWOT analysis (*Strengths, Weaknesses, Opportunities, Threats*), it is clear why transgenic plants are so well suited to being an expression system for the production of biopharmaceuticals. These drugs can be produced cheaply and comparatively quickly in large quantities.

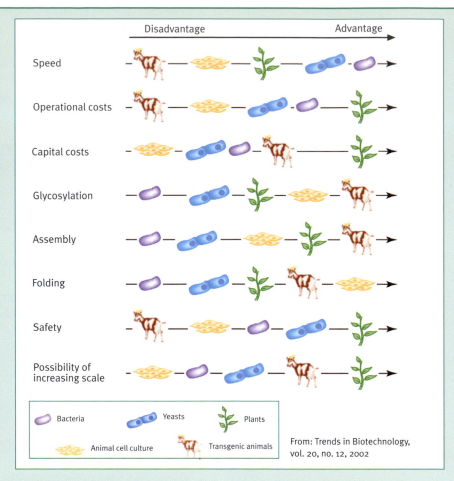

From: Trends in Biotechnology, vol. 20, no. 12, 2002

Strengths :

- development of additional production capacity
- high production and high protein yield
- short development time to protein
- safety advantages: no human pathogens are involved.
- stable cell lines: they have high genetic stability.
- simple medium: water, minerals, and light
- simple processing (e.g., ion exchanger)

Weaknesses :

- none of the drugs have yet been approved (Phase III).

- the authorities have not yet provided any final guidelines

Opportunities :

- lower costs and higher profit margins are to be expected
- prevention of production bottlenecks
- protein glycosylation similar to humans (in contrast to bacteria)

Threats

- contamination of the food chain
- escape of DNA

The "media" (basically water and nutrient salts) are cheap and free of germs (pathogens, viruses, prions, etc.) and the plants are very robust in relation to cultivation conditions. Transgenic moss, for example, produces the required protein in large quantities at temperatures of 15 °C to 25 °C and pH values of 4 to 8.

Schering's partner Icon Genetics in Halle, Germany, cultivates genetically modified tobacco plants for the production of three biopharmaceuticals that have until now been produced in bacteria or cell cultures. The first protein involves an active substance already on the market, the second should gain approval as a treatment for Crohn's disease, and the third is about to go into

Fig. 7.52 Transgenic tomatoes would have to be clearly recognizable with e.g., blue coloring.

1000 US dollars per m² for animal cell cultures. A 10,000 liter bioreactor for bacteria costs 250,000 to 500,000 US dollars and five years until it is up and running (Box 7.8).

Transgenic animals (Chapter 8) are certainly good at producing drugs in milk, but animal protectionists have severe reservations, and the latest setbacks make an expansion of the technology appear increasingly improbable. In contrast, there are no ethical or emotional problems with plants, so drugs

can be grown in pharmaceutical plantations. Plant cells have yet another crucial advantage:

The complicated proteins of higher living creatures continue to be modified in the cell after production. For example, complex sugars can be attached, without which many proteins are not active (Chapter 4). Bacteria cannot do this, so what is to be done?

So-called **plantibodies**, antibodies produced in plants, can already be produced in large quantities.

clinical development. The viral expression system of Icon Genetics is represented in the illustration and shows very clearly how the virus has infected the majority of leaf cells after only a few days. These cells appear green under UV light because they express green fluorescent protein (GFP, see Chapter 8).

Biopharmaceuticals are often too large and complex for chemical synthesis. In most cases, we are dealing with long-chain protein molecules that must be correctly folded and filled with sugars and other molecules in order to be effective as antibodies, enzymes, messenger substances or vaccines.

The Director of Chemical Microbiology at Schering believes that the great charm of moss, corn, and tobacco is that they have a more lavish synthetic apparatus than bacteria, production would be unbeatably good, and the drugs produced in this way would be guaranteed to be free of animal germs.

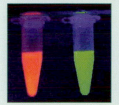

GFP (green) and DsRED (red) are both fluorescent proteins that can be used for controlling gene expression.

It is also true that with plants, the devil is in the detail. Plants definitely have other sugar preferences when it comes to the synthesis of proteins. Genetically-modified plants also conceal the basic danger that the changed genes will inadvertently be transferred to other plants.

Researchers at Icon Genetics have found an elegant solution to these problems. Gene distribution is brought to a halt by for example not permanently building into the plant genotype the growing instructions for the medicinal products. A happy side effect is that the production of active substances therefore runs at top speed.

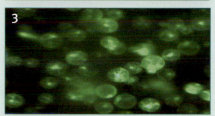
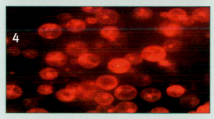

Experimental viral expression system: GFP (green) and DsRED (red) are both fluorescent proteins that are expressed together.
DsRED is a recently cloned fluorescent protein responsible for the red coloration around the oral disk of the coral *Discosoma*. DsRED has attracted tremendous interest as a potential gene expression marker.

The company's innovative procedure is directed at precision, speed, controlled expression, and safety in the processing of transgenes introduced into plants.

The magnICON System is suitable for rapid gene/vector assembly and optimization as well as for high throughput plant expression with optimum yield. ICON also offers the lexICON System which enables the specific marking of genetically modified plants ("DNA barcode").

The red fluorescent protein (DsRED) from the Indo-Pacific sea anemone *Discosoma* species and the green fluorescent protein (GFP) (see Chapter 8) are ideal markers for gene expression.

The expression is shown in the illustration. In Picture 1, you can see the plant just after viral infection and in Picture 2, after about a week. Under UV light, you can see that almost all the leaf cells are producing GFP. Pictures 3 and 4 show the same section. Under the microscope, you can see that the same cells express both GFP (green) and DsRED (red).

This means that a cell can be doubly transformed. We could, for example, produce both the light and heavy chains of an antibody in one cell. Such cells would also be able to functionally express a complete antibody consisting of light and heavy chains.

My expectations for the future are therefore positive. Assuming that the clinical trials also go to plan, the drugs of today can in the not too distant future be at least partially replaced by identical active substances that originate from plants.

Dr Jörg Knäblein is Head of Microbial Chemistry at Bayer Schering AG in Berlin.

Antibodies have already been produced in transgenic tobacco against the adhesin of the caries bacterium (*Streptococcus mutans*) (Fig. 7.53). Vaccine production in potatoes and bananas has already been successfully tested. Potatoes with the protein of some *E. coli* strains that lead to diarrhea (Enterotoxin B) have been ingested and have effectively protected human volunteers from the troublesome symptoms that accompany bacterial infections. Just imagine eating an **"anti-caries tomato"** (Fig.

7.52). But this immediately poses the question of safety (dosage). The plants would have to be conspicuously labeled; for example, the tomatoes would have to be colorless or have another color. What about blue?

7.20 Transgenic Plants – a Heated Debate

The discussion about transgenic plants and genetically manipulated food goes on. The main argu-

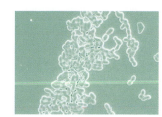

Fig. 7.53 The establishment of caries bacteria (*Streptococcus mutans*) could in the future be prevented with antibodies from transgenic tomatoes or bananas.

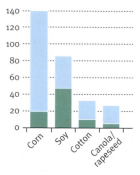

Fig. 7.54 The transgenic plant section (green) is growing continuously (in millions of metric tons).

Fig. 7.55 Genes from the daffodil (top) and the bacterium *Erwinia* were transferred to rice to create "Golden Rice" with a high provitamin A content. Its creator was Ingo Potrykus (bottom) (see Box 7.5).

Fig. 7.56 Canola/rapeseed is grown increasingly in Germany. With transgenic canola/rapeseed, PHB should be able to be produced. However, the technical problem is the processing of PHBs from the plant.

ments against it are of a rather political nature and difficult to refute. It is largely the interests of the large companies and big farmers that are served, making small farmers even more dependent, undermining renewable and alternative agriculture. However, this is countered by the fact that both large and small farmers, even in developing countries, decided in favor of the cultivation of these plants, because they promised economical advantages in the end in spite of the higher cost of seeds.

Polyhydroxybutyrate (PHB) – a biodegradable polymer (see Chapter 6) is a truly environmentally friendly product produced by transgenic *Arabidopsis thaliana* in rapeseed. This bioplastic material completely disappears from the environment after a certain period of time, because it is eaten by bacteria.

In the meantime, biosynthetic genes of *Ralstonia eutropha* have been expressed in thale cress (mouse-eared cress) *Arabidopsis thaliana* – a favorite subject for plant geneticists – and are making PHB production increasingly cost effective and attractive.

Almost 2.4 billion people live mainly on rice. If you eat rice almost exclusively, you risk severe symptoms of deficiency because vitamin A, iron, many trace elements and important proteins are missing from the white kernels. Approximately 800 billion people around the world suffer from acute **vitamin A deficiency**. Their vision, their immune system, blood formation, and skeletal growth become weaker. Scientists estimate that this is the cause of two million deaths a year and that half a million children per year go blind because of vitamin A deficiency. There are also widespread symptoms resulting from iron deficiency: for 1.8 billion women, especially in the third world, anemia is inevitable.

Ingo Potrykus (Fig. 7.55 and Box 7.5) of the Swiss University of Technology in Zürich and his German colleague **Peter Beyer** (Freiburg University) have succeeded in developing "**golden rice**". Both researchers handed over their results to the International Rice Research Institute (IRRI) in Manila for evaluation free of charge. The European public was very enthusiastic about the "golden rice". The shiny gold kernels contain a high concentration of carotene (provitamin A), a preliminary stage of vitamin A. The genes transferred for this purpose are those of the daffodil (*Narcissus pseudonarcissus*) and a bacterium, *Erwinia uredovora* (*Erwinia* is familiar to us from vitamin C synthesis in Chapter 4).

The company Zeneca had announced it would give seeds for this "golden rice" free of charge to small farmers in developing countries – a unique situation in the field of genetic engineering. Critics declared that this was a superficial solution, but it gave rise to an unexpectedly intense debate, which continues to this day.

A research team at the biotechnology company Syngenta has now developed a new variation on the "golden rice". It contains 23 times more β-carotene than the initially developed variation.

The scientists working with **Rachel Drake** of the British Jealott's Hill International Research Centre in Bracknell added a corn gene to the rice genome in 2005. With the help of the protein encoded within it, the rice plants produce provitamin A. "Golden rice II" could cover half the daily requirements of vitamin A, say researchers.

Other projects deserving mention here include caffeine-free coffee (since 20% of people all over the world drink decaffeinated coffee), poplar trees that collect heavy metals, and cultivated plants that are drought- and salt-resistant.

■ 7.21 Tropical Palms in Snow and Ice?

Will we in future be able to leave frost-sensitive tropical palms standing in the garden in the winter?

At 06:45 on the 24th of April 1987, the young scientist **Julie Lindemann**, was writing biotechnology history. She was tramping in a space suit across a Californian strawberry field spraying a concoction of **anti-frost bacteria**, genetically modified microbes which had been released with the approval of the state authorities for the first time. This event ended years of dispute over the release of genetically modified organisms into the environment. The history of anti-frost bacteria began in 1980 in a laboratory at the University of California at Berkeley.

The manner in which plants are damaged by frost had been known for a long time. Ice crystals form on the leaves and in the plant parts and these destroy living tissue. The new discovery was that bacteria play a key role in this. The organisms are only a thousandth of a millimeter in size but they often serve as crystallization nuclei for the ice crystals. Tap water freezes at 0 °C (32 °F) but purified distilled water can be cooled down to -15 °C (5 °F),

Box 7.9 Biotech History: Anti-frost bacteria – the story of their release

In April 1983, outdoor trials with anti-frost bacteria were given the green light by the health authorities in the USA whose advisory committees decide on genetic engineering experiments. However, they had not reckoned on the American public. Some residents went to court and criticized the fact that there had not been any comprehensive ecological studies. They asked who could guarantee that weeds and plant pests would not also benefit from the frost protection bacteria. They also asked whether the biological balance would be disturbed and pointed out that once microbes have been released into the environment, they cannot be recalled, nor could even climate changes be ruled out.

In May 1984, it was decided not to approve open-air trials with genetically modified lifeforms. There should be an ecological assessment of the balance between the benefits and risks for the environment. The researchers continued to experiment in the laboratory and the greenhouse and were able to refute some of the criticisms.

In the meantime, however, the University of California awarded a license to the related company Advanced Genetic Sciences (AGS). According to American law, the company was not affected by the legal ban, which only applied to government institutions such as universities. AGS planned to spray anti-frost bacteria on 2,400 strawberry plants in a field in Salinas and monitor them carefully for three months. If the bacteria should cross into the protection zone within a radius of 15 and 30 meters, they would be curbed with antibiotics.

The environmental authorities now intervened but approved the experiment in November 1985. The inhabitants of Salinas then protested. They rejected the field trials as too risky. Information also leaked out that AGS had already

secretly sprayed fruit trees on the flat roof of the company building with anti-frost bacteria before approval of the open-air trial. That made it into a real scandal. The environmental authorities withdrew the company's approval for an unlimited period of time and imposed a fine of 20,000 dollars.

The company now had to carry out additional experiments and work out further expertise. In one experiment, strawberries in the greenhouse were sprayed with various concentrations of Ice-minus bacteria and normal bacteria. None of the strains of bacteria subsequently showed any advantage, however. It also meant that uncontrolled distribution could also not be fully excluded.

The conditions for the US environmental authorities were finally fulfilled by AGS in 1987. Although some of the strawberry plants were taken by unknown sources shortly before the first test, in spite of modern safety assurances, the test was successful. **Julie Lindemann** tramped decoratively through the media. No manipulated microbes were found outside the 30 meter security zone. The manipulated bacteria had not even spread 15 meters.

Frost Technologies Corporation registered a mixture of three *Pseudomonas* strains ("Frostban B") with the EPA in 1992 for frost control. However, the EPA insisted on registering Frostban as a pesticide, since it inhibited bacteria. The cost seems to have brought the project to a standstill.

Bacteria serve as ice crystal nucleating germs.

In Europe, incidentally, genetically-modified viruses had already been released in September 1986, albeit in order to check the safety of genetic engineering experiments. Baculoviruses normally kill the caterpillars of the pine beauty moth (*Panolis flammea*), a forest pest. Oxford virologists inserted an 80 base-pair-long piece of DNA into the virus genotype, which served only to mark the virus genetically and did not change the metabolism. It was the fate of the microorganisms released in the field to be monitored further. Pest caterpillars were infected with "marked" viruses. The caterpillars and the hatched butterflies could not penetrate the close-meshed netting of the test field. If the experiments succeed, the gene for an additional insect poison and "suicide gene" will be introduced into baculoviruses to ensure that the viruses die after they have done their work.

In the future, will palm trees be growing in the graphic designer's garden in the winter?

as long as it does not contain any impurities to form nuclei for crystallization. One crystal-forming type of bacteria in particular is widespread in nature; this is *Pseudomonas syringae*. Biotechnologists **Steven Lindow** (Fig. 7. 57) and **Nikolas Panopoulos** carried out trials on specimens: by examining normal plants infected with *P. syringae* in a climatic chamber at temperatures below freezing point. At minus two degrees Celsius (28.4 degrees Fahrenheit), the first frost damage started to show, but plants where

the bacteria had been killed were able to tolerate -8 °C (17.6 °F) and even -10 °C (14 °F) without damage.

However, it is totally unrealistic to want to kill all bacteria on cultivated plants in the open countryside. Both researchers therefore sought the cause of the microbes changing into **frost bacteria**. They discovered that there was more to it than the size of the dust particles alone; there was a special protein

Fig. 7.57 Steven Lindow

Box 7.10 The Expert's View:
Crop Biotechnology in the United States – a Success Story

American growers continue to lead the farmers worldwide in the adoption of biotechnology-derived crops by planting 52 million hectares in 2006, the eleventh year of their commercial planting (James 2006). Benefits reaped from the technology have been the driving force in the expansion of the planted acreage from 2 million hectares in 1996 (first year of commercial planting) to 52 million hectares in 2006.

Corn (*Zea mays*).

American growers planted eight biotechnology-derived crops (alfalfa, canola, corn, cotton, papaya, soybean, squash, and sweet corn) and three traits (herbicide-resistance, insect-resistance, and virus-resistance) in 2005. Planted acreage was mainly concentrated in 13 different applications (herbicide-resistant alfalfa, canola, corn, cotton, and soybean; virus-resistant squash and papaya; three applications of insect-resistant corn, two applications of insect-resistant cotton, and insect-resistant sweet corn).

Trait information, trade names, and acreage planted to biotechnology-derived crops in 2005 is presented in the Table.

In general, **herbicide-resistant crops** were planted on a large scale compared with insect/virus-resistant crops. The rapid and widespread adoption of herbicide-resistant crops is mainly due to the ubiquitous nature of the weeds. The adoption of insect/virus-resistant crops varied each year based on the anticipated

level of target pest infestation. Adoption of these crops, Bt crops in particular, will continue to increase in future, as new varieties such as Vip-Cot cotton are commercialized and as more seed supplies are available for YieldGard Rootworm corn and WideStrike and Bollgard II cotton (see Table).

American experience from more than a decade-long planting of biotechnology-derived crops indicate that these crops have provided **simple, reliable, and flexible alternatives to traditional pest management choices, reduced the total amount of input costs in farming, and improved crop yields, all of which have translated to direct economic benefits to farmers. A recent report,** one of the annual updates to an initial study released in 2002 by the National Center for Food and Agricultural Policy, suggests that the widespread adoption of biotechnology-derived crops on 48 million hectares in 2005 **increased crop productivity** by 3.8 billion kilograms, **lowered crop production costs** by $1.4 billion, and **reduced the overall pesticide use** in production agriculture by 32 million kilograms (Sankula 2006). The numeric value of the benefits reaped from biotechnology-derived crops in 2005 was **improved grower net returns** of $2.0 billion.

The following discussion is focused on the pest management challenges encountered by growers in conventional crops and how biotechnology-derived crops offer solutions to address these challenges. Also, the impacts to growers and crop production of biotechnology-derived crops planted in the United States are discussed. Due to limited space, the following discussion focuses on only insect-resistant corn and cotton and virus-resistant papaya.

Corn insect pest problems

The most important insect pest problems in corn production in the United States are **corn borer, rootworm, armyworm, and cutworm**. European corn borer (ECB) and corn rootworm are the two most economically important insect pests of corn, costing growers billions of dollars each year in insecticides and lost crop yields. In fact, both ECB and corn rootworm are nicknamed "**billion dollar bug problems**" due to crop losses of at least one billion dollars each year from each of these insect pests.

Insecticide control of ECB is difficult for two reasons. First, European corn borer levels are difficult to predict and vary greatly from year to year. As a result, growers are usually reluctant to

incur costs on scouting to determine the feasibility and profitability of insecticide applications. Second, ECB control is complicated due to the feeding and survival behavior of the insect. Corn borer larvae feed in leaf whorls after hatching and eventually move into the stalks to pupate inside the stem burrows thereby avoiding insecticide applications. Insecticides need to be applied during the 2 to 3 day period between egg hatching and their burrowing in the stems. While carefully timed insecticide applications are the key for the successful control of ECB, control with the available insecticide options is only marginal to good.

The European corn borer (*Ostrinia nubilalis*) was introduced between 1910 and 1920 into North America from Europe: Butterfly and its eggs.

Similar to ECB, rootworm is an economically important and difficult pest problem in corn production. Excellent rootworm-control products have fallen by the wayside as corn rootworm has developed resistance to various insecticides. In addition to insecticide use, crop rotation is the most widely used cultural method to manage corn rootworms. Since a variant of the corn rootworm became the first pest ever to develop a way of foiling crop rotations, corn growers have been seeking a breakthrough in corn rootworm management. Biotechnology-derived Bt corn marked a new era for the management of both ECB and rootworm in the United States.

Cotton insect pest problems

The most damaging cotton pests are those that attack squares and bolls such as the cotton bollworm, tobacco budworm, pink bollworm, boll weevil, and lygus bugs. Yield losses due to bollworm/budworm complex, the most devastating of all the above pests, are typically higher in bloom stage. Without effective control, cotton bollworm and tobacco budworm can cause yield losses of 67%.

Biotechnology-derived crops planted in the United States in 2005

Papaya

Squash

Soybean

Canola

Corn

Cotton

Alfalfa

Trait	Crop	Resistance to	Trade name	Planted acreage as % of total
Virus-resistant	Papaya	Papaya ring spot virus	-	55
Virus-resistant	Squash	Cucumber mosaic virus Watermelon mosaic virus Zucchini yellows mosaic virus	-	12
Herbicide-resistant	Soybean	Glyphosate	Roundup Ready	88
Herbicide-resistant	Canola	Glyphosate Glufosinate	Roundup Ready Liberty Link	65 33
Herbicide-resistant	Corn	Glyphosate Glufosinate	Roundup Ready Liberty Link	31 4
Herbicide-resistant	Cotton	Glyphosate Glyphosate	Roundup Ready; Roundup Ready Flex Liberty Link	78 2
Herbicide-resistant	Alfalfa	Glyphosate	Roundup Ready	0.2
Insect-resistant	Corn	European corn borer/ Southwestern corn borer/ corn earworm	YieldGard Corn Borer	34
		European corn borer/ southwestern corn borer/ black cutworm/ fall armyworm/corn earworm	Herculex I	3-4
		Rootworm	YieldGard RW	4
Insect-resistant	Cotton	Bollworm/budworm	Bollgard	55
		Bollworm/budworm/	Bollgard II	2
		looper/armyworm	WideStrike	0.2
Insect-resistant	Sweet corn	European corn borer/ corn earworm	Attribute	Not available

Cotton is a major market for pesticide use in the United States. More than 90% of the entire cotton acreage in the United States is treated with insecticides. Insecticide costs for cotton bollworm, tobacco budworm, and pink boll-worm usually average around 60-70% of the total pesticide costs to American cotton growers. Moreover, bollworm/budworm complex has developed resistance to insecticides belonging to the classes of organophosphates, pyrethroids, and carbamates posing serious problems in cotton pest management. Biotechnology-derived insect-resistant cotton provided answers to the above challenges in managing cotton pests.

Insect-resistant corn

Insect-resistant crops or Bt crops were one of the first crops developed through biotechnology methods in the United States. These crops were developed to contain a gene from a soil bac-terium called *Bacillus thuringiensis*, and hence the name Bt (see lead text of the textbook, page 217 for details).

Although highly toxic to certain insects, **Bt is relatively harmless to humans** as the diges-tive enzymes that dissolve Cry protein crystals into their active form are absent in humans. Cry proteins from Bt have become an integral part of organic crop production in the United States for more than forty years in view of their safety and

See next page

effectiveness in controlling target insect pests. Three applications of insect-resistant corn were in commercial cultivation in the United States in 2005. They include Bt corn resistant to corn borer (trade names: YieldGard Corn Borer and Herculex I), Bt corn resistant to black cutworm and fall armyworm (trade name - Herculex I), and Bt corn resistant to rootworm (trade name YieldGard Rootworm). YieldGard Corn Borer has been on the market since 1996 while 2003 was the first year of commercialization for Herculex I and YieldGard Rootworm.

The corn borer destroys corn crops by burrowing into the stem, causing the plant to fall over. The larvae also thrive in the fruits.

Bt cotton

Three applications of Bt cotton are currently in commercial production in the United States: Bollgard I, Bollgard II, and Widestrike (see Table). The target pests for Bollgard I cotton, which express the Cry1Ac delta endotoxin, are tobacco budworm and pink bollworm. Bollgard II is the second generation of insect-protected cotton that offers enhanced protection against cotton bollworm, fall armyworm, beet armyworm, and soyabean looper while maintaining control of tobacco budworm and pink bollworm (similar to that provided by the Bollgard I).

Bollgard II contains two Bt genes, Cry1Ac and Cry2Ab, compared to the single gene in its predecessor, Bollgard I. The presence of two genes in Bollgard II provides cotton growers with a **broader spectrum of insect control**, enhanced control of certain pests, and increased defense against the development of insect resistance. The presence of the Cry2Ab gene in addition to the Cry1Ac in Bollgard II cotton provides a second, independent high insecticide dose against the key cotton pests. Therefore, Bollgard II is viewed as an important new element in the resistance management of cotton insect pests.

Using a strategy similar to Bollgard II, Dow Agrosciences developed 'WideStrike' cotton to simultaneously express two separate insecticidal Bt proteins, Cry1Ac and Cry1F. Similar to Bollgard II, the WideStrike cotton offers season-long protection against a broad spectrum of cotton pests such as cotton bollworm, tobacco budworm, pink bollworm, beet armyworm, fall armyworm, yellow-striped armyworm, cabbage looper, and soybean looper.

Impacts of insect-resistant corn and cotton

Bt crops provided high levels of protection against target insect pests, which is equal to, if not greater than the previously used conventional pest management options. Since the first planting of insect-resistant/Bt crops, growers noted that the most substantial impact has been improvement in crop yields. Unlike conventional insecticides, Bt crops offered in-built, season-long, and enhanced pest protection, which translated to gained yields.

Another significant impact of insect-resistant crops has been the **reduction in insecticide** use targeted against key pests because Bt crops eliminate the need for insecticide applications. Reduction in overall insecticide use and number of insecticide sprays has led to a reduction in overall input costs and improved returns for the adopters of Bt crops. Other benefits from Bt crops include reduced scouting needs, pesticide exposure to applicators, and energy use.

An indirect impact of Bt crops is the influence they exert on local target insect populations leading to an **overall reduction of insects** in a field. This effect is termed "halo effect" and has been noted with Bt corn and cotton. Volunteer crop plants have been reduced in the following season with Bt corn and cotton as dropped ears and bolls were significantly reduced.

By targeting specific insects through the naturally occurring protein in the plant, Bt crops reduced the need for and use of chemical insecticides. By eliminating chemical sprays, the **beneficial insects that naturally inhabit agricultural fields are maintained** and even provided a secondary level of pest control. **Local ecosystems** were impacted favorably since the planting of Bt crops as bird populations were reported to be higher in numbers in Bt cotton fields compared to conventional fields.

Papaya virus pest problems

Papaya ring spot virus is the most important disease that affects papaya. Papaya production in the United States, concentrated mainly in Hawaii, was declining in the 1990s due to epidemics of papaya ringspot virus. Hawaiian farmers relied on surveying and rouging the infected trees to keep the virus from spreading to other fields. This process of identification and destroying of infected trees turned out to be expensive and ineffective and led to a collapse of the papaya industry in Hawaii.

Biotechnology-derived virus-resistant crops were developed to express genes derived from the pathogenic virus itself. Plants that express these genes interfere with the basic life functions of the virus. Use of coat protein genes is the most common application of pathogen-derived resistance. The expression of the coat protein gene gives protection against infection of the virus from which the gene is derived, and possibly other viruses as well.

Two virus-resistant crops, papaya and squash, have been planted in the United States since 1998. Of these, biotechnology-derived papaya is a dramatic illustration of biotechnology success in the United States.

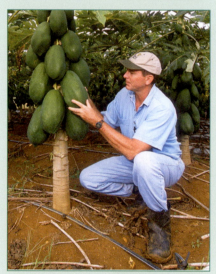

A specialist checking papaya trees for the Papaya ring spot virus.

Impacts of virus-resistant papaya

The papaya industry owes its continued existence in Hawaii to biotechnology. Virus-resistant papaya has facilitated strategic planting of conventional varieties in areas that were previously infested with the ringspot virus and also planting of conventional and biotechnology-derived varieties in close proximity to each other. Papaya production, which has fallen 45% from the early 1990s to 1998, rebounded to its original levels since 1999. Experts credit this increase in papaya production to planting of virus-resistant varieties. Biotechnology-derived papaya, overall, **has restored the economic viability of an industry that was on the verge of extinction.**

Conclusion

Every crop management decision has consequences, and the decision to plant biotechnology-derived crops is no exception. American growers have made the decision to choose biotechnology-derived crops because they realized clear and positive benefits from planting these crops. In addition to revolutionizing the way crops are produced, biotechnology provided best hope to growers by providing enhanced pest protection thereby improving yields with the use of minimal inputs. With that increased hope and confidence, American growers have increased planting of biotechnology-derived crops from 2 million hectares in 1996 to 52 million hectares in 2005. The fact that adoption of biotechnology-derived crops has continued to grow each year since their first introduction is testimony to the ability of these crops to deliver tangible positive impacts and to the optimistic future they hold.

Dr Sujatha Sankula has been the Director of Biotechnology Research Programs at the National Center for Food and Agricultural Policy, Washington, DC

Cited Literature:

James C (2006) *Global Area of Biotech Crops* (1996 to 2005) by Country.

Sankula S (2006) *Quantification of the Impacts on US Agriculture of Biotechnology* – Derived Crops Planted in (2005) Available at http://www.ncfap.org/whatwedo/biotech-us.php.

on the surface of the tiny organisms which stimulated the formation of ice crystals. If a section could be cut from a DNA strand of the frost bacteria, containing the command for formation of the frost protein, this would also destroy the capacity for forming ice crystals. Lindow and Panopoulos were able to use genetic engineering to achieve just that. The next thing they did was to populate some plants with the new **anti-frost bacteria**, to find that they did indeed protect their hosts from frost damage. Larger series of tests in the laboratory showed that spraying plants with an anti-frost bacteria liquid was sufficient, as the genetically modified microbes displaced the natural ones. It was possible to produce the spray liquid cheaply and in great quantities in bioreactors.

This opens up some attractive perspectives. For example, a number of cultivated plants that until now have only thrived in warmer regions can also be grown further north. Not least, frost damage could be limited. However, one crucial question remained open: how would the newly created microbes behave in the environment? (Box 7.9) In the end, the project was halted.

■ 7.22 Bacteria in Snow Guns Safeguard Skiing Holidays

In the meantime, however, managers at AGS had the bright idea of selling the natural frost bacteria mass-produced in bioreactors and then destroyed under the name "Snowmax" for the production of artificial snow.

The dead bacteria are added to water in snow guns and with "Snowmax", snow production increases by 45%. "Snowmax" also saves on the energy required for refrigeration. None of the authorities in the USA forbade the release of non-manipulated natural microbes.

In the meantime, the artificial snow business *made by Snowmax* is booming worldwide. The dead frost bacteria incidentally saved the Winter Olympics in 1988 in Calgary (Canada) during an unexpected warm spell.

By the beginning of the 1990s, studies from Switzerland had brought into question the profitability of many low-lying ski regions.

In view of climate changes, this situation was likely to get worse in the future. Only 44% of Swiss ski regions can be sure of getting snow in the future.

Fig. 7.58 The controversial transgenic corn (above).

In June 2005, the EU Committee failed to lift a ban on the planting of Bt corn varieties because of health reservations based on trials with rats. Less insecticide is generally needed when growing bt corn.

Middle: the caterpillars of the popular monarch butterfly (*Daunus plexippus*) were intensively tested for five years in the USA for damage caused by bt corn pollen.

Bottom: Under extreme laboratory conditions, there was damage to individual caterpillars but not to the whole population. However, all were affected by insecticides.

Fig. 7.59 Ernst Haeckel (1834–1919), was an eminent German biologist, naturalist, philosopher, physician, professor, and artist. Ernst Haeckel named thousands of new species, mapped a genealogical tree relating all lifeforms, and coined many terms in biology (phylum, phylogeny, ecology) and the kingdom Protista. Haeckel promoted Charles Darwin's work in Germany and developed the controversial "recapitulation theory" (see end of Chapter 8).

The published artwork of Haeckel includes over 100 detailed, multi-color illustrations of animals, plants, and sea creatures (Kunstformen der Natur, "Artforms of Nature"). The plates shown here are reprinted with kind permission of Dr Kurt Stüber of Max-Planck-Institut für Züchtungsforschung in Köln (Germany) (for complete collection, see *http://www.biolib.de/*)

In the United States, Mount Haeckel, a 13,418-ft summit in the Eastern Sierra Nevada, overlooking the Evolution Basin, and another Mount Haeckel, a 2,941-m (9,649-ft) summit in New Zealand, are named in honor of Ernst Haeckel, as is the asteroid 12323 Haeckel.

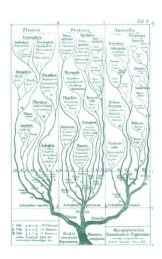

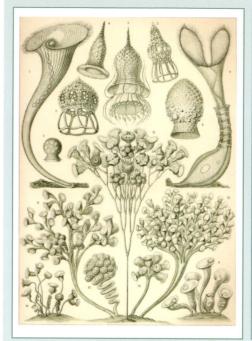

Worth Thinking About

But it is illusion to think that there is anything fragile about the life of the Earth; surely this is the toughest membrane imaginable in the universe, opaque to probability, impermeable to death.

We are the delicate part, transient and vulnerable as cilia. Nor is it a new thing for man to invent an existence that he imagines to be above the rest of life; this has been the consistent intellectual exertion down the millennia.

As illusion, it has never worked out to his satisfaction in the past, any more than it does today.

Man is imbedded in nature.

Lewis Thomas

Cited and recommended Literature

- **Thiemann WJ, Palladino** MA (2004) *Introduction to Biotechnology.* Benjamin Cummings, San Francisco

- **Barnum S R** (2006) *Biotechnology: An Introduction*, updated 2nd edn., with InfoTrac® Brooks Cole, Belmont
  Two recommended beginners' textbooks. Susan Barnum is author of a Box in this book.

- **Hopkins WG** (2006) *Plant Biotechnology* (*The Green World*) Chelsea House New York

- **Heldt H-W** (2004) *Plant Biochemistry* (3rd edn.) Academic Press, New York
  Recommended as a text in plant biochemistry, physiology and biotechnology courses

- **Rifkin J** (1998) *The Biotech Century*. Harcourt Boston:
  Jeremy Rifkin is the most prominent critic of GMOs. Worth reading.

Useful weblinks

- Wikipedia, for a start, for example algae:
  http://en.wikipedia.org/wiki/Alga

- AlgaeBase – a comprehensive database of over 35,000 algae, including seaweeds, with over 5000 images and some 40,000 references.
  http://www.algaebase.org/

- The European Union s website on GMO, the work of independent science journalists
  http://www.gmo-compass.org/eng/home

- Everything you wanted to know about GM organisms, provided by *New Scientist*.
  http://www.newscientist.com/channel/life/gm-food

- Internet portal providing up-to-date clear and intelligible information about current and past biosafety research into genetically modified plants
  http://www.gmo-safety.eu/en/

- Greenpeace about GMOs:
  http://www.greenpeace.org/international/footer/search?q=GMOs

- A fantastic collection of historic and modern biology books: Kurt Stüber's Online Library. Now over 100,000 scanned pages available:
  http://www.biolib.de/

Eight Self-Test Questions

1. Why don't famine regions "just" look after themselves and set up algae farms?

2. Why did the bright idea of producing protein from oil fail? Which bioproduct is a health food and a recent result of this Single Cell Protein research?

3. What is the name of the bacterium which causes root gall and how can it be used for gene transfer in plants?

4. Does the gene transfer in question 3 also apply to corn or rice? What is the way out?

5. Is the cloning of plants completely new? Were genetic engineers the first to discover it?

6. What is the principle that was used to create anti-mush tomatoes? Which enzymes are blocked by this?

7. Name three "well intentioned" transgenic plant products created by scientists.

8. What is the connection between frost damage in plants and infestations with bacteria? What are anti-frost bacteria?

EMBRYOS, CLONES AND TRANSGENIC ANIMALS

CHAPTER **8**

Fig. 8.1 Hippocrates (460 – 370 B.C.)

Fig. 8.2 Artificial insemination of dogs was possible already in Goya's (1746 – 1827) days.

Fig. 8.3 Artificial insemination in dogs. From top to bottom: A golden retriever bitch in Germany is artificially inseminated with semen from a male in Finland. The semen was deep frozen and sent by courier. The insemination process is monitored on screen. The happy mother and her puppies.

■ 8.1 Artificial Insemination

Artificial insemination (AI) of cattle has become standard practice in nearly all industrial countries and is a highly successful conventional method of propagating the traits of a few of the most precious breeding bulls.

Artificial insemination in **dog breeding** had already been described by **Lazzaro Spallanzani** (1729-1799, Ch. 2) around 1780 (Fig. 8.2).

Until this point scientists had a rather primitive understanding of conception, largely based on how plants grew. They speculated that the embryo was the product of male seed, nurtured in the soil of the female.

Spallanzani's experiment on dogs proved for the first time that there must be physical contact between the egg and sperm for an embryo to develop. With this new knowledge, Spallanzani experimented on frogs, fish, and other animals.

Most European dog breeding clubs require that a bitch be naturally inseminated before the clubs agree to artificial insemination. This is to prevent what generally happens to cattle and has been observed in dogs in the US: animals that never experienced natural copulation are unable to breed naturally later (Fig. 8.3).

By the mid-1940s, artificial insemination in cattle had become an established industry.

During the 1950s, semen storage methods in **liquid nitrogen** (-196°C or -320°F) were developed, which revolutionized animal breeding. It was now possible to send **frozen sperm** to other countries. It certainly makes a difference whether a hot-tempered bull or a pack of frozen bull semen (Figs. 8.11 to 8.13) is shipped across the Atlantic. Today, most dairy cows and pigs in industrial countries stem from artificial insemination.

It is a **cost-efficient method**, as the ejaculate of a stud bull "copulating" with a fake cow yields 400 portions of semen, each containing 20 million spermatozoa. One Artificial Insemination bull replaces approximately 1,000 traditional breeding bulls.

In the past 40 years, the **performance of dairy cows** increased dramatically, without any genetic engineering. In the 1950s, a dairy cow would produce 1,000 liters of milk per year, while today, 8,000 liters or more per year are normal.

■ 8.2 Embryo Transfer and *In Vitro* Fertilization

Artificial insemination makes it possible to exclusively **use semen of top-breeding bulls**. However, even the best breeding cow, when inseminated, takes nine months to produce one or sometimes two calves. Of course, the breeder would like to have more progeny from such a cow within a shorter space of time (Fig. 8.11).

This can be done thanks to hormones. By injecting gonadotropin, **superovulation**, i.e., the maturation of several eggs at the same time, is initiated. The eggs are artificially inseminated, and once embryos have developed, they are flushed out of the uterus using a catheter or recovered by ultrasound-guided follicle aspiration. Up to eight transferable embryos can be obtained in this way and **implanted into surrogate mothers**, resulting in four calves on average.

Just like sperm, **embryos can be deep-frozen** in liquid nitrogen for almost unlimited storage, and two thirds of them will be viable for embryo transfer.

Again, the stress and fuss of getting a premium cow and a prize bull together is thus avoided. Since the calf develops inside a surrogate mother, she passes on her **immune protection to the calf**. The calf will be born with the antibodies against local diseases in the blood.

In *in vitro* **fertilization**, egg and sperm are brought together outside the animal. The growing embryo is then implanted into surrogate mothers.

Thus, the genetic material of premium cattle can be passed on to a large number of offspring.

Having control over the gender of offspring is an idea that **Hippocrates** (460-370 B.C.) (Fig. 8.1) found quite attractive. Nowadays, it is possible to determine the gender of the embryos created through the **polymerase chain reaction** (PCR, Ch. 10). When bovine embryos have grown to the eight-cell stage, one cell is removed under the microscope, while the other cells continue to grow and can be implanted.

DNA is extracted from the separated single cell, and a PCR assay is used that replicates a specific region on the **Y-chromosome** (the male sex chromosome) millions of times (Fig. 8.53).

The DNA is dyed with ethidium bromide (Ch. 10) and shows up in electrophoresis. If the Y region shows, it is a male cell. If it does not, the calf must be female. The use of this technique in human **pre-implantation diagnosis** (**PID**) is highly controversial (see end of this Chapter!).

Dairy farmers can use PID to select female calves, while meat cattle farmers will go for males. It is already possible to buy gender-assorted cattle embryos online, ready to be implanted into surrogate mothers (Fig. 8.13).

■ 8.3 Animals Threatened With Extinction Could be Saved by Embryo Transfer

The Cincinnati Zoo in Ohio used Holstein cows as surrogate mothers for the threatened Malaysian gaur (*Bos gaurus*) in 1984.

In Kenya, Oryx antelopes were used as surrogate mothers to rebuild the bongo antelope population (*Tragelaphus euryceros*, Fig. 8.6). Deep frozen bongo embryos were flown to Kenya, and embryos of wild bongos from Mt. Kenya were flown to Oryx and to cow surrogate mothers in Ohio. Simultaneously, US bongo embryos were implanted into wild bongos in Kenya. Once they are born and released into the wild, they are tracked by behavioral scientists via a miniature radio transmitter.

The Audubon Institute in New Orleans has been very successful in breeding the African lynx or caracal (*Felis caracal* or *Caracal caracal*, Fig. 8.7), also seen on the cover of this book.

The list of threatened species is long. The first successful *in vitro* fertilization of dolphins in Hong Kong received a lot of attention (Fig. 8.5). Later in this chapter, we will describe in detail the much-

Fig. 8.5 Dolphins, highly intelligent marine mammals, can be reproduced through artificial fertilization.

Fig. 8.6 Bongo antelope (*Tragelaphus euryceros*).

Fig. 8.7 An African lynx or caracal (*Felis caracal*) was created through *in vitro* fertilization at the Audubon Institute in New Orleans. The embryo was frozen, thawed, and successfully implanted into a surrogate mother. The offspring, named Azalea, is hale and hearty.

Fig. 8.8 Goat-sheep chimera.

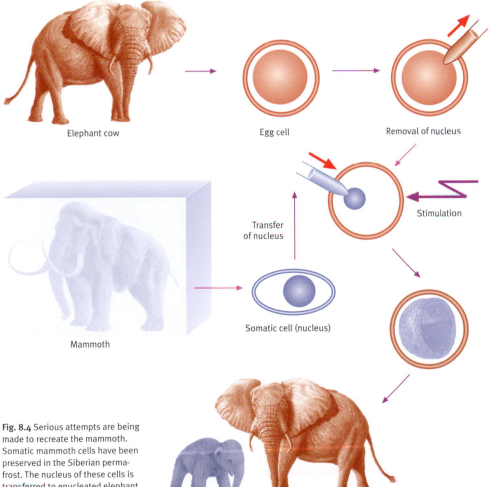

Elephant cow

Egg cell

Removal of nucleus

Transfer of nucleus

Stimulation

Somatic cell (nucleus)

Mammoth

Surrogate mother with mammoth calf

Fig. 8.4 Serious attempts are being made to recreate the mammoth. Somatic mammoth cells have been preserved in the Siberian permafrost. The nucleus of these cells is transferred to enucleated elephant egg cells. The embryo is then gestated by a surrogate mother.

Genome **Genetic Engineering**

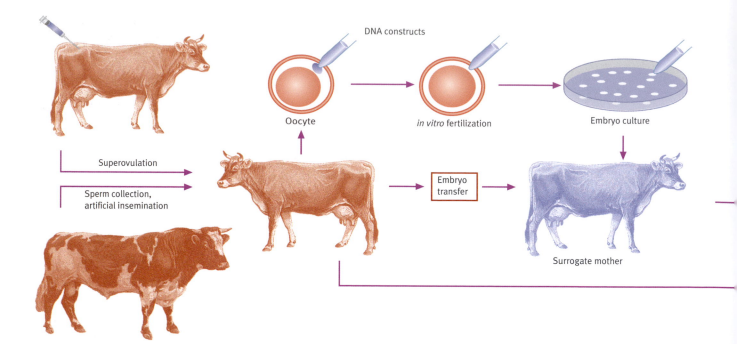

DNA constructs

Oocyte

in vitro fertilization

Embryo culture

Superovulation

Sperm collection,
artificial insemination

Embryo
transfer

Surrogate mother

Fig. 8.9 A qilin, mythical creature with dragon head, lion tail, ox hooves, and deer antlers, guarding the Summer Palace in Beijing.

Fig. 8.10 The Minotaur is a Greek mythological figure, a monster with a human body and a bull's head. King Minos of Crete kept the Minotaur confined in the labyrinth where the creature was defeated by Theseus.

debated animal cloning, i.e., the production of genetically-identical copies of individual animals.

■ 8.4 Chimeric Animals Have at Least Four Genetic Parents

Another technique is the production of chimeras (Latin *chimaera*, a flame-spewing creature consisting of parts of several creatures). These animals have at least four different genetic parents. The most famous chimera in Greek mythology is Minotaur on the island of Crete (Fig. 8.10).

Fusion chimeras (Fig. 8.16) are the result of the fusion of two embryos in the two- to-eight-cell stage. The embryonic membrane (*Zona pellucida*) is previously removed. The embryos must be at the same division stage.

Injection chimeras, by contrast (Fig. 8.18), are obtained by removing some cells (blastomeres) from a blastocyst and injecting them into the blastocyst cavity of another embryo. Within 24 hours, the injected blastomeres merge into a uniform cell complex. The created chimeras are ready for implantation into surrogate mother animals.

So far, the creation of chimeras in mice has been a way of exploring the terrains of embryology, cancer research, and evolutionary genetics. **Chimeras of sheep and goat** were also created (Fig. 8.8). Whether such chimeras, which cannot be created naturally, could one day become useful for farming, remains to be seen.

The most intriguing development seems to be the creation of injection chimeras from genetically engineered stem cells (see Ch. 9).

Cells from the inner cell mass of an embryo at the blastocyst stage are removed and grown in tissue culture. These are known as **embryonic stem cells** (**ES cells**) (Fig. 8.18).

Differentiated cells may divide (e.g., liver cells) or not (e.g., nerve cells), but it is impossible that they will convert into different types of cells. Nerve cells are considered to be terminally differentiated. It does not mean they have a short lifespan – on the contrary, they seem to function over years.

Stem cells are very different indeed. Embryonic stem cells taken from a blastocyst can turn into all types of cells. They are called **pluripotent**. They are not considered to be totipotent because they cannot develop into a whole new organism.

They multiply in a feeder layer of mouse fibroblasts. Foreign genes can be introduced into these cell lines by transfection. Almost all known DNA methods can be applied to ES cells.

The main advantage is that the cells into which the new gene has been inserted can be selected before they are injected into blastocysts (embryos). They can then be retransferred into blastocysts to create transgenic chimeras.

Cells up to the morula stage are totipotent (sometimes also called omnipotent), whereas a blas-

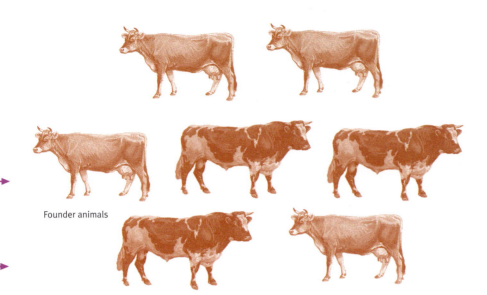

Fig. 8.11 Due to *in vitro* fertilization (IVF) and embryo transfer (ET), the reproduction rate of high-performance cattle has been boosted. In cows, hormones induce superovulation. The cows are then artificially inseminated (center) or fertilized *in vitro*. The embryos are then implanted into pseudopregnant surrogate mothers.

Founder animals

Fig. 8.12 Fertilization of an egg cell. Drawing by Ernst Haeckel in his book *Anthropogenie*.

Fig. 8.13 Fertilization centers offer bull sperm and embryos via the Internet.

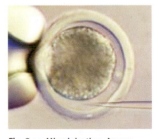

Fig. 8.14 Microinjection. An egg cell is taken up by vacuum pipette. A micromanipulator guides the injection needle to its target under the microscope.

Fig. 8.15 A cloned bull.

tocyst containing ES cells is probably only pluripotent. However, nothing that has been said on this matter is set in stone (more in Ch. 9).

This is a **highly controversial** topic because even embryos that produce ES cells for therapeutic cloning (see end of this chapter) could be misused for **reproductive cloning** and give rise to complete organisms. The genetically engineered cells are injected into intact mouse blastocysts where they mingle with the ICM cells and produce a chimeric animal that has four parents (Fig. 8.18). If, for example, the ES cells came from a brown mouse (the *agouti* gene that is expressed as brown fur color even if only a single copy is present) and are injected into the blastocyst of a white mouse, the color of the newborn mice indicate if they contain transformed ES cells.

■ 8.5 Transgenic Animals – From Giant Mouse to Giant Cow?

Just as in plants (Ch. 7), it is possible in animals to introduce DNA to insert new genes into cells or switch off existing genes. The new genes are passed on. Such **transgenic animals** are the most high-profile products of biotechnology.

This is done through a variety of methods:

- **Gene guns** (Ch. 7) shoot bullets carrying adsorbed DNA into cells
- In **retrovirus-mediated transgenesis**, mouse embryos at the eight-cell stage are infected with defective retroviruses which serve as vectors for the foreign genes. The virus material is defective to the extent that it does not produce viral envelopes or infectious viruses. However, the method is limited by the size of the gene to be transferred (eight kilobases of foreign DNA only). It was used to produce fluorescent piglets in Munich, using lentiviruses carrying the gene for green fluorescent protein (GFP) (Box 8.3).

- **Pronuclear microinjection**: When sperm and egg unite to produce a zygote, foreign DNA is injected directly into the pronucleus of sperm or egg (Fig. 8.14). No vector is needed.

- **Embryonic stem cell (ES) method**: Embryonic stem cells are taken from the inner cell mass of blastocysts and mixed with foreign DNA. Some ESs absorb the foreign DNA and are thus transformed. These cells are injected into the inner cell mass of a blastocyst. This method has now become the most relevant technique, although the resulting young animals are chimeras (see Sect. 8.4), i.e., only a proportion of cells carry the foreign gene. Only in the second generation will there be transgenic animals that carry the foreign genetic material in all their cells. Further crossbreeding results in homozygous transgenic lines.

- **Sperm-mediated transfer** uses linker proteins to bond DNA to sperm cells. They act as Trojan horses, introducing the DNA into the egg.

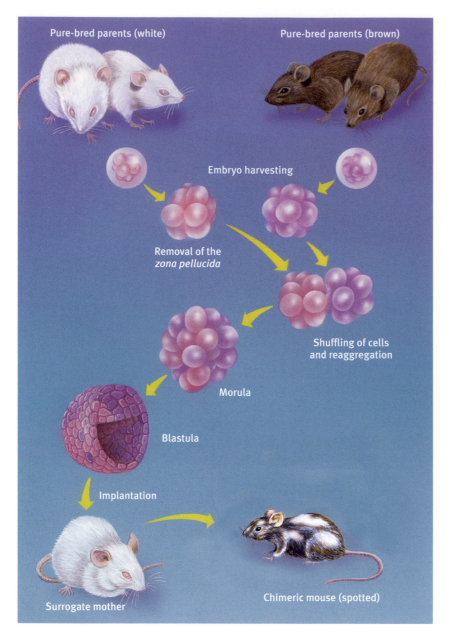

Fig. 8.16 The creation of fusion chimeras.

Fig. 8.17 Summer 2005: the latest success in cloning, i.e., the Afghan dog Snuppy (with donor, whose cells were taken from the ear). An apparently genuine success of South Korean researcher Hwang who became infamous for his fraudulent claims to have cloned human stem cells.

The **Giant mouse** (Box 8.1) created in 1982 by **Richard Palmiter** (University of Washington) and **Ralf Brinster** (University of Pennsylvania) provided an idea of things to come.

Of two mice, both two weeks old, one weighed 44 grams, the other 29 grams. Due to an **inserted rat growth hormone**, the transgenic giant baby mouse grew twice as fast as its non-transgenic siblings and ended up twice as big.

Physiologically, mice (*Mus musculus*) are surprisingly similar to humans, although pigs are still more closely related. As mice are easy to breed and keep, murine models (mouse models) are often used in medical research on human diseases.

It was an ingenious idea to inject the rat growth hormone gene into a non-differentiated mouse egg (Box 8.1 and Fig. 8.20). If successful, the gene is integrated into the genome of the embryo. The gene is linked to another gene and its promoter gene. Thus, the hormone, the production and secretion of which would normally be controlled in the brain, is produced in a different type of tissue, e.g., the liver, while the pituitary gland has no control. A resultant mouse became twice as heavy as untreated mice of the same age.

Its offspring also grew faster and became bigger, suggesting that the rat gene had been stably inserted into the mouse genome.

■ 8.6 Growth Hormones For Cows and Pigs

In cattle, size is not really the issue – who wants to build bigger cowsheds anyway – but milk quality, rate of growth, fertility, and disease resistance are important.

The **composition of milk** can be modified. A higher production of κ-casein, a phosphoprotein, is desirable for better cheese production. Millions of people with lactose intolerance would benefit from lactose-free milk and forget about indigestion (Ch. 2. So far, lactose has been enzymatically degraded by galactosidase).

Resistance to bacterial mastitis is another desirable trait. In the U.S. alone, mastitis causes annual losses of $2 billion in milk production. Fewer animals would need vaccination if they had this trait.

Fully grown dairy cattle have already been successfully treated with genetically engineered **bovine growth hormone** (recombinant bovine somatotropin, rBST). It led to a rise in milk production by 10 to 25 percent, while the cows were fed no more than 6 percent more feed – in other words, production increased while the financial input remained the same. Other researchers say more concentrated feed is required. What remains unclear is the long-term effect of hormone treatment.

Even without hormone treatment, many dairy cows seem to be struggling with health problems. Although the idea of burnt-out hormone cows may be exaggerated, the final verdict about this treatment is still not in. These methods are viewed with a good deal of skepticism, and there has been **embittered controversy** about the injection of growth hormone into cattle. This could soon come to an end when the growth hormone gene will be inserted into the animal genome instead of being injected, and **transgenic calves** will produce their own growth hormone.

The use of growth hormones makes more sense in **pig breeding** where they encourage growth of muscle rather than fat. A certain amount of fat is always desirable to give the meat some taste. However, injecting pigs with growth hormone on a daily base is simply not practical. The idea is to use slow-release medication that only needs to be repeated every few weeks.

What about **transgenic low-fat, muscle-packed pigs**? The relevant gene transfer was successful, but the results were far less impressive than in mice. It was disappointing to see that, although the muscles of transgenic pigs developed quickly and the fat content was low, the animals suffered from kidney and skin problems as well as inflamed joints. Transgenic pigs are often lame and are not particularly fertile, which is not surprising, given their general state of health.

Transgenic goats and sheep seem to be free from health problems. Experiments are being carried out to use their milk glands as bioreactors for the production of human proteins (Factor VIII and IX, plasminogen activator, Ch.9) – a kind of gene "pharming".

■ 8.7 Gene "Pharming" – Valuable Human Proteins in Milk and Eggs

Milking mice is serious business for biotechnologists. In normal mice, it is hardly worth the effort, whereas transgenic mice have been producing in their milk high concentrations of **human tissue plasminogen activator** (**t-PA**) (Fig. 8.20) as early as 1987. t-PA is an important agent that dissolves thrombi (blood clots) in heart attacks.

The basic idea was to inject the gene for human t-PA (Ch. 9) into fertilized mouse eggs, as had been successfully done with other mammalian genes. In some cases, the foreign hereditary material was stably inserted into the genome.

In order to give easy access to products like t-PA in milk (Fig. 8.20), the **promoter gene for acidic whey** protein, the most common protein in mouse

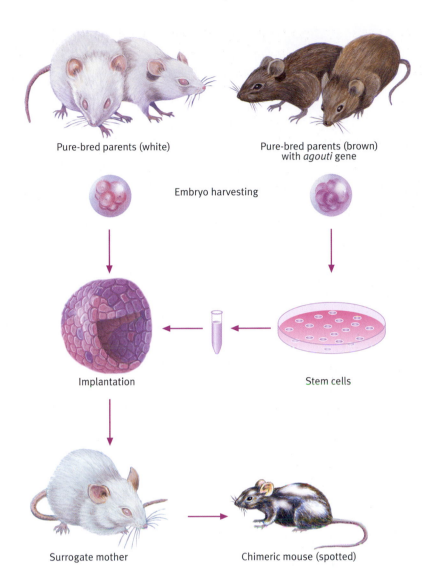

Pure-bred parents (white)

Pure-bred parents (brown) with *agouti* gene

Embryo harvesting

Implantation

Stem cells

Surrogate mother

Chimeric mouse (spotted)

milk, was combined with the isolated gene for human t-PA. As expected, human t-PA was found exclusively in the milk gland tissue of the transgenic mice and was excreted into their milk. No human t-PA was found in the blood of the mice (Fig. 8.20).

A lot of other scenarios are possible. Thus, **low-cost rodent farms** could replace expensive mammalian cell bioreactors to produce genetically engineered products. Every mouse that is milked for 15 minutes with an especially developed mouse-milking machine provides 4 milliliters of milk per day. Large colonies of such transgenic animals could be bred within a very short time.

In the late 1980s, the Dutch company GenePharming began to engineer **cattle embryos**. The aim was to create cows that would produce large quantities of the human protein **lactoferrin** in their milk. Lactoferrin was intended to become an addi-

Fig. 8.18 Injection chimeras, obtained using embryonic stem cells (ESC). ESCs can also carry genetically engineered material.

Fig. 8.19 Top: cloned goat. Transgenic cows can secrete human proteins into their milk.

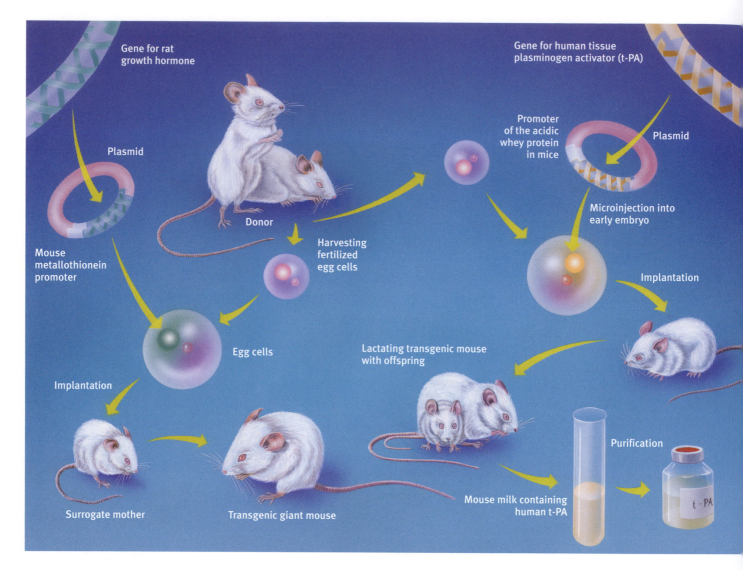

Gene for rat
growth hormone

Plasmid

Gene for human tissue
plasminogen activator (t-PA)

Promoter
of the acidic
whey protein
in mice

Plasmid

Microinjection into
early embryo

Mouse
metallothionein
promoter

Donor

Harvesting
fertilized
egg cells

Implantation

Implantation

Egg cells

Lactating transgenic mouse
with offspring

Surrogate mother

Transgenic giant mouse

Purification

Mouse milk containing
human t-PA

t-PA

Fig. 8.20 Left: Creation of the giant mouse. Right: Transgenic mice produce human tissue plasminogen activator in their milk.

Fig. 8.21 One liter (2.1 pts) of cow's milk weighs approximately 1030 grams and contains 3 to 8 grams of vitamins and minerals, 15 to 35 grams of protein, 50 to 70 grams of lactose, 40 grams of milk fat, and the rest is water. In future, the proteins contained in milk could be therapeutics for humans.

tive to formula milk. In a first step, 2400 embryos were genetically modified. 128 of these embryos developed further and were transplanted into normal cows. However, the transgenic animal that was born as planned in December 1990 was not female and thus unable to produce milk containing lactoferrin.

The male calf was called **Hermann**, and the Dutch government gave permission in March 1993 to use transgenic sperm to fertilize non-transgenic cows. Hermann's progeny were born from the end of 1993 onwards.

Out of 55 calves that were born, eight were female and transgenic. Three of them died, and the remaining five became sexually active in spring 1995. They were fertilized with non-transgenic sperm.

In March 1996, Hermann's daughters gave birth and began to produce milk. Four daughters secreted

human lactoferrin in their milk in concentrations between 0.3 to 2.8 grams per liter.

The lactoferrin produced by these cows was identical to the lactoferrin found in breastmilk. It improves the absorption of iron and protects against intestinal infections.

Goats have become even more desirable animals for genetic engineering, as they reproduce faster than cattle and are cheaper to keep. A blood-clotting factor has already been successfully produced in goat milk. A herd of 100 goats can produce $200 million worth of genetically modified protein.

Chickens lay on average 250 eggs per year. The egg white contains about three to four grams of protein. Even if only a fraction of this quantity consisted of therapeutic proteins, it would be well worth the effort. The production of **monoclonal antibodies** in eggs (Fig. 8.22) is already successful.

■ 8.8 Transgenic Fish – From GloFish to Giant Trout

The wise **Laotse** (Fig. 8.24) said: "Give a man a fish and you feed him for a day. Teach him how to fish and you feed him for a lifetime."

Fish consumption is rising at an incredible rate. Every year, around 80 million tons of fish are caught, and fish stock is increasingly under threat (Fig. 8.30) and, as a result, catches are declining.

Could **giant fish in fish farms** be the way ahead? Currently, approximately 40 million tons of fish are bred in aquaculture every year. Young rainbow trout were injected with genetically engineered salmon growth hormone. The fish reached twice their normal size. However, catching and injecting thousands of trout is difficult and expensive. Experiments with transgenic trout have begun.

The **genetic modification of fish** is much easier than of mammals. The eggs are fertilized outside the fish body, and the embryos develop not inside the mother, but at the bottom of the sea, lakes, or rivers. The eggs can be easily collected, and there is no need for implantation into surrogate mothers. There is also an abundance of eggs compared to other animal species, permitting several experiments to be run in parallel, which shortens research time considerably.

Since fish eggs are huge, the growth hormone can be easily injected. The transgenic Pacific trout are approximately ten times bigger than normal trout. Some were even 37 times larger, dwarfing wildtype trout (Fig. 8.28).

Trout breeders also aim for better looks as well as taste. In **salmon**, consumers like a pink color that used to come from small plankton crabs eaten by wild salmon. Expanding **aquaculture**, particularly in Norway, has brought about a massive salmon glut. Salmon has ceased to be gourmet food and has become an everyday item. Gene technology does not change the overall aims of breeding, but opens

Fig. 8.22 Human insulin, produced in the eggs of transgenic hens – a realistic dream.

Fig. 8.23 Zebrafish originate from the river Ganges in India and have become a favorite subject of biological research (See Box 8.2).

Box 8.1 **The Giant Mouse**

The gene for the growth hormone in mice (and men!) is usually switched on in the pituitary gland only. This is where the hormone is produced and where its release into the bloodstream is controlled. Thus, the growth process of young mice is intelligently controlled.

Ralf Brinster had the ingenious idea of producing growth hormone in mice outside the control of the pituitary gland, e.g., the liver. This could speed up growth considerably. The liver synthesizes a protein called metallothionein which protects it from heavy metals. The presence of zinc stimulates its production. Could the gene coding for this protein be used to introduce the growth hormone into liver cells?

When the DNA of two genes are combined with a gene that switches one of them on – in this case, the growth hormone gene and the metallothionein gene plus metallothionein promoter gene – and introduces them into an embryonic mouse, at least part of the growth hormone gene can be expected to be expressed in the liver and the production of growth hormone to begin.

The direct approach of introducing a new gene into a cell involves the injection of the DNA into the nucleus of an oocyte under the microscope, in the hope that it will be inserted into the genome. This works very well in mice.

First, the DNA containing the growth hormone, metallothionein and inducible promoter genes is prepared and inserted into a plasmid which, in turn, is introduced into bacteria. The bacteria are replicated by millions. The relevant DNA is cut out in large numbers using restriction nucleases. Nowadays, it is also possible to replicate the DNA using PCR, a technique not yet available when the giant mouse was created.

The embryos are viewed under the microscope with a special lens of 100x or 200x magnification which gives a high-contrast image of the cell structure. Eight to twelve hours after fertilization, the two circular pronuclei are clearly visible. The paternal pronucleus is larger than the maternal one.

Using an aspiration pipette, the zygote is gently held, while a microinjection needle introduces 50 to 500 gene copies into the male pronucleus. Not all embryos survive the procedure, but generally, 60–80% remain viable.

The prospective surrogate mothers are mated with sterile males to induce a false pregnancy, i.e., to activate the pregnancy hormone cycle although the mice do not carry an embryo.

The embryos are implanted into the surrogate mothers and develop into normal fetuses. After birth, the newly born mice spend another three weeks with their surrogate mothers until they are weaned.

Is the newborn mouse transgenic? Does it carry the injected DNA in its genome? Does the DNA replicate at each cell division during the growth of the embryo?

The answers lie in a tissue sample taken from the tail of the mouse. Usually, PCR will show that in around 27% of the offspring, the foreign DNA has been integrated and activated. It will be passed on to their progeny through their gametes.

When exposed to small amounts of zinc, the giant mouse went through amazing growth spurts, as the liver cells produced both metallothionein and rat growth hormone. Switching on the metallothionein gene had switched on both genes, and the liver had no control mechanism to regulate growth hormone secretion. Thus, the growth stimulus remained unabated.

Currently, 2,000 different strains of transgenic mice are available for research, facilitating the functional analysis of the human genome. Mice are good human disease models, e.g., for arthritis, Alzheimer's disease, and coronary disease.

Ralf Brinster and the giant mouse.

Box 8.2 Biotech History:
GloFish® – The First Transgenic Pet

Fish have been all the rage in the US recently – first in "Finding Nemo", and then real aquarium fish that glow an attractive bright red when exposed to fluorescent light. The first genetically modified pet was created.

Biotechnologists became interested in black light about 20 years ago when the firefly enzyme **luciferase** was introduced into cells of tobacco plants. When its substrate luciferin which had been added to the water in the watering can was taken up by the transgenic plants, they gave off a greenish-yellowy glow. This was done not to make tobacco harvesting at night possible or perhaps to add a gentle glow to Christmas trees, but it was an easily recognizable marker to show which genes had been switched on in which parts of the plant (see Ch. 7).

The **green fluorescent protein** (**GFP**) converts UV light into low-energy green light. Under normal electric light, a solution containing purified GFP looks yellow, but glows green in sunlight.

Whatever is coupled with GFP can thus be made visible, e.g., the journey of proteins through a cell or of viruses through the body. A GFP gene that has been introduced together with other genes can be seen wherever it is expressed.

The transgenic zebrafish created in Singapore, which were designed with the long-term intention to make them fluoresce if under stress or in the presence of heavy metals in the water. Right: The GFP molecule with its active fluorescent group.

Researchers in Singapore tried to engineer the black-and white zebrafish from the Indian river Ganges in such a way that the fish would glow red or green when under stress. Female sexual hormones (estrogens) or heavy metals in water can thus easily be detected.

In Taiwan and the US, this bright idea became big business. From November 2003, it has been possible to buy **GloFish**® for just five dollars in the United States. This rapid commercialization initially caused a lot of concern among the scientific community due to a perception that the fish were not adequately reviewed. However, this concern was almost entirely eliminated as it was

GloFish®, the first transgenic pet, glows red when exposed to black (UV) light.

learned that the U.S. Food & Drug Administration did review the fish. Although they decided not to formally regulate the fish, they did find the following:

"Because tropical aquarium fish are not used for food purposes, they pose no threat to the food supply. There is no evidence that these genetically engineered zebra danio fish pose any more threat to the environment than their unmodified counterparts which have long been widely sold in the United States."

Although GloFish® have been marketed without incident, and with broad public acceptance, other fish have been waiting to be licensed for years, e.g., **transgenic Pacific trout** (with additional growth hormone).

Even where gene technology is not involved, the release of a non-native species can have disastrous effects, as the example of the **Nile perch** (*Lates niloticus*) has shown. It was released into Lake Victoria with the best intentions, but has displaced nearly all local fish species, most importantly the famous Victoria cichlides, and driven them to extinction. And for all that, its food value is nothing to write home about. Short-sightedness and carelessness interfered with an ongoing evolution process, and, by introducing a foreign competitor, terminated it irrevocably.

Researchers at Purdue University (West Lafayette, USA) have been experimenting with transgenic fish for a long time and found that in a Japanese relative of the salmon called **medaka**, transgenic males have the edge over wildtype medakas, as they are able to fertilize eggs four times as frequently. They conclude from their computer simulations that it would take no more than fifty generations of interbreeding to wipe out the wildtype completely. It is important to note, however, that in this simulation, the transgenic medaka reaches a larger size than the wild-type fish. Since this does not apply to the growth-enhanced Atlantic Salmon, it is thought that it would not raise the same ecological concern.

According to Internet sources such as Wikipedia, GloFish® have continued to be successfully marketed throughout the United States. After more than three years of availability, there are no reports of any ecological concerns associated with their sale. In addition to the red fluorescent zebrafish, trademarked as "**Starfire Red**", Yorktown Technologies released a green fluorescent zebrafish and an orange fluorescent zebrafish in mid-2006. The new lines of fish are trademarked as "**Electric Green**" and "**Sunburst Orange**", and incorporate genes from sea coral (see running text, page 245). Despite the speculation of aquarium enthusiasts, it has been found GloFish are indeed fertile and will reproduce in a captive environment. However, the sale of any such offspring is restricted by a US Patent.

As of January 2007, the sale or possession of GloFish is illegal in California, on the basis of a regulation that restricts all genetically modified fish. The regulation was implemented before the marketing of GloFish, largely due to concern about a fast-growing biotech salmon. Due to the State's interpretation of the California Environmental Quality Act, Yorktown Technologies was informed by State attorneys that it would first need to complete an extremely expensive study, which could cost hundreds of thousands of dollars and take years to complete. According to the company's web site, they have thus far declined to undertake this study.

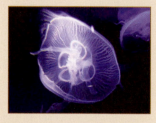

Luminescent jellyfish

As of January 2007, Canada also prohibits import or sale of the fish, due to what Government considers a lack of sufficient information to make a decision with regard to safety. Despite the relatively extreme cold of the Canadian climate for a tropical fish, it has been hypothesized that the fish could theoretically survive in certain warm springs. However, in light of the fact that there are no known populations of wild-type zebra fishes, this eventuality is seen as extremely unlikely by those familiar with the species.

The import, sale, and possession of these fish is not permitted within the European Union. In November 2006, however, it has been rumored that the Dutch Environmental Inspection found 1,400 fluorescent fish, which were sold in various aquarium shops.

up an opportunity to insert specific genes and thus achieve the desired changes much more quickly and with more precision. Even genes foreign to the species can be used.

To date, up to **35 fish species have been genetically modified** in various labs, mainly in the U.S., Canada, the United Kingdom, Norway, and Japan. The most important species has been Atlantic salmon, followed by trout and carp. Salmon is the most widely farmed fish species.

Sea fish such as **cod, turbot, and halibut** will be more widely farmed in the future and have been genetically modified. When the technology will be commercially viable is not clear yet. Intensive research has been dealing with the genetic modification of the freshwater fish *Tilapia* (Fig. 8.25). Originally from Africa, it is has become one of the most important farmed fishes in tropical countries and is also exported to Europe.

Currently, the commercial focus is on **increasing the size** of farmed fish (Fig. 8.26). Many projects are still in the first development stages and have a long way to go until transgenic fish is market-ready. It is a matter of finding suitable genes and developing suitable methods to insert them into the fish genome.

Initially, attempts were made to transfer **growth hormones** from various other species (other fish, rats, cattle, even humans) to speed up natural growth in fish. However, experiments showed that the modification of the species' own growth hormone genes was more reliable and successful. Thus, for example, if the Pacific salmon growth hormone is transferred into Atlantic salmon, the hormone is also secreted in winter, making the Atlantic salmon grow all year round, instead of just in spring and summer. A thus modified salmon reaches its final weight far more quickly (Fig. 8.28).

The yield of fish farms is often considerably reduced through disease, as the crammed conditions of cages and basins favor the spread of bacterial and viral pathogens. Genetically transferred **resistance to pathogens** is therefore desirable for the fish-breeding industry. One approach to make fish more resistant is to find the relevant resistance genes; another, to enhance the production of the antibacterial enzyme lysozyme (Ch. 2).

In 1974, a tank full of flounders froze at the University of Newfoundland. Most of the fish were frozen solid. They even had ice in their hearts. Young Assistant Professor **Choy L. Hew** (Fig. 8.29) was distraught, but then became very popular with his colleagues by giving the frozen fish away. Finally, when taking out the last dead fish, he suddenly discovered that some of them had survived the deep freeze.

Later, Hew discovered why they were able to survive – due to an **antifreeze protein**. Once the gene coding for the protein had been found, scientists tried to transfer it to salmon. It was hoped that it would be possible to breed salmon in waters at near freezing temperatures. Hew is now very successful in producing transgenic fish at the University of Singapore.

The transfer of antifreeze genes is supposed to enhance the cold-water tolerance of salmon or other farmed fish and would allow the fish to be farmed in regions where it had been so far impossible due to harsh winter temperatures.

Ecologists have general reservations about the production of transgenic fish. There is no guarantee that they will not escape from fish farms and spread uncontrollably. When the Nile perch (*Lates niloticus*) was introduced into Lake Victoria, it devastated the native cichlid population and half the native species became extinct.

The greatest fear associated with the intentional or unintentional release of transgenic fish into the wild is **crossbreeding with the wild population**. The escape of farmed fish into the open sea cannot be avoided. Recent research therefore explores the creation of sterile transgenic fish that will not be able to spread their genes.

It is far less likely for transgenic cattle and pigs to become wild in densely populated developing countries. They are simply caught and eaten. There are exceptions, however. On the remote Galapagos islands, wild domestic pigs ate the eggs of the famous tortoises that inspired Darwin. The only natural predators the tortoises had known until then were seagulls.

In Taiwan and the US, it costs only five dollars to buy **transgenic zebrafish** that are fluorescent under UV light. They are called GloFish® and have become the first transgenic pet. GloFish® carries the **Red Fluorescent Protein (DsRED) gene** from the coral-like sea anemone of the genus *Discosoma* in his genome (Box 8.2, see also Ch. 7, Box 7.8 for

Fig. 8.24 Laotse, also known as Li Er. He is supposed to have been an older contemporary of Kung fu tse (551–479 B.C.).

Fig. 8.25 *Tilapia* is a fish that has its home in the warm waters of Africa and the river Jordan. They are also known as St Peter's fish and are suitable for breeding in areas where protein is at a premium. They are not choosy and will feed on any organic matter. They reproduce rapidly, are disease-resistant, and produce valuable food.

Fig. 8.26 Transgenic *Tilapia* (left) in comparison with the wildtype species.

Fig. 8.27 Norman Maclean of the University of Southampton is working on transgenic *Tilapia* and medakas.

Fig. 8.28 Wildtype trout (top); size comparison between transgenic Pacific trout and wildtype trout (bottom).

Fig. 8.29 Choy L. Hew of the University of Singapore worked on transgenic antifreeze fish and is now cooperating with Norman Maclean on *Tilapia*.

Fig. 8.30 Coral reefs suffer from overfishing, and often, illegal methods are used to improve fish collection such as attacks with cyanide and dynamite. The fish are then sold in Hong Kong's restaurants.

Fig. 8.31 Knockout mice.

Fig. 8.32 Transgenic pigs produce organs for xenotransplants.

application of GFP and DsRED in plants). In GloFish®, the gene was introduced into the genome of the tropical zebrafish (*Danio rerio*), which is normally just black and silver (Fig. 8.23). Even just a trace of light will make the fish glow bright red.

The glowing fish had been developed by Singapore University to use them as **detectors of water pollution**. A later form of the genetically engineered fish glowed red or green only when the surrounding water contained toxins.

ANDi, the first genetically engineered primate, was created in 2000 and is also carrying jellyfish genes. ANDi is the reversed acronym for inserted DNA. Here, GFP is used as a reporter protein to label the activated genes. The gene has been successfully transferred, but not activated.

In 2003, by contrast, the activation of fluorescent genes introduced into **transgenic pigs** was successful.

30 piglets emanated a green glow in the Munich piggery of reproduction biologist **Eckhard Wolf** and pharmacologist **Alexander Pfeifer**. Lentiviruses had been used as vectors of the jellyfish genes (Box 8.3) – a step on the route of gene pharming from vision into reality.

The **ethical reservations** about gene pharming must be taken seriously, in spite of the obvious advantages the method has to offer – after all, milking transgenic cows, sheep, or goats is much easier than milking mice ...

Various pharmaceutical products have already been obtained in this way, with yields of 35 grams per liter. The products can be either isolated from the milk or consumed with it. One good dairy cow can produce 10,000 liters (2,600 gal) per year. This could cover the **total US demand for factor VIII** (120 grams), for the treatment of hemophiliacs. However, the production method has not yet been licensed (see also Ch.9).

Other proteins on the wish list are insulin, erythropoietin (EPO, Ch. 9), fibrinogen, hemoglobin, interleukin-2, growth hormone, and monoclonal antibodies.

It remains to be seen whether transgenic animals (a very emotion-fraught topic for animal lovers) or their mute relations, **transgenic plants** (Ch. 7), will be ahead in the Gene Pharming Race (see also Ch. 7, Box 7.8 for comparison).

■ 8.9 Knockout Mice

Mice reproduce up to eight times per year, having three to eight young ones each time. After four to six weeks, they become sexually active and have a lifespan of two years. Thus, a mouse can produce up to 150 offspring during its lifetime. This makes mice ideal lab animals.

The nude mouse has no hair and is used for skin sensitivity tests. It is known as a **knockout mouse** (Fig. 8.31) and caused an uproar in the media, but it is worth keeping in mind that research into human diseases such as SCID (immune deficiency), cancer, and hypertension would not have made much progress without specific knockout mice. More than 5,000 human diseases are caused by defective genes.

In **gene targeting**, specific genes are switched off in order to clarify the role of those genes in the phenotype. As 99 % of mouse genes have human equivalents, this can be very instructive.

A **nonfunctional (knockout) gene** is transferred into embryonic stem cells. This is done by cutting out part of the DNA in a first step, or some of the DNA is replaced by nonfunctional DNA. The introduced gene associates physically with the corresponding gene in the mouse chromosome and sets off an exchange process (homologous recombination) that is not yet well understood. The embryonic stem cells that have been thus modified are injected into an early mouse embryo, in the hope that they will be integrated. The embryo is then implanted into a surrogate mother. The next generation mice are tested as to whether the knockout gene has been passed on. In this generation, each individual has two gene copies, a normal and an inactivated one – in other words, the mouse is heterozygous regarding that specific gene. Through crossbreeding with other heterozygous mice, homozygous mice arise that do not produce the protein the knockout gene was supposed to block. It takes two generations of interbreeding of these animals to end up with pure-bred knockout mice.

Knockout mice have helped to make dramatic progress in research into **human cystic fibrosis**, for example. Cystic fibrosis is the most widespread hereditary disease among Europeans and Ashkenazi Jews. Approximately one in 2,000 newborns is affected. One in 20 carries the defective gene. The patients' lungs clog up with large amounts of sticky mucus, which leads to breathing difficulty and susceptibility to lung infections. A cure has not yet been found. It is not difficult to see that the knockout approach could also be very useful in studying cancer.

8.10 Xenotransplantation

Demand in transplant organs is growing steadily. In the U.S., 45,000 people under 65 are waiting for a heart transplant. Only 2,000 suitable donor hearts are available. Demand is rising, while willingness to donate stagnates. In this situation, **xenologous organs** obtained from animals or organs grown in tissue culture could be the way forward.

Pigs are the most suitable animals to provide human organs, due to their size, physiology, and anatomy. However, the major problem that needs to be overcome is **immune rejection**. Human antibodies react to antigens on the surface of a porcine organ as they would to any foreign molecule. It is therefore a matter of outwitting the immune defenses through genetic engineering.

The first **transgenic pig** named Astrid was born in 1992. Transgenic pigs produce human immune defense regulators that prevent the organism from recognizing the organ as foreign. In the first transgenic pigs that were developed specifically for xenotransplantation by the Scottish company PPL Therapeutics, the gene for α-1,3-galactosyl transferase (an important enzyme that builds sugars in porcine cell membranes) was silenced.

Sugars play an important role in the immune response. Their absence improves the chance of the transplant tissue not being rejected by the recipient organism.

The 30to 60-day survival of pigs' hearts in primates is somewhat of a success, while non-transgenic control hearts were attacked by the recipients' immune system within minutes. All recipients were given additional **immunosuppressants** (cyclosporin, Ch. 4).

There is, however, the risk that animal viruses (e.g., the porcine endogenous retrovirus PERV) that are innocuous to pigs could be passed on to humans via transplanted organs and prove pathogenic to humans.

Human insulin-producing islet cells from knockout pigs that do not produce porcine insulin will probably be transplanted to treat serious **diabetes**.

Tissue engineering involves fewer risks. Cells such as human nose or ear cartilage cells are grown on polymer scaffolding in nutrient broth. In order to avoid rejection problems, the cultured cells are taken from the recipient. They can be modelled and implanted to give a nose a better shape.

Could one regrow an ear? Today, using tissue engineering techniques, we still cannot regrow an ear. However, science is reaching the level where one can regrow pieces of cartilage, pieces of bone, pieces of muscle and bring all these tissue types together into a coordinated functional organ. That is the next task (Fig. 8.33).

8.11 Cloning – Mass Production of Twins

The word cloning sends a shower down the spines of many people, although the meaning of the **Greek word "klon" is a shoot or twig**.

A gardener who grows a plant from a cutting or puts a graft on a fruit tree produces a clone. Biologists call it asexual reproduction.

Greenflies are masters of cloning, and male bees or drones develop from unfertilized eggs (Fig. 8.36).

The more highly developed an organism, the lower the chances of **asexual reproduction**. It is assumed that sexual reproduction had the evolutionary edge when it came to adapting to environmental conditions. The choice of a sexual partner involves competition and results in the selection of the best-adapted individuals.

At the same time, the genetic material is rearranged when gametes form, offering unlimited variation possibilities and ensuring that there will always be individuals flexible enough to adapt to further changes in living conditions. Cloning does not offer these options.

Apparently, sexual reproduction promotes variation and selection, which are the all-important evolutionary factors, according to **Charles Darwin** and **Alfred Wallace**.

However, this does not exclude the possibility of genetically identical individuals arising from sexual reproduction. Identical twins make up 0.3% of human births.

From a genetic point of view, according to bioethics professor **Jens Reich** (Fig. 8.52), **a clone is just a twin being born with a time lag**.

Neither genetically engineerered, nor a human ear!

Fig. 8.33 Back in 1997, a rather bizarre photograph suddenly became very famous. It showed a totally hairless mouse, with what appeared to be a human ear growing out of its back. That photograph prompted a wave of protest against genetic engineering, which has not yet abated. But there was **absolutely no genetic engineering involved**, says Australia's most popular science writer Karl S. Kruszelnicki (see www.abc.net.au/ science/ k2/aboutkk.htm).
In August 1997, Joseph Vacanti and his colleagues wrote their groundbreaking paper in the journal *Plastic and Reconstructive Surgery*. The publicity was enormous. The "mouse-ear" project began in 1989, when Vacanti managed to grow a small piece of human cartilage on a biodegradable scaffold. The scaffold was the same synthetic material (99% polyglycolic acid and 1% poly-lactic acid), PLA, see Ch. 6) used in dissolving surgical stitches.
After 8 years, Vacanti's team were able to mold their sterile biodegradable mesh into the exact shape of the ear of a 3-year-old. The next step was to seed this ear-shaped scaffold with cartilage cells from the knee of a cow. Where does the mouse come in? Its only purpose in this project was to **supply nutrients and energy** to make the **cow cartilage cells** grow. The team used a Nude Mouse. The Nude Mouse got its name thanks to a random mutation in the 1960s that left the mouse with no hair, and virtually no immune system, so the mouse would not reject the foreign cow cartilage cells. The cartilaginous ear was implanted under the skin of the mouse. Over some three months, the mouse grew extra blood vessels that nourished the cow cartilage cells, which then grew and infiltrated into the biodegradable scaffolding (which had the **shape of a human ear**). By the time the scaffolding had dissolved, the cartilage had grown strong enough to support itself.

That cartilaginous structure that looked like a human ear was never transplanted onto a human because it consisted of cow cells and would have been rejected by any human immune system.

247

Box 8.3 The Expert's View:
Animal Cloning in Germany

On December 23rd 1998, Uschi was born in Munich – the first transgenic calf in Europe. Like Dolly, it had been cloned from cells of an adult animal.

Uschi is the first cloned calf in Europe.

What is Cloning by Nuclear Transfer?

The basic technique of nuclear transfer is easy to understand. To put it simply – the nucleus of a donor cell is introduced into an egg cell that has been previously enucleated. Thus, a new embryo is created – an embryo with a difference because in Dolly's as well as in Uschi's case, the nucleus came from the gland cells of a mammal.

After nuclear transfer, something fascinating happens that we reproductive biologists have not yet fully understood. The nucleus that heretofore had been totally specialized to take over the function of a mammal gland cell is reprogrammed. It is rejuvenated and transformed into a nucleus able to activate the programming for all potential organ functions. The current genetic program is temporarily brought to a halt and then reactivated by complex mechanisms. The egg cell and the nucleus that were brought together artificially begin to work as a unit, in a way similar to what happens after an egg has been fertilized; the success of a nuclear transfer depends on it.

A Dogma of Cell Biology Has Been Overturned

Pioneering experiments at the Roslin Institute near Edinburgh showed for the first time that under certain conditions, even cells of adult animals can be used for nuclear transfer. The experiments that culminated in the creation of Dolly refuted a central dogma in cell biology according to which whole organisms can only develop from embryonic cells.

It was no surprise that these results were first met with skepticism by many well-known scientists. Meanwhile, such doubts have been scattered by the successful repetition of the experiment in various animal species. Alongside research teams in Japan, the US, and New Zealand, our team in Munich succeeded in nuclear transfer cloning from adult cells as early as 1998. Out of 93 successfully fused egg cells, 32 (24%) developed into blastocysts – embryos ready for implantation into surrogate mothers.

In our pilot study, only four of them were implanted into two recipient animals. Both became pregnant, but one of them had a miscarriage during the fifth month of gestation. The fetus did not exhibit any abnormalities, but the placenta showed signs of disruption at the fetal-maternal interface. The other pregnancy continued without any problems and resulted in the birth of a healthy calf. Uschi, the first calf cloned from adult cells in a German-speaking country, has developed normally and given birth to two calves of her own.

Reproductive Biotechnology – Key to Fast-Track Breeding

In the past, all knowledge about the heritability of traits was derived from the phenotype, i.e., the visible traits of an animal. Through genetic engineering methods, it is now possible to access the genetic material directly. Genome analysis can provide a better understanding of the structure and function of genetic material in order to identify and select relevant genetic traits for breeding.

This, however, is only one aspect. In order to make full use of the advantages of the direct identification of genotypes, the reproduction rate of genetically desirable animals must be increased. In cattle breeding, artificial insemination has been established since the 1950s and made the rigorous selection of breeding bulls possible. Embryonic transfer as a biotechnological method includes the hormonal enhancement of the ovulation rate in genetically valuable donor animals, the harvesting of embryos and their implantation into recipient animals. Thus, the selection process has also been improved in females. It is, however, a labor-intensive and costly process, and often, the success rate is insufficient. Alternative or complementary techniques are therefore more than welcome.

For the *in vitro* production (IVP) of bovine embryos, egg cells are harvested from the ovaries of slaughtered cows. These are grown in culture until maturation, then fertilized and cultured for

another six to eight days until they are ready for transfer. The IVP of bovine embryos has immediate practical use because genetically valuable or threatened races can be preserved when the animals must be slaughtered because of age or for other reasons, e.g., the foot-and-mouth disease epidemic in Great Britain in 2001.

Egg cells are increasingly harvested from living heifers and cows (ovum pickup, OPU) and used for the IVP of bovine embryos. Egg harvesting is an important technique for boosting the numbers of valuable breeding stock.

Cloning – The Biotechnological Way Forward in Animal Breeding

Cloning is still largely in the stage of basic research. Before routine applications can be used in animal breeding, the biological mechanisms should be understood as far as possible in order to be able to resolve problems during pregnancy or in calves that have been caused by cloning. Research is now focused on the synchronization of nucleus donor and recipient egg cell in order to ensure a functioning genetic program. Another research focus is the optimization of culture conditions for bovine embryos.

Nuclear transfer cloning (Fig. 8.51) is a technique mainly relevant to basic research, but, it undoubtedly also opens up new perspectives, even for practical application in animal breeding. The complex reprogramming mechanisms involved in cloning have intrigued research teams all over the world. Clones produced by nuclear transfer provide excellent models for the study of epigenetic mechanisms.

In biotechnology, cloning comes into its own where the targeted transfer of genes is paramount. Heretofore, gene transfer in farm animals involved the injection of many copies of genetic information into fertilized egg cells – a very inefficient method, as only a small proportion of the egg cells will insert the injected information into their genome. In order to produce one or two calves from the desired additional or modified gene, hundreds of fertilized egg cells need to be injected.

If cloning is the chosen strategy, the actual gene transfer can be carried out in a cell culture. Cells into which the desired information has been stably inserted are then selected and their nuclei transferred into enucleated egg cells. A viable nuclear transfer embryo arises, ready for implantation into a recipient animal where it may develop into a transgenic calf.

The cloning of pigs from cells that have undergone targeted cloning is essential for the creation of genetically modified pigs as organ donors for xenotransplants. It is thus possible, for example, to remove from the surface of pig cells sugar residues that would prompt violent rejection of pig tissue in primates. This, however, is only the first hurdle to overcome in a complex rejection reaction. In order to ensure the long-term survival of xenotransplants, several strategies must be combined to produce multitransgenic pigs. We have developed a very efficient method to do just that.

Success in Transplant Medicine

Together with **Alexander Pfeifer**, Department of Pharmaceutical Science, Munich, we successfully inserted foreign hereditary material into the genome of higher mammals, developing fluorescent piglets. The vector we used for the foreign genetic material is a virus, able to penetrate mammalian cells.

Alexander Pfeifer

The foreign genetic material is a gene that codes for Green Fluorescent Protein (GFP). It was selected because of its suitability for labeling. In the majority of the piglets born, the active gene

Transgenic pigs into which GFP genes have been inserted using lentiviruses have a green glow.

was found throughout all tissue and was even passed on to offspring. This marks a step forward on the route to the targeted insertion of genes carrying desired traits into farm animals and the use of animal organs as transplants for humans. The targeted transfer of genetic material could be tailor-made for the individual recipient to enhance immune compatibility.

Until recently, there had only been very inefficient ways of inserting foreign genes into higher mammalian cells. The genetic material is usually injected into embryos – a very labor-intensive method with a low success rate.

The use of viruses as vectors for the foreign genes seems to be the most promising alternative, as they have the ability to penetrate into the mammalian cells and insert their genes – as well as additional foreign genes – into the DNA of the infected organism. This method, however, often failed because the viral genetic material had been silenced by the mammalian cells and could no longer be activated.

At the Center for Pharmaceutical Research, Alexander Pfeifer and his group used cutting edge viral technology to circumvent the problem. They used a lentivirus to infect pig embryos at the very early one-cell stage. A total of 46 piglets were born, of which 32 animals or 70 percent carried the GFP gene. In 30 pigs or 94 percent of this group, the gene was even active. Not only

did all tissue, including gametes, emanate a green glow, but it was also passed on to their offspring.

Further experiments were carried out to test whether it was possible to activate the foreign DNA in certain pig tissues only. Again, a GFP gene was introduced into the embryos, but this time, it was preceded by a piece of human DNA, usually responsible for the activation of a gene in skin cells.

The GFP gene was again identified in all tissues of the piglets, but it was only active in skin cells. The same procedure has now also become an established method in cattle.

Therapeutic and Reproductive Cloning

Therapeutic and reproductive cloning are hotly debated topics, and in my view, they cannot be discussed separately, as both go through the same stages until a blastocyst (an embryo on the seventh day after fertilization) has developed.

The divide only comes when a decision is taken to determine what to do with the embryo – whether to culture it to grow pluripotent cells or to implant it into a woman at the critical moment in her cycle. At least in theory, it would then be able to grow into a child.

It is conceivable that sensible legislation could deal with the situation, although there are bound to be borderline cases. I am convinced that in the long run, by making therapeutic cloning possible, we are preparing the ground for reproductive cloning. I am also fairly certain that some scientists in reproductive medicine would soon demand that reproductive cloning should be reclassified as therapeutic cloning for the treatment of childlessness.

I advocate a ban on therapeutic cloning, as it exists *de facto* in Germany under the embryo protection legislation. It should remain in place until all other possibilities for cell replacement treatment have been exhausted.

Prof Eckhard Wolf is Professor for Molecular Animal Breeding and Biotechnology at the University of Munich, Germany.

Box 8.4 Biotech History: "Clonology"

The chronology of cloning – dubbed "clonology" by Hwa A. Lim in his book *Sex Is So Good, Why Clone?* – probably began with parthenogenesis (Greek for virgin birth) when German embryologist **Oskar Hertwig** (1848 - 1922) added strychnine or chloroform to sea urchin eggs and got them to develop without the use of sperm.

Jacques Loeb (1859 - 1927) repeated the experiment three years later. In 1900, Loeb pricked unfertilized frog eggs with a needle and set off their embryonic development. In 1936, **Gregory Goodwin Pincus** (1903 - 1967) used a temperature shock to start off embryonic development. It was not until 2002 that **Jose Cibelli** of Advanced Cell Technology reported the successful parthenogenesis of the first primate, a macaque called Buttercup.

Paul Berg was the first to clone genes successfully. He later called for a moratorium in order to keep DNA experiments under control. He was awarded the Nobel Prize in 1980.

In 1972, **Paul Berg, Stanley Cohen, Annie Chang** (Stanford), **Herbert Boyer,** and **Robert B. Helling** (UC San Francisco) made significant progress in the cloning of genes (Ch. 3) – foreign DNA was inserted into bacteria, for example, and was replicated, i.e., copied or cloned, millions of times.

In 1976, **Rudolf Jaenisch** (Salk Institute, La Jolla) injected human DNA into just fertilized murine egg cells to obtain mice with human DNA in their genome. These founder mice passed on their genetic material to their offspring. Transgenic mice had been created.

Cloning has remained a contentious issue in the public debate.

Two years later, on July 25th 1978, **Louise Joy Brown** was born. She was the first test-tube baby in history. The British doctors **Bob Edwards** and Patrick Steptoe went down in history as the founder fathers of *in vitro* fertilization (IVF). In 1983, **Kary Mullis** (Ch. 10) invented the polymerase chain reaction (PCR), a method that allowed billions of copies to be made from DNA and RNA sections – to clone them.

Steen Willadsen succeeded in 1984 in cloning a living lamb from immature embryonic sheep cells. A year later, he moved on to Grenada Genetics to clone cattle for commercial purposes. Animals developed from embryonic cells, however, still contain the genetic material of both parents, as they arise from sexual reproduction. In 1995, **Ian Wilmut** and **Keith Campbell** let embryonic sheep cells go through a dormant phase before their nuclei were implanted into sheep cells. In 1996, Dolly was born.

In 1998, **Teruhiko Wakayama** of Hawaii developed what became known as the Honolulu technique, creating several generations of genetically identical mice. At the Kinki University in Japan, eight calves were cloned from one cow, and in Germany, the cloned calf Uschi was born in 1998 (Box 8.3). In South Korea, cells were taken from an infertile woman in 1999 and cloned into an embryo where they reached the four-cell stage. The experiment was then terminated for ethical and legal reasons. The embryo was never implanted.

At the same time, at Texas A & M University, the calf Second Chance was cloned from 21-year-old Brahman bull Chance. It is the oldest cloned animal so far. In 2000, the female rhesus macaque Tetra was created by cloning, whereas in Japan and Scotland, pigs were cloned.

Steve Stice trebled the success rate in cattle cloning in 2001, using skin and kidney cells. For the first time, cells of a cow that had been slaughtered 48 hours previously were used.

Prometea, the first cloned Haflinger horse was cloned in Italy from skin cells in 2003.

Top: cloned pigs; bottom: Second Chance, the Brahman bull created from a nucleus the donor of which, the bull Chance, was 21 years old and, due to testicular surgery, no longer fertile.

Fig. 8.34 Polyps can reproduce through budding. Drones develop through parthenogenesis.

蜂后 Queen 雄蜂 Drone 工蜂 Worker

■ 8.12 Clones of Frogs and Newts

German zoologist **Hans Spemann** (1869 - 1941) carried out experiments with newt embryos, partially tying off a fertilized oocyte (zygote). Through this ligature, the nucleus ended up in one side, which was the side that began to cleave. When one of these cleaved cell nuclei was allowed to pass through the ligature into the uncleaved cytoplasm, it began to cleave as well. This is why the procedure was named "delayed nucleation". Both halves developed into complete embryos. This experiment inspired Spemann to suggest a further experiment in which nuclei of developed somatic (body) cells should be used to promote the normal development of oocytes. He was the first zoologist to be awarded the Nobel Prize for Physiology or Medicine in 1935 (Fig. 8.35).

Nowadays, Spemann's suggestion is described as **nuclear transfer** into an enucleated egg cell.

It was first successfully performed in 1952 at the Cancer Research Institute in Philadelphia by **Robert Briggs** and **Thomas King**. They destroyed the nucleus of a just fertilized egg cell using UV light and replaced it by the nucleus of a blastocyst cell of a leopard frog. Blastocysts are liquid-filled hollow spheres consisting of approximately 100 cells, which form a week after fertilization. The thus activated egg began to turn into a "fatherless tadpole". The same technique, however, did not work in embryos that had developed into the gastrula or neurula stage, nor was it successful in mature somatic cells.

The experiments that British biologist **John Gurdon** carried out with the South African **clawed frog** (*Xenopus laevis*, Ch. 3) in the early 1960s came to a wider attention.

Using a fine glass capillary, he punctured a cell from the intestinal wall of a tadpole and vacuum-extracted the diploid (containing a double set of chromosomes) nucleus. He used the same capillary to puncture the oocyte of a frog and insert the nucleus of the intestinal wall cell.

The haploid (containing a single set of chromosomes) nucleus of the unfertilized egg cell had been either irradiated with UV or completely destroyed by vacuum extraction. There was no way for DNA fragments to reorganize.

It was a **limited success**, as only a few of the thus created diploid egg cells behaved like fertilized egg cells, going through cell division and tadpole stage to end up as healthy frogs. There were also malformed and diseased animals, but the method worked in principle – proof of the **omnipotence** of the nucleus of a differentiated cell.

Critics would still argue that an intestinal cell of a tadpole is **not a terminally differentiated cell**. So John Gurdon repeated the experiment using cells from the webbed feet of fully grown clawed frogs. He obtained a sizeable number of adult animals through serial cloning.

Serial cloning is the nuclear transfer into an enucleated oocyte which grows into a blastula which, in turn, is transferred into enucleated oocytes. This is the nucleus conditioning procedure used in Dolly, the cloned sheep (see sect. that follows). It would seem that the DNA of a terminal nucleus unpacks for transcription only if certain experimental conditions are fulfilled.

These transfer experiments showed that a **mature somatic cell contains the full information** to create an entire organism and – if the conditions are right – has the ability to activate the development program. This, however, is a complex process that mostly ends in failure.

8.13 Dolly – The Breakthrough in Animal Cloning

In the decades after these experiments, various attempts to clone mammals failed, such as with mice in the 1970s. A mammalian oocyte is 4,000 times smaller than a frog egg. The egg cell of a mouse has a diameter of a tenth of a millimeter.

A breakthrough came in Cambridge in 1986 when Danish biologist **Sten Willadsen** denucleated the egg cells of sheep. He introduced the nuclei of sheep zygotes into the enucleated egg cells, and embryos developed. However, experiments with mature somatic cells failed.

In the mid-nineteen-nineties, collaborators of **Ian Wilmut** (Box 8.5) and **Keith Campbell** in Roslin, Scotland, took cells from the udder of an adult Finn-Dorset sheep called **Tracy** and cultured them. Tracy had already died when her cell nuclei were injected into the enucleated egg cells of sheep of a different race. Egg cell and nucleus were stimulated in a nutrient broth, and the development of an embryo began. The cloned sheep Dolly was born on July 5, 1996.

Dolly was living proof that it was possible to clone adult mammals and that a normal somatic cell can "forget" all of its specifications, behaving like a totipotent fertilized egg cell. A short letter to the journal *Nature* overturned one of the most enduring dogmas in biology.

To be fair, Dolly was lucky (see Box 8.5 for the story). Out of the **277 cloning attempts, only 29 reached the stage of transferable embryo**. The few sheep that became pregnant had early miscarriages – Dolly's mother being the only exception. Anybody speculating about cloning humans should bear these figures in mind. Malformed sheep embryos are bad enough!

Later, Dolly herself became a mother – no cloning involved! On April 13th, 1998 the healthy lamb **Bonnie** was born. Dolly was by no means the only cloned sheep. There had been predecessors: The Welsh Mountain sheep Megan and Morag had been cloned directly from embryonic cells, and two male Black Welsh sheep, **Taffy** and **Tweed**, were cloned from cultured fetal cells around the same time as Dolly. These predecessors had shown that cultured cells can be successfully used in cloning. What was new in Dolly was that the nucleus donor cells had been adult somatic cells.

Fig. 8.35 Hans Spemann (1869 - 1941).

Fig. 8.36 Developmental stages of clawed frogs, from the work of embryologist E.J. Bles, 1905.

Fig. 8.37 Frogs were the first animals to be cloned.

Fig. 8.38 Lucky student: Stanford student Stephen Lindholm went to Scotland and Ireland to visit his high-school friend Dale, who had a summer job at the Roslin Institute where Dolly the sheep had been cloned in collaboration with PPL Therapeutics. One day, Dale took Steve to work and he got to meet the inhabitants of the Large Animal Unit. At that time, Dolly was still being kept outdoors and very much alive and kicking!

Box 8.5 Biotech History:
Dolly the Sheep

Few celebrities have so besotted the world's media as she did. Her arrival caused a sensation, triggering an orgy of speculation, gossip, and hype. She posed for People magazine, became a cover girl, and even caught eye of **Bill Clinton**. Plays, cartoons, and operas were inspired by her story, advertisers traded on her images, and her name seemed to be on everyone's lips. Those closest to her said the attention rather went to her head, and she soon started putting on airs.

Dolly the superstar

Nor was her career slowed by motherhood; even when she had six young ones to care for, she continued to be in the news. But, as with so many global superstars who live life in the fast lane, it was destined to go wrong. She developed a cough and began to go into decline. After a few weeks, she died tragically young of lung cancer, even though she had never smoked. Her passing marked the end of a great celebrity life. But this was no ordinary diva, for she was a sheep.

So much for the breathless media account of the life and times of Dolly the sheep. Her story is now as much a part of the history books as of the annals of science. When she was born in the **Roslin Institute**, near Edinburgh, on 5 July 1996, Dolly marked the beginning of a new era of biological control. With a large team, I was the first to reverse cellular time, the process by which embryo cells differentiate to become two hundred or so cell types in the body.

We had defied the biological understanding of the day: development runs only in one direction in nature: cells in tissues as different as brain, muscle, bone and skin are all derived from one small cell, the fertilized egg. Until Dolly was born, it was thought that the mechanisms that picked the relevant DNA code for a cell to adopt the identity of skin, rather than muscle, brain, or whatever, were so complex and so rigidly fixed

that it would not be possible to undo them. This deeply held conviction was overturned by Dolly. She was the first mammal to be cloned from an adult cell, a feat with numerous practical applications, many of which will raise profound moral and ethical issues.

The term "clone", from the Greek for "twig", denoted a group of identical entities. In the case of Dolly, she was (almost) genetically identical to a cell taken from a six-year-old sheep, the nucleus from which had been transplanted into an egg cell from a second sheep, and then inserted into the uterus of a third sheep, and then a fourth, to develop. It was because the process started with a mammary cell from an old ewe that she was called Dolly-our affectionate tribute to the buxom American singer **Dolly Parton**.

While we now take for granted that it is possible to clone adult animals, the birth of Dolly shocked those in the general public who dwelled on the implications of reproduction without the act of sex. The feat shocked many in the research community too. Scientists were apt to declare that this or that procedure would be "biologically impossible," but with Dolly that expression lost all meaning. Some pundits have even said that Dolly broke the laws of nature. But she revealed, rather than defied, those laws. She underscored how, in the twenty-first century and beyond, human ambition will be bound only by biology and society's sense of right and wrong.

Megan and Morag, the first sheep to be cloned from cultured embryonic cells.

When we created Dolly, we were not thinking about rooms full of clones, or creating hillock upon hillock of identical sheep to guarantee a good night's sleep. We were not thinking about helping lesbians to reproduce without the help of a sperm bank or multiplying movie stars. We were certainly not thinking of duplicating dictators. The tortuous tale of Dolly does not fit the storyboard of a traditional Hollywood movie: my single-minded struggle, against all odds, to fulfill my dream; strange goings-on in a subterranean laboratory one dark and stormy night; or how a tight-knit group of hirsute underdogs toiling in an obscure Scottish lab snatched the cloning

prize from arrogant, clean-shaven, lantern-jawed scientists working in a well-funded North American powerhouse of genetics.

Dolly hiding from the paperazzi.

True, my team did sport a fair amount of facial and long hair, but it was a special combination of factors that led to success, along with a spark of serendipity. The right people with the right skills and understanding came together at the right time in the right place-the Roslin's collection of pristine molecular biology laboratories, haystrewn barns, and up-to-date facilities for surgery, whether on livestock or on living cells.

Dolly's story is all the more remarkable for other reasons. The classic scientific tale of rival teams racing for a prize, as vividly depicted in **Jim Watson's** The Double Helix, did not apply in our case. We were mostly spurred on by pure curiosity, though we did have in mind practical applications in research and agriculture. As in any government laboratory, research was a grind because money was always in short supply. Even more so in those days, when the science establishment was subjected to punishing cutbacks: on the day Dolly was unveiled, the government withdrew its funding for my project. The company that helped us create her, PPL Therapeutics, was our ally but not our friend in the sense that we shared information only on a need-to-know basis.

The group was led by myself with **Keith Campbell**, a cell biologist ten years my junior, who had built up a profound understanding of what makes cells tick. It is still a surprise that we were so effective together because neither of us is well organized. His adventurous nature and knowledge of cells complemented my own scientific curiosity, my experience with cloning, and my strategic, step-by-step approach to achieving the goal of the genetic transformation of livestock. I conceived the project, but it was thanks to Keith's inspiration that it succeeded.

Roslin the place, not the institute, held us tight during those difficult days. Keith and I would continue to drive to the laboratory together and have animated chats about the details of cloning. We were both excited by the science. We both hailed from the Midlands. But we were no Watson and Crick engaged in a race with rivals. We had different styles and outlooks. We did not socialize much.

Our shared aim was to develop methods for nuclear transfer in sheep that could be used to introduce precise genetic changes. Superficially, it may seem paradoxical that the wherewithal to create genetically identical offspring can also be used to introduce genetic changes into animals: the answer from the one cell where you have cloning to make a whole animal from the one cell where you have succeeded in carrying out successful genetic surgery among the billions of cells in which the surgery was botched or incomplete or had failed. That would offer huge advantages over the old-fashioned brute-force method in which we had to insert endless embryos into sheep without any clear idea of whether the embryos had been successfully modified.

Dolly, now in the National Museums Scotland, Chambers St, Edinburgh.

We had just £ 20,000, which a Roslin committee had offered us to do some more experiments because of the good progress we were making with cloning. Of course, because of our previous experience we would use sheep. That meant we had to deal with their notorious obstetric problems (shepherds say that sheep spend their time dreaming up new ways to die).

We were slave to their breeding cycles, which meant intense work during the winter, when sheep mate and conceive. The point was, however, sheep were astonishingly cheap. In those days, it was possible to buy one at the market for less than a bottle of mineral water in a posh hotel and for around one five-hundredth the cost of a cow. However, we were confident that because of their similar reproduction and embryology any method developed in sheep would also prove suitable for cattle. In effect, the sheep were small cheap cows.

It all seemed so simple. We would use special cells from embryos-stem cells-for cloning. These grow in the laboratory, but retain many of the characteristics of embryo cells. This was the reason why I was optimistic that this approach to

Dolly with her first lamb, Bonnie.

making genetic changes in animals would work. But there was a problem. We could not multiply those sheep stem cells in the laboratory (we still cannot). However, in our failed attempts we did grow more mature cells that had, as scientists say, begun to differentiate.

Thanks to the insights of Keith Campbell into the mechanics of the cell cycle, we went on to make a remarkable find: it was possible to put these more differentiated cells into a special state (a resting state, called quiescence, which is unattainable by embryonic cells), and we discovered that we could clone them.

Megan and **Morag**, Welsh Mountain sheep, were born as a result of this crucial insight. They were the first clones from differentiated cells. We had failed to grow stem cells, but Keith's new method of cloning had proved more powerful than we had imagined. Keith had always believed that cloning with adult cells would become possible, and with this new encouragement we set out to test the idea. The birth of Dolly confirmed the point emphatically.

As I waited for Dolly to be born during the summer of 1996 my mood would swing between elation and fear. Keith and I were confident of success, but, with such a complex process, involving so many steps and people, failure was always a strong possibility. And if we succeeded, the thrill of it all would be tempered by the thought of the fuss that would follow. I am a private person, and I knew my life would not be quiet anymore.

And succeed we did, but only just-in 1 of 227 attempts. At birth, Dolly weight in at 6.6 kilograms, heavy for a Finn-Dorset but not surprisingly so. Perhaps in another season we would have had twenty Dollys. But it was much more likely that we would have failed. Despite the emphasis on objectivity and rationality, even in science it helps to be lucky. For Keith, however, Megan and Morag were the real stars of the scientific story. He considered Dolly merely "the gilt on the gingerbread." We had agreed that he would hold the coveted position of first author of the Megan and Morag paper, but that I would be

first author of the paper that described Dolly's difficult passage into the world.

We kept Dolly's existence secret for the first six months, as paper's detail of how we had created her were scrutinized by other scientists before they could be published in an academic journal. Once the news of her birth leaked out, in February 1997, she made headlines, capturing the imagination of commentators, columnists, and opinion writers across the planet.

By the standards of her fellow sheep on the farm, many of which were slaughtered as early as nine months, Dolly lived a very long and full life. Some scientists feared, though for no clear reason, that she was sterile, but she defied her critics and went on to breed with a Welsh Mountain ram called David and have a daughter in April 1998. Her firstborn was named **Bonnie** because, as her vet **Tim King** commented, "She is a bonnie wee lamb". Dolly had two more lambs in 1999. In all, she gave birth to six lambs, all conceived naturally and all born healthy. She was the living, bleating, and woolly proof that new life – in the full sense of being able to reproduce – could come from a cloned adult cell.

Given the unusual way she came into being, Dolly's every bleat and every baa were by then being analyzed for their biological significance and the merest hint of decrepitude. She showed no real signs of being any different from the rest of the flock, however. But there were limits to what we could do to check she was really "normal". For example, we could not scan for aberrations of ovine mind or mood that could be linked with cloning ...

No one knows the mind of a normal sheep, after all.

Cited and shortened with permission from **Wilmut I and Highfield R** (2006) *After Dolly: the uses and misuses of human cloning.* WW Norton & Co, New York, London pages 11-18 and 23-25

Superstar Dolly with Ian Wilmut.

Fig. 8.39 The creation of Dolly.

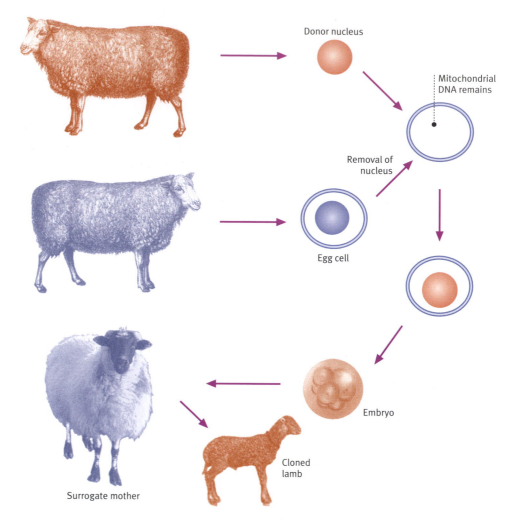

Donor nucleus

Mitochondrial DNA remains

Removal of nucleus

Egg cell

Embryo

Cloned lamb

Surrogate mother

Fig. 8.40 Dolly with her surrogate mother soon after birth. The ewe is a Scottish Blackface and obviously cannot be the natural mother of Dolly.

Master, what happened to my thousand twins?

clone

Fig. 8.41 A foal cloned in Texas, named after the Wim Wenders movie Paris-Texas.

Fig. 8.42 Cloned roe deer.

Meanwhile, after six years in the limelight, Dolly has died of a lung disease. Before Dolly, there had only been one systematic study on cloned mice, at the National Institute of Infectious Diseases in Tokyo. These mice met a premature death.

We need to wait until the 24 calves cloned by Advanced Cell Technology have grown up to see the effects. In February 2003, Australia's first cloned sheep died of unclear symptoms at the age of two years and ten months. Lung diseases such as Dolly's are typical in older sheep. Her donor mother had been six years old when the cells were taken out.

Soon, a debate began about the life expectation of cloned animals. In January 2002, it was found that Dolly suffered from arthritis which usually only affects old sheep. **Harry Griffin** from the Roslin Institute commented after Dolly's death: "Sheep can live to 11 or 12 years of age and lung infections are common in older sheep, particularly those housed inside. A full post-mortem is being conducted and we will report any significant findings."

Meanwhile it has been found that some cloned animals have shorter **telomeres** than their normal counterparts. Telomeres are chromosome caps at the end of chromosomes. They carry a characteristic DNA repeat sequence that protects the chromosomes (a bit like plastic ends on a shoelace). Each time a cell divides, the telomeres become shorter and are thus indicators of the age of a cell. (At some stage, there is no plastic end left on the shoelace).

Intense research is going into the activation of **telomerase** in aging human cells. Telomerase is an enzyme that increases the number of possible cell divisions from 50 to 300. So far, it has only worked in cell cultures. Could this be a possible fountain of youth?

■ 8.14 Difficulties in the Cloning Process

The difficulties that arise during the cloning process cannot all be put down to insufficient technique, but also to biological causes.

When a cell nucleus comes from a fully developed somatic cell, certain gene regions have been blocked. Thus, an islet cell that produces insulin will not produce the same substances as a nerve cell, whereas a nerve cell does not need insulin.

In Dolly, Ian Wilmut **removed the gene blockade** by starving the somatic cells in a broth that was poor in nutrients. This probably reversed the DNA packaging to its original state, and the blocked DNA was reprogrammed.

Most donor cells have suffered some previous damage such as UV irradiation, reactive oxygen radicals and toxins that may have damaged some regions in the genome. In humans, added damage can come from alcohol, drugs, X-rays, and barbecued food. Normally, such damage does not affect somatic cells, as long as they do not rely on the information that has been damaged. The damage only shows once the cell has been reprogrammed and a clone develops. This would explain why a clone ages prematurely.

Finally, the interaction between denucleated cell and injected nucleus may be faulty. The egg cell's cytoplasm produces active substances that control the functioning of the nucleus. Faults in the reprogramming of the nucleus may result in faulty communication. During the normal fertilization process, paternal chromosomes are demethylated (CH_3 groups are removed) shortly after their entry into the egg cell. An egg cell at a later stage is unable to do that because it is no longer needed.

The copy hype in the mass media has no scientific foundation. **Most existing clones are by no means exact copies of their donors!**

Like all animal clones, Dolly had a nucleus from her donor mother, but the cytoplasm came from the egg cell donor, thus passing on her **mitochondrial DNA**. In normal fertilization, mitochondria are always passed on down the maternal line. Mitochondrial DNA is important and can also be used in DNA fingerprinting (Ch. 10).

Thus, Dolly is the product of the **nuclear DNA of the donor plus mitochondrial DNA from the egg cell**, with the added impact of the hormones of the surrogate mother. Ian Wilmut says in his book *The Second Creation: The Age of Biological Control by the Scientists Who Cloned Dolly*: "So Dolly and the ewe who provided the original nucleus have identical DNA but they do not have identical cytoplasm. In fact Dolly is not a 'true' clone of the

original ewe. She is simply a 'DNA clone' or a 'genomic clone'. "

Wilmut chooses his wording carefully when he continues: "So, in fact, although the cells in Dolly's body are descended from a cell which mainly contained Scottish Blackface cytoplasm, that cell also contained some Finn-Dorset cytoplasm surrounding the donor nucleus. To be sure, the oocyte that came from the Scottish Blackface is far, far bigger than the cultured body cell that supplied the Finn-Dorset nucleus, so we might suppose that the meagre Finn-Dorset contribution would simply be swamped. In fact, this does seem to be the case…"

How much of the phenotype can be attributed to the cytoplasm and how much to the surrogate mother became apparent when cats were cloned.

8.15 Cloning Cats – Parental Variations

The cloning process of a cat is given as an example to clarify the parts that the various parents play:

1. A female cat can be an egg cell donor. The nucleus is removed and the nucleus of one of her body cells is injected. The resulting embryo is implanted into the womb of the same cat who is egg cell donor, somatic cell donor and surrogate mother rolled into one. One parent.

2. The nucleus of the somatic cell can come from another cat (male or female), and the egg cell is re-implanted into the egg cell donor cat. Two parents.

3. Egg cell and somatic cells come from two different cats, and a third cat is used as surrogate mother. Three parents.

The next steps are the same as in the creation of Dolly. Here they are:

- A somatic cell is taken from a cat and starved in low-nutrient broth. The nucleus is chemically or mechanically removed.

- The donor cat undergoes hormone treatment to produce several egg cells simultaneously (superovulation). The egg cells are obtained through vacuum aspiration (as during artificial fertilization) and enucleated. What is left is the cytoplasm containing mitochondrial DNA (which has an important role to play, see above).

Fig. 8.43 Mitochondrion. In normal fertilization, mitochondria always come from the mother, containing mitochondrial DNA. (In nuclear transfer, the mitochondria come from the enucleated cell).

Fig. 8.44 Cats have 19 chromosome pairs. Only the female X-chromosome contains color information. Male cats (XY) can thus pass on their color only to female offspring (XX), as semen carrying the Y-chromosome does not contain color information. Therefore, male cats always inherit their color from their mother.

Tortoiseshell and calico describe a coloring caused by a combination of specific genetic traits. In female cats, it is a result of X-chromosome inactivation, in which different patches of fur receive coding for different hair color due to the activation of an X chromosome from either the mother or the father.

Calico coloring of my female cat Fortuna is a mix of phaeomelanin-based colors (red) and eumelanin-based colors (black, chocolate and cinnamon). Calico cats are believed to bring good luck!

The lower photo shows two interesting phenomena – my female calico cat Fortuna, having caught the tail of a lizard. Shedding the tail – which will regenerate – saved the lizard's life.

Box 8.6 The Expert's View:
"Social Genes" – RNA interference and the Life of Honeybees

Bees and ants have always fascinated me. These insects live in groups often specialized in different tasks, creating a division of labor.

One extreme example can be seen in honeybees, in which most tasks are performed by thousands of worker females that are essentially sterile helpers. Workers start out as nurse bees that care for larvae in the nest. Later they embark on foraging trips, specializing in either pollen or nectar collection, and continue to forage until they die. The age at which workers switch to foraging and collecting pollen or nectar seems to be determined by their rudimentary reproductive physiology and the protein vitellogenin in particular.

Vitellogenin is normally involved in the production of egg yolk, but it may affect behavior and lifespan in workers. We tested this hypothesis by knocking down the vitellogenin gene in worker bees. When investigating the genome of the honeybee, we expected to find many new genes responsible for social behavior. However, we didn't find much diversification of such social genes. In fact, the honeybee genome was smaller than that of a fly, which has a solitary lifestyle.

The genetic and hormonal mechanisms that mediate the control of social organization are not yet understood and remain a central question in social insect biology.

To test our hypotheses, we injected vitellogenin double-stranded RNA (dsRNA) into a subgroup of bees and compared the bees' behavior and lifespan to two other groups that had received either dsRNA derived from a different gene (green fluorescent protein) or no injection (reference group).

First of all, we found that knockdowns expectedly showed reduced hemolymph levels of vitellogenin. However, they also showed high levels of juvenile hormone.

This combination is typical of bees in the foraging stage, in contrast to bees in the earlier nursing stage, which have high levels of vitellogenin

and low levels of juvenile hormone. Therefore, we expected that in field experiments, the knockdown bees would become foragers earlier in life than controls and the shift from nest care to foraging would occur earlier.

Because selectively bred bee strains specialized in nectar collection have lower vitellogenin hemolymph levels than those bred to specialize in pollen collection, we predicted that vitellogenin knockdowns would more likely become nectar collectors. We also predicted that knockdowns would live shorter than controls, as it is known that honeybee vitellogenin enhances longevity. This is due to its ability to scavenge free radicals, acting as an antioxidant shield.

The vitellogenin gene, which is involved in egg production in all egg-laying animals, coordinates three core aspects of the social life of bees. It paces the onset of foraging behavior, primes bees for specialized foraging tasks, and enhances longevity.

Workers with suppressed vitellogenin levels foraged earlier, preferred nectar, and lived shorter lives. Thus, vitellogenin has multiple effects on the social organization of honeybees. By using gene knockdown to understand insect social behavior, our study supports the view that social life in bees evolved by co-opting genes involved in reproduction.

The results also confirmed predictions drawn from the sequence of the honey bee genome, which was published in 2006. The sequence suggested that new, distinctive social behaviors likely developed from old genes and mechanisms.

It's the first mechanistic study to look at the genetic and hormonal regulation of complex social behaviors by a single gene using honeybees as a model organism.

It seems that vitellogenin, which is typically used as a yolk precursor protein in reproduction by insects, including the nonsocial *Drosophila*, has been co-opted in social honeybees for the regulation of social behaviors. It makes sense to assume that evolution has used a previously existing mechanism for a new function, which is very economical.

We also had an important technical breakthrough, because RNA interference (RNAi) is used to manipulate the behavior of bees out in the field, as opposed to in the laboratory. This technique promises to be extremely helpful in identifying the no-doubt many other genes involved in regulating division of labor.

And what is the conclusion for us humans? Nature rewards intelligent laziness…

Gro V. Amdam is Professor at the School of Life Sciences at Arizona State University, Tempe, and the Department of Animal and Aquacultural Sciences, Norwegian University of Life Sciences, Aas.

Cited Literature:

Nelson CM, Ihle KE, M. Kim Fondrk MK, Page Jr. RE, Amdam GV (2007) *The gene vitellogenin has multiple coordinating effects on social organization* PLoS Biol 5(3) http://biology.plosjournals.org/perlserv/?request=get-document&doi=10.1371/journal.pbio.0050062

Phillips ML (2006) *Honeybee genome sequenced*. The Scientist, October 25, 2006. http://www.the-scientist.com/news/display/25318/

Amdam GV et al. (2003) *Disruption of vitellogenin gene function in adult honeybees by intra-abdominal injection of double-stranded RNA. BMC Biotechnol.* 3:1 http://www.the-scientist.com/pubmed/12546706

- The donor nucleus is introduced into the enucleated egg cell. This can be done using micropipettes or through subtle electric impulses. The egg cell now contains a diploid set of chromosomes, as it would if sperm and egg had met. Electric impulses are used to stimulate cell division. Once the embryo has reached the eight-cell stage, it is examined for genetic damage.

- The growing embryo is implanted into the surrogate mother who gestates the embryo.

If the cat to be cloned is female, the nucleus donor cat could also be used as surrogate mother instead of another cat. In a sense, the cat is producing its own clone. The similarity between mother and kitten would be greatest. It would be a genuine clone. If a male cat is to be cloned, the help of females is indispensable, whereas a female cat could simultaneously be twin sister, egg cell mother, and surrogate mother of a clone.

The white-and-tortoiseshell cat **Cc or Carbon copy** was born in 2001, just two days before Christmas eve. Texan researchers working with **Mark Westhusin** (Fig. 8.45) declared that the first cat in history had been cloned.

Why then is Cc's **coat pattern not identical** with that of the donor of the somatic cell and egg cell donor? Basically, the coloring is similar to that of Rainbow. Obviously, the coloring will not resemble that of the surrogate mother because she is a different cat altogether (Fig. 8.46).

We cannot expect the color pattern of the clone to be identical with that of the donor cat, as the color pattern is a result of genetic as well as environmental input.

For example, the position of the embryo in the uterus of the surrogate mother decides which hair follicles are reached by the pigment-producing cells. Other environmental factors can cause minor differences between clone and donor. The food intake of the surrogate mother has a bearing on the size of the newborn.

It should be kept in mind that a clone may be genetically identical with the donor, but is not the same animal! There are differences in appearance, but they are not dramatic.

The cells used for cloning Cc are known as **cumulus cells**. Cumulus cells surround the egg cell while it is maturing in the ovary. They feed the egg cell before and after ovulation. The Texan researchers believe that cumulus cells work better than fibroblasts. First, 82 cloned embryos from skin fibroblast cell nuclei were created and implanted into seven surrogate mothers – no success! Then Rainbow's cumulus cells were used and cloned embryos produced. These were implanted into the surrogate mother, and 66 days later, Cc was born.

Can cats only be cloned from cumulus cells? No, Cc just happens to be the first cloned cat. The six other cloned animals (sheep, goats, cows, pigs, mice, and a gaur) were produced from frozen and re-thawed skin fibroblast cells.

■ 8.16 What About Humans? Cloning, IVF, and PID

Cats, mice, cats, sheep, horses, and goats have all been cloned, but in spite of headline-grabbing statements from esoteric sects, the cloning of humans through transferring the nucleus of a mature somatic cell into an enucleated egg has not yet overcome technical flaws. Take just the **high rate of malformation**! Science would be thoroughly discredited if malformed babies resulted from cloning.

What is more, embryo protection legislation made the cloning of humans illegal.

The Charter of Fundamental Rights of the European Union explicitly prohibits reproductive human cloning, though the Charter currently carries no legal standing. The proposed European Constitution would, if ratified, make the charter legally binding for the institutions of the European Union.

The British government introduced legislation in order to allow licensed therapeutic but not reproductive cloning in a debate in January 2001. Currently **therapeutic cloning** is allowed under license of the Human Fertilisation and Embryology Authority. The first known licence was granted in August 2004 to researchers at the University of Newcastle to allow them to investigate treatments for diabetes, Parkinson's disease, and Alzheimer's disease.

In 1998, 2001, and 2003, the U.S. House of Representatives voted whether to ban all human cloning, both reproductive and therapeutic. Each time, divisions in the Senate over therapeutic cloning prevented a competing proposal (a ban on both forms

Fig. 8.45 Mark Westhusin with Cc (Carbon copy).

Fig. 8.46 Cc's genetic mother Rainbow (top) and her surrogate mother (center) with Carbon copy (bottom).

Fig. 8.47 The coloring of the cloned cats came as a surprise to researchers.

Fig. 8.48 This is how Ernst Haeckel saw the human egg cell in his book The Evolution of Man.

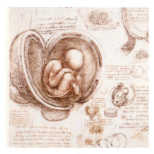

Fig. 8.49 Renaissance genius Leonardo da Vinci (1452-1519) collaborated with anatomy expert Marcantonio della Torre. For this drawing, however, Leonardo relied on knowledge gained from the dissection of a cow fetus.

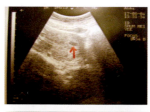

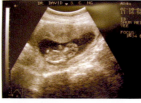

Fig. 8.50 From ultrasound image to a healthy happy boy. Baby Theo Alex Kwong, however, did not need PID or IVF.

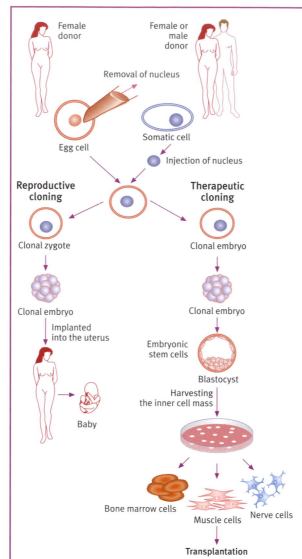

Female donor

Female or male donor

Removal of nucleus

Egg cell

Somatic cell

Injection of nucleus

Reproductive cloning

Therapeutic cloning

Clonal zygote

Clonal embryo

Clonal embryo

Clonal embryo

Implanted into the uterus

Embryonic stem cells

Blastocyst

Baby

Harvesting the inner cell mass

Bone marrow cells

Muscle cells

Nerve cells

Transplantation

Box 8.7 *In Vitro* Fertilization in Humans

In a first step, egg cells are taken from the tubes through aspiration. Although this is a highly advanced technique, it is still associated with problems. The woman must undergo hormone treatment to make several egg cells in the ovary mature simultaneously. This can lead to hormonal imbalances. Ultrasound is used to establish beforehand whether enough follicles have matured. The whole procedure of taking out the egg cells using a syringe is visualized through ultrasound. This is normally done through the vagina. Usually, a local anesthetic is given. There is a potential risk of infection and bleeding. While eight to ten egg cells are thus obtained, male sperm (resulting from masturbation) is collected.

In the lab, an egg cell is fertilized with the sperm. Both cells are united in broth in an incubator – outside the female body. Between the third and fifth day, in the multiple-cell stage, the embryo is transferred into the womb.

If successful, a normal pregnancy begins. However, success cannot be taken for granted. Only 10 to 15 percent are successful. If several egg cells are fertilized and implanted in the same cycle, a multiple pregnancy can be the result. For this reason, it is illegal in Germany to implant more than three fertilized egg cells at once, whereas in the US, six are allowed.

or on reproductive cloning alone) from passing. President Bush is opposed to human cloning in any form. Some American states ban both forms of cloning, while others outlaw only reproductive cloning.

In his book *Es wird ein Mensch gemacht* (A man is in the making) Jens Reich, former human rights activist in East-Germany, an expert in bioinformatics, medicine, and bioethics says: "I have no intention to start a moral debate about the cloning of animals, although I can see its problems. However, in humans, I find it an obscene suggestion, for two fundamental reasons – one is that the making of humans is turned into a technical process, the other the idea of creating a human in one's own image."

Pre-implantation diagnosis (**PID**) is currently as hot a topic as gene food (Ch. 7) and stem cell research (Ch. 9).

In order to test the genome of an embryo, at least one cell must be taken from the cell agglomeration and destroyed. One usually waits until the third day after fertilization when the eight-cell stage has been reached. One cell is removed (plus another one to back up the diagnosis). At this stage, the cell contains all the information and is totipotent. No harm is done to the embryo, as it has not yet entered the compact stage.

What can be recognized within the cell?

The **gender** of the embryo (XX female, XY male) and an abnormal chromosome count (the normal number is 23 pairs) are easy to identify. Specific chromosome sections can be visualized through microscopy to identify faulty regions (Ch. 10). It is even possible to identify pathogenic gene mutations. In theory, it would be possible to read the entire DNA, but the cost would be astronomical.

Box 8.8 Biotech History:
Embryos, Haeckel, and Darwin

Vertebrate embryos (not to scale) at three arbitrary stages of development: from early (approximately the tailbud stage) through late (when the definitive adult form is visible). No evolutionary sequence is implied in the way the specimens are arranged. (reprinted with permission of Michael K .Richardson).

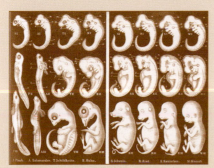

Haeckel' s embryo drawings.

Darwin commemorated on an English ten-pound note.

It has been claimed that some of **Ernst Haeckel**'s famous embryo drawings of 1874 were fabricated. There were multiple versions of the embryo drawings, and Haeckel rejected the claims of fraud made against him, but did admit one error which he corrected.

Some creationists have claimed that Charles Darwin relied on Haeckel's embryo drawings as proof of evolution, implying that Darwin's theory is therefore illegitimate and possibly fraudulent. This claim ignores the fact that Darwin published the *Origin of Species* in 1859, and *The Descent of Man* in 1871, whereas Haeckel's famous embryo drawings did not appear until 1874.

Michael K. Richardson (University Leiden, The Netherlands) comments:

Our work has been used to attack evolutionary theory, and to suggest that evolution cannot explain embryology. We strongly disagree with this viewpoint. Data from embryology are fully consistent with Darwinian evolution. Haeckel's famous drawings are a Creationist cause célèbre.

Early versions show young embryos looking virtually identical in different vertebrate species. On a fundamental level, Haeckel was correct: **All vertebrates develop a similar body plan** (consisting of notochord, body segments, pharyngeal pouches, and so forth). This shared developmental program reflects shared evolutionary history. It also fits in with overwhelming recent evidence that development in different animals is controlled by common genetic mechanisms.

Unfortunately, Haeckel was overzealous. When we compared his drawings with real embryos, we found that he showed many details incorrectly. He did not show significant differences between species, even though his theories allowed for embryonic variation. For example, we found variations in embryonic size, external form, and segment number which he did not show. This does not negate Darwinian evolution. On the contrary, the mixture of similarities and differences among vertebrate embryos reflects **evolutionary change in developmental mechanisms** inherited from a common ancestor.

We show here a more accurate representation of vertebrate embryos at three arbitrary stages, including the approximate stage, which Haeckel showed to be identical. We suggest that Haeckel was right to show increasing difference between species as they develop. He was also right to show strong similarities between his earliest embryos of humans and other eutherian mammals (for example, the cat and the bat). However, he was wrong to imply that there is virtually no evolutionary change in early embryos in the vertebrates.

These conclusions are supported in part by comparisons of developmental timing in different vertebrates. This work indicates a strong correlation between embryonic developmental sequences in humans and other eutherian mammals, but weak correlation between humans and some "lower" vertebrates.

Haeckel's inaccuracies damage his credibility, but they do not invalidate the mass of published evidence for Darwinian evolution. Ironically, had Haeckel drawn the embryos accurately, his first two valid points in favor of evolution would have been better demonstrated.

Charles Darwin's 1837 sketch, his first diagram of an evolutionary tree from his First Notebook on Transmutation of Species.

Cited Literature:

Richardson MK et al. (1997) *Anat. Embryol.* 196, 91
Richardson MK (1998) *Letters to Science*
15 May 1998: Vol. 280. no. 5366, p. 983

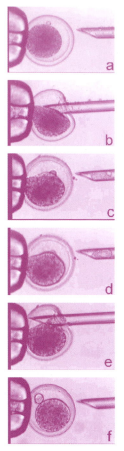

Fig. 8.51 Nuclear transfer in cattle.

Fig. 8.52 Jens Reich, medical doctor, bioinformatician, and expert in bioethics, member of the German government Ethics Commission.

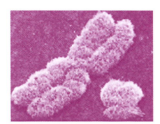

Fig. 8.53 A small difference, magnified 10,000 times. On the left the female X-chromosome, on the right, the significantly smaller Y-chromosome.

Future technological developments could change the situation.

Those who take a **skeptical view of PID** may be justified in arguing that the technique could be used not only to identify severely damaged embryos, i.e., for negative selection, but also for **positive selection** such as gender and certain traits.

■ 8.17 The Embryo Yielding Its Secret

In our coauthored book *Liebling, Du hast die Katze geklont* (Honey, you have cloned the cat) **Jens Reich** (Fig. 8.52) says:

"The 0.1 percent of our genome that differs from that of other people makes it our very own unique genome. However, to think that this **0.1 percent** would tell you how tall or intelligent a person will become, what eye color they will have, what food they will prefer, and how musical or sporty they will be, remains wishful thinking.

Although many human qualities have been genetically hardwired, the fulfillment of their potential largely depends on other life circumstances. Size and health depend on nutrition, and intelligence can be significantly boosted during the first three years of a child's life. A talent such as playing the piano may remain undiscovered, however, if the person in question does not get the chance to open a piano or lacks the perseverance to practice.

Similarly, most forms of diabetes only develop in an environment of convenience food and stress.

On the other hand, it is also true that all these traits cannot be put down to just one single gene, but several – sometimes even several thousand. Which genes are responsible for what can usually be found out only in animal experiments. Since mice are unable to play the piano and their idea of beauty differs from ours, this is a route that we cannot go down to find out which gene combinations are responsible for musicality or beauty.

So far, little is known about the function of single genes, and the individual portion of the genome does not give away much about the individual. Does this mean that there is no need to be worried about the privacy rights of the individual citizen?

The use of data that have been gathered with mutual consent for medical or research purposes in any other context must be strictly prohibited. There must be a clear understanding that contract partners (employers, insurance companies) have no right to the revelation of such data, and the individual cannot use them for their own purposes. These data must be treated as nonexistent for all other purposes except for the one they were collected for in the first place. A DNA profile does not give a personality profile.

Each individual has 1,000 billion nerve cells, each of which has about 1,000 links to other nerve cells. It is impossible that this highly individual and specific network could be coded in just three billion DNA letters.

Genetic data on their own would not be too compromising, but the danger lies in their combination with other data – with health records, with records of websites visited, with shopping behavior, with a history of telephone contacts or bank account activity or even credit card records to track down the movements of a person. "

Homunculus?

It flashes, see! Now truly we may hold
That if from substances a hundredfold,
Through mixture – for on mixture all depends –
Man's substance gently be consolidated,
In an alembic sealed and segregated,
And properly be cohobated,
In quiet and success the labour ends.
Twill be! The mass is working clearer,
Conviction gathers, truer, nearer.
What men as Nature's mysteries would hold,
All that to test by reason we make bold,
And what she once was wont to organize,
That we bid now to crystallize.

J.W. Goethe, Faust II

Cited and recommended Literature

- **Thiemann WJ, Palladino MA** (2004) *Introduction to Biotechnology*. Benjamin Cummings, San Francisco. Chapter about Animal Biotechnology is a fine introduction.

- **Watson JD** (2004) *DNA. The Secret of Life*. Arrow Books, London. This book is a MUST HAVE, "an immediate classic" (E.O Wilson)

- **Brown TA** (2001) *Gene Cloning and DNA Analysis- An Introduction*. 4th edn. Blackwell Science, Oxford. A companion to this book, covering genetic engineering.

- **Beaumont AR, Hoare K** (2003) *Biotechnology and Genetics in Fisheries and Aquaculture*. Blackwell Science, Oxford. Biotech goes to sea... Good overview!

- **Wilmut I, Campbell K** (2001) The Second Creation: Dolly and the Age of Biological Control. Harvard University Press, Cambridge.

- **Wilmut I, Highfield R** (2006) After Dolly: The Uses and Misuses of Human Cloning . W. W. Norton, New York.

- **Bains W** (2004) *Biotechnology from A to Z* (3rd edn.) Oxford University Press, Oxford. A great reference book covering the entire field of biotechnology

- **Barnum S R** (2006) *Biotechnology: An Introduction*, updated 2nd edn., with InfoTrac® Brooks Cole, Belmont. Reliable beginners' textbook. Susan Barnum is author of an Expert's View Box in this book.

- **Lim H A** (2004) *Sex is so good, why clone*? Enlighten Noah Publishing, Santa Clara Entertaining history of "Clonology"

- **Scientific American Reader** (2002) *Understanding Cloning* (Science Made Accessible) Warner Books, New York. Good reviews, nicely illustrated.

- **Rifkin J** (1998) *The Biotech Century*. Harcourt Publishers, Boston: Worth learning from Jeremy Rifkin, the prominent critic of GMOs, about the concerns about biotech.

Useful Weblinks

- Wikipedia, for a start, e.g., cloning
  http://en.wikipedia.org/wiki/Cloning

- "Cloning" Freeview video by the Vega Science Trust and the BBC/OU
  http://www.vega.org.uk/video/programme/15

- Clone Guide – good resource to human reproductive cloning information
  http://www.reproductivecloning.net/

- Click and Clone. Try it yourself in the virtual mouse cloning laboratory, from the University of Utah's Genetic Science Learning Center
  http://learn.genetics.utah.edu/units/cloning/clickandclone/

- World Wide Web Virtual Library: Biotechnology, catalog of WWW resources:
  http://www.cato.com/biotech/

- DNA interactive, mainly for teachers: *http://www.dnai.org/*

- Everything you wanted to know about GM organisms, provided by New Scientist.
  http://www.newscientist.com/channel/life/gm-food

Eight Self-Test Questions

1. How many natural breeding bulls could in theory be replaced by one artificial insemination bull?

2. Apart from chromosomal donor DNA, which kind of DNA could have an impact on a cloned animal?

3. How did Professors Wolf and Pfeifer in Munich make piglets glow green?

4. In cat cloning, is the coloring of the cloned kittens the same as that of the donor cat?

5. How can a jellyfish gene be useful in transgenic experiments?

6. What are the specific opportunities that open up for the pharmaceutical industry in transgenic cattle, goats, and chickens? What advantage would they have over traditional production methods?

7. How is it possible to show which cells are omnipotent in frogs?

8. How could threatened animal species be saved from extinction through embryo transfer?

MYOCARDIAL INFARCTION, CANCER, AND STEM CELLS

BIOTECHNOLOGY IS A LIFE SAVER

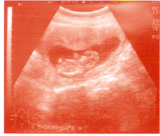

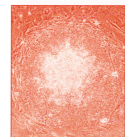
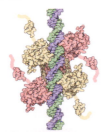

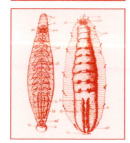

CHAPTER **9**

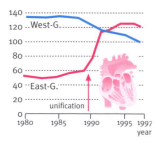

Fig. 9.1 Deaths caused by coronary infarction in both parts of Germany before and after reunification (per 100,000 inhabitants).

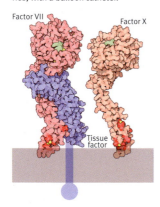

Fig. 9.2 If someone has had a heart attack, rapid aid is crucial, involving police and ambulance in a coordinated rescue operation. Once a coronary infarction has definitely been diagnosed, there are two possibilities: enzymatic lysis (see main text) and, increasingly, percutaneous transluminal coronary angioplasty (PTCA), a procedure to dilate the restricted coronary vessels, the coronary arteries, with a balloon catheter.

Fig. 9.3 Tissue factor (above) binds to factor VII and activates it a thousand-fold; the activated factor VIIa activates factor X and as factor Xa, this then activates thrombin.

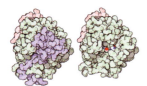

Fig. 9.4 Hirudin (here in blue) from the leech binds perfectly to thrombin and blocks the active center (shown right without hirudin).

Fig. 9.5 Clotting cascade (Figure to the right).

9.1 Myocardial Infarction and Anticoagulants

Myocardial infarction (Fig. 9.1) and stroke are among the most frequent causes of death in developed countries (more detail in Ch. 10). The enzyme **thrombin** is the key to **blood clotting** (Fig. 9.3), a process which starts with molecules that indicate that something is "going wrong". In vascular injuries with damage to the endothelium (the thin, single layer coating of smooth cells on the inside of blood vessels), the underlying **tissue factor** is released by the endothelial cells. This starts a cascade of enzyme activation, beginning with a few tissue factor molecules and building up like a pyramid (Fig. 9.4).

The tissue factor first activates a few molecules of inactive **factor VII** to become active factor VIIa, and these proteolytically activate a large number of **factor X** molecules into Xa in tissue factor complexes. The tissue factor itself is not a protease but a transmembrane glycoprotein; it forms a complex with circulating factor VIIa, which finally enzymatically activates factor X.

In its turn, factor Xa activates the preliminary stage thrombin, prothrombin (factor II), to thrombin (factor IIa). The protease thrombin now splits off a part of the **fibrinogen**: this creates fibrin and forms a

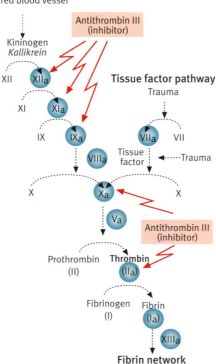

Contact activation pathway
Injured blood vessel

Fibrin network

fibrin network which encloses blood cells within it. This is how a dark red scab is formed on a wound.

Anticoagulants prevent blood clotting and the formation of a blood clot (thrombus). **Heparin** and cumarin derivatives are known for their ability to inhibit blood clotting (Fig. 9.7). **Cumarin** comes from woodruff (*Asperula odorata*) and is responsible for woodruff's typical smell. It is also used in the form of warfarin as a rat poison, which causes rats to bleed internally. Cumarin works as a vitamin K antagonist.

Acetylsalicylic acid (ASS; Aspirin) is an ancient and unsurpassed drug for inhibiting blood clotting. It is a derivative of salicylic acid, originally isolated from willow bark (Lat. *salix*, willow) (Fig. 9.21). Aspirin thins the blood and is recommended for patients at risk of coronary infarction. At the onset of a coronary infarction, it is recommended that the emergency doctor be called immediately. The physician may give the patient Aspirin tablets which can be quickly chewed and swallowed. New rapid immunological tests help to make a definite diagnosis quickly (Ch. 10 and end of this chapter).

Hirudin or hirundin is also interesting. It comes from the medicinal leech (*Hirundo medicinalis*, Figs. 9.4 and 9.6) and is the only anticoagulant that is genetically engineered at present (trade name Lepirudin), making the leech one of the few animals that we use in medicine today.

9.2 Fibrinolysis Following Coronary Infarction: Using Enzymes to Dissolve Thrombi

Thrombolytics (or **fibrinolytics**) dissolve thrombi that have already formed. They are proteases (protein splitting enzymes, see Ch. 2). In fibrinolysis, fibrin clots are dissolved after vascular occlusion by a group of proteins. The main component of clots in thrombosis or embolism is the protein fibrin. **Plasminogen activators** first change plasminogen into the active fibrin-splitting enzyme plasmin. Plasmin is a serine protease (Ch. 2). An endogenous plasminogen activator is known as **t-PA** (tissue plasminogen activator).

If, for example, a coronary vessel is blocked, that area of heart muscle is cut off from its oxygen supply, leading to coronary infarction with possibly fatal consequences. In the search for a treatment for the causes, researchers have concentrated on dissolving thrombi as effectively and with as little risk as possible within the first few hours – i.e., before the final death of the tissue.

Until recently, **streptokinase** has been an all-time favorite for thrombolysis. This is an exogenous, foreign plasminogen activator. In spite of the "-ase" ending, this is not an enzyme, but a nonenzymatic protein produced by streptococci, which indirectly activates the clotting system.

As soon as streptokinase is injected, it forms an activator complex with the plasminogen which activates further plasminogen molecules. Streptokinase is derived from streptococci and is very effective, but it can cause allergic reactions in the patient's immune system because of its bacterial origin.

The enzyme **urokinase**, on the other hand, is derived from human urine or human kidney cell cultures; it has recently even been genetically engineered with *Escherichia coli* bacteria. If urokinase or streptokinase is injected in sufficient quantity, it activates the plasminogen to active plasmin and eliminates blood clots.

However, both proteins work throughout the vascular system, i.e., the effect is not limited to the "place of damage". This has a strong effect on the whole clotting system. Compared to modern fibrinolytics, there is at least a tendency to a higher risk of internal bleeding (e.g., if the patient suffers from stomach ulcers).

With modern drugs, only the fibrin-bound plasminogen is activated. The effect is predominantly limited to the damage site. **t-PA is a natural tissue type plasminogen activator** and only works where it is needed, at the thrombus. Therefore, it does not reduce the clotting ability of the blood throughout the whole body. However, with a coronary infarction, this is generally too much to ask of the body's own t-PA and it must be supported with additional recombinant t-PA molecules (rt-PA). rt-PA is the gold standard in fibrinolytic **treatment of coronary infarction**. If treated promptly, the patient has a realistic chance not only of living but of maintaining the full function of the heart muscle tissue under threat.

Trials have shown that in the first few hours, 60 to 80 extra lives per 1000 of those treated have been saved. Treatment commencing later than three to four hours after the infarction can still relieve the pain (by opening the vessel again) but the probability of it being too late to preserve the heart muscle tissue increases. Modern fibrinolytics can be injected immediately in the emergency ambulance.

Human t-PA was first cloned in 1982 and was approved about 5 years later. As the molecule contains sugar as well as protein, it cannot be "simply" produced by recombinant bacteria; it is produced at a much higher cost in **mammalian cell culture** (ovarian cells from the Chinese hamster, CHO cells) and also in the milk of transgenic sheep and goats (Ch. 8).

Even with rt-PA, the tactics of biotechnologists revolve entirely around fighting human diseases with biotechnologically produced human substances. By targeting molecular changes, the efficacy and the time for the t-PA to take effect is increased. This has led to the development of rt-PA/Reteplase (Rapilysin® from Hoffman-La Roche), which is produced in *E. coli* cells, and TNK-t-PA (from Boehringer Ingelheim), which is made in CHO cells. TNK-t-PA is an excellent example of protein design.

■ 9.3 Stroke: Help from the Vampire Enzyme

Stroke is the third most frequent cause of death in industrial countries after coronary disease and cancer and is also one of the most frequent causes of severe long-term disability. In industrial countries, 300 to 500 out of every 100,000 people suffer a stroke every year. In the USA, that is 700,000 victims and in Germany, 200,000.

20% of stroke victims die within the first four weeks. Only about one third of survivors can return to a normal working life. Another third remain dependent on the help of other people for the rest of their lives. The financial strain on the health services resulting from stays in hospital caused by stroke and long-term care programs for patients is enormous. In 2004 in the USA alone, it amounted to about 54 billion dollars. At the same time, few stroke patients today receive direct treatment for the cause.

A stroke occurs if one of the blood vessels in the brain is blocked by a blood clot (ischemic stroke) or ruptures (hemorrhagic stroke). In either case, the affected part of the brain is no longer supplied with sufficient oxygen or nutrients. This causes the nerve cells in the affected area of the brain to die. The type

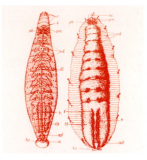

Fig.9.6 In the past, leeches (*Hirundo medicinalis*) were used as a living cupping glass for lowering blood pressure. Leeches bite into the skin with their hundreds of tiny pointed teeth and inject hirudin which prevents mammalian blood from clotting and enables them to suck better. The leech grows by up to ten times in size while it is feeding but is also able to fast for up to 18 months.

Fig. 9.7 A heparin injection before a long journey helps to prevent thrombosis.

Fig. 9.8 Vampire bats have tPA, the model for genetically engineered rtPA.

Of course, it's cheaper to train on drugs than in the mountains!

Box 9.1 The Expert's View: Perspectives on "Red" Biotechnology

Biotechnology, which predominantly refers to the development of drugs and diagnostic tests, is opening up all sorts of possibilities for the pharmaceutical industry.

First of all, the drugs market is an **attractive growth market**, which should also increase dramatically with the demographic development of the Western industrial nations; secondly, the **social acceptance** of biopharmaceuticals is very high. However, we may not be able to make use of all biotechnological opportunities because of the serious ethical connotations in this area. This is especially the case in Europe where there has been great suspicion about the use of recombinant plants for agriculture. However, this problem is on the decrease according to recent developments there.

Gene Therapy and Stem Cell Research Raise Hopes

The decoding of the **human genome** in 2000 represented a milestone in improving the diagnosis and treatment of diseases. In the future, rapid breakthroughs in the development of new drugs and a high rate of growth for sales and profits for the biotech companies may be expected. However, the sense of euphoria created by this does not take into account the fact that the development of new drugs takes on average 12 years and only a few drugs survive long-term testing and achieve eventual market approval.

An important field of activity for Red Biotechnology is the fight against **cancer** – one of the most lucrative markets for new drugs. This disease is one of the most frequent causes of death in industrial countries. As the likelihood of developing cancer increases as we get older, the increasing age of the overall population also means that the number of patients increases significantly. The biotech products that may be expected might be able to stop or slow down the reproduction of diseased cells, without the severe side effects that normally accompany such treatments as chemotherapy. Researchers have given up hope of overcoming all forms of cancer with just one drug, and the selective treatment of tumors is a subject of great interest and activity.

With the aid of **stem cells**, it may in the future be possible to cure diseases caused by tissue defects. These include injury to the bone marrow, diabetes, the consequences of a heart attack, and various hemopoietic diseases.

It is hoped that red biotechnology will lead to a breakthrough for **individualized drugs**, i.e., to the development of tailor-made drugs for specific patient groups. Human genome research offers the possibility of creating a precise genetic profile for each patient. This is a crucial prerequisite for individually targeted treatment. Based on the individual genome, biochips should be able to find out whether and in what connection drugs would work best in an individual patient.

Legal Framework

Research with embryos was banned in a number of developed nations in the 1990s. This led to a reduction in research activities in this field. The import and use of embryonic stem cells for research purposes has been governed by laws on stem cells which came into force in more recent years. Thus, the import and use of **embryonic stem cells** is prohibited, although there are certain exceptions.

Discoveries in Red Biotechnology involve a high level of investment, long periods of development and complex approval procedures. However, pharmaceutical and biotech companies normally only make these investments and accept the risks linked with them if their research results are protected by **patents** under the certainty of the law. Effective patent protection is therefore crucial.

The EU guidelines on the protection of biotechnological discoveries were adopted in 1998 (BioPatent guidelines). After that, no patents could be granted for the cloning of human beings or for the use of human embryos for industrial or commercial purposes. Neither could the creation and development of the human body be patented.

Many small German biotech companies were overtaxed with coping with the whole process from research to development and then marketing of the drugs. Almost half of the companies questioned by Ernst & Young therefore aspired only to development as far as clinical phase II (first test on patients).

It is predominantly the **large pharmaceutical companies** that currently need the innovative strength of Red Biotechnology. One of the main reasons is that many patents for drugs that sell well will be running out in the next two to four years; there are also gaps in the development pipelines of these companies.

A series of biotech companies have already positioned themselves as **research services providers** for established pharmaceutical companies. This is an excellent way for biotech companies and large pharmaceutical companies to become connected. While the pharmaceutical companies have strengths in the later clinical phases and in production, approval, marketing, and sales, the biotech companies specialize predominantly in development of the active substance and applications research.

We estimate that German Red Biotechnology will continue its **rapid growth in sales** until the end of the decade. The $57 billion market for therapeutic proteins in 2005 should continue to increase. The well-filled development pipelines of the biotech companies will play a large part in this.

In general, the medical and pharmaceutical industries benefit from the **increasing age of the population**. From 2010 onwards the "Baby Boomer" generation (the high birth rate in the years between 1946 and 1964) will reach retirement age. The increasing average life expectancy also contributes to this. The number of elderly people in the total population will increase in all industrial countries and this population group is particularly indicated for medical care.

To sum up

Biotechnology will remain a **very dynamic growth market**. The number of possibilities for tackling and curing previously untreatable diseases and closer cooperation with the traditional pharmaceutical industry suggests that this will be the case. All phases of the supply chain will be used in biotechnology, from development to marketing of the drugs.

Companies in the industry will not be able to deal with all the research possibilities, as the legislation in certain countries is very restrictive and they will have to take into account the ethical aspects more than in other industries. The results of genomic and cell research will only be reflected in products that will sell strongly in the medium and long term. However, the course for this must be set today.

Dr. Uwe Perlitz is Senior Economist at Deutsche Bank Research, Frankfurt am Main.

of treatment required depends on whether the stroke is caused by vascular occlusion (88% of all strokes) or intracranial bleeding (12%). An **ischemic stroke** (vascular occlusion) is treated by removing the blood clot that caused the stroke and restoring the blood supply.

The only drug that has so far been approved for treating the cause of an acute ischemic stroke is the **thrombolytic rt-PA** (see above), the effect of which is based on dissolving the blood clot. However, the use of rt-PA is limited, as it must be administered within the first three hours after the symptoms of a stroke appear. In view of this very narrow window of time, only a small number of acute ischemic strokes are treated with rt-PA.

Desmoteplase, a highly selective plasminogen activator, is a new active substance from the German company, Paion AG. This genetically engineered protein was originally found in the saliva of the vampire bat (*Desmodus rotundus*) (Fig. 9.8) whose source of nutrition is the blood of mammals.

Desmoteplase activates fibrin-bound plasminogen. After activation, plasminogen is converted into the enzyme plasmin, which dissolves the thrombus by breaking up the fibrin matrix. Blood flow to the affected tissue is restored and the ischemic damage, which would otherwise have continued, is minimized.

The advantage of Desmoteplase is that it is not neurotoxic. Positive clinical data from a phase-II study which has been concluded indicate that Desmoteplase can still be effective after the appearance of stroke symptoms in a time window of up to nine hours.

■ 9.4 Genetically Engineered Factor VIII – Safe Help for Hemophiliacs

"Kings are carefully protected from unpleasant realities. The hemophilia of Zarewitsch (**Alexej of Russia**, Great Grandson of **Queen Victoria**) was really just a symptom of the chasm between royalty and reality", wrote the English biologist J. S. B. Haldane.

The blood disease **hemophilia** A (Fig. 9.10) affects only men but can be passed on through a mother to her sons. The gene for clotting factor VIII must therefore lie on the X-chromosome.

Like the genetic disorder of the "Habsburg lip" [mandibular prognathism], the case of **Queen Vic-**

toria of England (1819-1901, Fig. 9.10) and her numerous offspring has been well documented in the history of the European aristocracy. The fact that so many of her male offspring became old enough to have children can only be attributed to their uniquely protected life "under a bell jar".

The number of hemophiliacs is small compared to the number of cases of infarction, only affecting one man in 5,000, and can be increasingly treated by biotechnology. Hemophiliacs are lacking an important component of the blood clotting system, **factor VIII**. 80% of hemophiliacs have hemophilia A, i.e., factor VIII deficiency, while the rest have a lack of factor IX (hemophilia B). Worldwide approximately 400,000 people are **hemophiliacs**.

The symptoms of both types of hemophilia are the same. They are more dramatic the greater the clotting factor deficiency. In severe forms, minimal injuries can lead to unstoppable life-threatening external bleeding or hemorrhages into the tissues or the joints. With milder forms, hemorrhage needs to be taken into account predominantly during surgical procedures. Treatment consists of the intravenous administration of the missing clotting factor, and the dose depends on the risk of bleeding, i.e., it needs to be relatively high for children and before operations.

Less than 500 grams (1.10 lb) of factor VIII should cover the needs of the world, but this small amount had so far cost 170 million dollars without biotechnology!

The factor VIII protein consists of 2,332 amino acid components. It is therefore one of the largest proteins whose structure has been elucidated so far. Each person has one milligram of factor VIII in his approximately ten pints (6 liters) of blood.

A hemophiliac must be injected with one milligram of the factor twice a week in order to live a normal life. If one donation of blood is one pint (½ liter), we theoretically need 24 blood donors a week for a single hemophiliac! However, the extraction and processing of donated blood is not just expensive, it is also extremely risky: the main danger used to be infection with hepatitis B viruses which cause jaundice, but today there is the risk of infection with the AIDS virus or hepatitis C (see Ch. 5).

It is estimated that 60% of hemophiliacs have been tragically infected by donated blood, some of those in France criminally so. In the USA, nearly 90% of

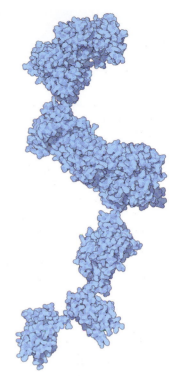

Fig. 9.9 Structure of blood clotting factor VIII.
This huge structure is not yet available. To give an impression, it was composed by David Goodsell (Scripps, La Jolla) from several Protein Databank data.

Fig. 9.10 Queen Victoria of the United Kingdom commemorated on a Hong Kong stamp. Born in 1819 and ascending to the throne in 1837, Victoria was Britain's longest reigning monarch. Prior to her death in 1901, she had presided over a massive expansion of the British Empire and the continued rise of Britain as an industrial power.
Below: The family of the Tsars: little Alexei can be seen at the front on the left of the photo.

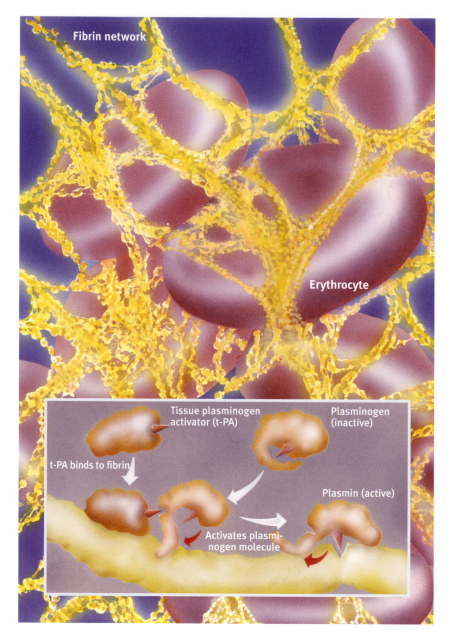

Fibrin network

Erythrocyte

Tissue plasminogen activator (t-PA)

Plasminogen (inactive)

t-PA binds to fibrin

Plasmin (active)

Activates plasminogen molecule

Fig. 9.11 How t-PA dissolves thrombi: the proenzyme is changed by fibrin into active serine protease which in its turn activates plasminogen in blood clotting. This essentially ensures local fibrinolysis.

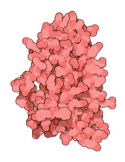

Fig. 9.12 Erythropoietin (EPO)

Americans with severe hemophilia became infected with AIDS in the 1980s when the nation's blood supply was contaminated. More than 50% of those infected with HIV have died.

On the other hand, the genetically engineered factor VIII is virus-free and safe. As **factor VIII is a glycoprotein**, it is created in mammalian cell lines and used for the treatment and prophylaxis of bleeding in hemophilia A patients.

Factor VIII has also been produced recently by **transgenic pigs** (Ch. 8). The complete human cDNA was combined with the promoter for whey acid protein from the mammary glands of pigs and factor VIII was expressed in the milk.

9.5 EPO for Kidney Patients and in Sport

Other genetically engineered products have followed factor VIII, such as the hormone **erythropoietin** (**EPO**) which is formed in the kidneys and stimulates the body's own erythrocyte production (erythropoiesis) in the bone marrow. The glycoprotein EPO is a **growth factor for red blood cells** and induces the formation of hemoglobin in the bone marrow.

EPO is important for the lives of dialysis patients, who suffer anemia as a result of dialysis.

In the past, athletes were sent to train at high altitude, since the "thin air" promotes the formation of red blood cells for a perfect source of oxygen in subsequent competitions. That can now be achieved more cheaply. EPO is produced, like factor VIII, in mammalian cells. The EPO sugar chains are important, since they protect EPO from rapid degradation in the liver and EPO is inactive without sugar.

In the Tour de France in 1998, whole teams dropped out of the race when they discovered testing was to be carried out by the police and independent doping experts, instead of the internal sport control bodies. The cyclists had supposedly been doped with EPO.

Hundreds of thousands of patients have used the life-saving growth factor, **GM-CSF** (**granulocyte macrophage colony stimulating factor**) which stimulates the formation of eosinophils, neutrophils, and macrophages.

EPO has a world market value of about twelve billion US dollars each per year. GM-CSF's market is about 2 billion dollars.

9.6 Interferons for Fighting Viruses and Cancer

Why is it that when a human being or an animal is infected with a virus such as influenza, they almost never catch a second viral disease at the same time? With bacterial diseases, one type of bacteria often clears the way for another because the body's powers of resistance have been weakened. Why should it be different with viral diseases?

This question was posed by **Alick Isaacs** from England and **Jean Lindemann** from Switzerland in 1956 when they were working at the National Institute for Medical Research in London. They

Box 9.2 What are Interferons?

Interferons are endogenous proteins which have a rapid and nonspecific guarding mechanism against viruses. While a virus is growing in a cell, a by-product of virus multiplication evidently induces the production of interferon in the cell. The interferon is released and migrates to other cells in the body. It binds to specific **interferon receptors** on the cell surface and sends out a particular signal (Fig. 9.11). This signal causes the cell to produce several new proteins, but these proteins alone are not sufficient to disturb the cell mechanism. They are passive until a cell is infected by a virus. The proteins are then activated and inhibit further growth of the virus.

A number of interferons have been found so far; the three most important are a, β, and γ interferons.

α-Interferon is mainly formed by white blood corpuscles (leukocytes) after a virus attack. It moves freely from cell to cell through the body in blood serum. At least 16 subtypes of α-interferon are known so far. They are differentiated only by slight changes in their chemical structure – in the amino acid sequence.

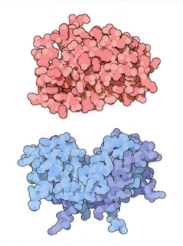

α-Interferon (above) and γ-interferon

α-Interferon acts as a unique drug against a rare blood cancer, hairy cell leukemia. It is also effective against melanoma – a cancer that comes from the pigment cells of the skin and is known to be malignant. It also works against myeloma (cancer of the bone marrow) and hepatitis B and C.

β-Interferon is formed from a series of cell types, including fibroblasts (connective tissue cells), which are responsible for the building of connective tissue. β-Interferon is used predomi-

nantly against multiple sclerosis (MS, an inflammatory disease of the central nervous system). Multiple sclerosis is a chronic, inflammatory condition of the nervous system and is the most common nontraumatic neurological disease in young adults. MS affects approximately 350,000 Americans.

γ-Interferon is formed by white blood corpuscles in the spleen during an immune reaction (response of the organism to a foreign body). This simultaneously creates antibodies or specific cells to kill cells that have been attacked by a virus or to reject foreign tissue.

γ-Interferon is known by its **antiviral properties**. It is a stronger modulator of the immune system than the two other types of interferon and was approved for the treatment of granulomatosis. Granulomatosis is a congenital defect of the oxidative metabolism of granulocytes, a subgroup of white blood cells. The granulocytes of a diseased person do not form aggressive free radicals and cannot kill pathogens. Patients must therefore receive permanent treatment with antibiotics.

Interferon may also be effective against arthritis (inflammation of the joints) and asthma.

---

found the substance responsible for this in 1957. Protein is secreted by cells affected by a virus and makes other cells resistant not only to infection with this virus, but also to other types of virus. Isaacs and Lindemann named this substance **interferon**, because it apparently prevents or interferes with viral reproduction (Box 9.2).

It was clear from the moment it was discovered that interferon was very promising as an antiviral agent, as it not only directed itself against any virus, it also protected cells from a whole series of viruses. However, interferon can do much more; it influences various cell activities in a manner that suggests other therapeutic possibilities. Interferon is also a highly effective drug, as a tiny quantity is sufficient for long-term protection. Assuming the correct dose, interferon is a natural cell product and thus it is probably safer than most chemical products that have been tried as drugs. However, interferon is expressed in such tiny quantities by the cells that produce it that many researchers even doubted the existence of the substance.

The situation began to change when **Kari Cantell** from the Finnish Red Cross developed a production

technique in the 1970s that could produce interferon from human blood. Cantell collected sufficient blood from blood donors in Finnish blood banks to produce interferon for clinical tests. However, this process was complicated and costly: it involved infecting leukocytes from blood donors with a virus, then collecting and purifying the released interferon. Using this technique, a total of only half a gram of partially purified interferon could be collected from more than 1,300 gallons (50,000 liters) of blood plasma!

Although the medical profession showed great interest in interferon, the research had to stay on a back burner because of the extremely tiny quantities available. The **blood of 100,000 donors would have yielded one gram** of interferon which would have contained at best 1% of the pure substance. It was calculated that this single gram of pure interferon would have cost a billion dollars – although such a quantity was never obtained.

In spite of the difficulties, the evidence began to mount that interferon could be successful in the battle against viral diseases and perhaps some forms of cancer. In January 1980, the news came from

Fig. 9.13 γ-Interferon (in red) binds to receptors (yellow) on the cell surface and the signal is transferred into the cell.
The one interferon molecule outside the cell brings together four receptors. Various protein kinases (orange) are bound in the inside and are phosphorylated and activated. They add a phosphate group to the STAT proteins (pink). The proteins now form dimers which adhere to the target genes (inserted picture) in the cell nucleus and trigger synthesis.

Types of cancer

More than 200 types of cancer are known in humans. The main ones are as follows:

- Lung cancer is at the top of the list for both sexes (it is often caused by smoking, environmental pollution, or asbestos).
- Breast cancer is the second most frequent fatal cancer in women (it is age-dependent and there is often a possible hereditary predisposition).
- Prostate cancer is the second most frequent cancer in men (it is age-dependent).
- Colorectal cancer (bowel cancer) is the third most dangerous form of cancer in either sex (encouraged by a diet including a great deal of fat and little fiber, although there can also be a hereditary element).
- Skin cancer is caused by UV rays or chemical mutagens.
- Ovarian cancer (affects women over 40 and risk factors include a high level of estrogen and diet).
- Leukemia, uncontrolled growth of white blood cells.
- Ovarian (cervical) cancer in women over 30 (increased risk caused by smoking, early sexual activity and multiple sexual partners. The role of sexually transmitted diseases such as herpes is under discussion).
- Testicular cancer is the most frequent cancer in young men.

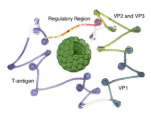

Fig. 9.14 SV40 (Simian virus 40) is a polyoma (DNA) virus found in both monkeys and humans.

It is enclosed by a spherical capsid composed of 360 copies of one protein, and a few copies of two others. This capsid encloses a small circle of DNA 5243 nucleotides long. To replicate its DNA and to package it inside new viral capsids, amazingly, SV40 only needs one protein, the T-antigen. T-antigen is formed and can occasionally transform the cell into a cancer cell. SV40 has not been proven to cause disease in humans, but several studies have suggested a link to cancer based on the presence of relatively large amounts of what may be SV40 DNA fragments in some tumor tissues, particularly non-Hodgkin's lymphoma.

Switzerland of an "interferon boom", which provided a crucial breakthrough for the further development of biotechnology. Precisely on Christmas Eve, researchers at the laboratory of the Zurich molecular biologist **Charles Weissmann** found that laboratory strain HiF-2H of *E. coli* began to produce a protein with biological properties that precisely resembled interferon from human leucocytes. This was the first obviously successful biological synthesis of the human antiviral substance interferon.

In contrast to attempts at the bacterial synthesis of insulin, Weissmann and his group were dealing with human leukocyte interferon, a substance where the precise structure was not yet known in detail. The news of the interferon-producing bacteria in Zurich therefore came unexpectedly early, although the possibility had long been known.

It was not until the successful synthesis of this protein in *E. coli* that its properties also became known. Instead of 1,300 gallons (50,000 liters) of blood, now 20 pints (10 liters) of bacterial culture solution were sufficient to produce half a gram of impure interferon. However, it also became clear that the name **"interferon" included a whole family of substances** (Box 9.2).

These days, many of the major pharmaceutical companies all over the world are involved with interferon research. The current worldwide market for α-interferon is 3.8 billion US dollars. The three worldwide ß-interferon manufacturers, Serono, Schering, and Biogen, made two billion US dollars in sales of β-interferon in 2001 alone. γ-Interferon has a market volume of 200 million dollars.

Interferons served as a model for the development and production of a series of mediators, such as **interleukin-2**, which causes the immune cells of the body to divide and is helpful in the treatment of the immune deficiency disease AIDS. The experiences with interferons were also promising in a series of viral and other diseases. Efficacy against hepatitis B and C as well as against various forms of herpes (virus diseases of the skin and mucosa) has already been proved. Interferons are now mainly used for indications where it is as yet unclear why they are effective – diseases that do not have a great deal to do with viral infections, particularly cancers such as cancer of the bladder, myeloma, melanoma, and lymphoma.

At the moment, interferon is the only drug available for the relatively rare **hairy cell leukemia**, a cancer of the blood. Three milligrams of interferon are sufficient to treat one patient. Patients with multiple sclerosis, rheumatoid arthritis and chronic granulocytomatosis are being given new hope because of interferon. It is interesting that interferon is used as a prophylactic measure against viral diseases in the elderly or infirm in risk situations.

■ 9.7 Interleukins

Interferon belongs to the **cytokine** family. These are polypeptides (often glycoproteins) which are formed from activated cells from the immune system, the hematological system, or the nervous system. Cytokines also include the interleukins, colony stimulating growth factors, chemokines, inflammatory cytokines, and anti-inflammatory factors. Cytokines are always of cellular origin, and however different they may be, they are always freshly produced as required and never stored.

The most important cytokine (**lymphokine**) formed from lymphocytes so far is **interleukin-2** (**IL-2.**). It was first found in 1976 at the National Institutes of Health (NIH) in the USA. IL-2 promotes the **activity and growth of T-cells** (a subgroup of lymphocytes) and other immune cells. As the vital helper T-cells of the immune system are destroyed by the AIDS virus, interleukin-2 is extremely interesting for the treatment of AIDS. It works like a messenger molecule, which uses the leucocytes with other mediators to regulate the immune response. IL-2 is the first of more than 20 interleukins, the "**hormone of the immune system**", was approved as Proleukin® for the company Chiron (now Novartis) for the treatment of renal cell cancer. Future clinical applications for interleukins include asthma, AIDS, lung cancer, and inflammatory diseases.

■ 9.8 Cancer: Abnormal, Uncontrolled Cell Growth

Cancer is the general term for any disease that involves **abnormal and uncontrolled cell division** (proliferation) (see this page, left border column). The following is only an extremely simplified introduction into this complex area.

New cells which do not have any useful function in the body and form a tumor are called neoplastic

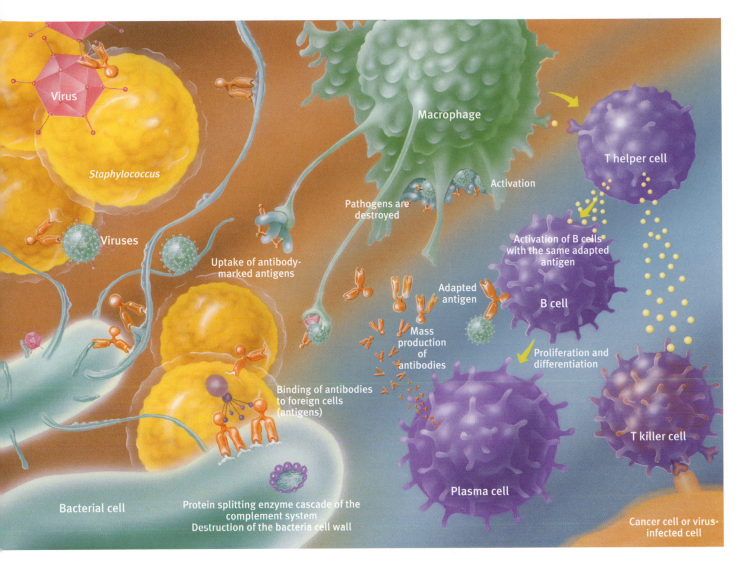

Within the figure:
Virus
Staphylococcus
Viruses
Uptake of antibody-marked antigens
Macrophage
Pathogens are destroyed
Activation
Adapted antigen
Mass production of antibodies
Binding of antibodies to foreign cells (antigens)
T helper cell
Activation of B cells with the same adapted antigen
B cell
Proliferation and differentiation
T killer cell
Plasma cell
Bacterial cell
Protein splitting enzyme cascade of the complement system
Destruction of the bacteria cell wall
Cancer cell or virus-infected cell

cells. **Benign tumors** contain cells that resemble normal cells. Because they are enclosed in a capsule of fibrous extracellular material, they remain in one place and do not form secondary tumors. **Malignant tumors**, on the other hand, contain abnormal cells which can become detached. They migrate to other parts of the body, spreading the tumor and forming secondary tumors (**metastases**). Since it is difficult to control this metastasis, it is essential to diagnose and treat cancer as early as possible, before the tumor cells have spread throughout the body.

Carcinomas are tumors that grow from the epithelial cells of the inner and outer surfaces of the body such as skin, gastric mucosa, mammary glands, and lungs. **Sarcomas** arise from cells in support structures such as bones, muscles, and cartilage. **Neoplasms** of the immune and hematopoietic system develop from blood cells or their precursors and include leukemia, lymphoma, and myeloma.

Cancer is occasionally inherited or caused by infection. Familial adenomatous polyposis and breast cancer can be inherited, for example. Cancer of the neck of the womb (cervix) (human papilloma viruses), certain forms of cancer of the blood (retrovirus HTLV-1) and liver cancer (hepatitis) can be encouraged by viruses.

Cancer is a genetic disease because it always begins with a change in the DNA. It is acquired by somatic cells during the life of an individual and is not passed on to the next generation, but a familial predisposition can favor the disease. However, a single gene mutation is insufficient to cause cancer in humans, as there are complex regulatory networks to counteract this. Cancer is therefore a **disease of many stages**, as individual neoplastic cells change progressively into abnormal and aggressive cells.

Two important classes of genes are directly linked with the loss of growth control. **Proto-oncogenes**

Fig. 9.15. How the immune system fends off infection (simplified, see Ch. 5)

Infection mobilizes the immune cells so that, first of all, the antigen of an antigen-presenting cell (macrophage) must be taken up, which processes the antigen. It appears on the cell surface and is presented to a T-helper cell which is stimulated to form interleukin 2 which activates the B-cells which have also had previous antigen contact.

These B-cells proliferate rapidly and become differentiated into plasma cells which form antibodies which mark the antigen for destruction.

The complement system (below left) and killer T cells (below right) work alongside, continuously searching the surface of all cells and killing those cells with signs of foreign bodies.

Box 9.3 Yew Trees, Paclitaxel Synthesis, and Fungi

In August 1962, the 32-year-old botanist, **Arthur S. Barclay**, peeled off the bark of a **Pacific Yew tree** (*Taxus brevifolia*) under contract to the US Department of Agriculture. This took place in the Gifford Pinchot National Forest near the famous Mount St. Helens. He sent the samples to a laboratory in Maryland to be tested for their biological activity. The tests included inhibition of growth in the cancer cell culture "9KB". By September 1964, it was clear that yew extract inhibits cancer cells.

There is now a commemorative plaque at the site to celebrate the 40th anniversary. In 1967, **Monroe Wall**, who obtained the extract, informed the specialist world that **paclitaxel** showed an extraordinarily broad spectrum of anti-tumor activity. It was another three years until the crazy structure of the substance was evident: there were two molecules in one!

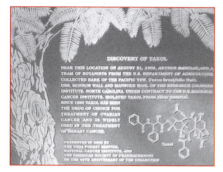

Plaque at the site where the active substance in yew was discovered.

The worldwide competition to synthesize the substance began in 1971. At the same time, the National Cancer Institute (NCI) was ordering ever greater quantities of yew bark reaching a peak in July 1977 of 3,500 kilograms (7,700 pounds) of bark from 1,500 felled yew trees.

The mechanism of action of paclitaxel on cancer cells was explained by **Susan B. Horwitz** (Yeshiva University, New York) in "Nature" magazine in 1979. It stimulated stability of the microtubules, not direct inhibition of cell division. It disturbed the essential dynamics of the otherwise lightning fast construction and disassembly of the **microtubules**, thus inhibiting cell division.

Paclitaxel and docetaxel impair cell reproduction and work predominantly on cells that divide quickly, like tumor cells. They stabilize the

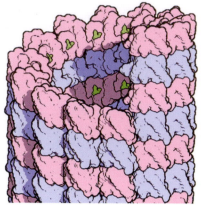

Microtubules are hollow, cylinder-shaped polymers of the cytoskeleton, are crucial in helping to decide the cell form and are of great significance in the separation of daughter chromosomes in cell division. A typical microtubule has a lifetime of only ten minutes before it is disassembled, after which the components are used for new microtubules. Microtubules are therefore excellent target sites for chemotherapy in cancer.

microtubules and block their necessary reorganization. Vinca alkaloids such as vinblastin and vincristin, on the other hand, bind to the ends of the microtubules and inhibit dimer uptake. Paclitaxel is shown in green in the illustration: It binds to tubulin and blocks the normal dynamics of disassembly and construction. The **cell activates its suicide program** – apoptosis – and dies.

Paclitaxel was released by the Food and Drug Administration (FDA) for phase I tests on 6th April 1984 after 22 years! The results of the phase-II tests for the most aggressive form of ovarian cancer were also published in 1984: The tumors shrank in three out of ten patients. However, it would have been necessary to fell 360,000 trees to treat all ovarian cancer patients in the USA for one year and tree cultures were not an option, as yew trees take hundreds of years to yield a good amount of paclitaxel.

The **environmentalists on the Pacific coast** had already been protesting for some time, so a means of synthesis had to be found! In 1980, a French group extracted the substance 10-deacetyl-baccatin III (10-DAB) from the needles of the **European yew tree** (*Taxus baccata*). It was possible to harvest the needles from the trees by the ton, without negative consequences for the environment. The 10-DAB taxane ring is the source substance for paclitaxel. **Pierre Potier** and **Andrew Greene** took six years to bind 10-DAB with the smaller paclitaxel part, a cluster of 34 atoms which looked like a

snowflake. The yields were slim, but from now on the aim was clearly to change 10-DAB into paclitaxel. The best synthesis chemists in the world competed for semisynthesis.

Synthesis chemistry intervenes

Robert Holton (Florida State University, FSU) produced his first paclitaxel publication in 1982. With one million dollars of research money the workaholic Holton managed to synthesize taxusin in 1988, a related compound that also occurs naturally in yew.

The paclitaxel crisis reached its climax in 1988 when the huge requirement had to face up to the protests of the Greens. The NCI wanted to withdraw from the project, even though it had already swallowed 25 million dollars! Other hopeful candidates had been deferred in favor of this substance, so Holton received an historic call from NCI: "Invent finally a damn semisynthesis!"

Robert Holton

At the end of 1989 Holton had developed a process that was double the productivity of the French; he patented it and began to write to the large pharmaceutical companies. Congress had already passed the Federal Technology Act in 1986, allowing the commercial use of state research results by private companies.

Rhone-Poulenc held the French patent and was competing for paclitaxel with Bristol-Myers (before its merger with Squibb) and two other giants. Bristol-Myers made a deal and Florida State University signed the agreement of its life on 1st April 1990, gaining an excess of money for research.

Bristol-Myers also took over the patent costs for all paclitaxel derivatives. Some would prove more effective than the mother substance itself. In 1992 the company patented a new semisynthesis set up by Holton, the metal alkaloid process. The source material for this process was 10-DAB, the complex precursor of paclitaxel, which was derived by extraction from renewable yew needles.

Production began in Ireland and took 31 years and costed 32 million dollars "from bark to business".

At the end of 1993, one gram of paclitaxel costed 5,846 dollars, bringing to an end the need to supply yew bark. It was a triumph for the environmentalists.

By 1995, innumerable women with cancer had had their lives extended by paclitaxel. In 1995 paclitaxel was the fastest selling anti-cancer product in the world. However, it also had toxic side effects caused particularly by the castor oil/ricinus oil used to make the paclitaxel soluble. Bristol-Myers recorded annual growth of 38 % through sales reaching a peak in 2000 of 1.6 billion. Paclitaxel was being sold at 20 times the price of the raw product. Bristol used the argument that development had cost a billion.

Bob Holton celebrated his triumph over almost one hundred other synthesis groups all over the world on 9th December 1993, having achieved total synthesis of paclitaxel – 40 synthesis steps with 2% yield. Holton had also found hundreds of paclitaxel analogues which were now a potential goldmine and were being feverishly protected by the FSU in 35 patents. As the five-year agreement with Bristol ran out, Holton accused the company of not really testing the analogues, so as not to provide competition with their blockbuster paclitaxel. It was agreed that the FSU would use their own metal oxide process for the paclitaxel analogues, but would pay 50% of the licensing fee.

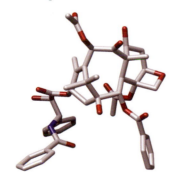

Structure of paclitaxel

The FSU had earned about 200 million dollars from paclitaxel research by 2000, unique in the history of the university. Even Bob Holton was now wealthy and founded his own company with his own money, Taxolog, Inc.

It seems that the analogue TL-139 will be successful, as it is 1000 times better than paclitaxel was at the same stage of development, according to Taxolog, Inc..

Holton's motto: "*If you have the opportunity to do something that could save someone's life, you have to do it!*"

… … and now the biotechnology!

The microbiologist **Gary Strobel** lives in Montana, a region where yew trees grow. He is an enthusiastic rambler and spends a lot of time in the mountains and forests of Montana. His passion is research into endophytes, microorganisms that use plants as hosts.

Gary Strobel

Strobel took samples from various plants on his treks with colleagues **Andrea** and **Don Stierle** from the Montana Institute of Technology. They found a fungus in 1993 on the bark of the Pacific yew in North West Montana. Strobel was possessed by the idea that microbes on plants produce similar products to their hosts. The fungus was therefore isolated and named *Taxomyces andreanae* after the host yew (*taxus*) and according to gentlemanly custom after the lady in the team, Andrea. After culture of the fungus and analysis of the product, Strobel had still not found what he was looking for but continued studying the structure he had found. He was proved right in the end, when he found the little fungus itself actually produced paclitaxel. Unfortunately, it produced it in quantities that were too tiny to be competitive. Nevertheless, Strobel had made the brilliant basic discovery that **microbial parasites imitate the chemistry of their host** and made the same chemical compound as the host plant. If it could be isolated and cultured it could be a discovery equivalent to penicillin, and there are thousands of other fungi!

Paclitaxel is the most promising anti-tumor drug of the last three decades. The worldwide market in 2001 peaked at billion dollars. The sad thing is that, although it has dramatically increased the survival time of patients, it is not a cure for cancer. Many patients develop resistance to it and many tumors do not respond to it, so the search goes on for modifications.

Rainer Zocher in front of the yew trees on the TU campus in Berlin from which 10-DAB acetyl transferase was isolated.

Paclitaxel is synthesized in plant cells by specific enzymes. This caused biochemist **Rainer Zocher** at TU Berlin to try extracting paclitaxel in an enzyme reactor towards the end of the preliminary stage 10-DAB, i.e., carrying out Holton's semisynthesis enzymatically. This has the advantage of not producing by-products and unchanged 10-DAB can flow back into the process.

Enzymes are also stereospecific and work in a watery milieu at room temperature without the use of poisonous chemicals or solvents, unlike chemical synthesis. Enzymes are therefore tough competitors for organic chemistry methods. The key reaction of semisynthesis, 10-position acetylisation, was tackled first. There was actually evidence of relevant **10-DAB acetyl transferase** activity in the needles and roots of yew trees on the campus at TU Berlin. They then succeeded in cleaning the enzyme and partially sequencing it. DNA probes (see Ch. 10) could be produced from internal sequences and helped to isolate the gene from the yew. The next step was the expression of the plant enzyme in *E. coli*, leaving the way open for enzymatic extraction of baccatin III, the paclitaxel precursor, in a membrane reactor.

The technical realization of the process is currently being carried out in cooperation with IFB Halle and IBA Heiligenstadt (Germany).

The project is a model system for long-term future technologies in which the organic synthesis of complex natural materials will be replaced by enzymes.

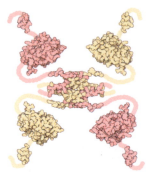

Fig. 9.16 The p53 tumor suppressor protein is a flexible molecule made of four identical subunits. P53 mutations are probably co-responsible for half of all human cancers. The trigger can be a single incorrect amino acid in p53.

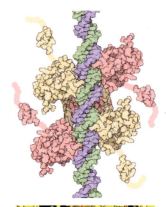

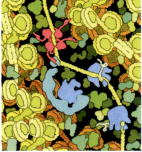

Fig. 9.17 p53 binds with all four "arms" (above) and activates the neighboring proteins involved in sensing the DNA.
Below: the p53 tumor suppressor (red) in action. It activates transcription.
DNA is shown in yellow and two RNA polymerases in light blue. These have begun to synthesize mRNA.

Fig. 9.18 Fighting the monster of cancer by Science on a French stamp.

code for proteins which promote cell growth and proliferation. Under normal conditions, these proteins are only expressed in rapidly dividing cells and must be activated via external signals. If such genes are overexpressed by mutation or their protein products are overactive, this can lead to uncontrolled proliferation. Mutated versions of proto-oncogenes which increase the risk of cancer are called **oncogenes**.

On the other hand, **tumor suppressor genes** suppress cell growth and proliferation in the body. These genes can act like a brake on cell division or they can promote cell differentiation or **programmed cell death** (apoptosis). If one of these functions goes wrong, there is an excess of proliferating cells.

The change from proto-oncogenes to cancer-causing oncogenes can occur in a variety of ways; it is linked with a **gain-of-function mutation**. On the other hand, the tumor suppressor genes lead to cancer through a **loss-of-function mutation**. In many types of cancer, they are silenced epigenetically, i.e., by DNA methylation and alteration of the chromatin structure.

The **p53-tumor suppressor** is shown in Fig. 9.16. Its concentration is normally low. If however DNA damage occurs, the p53 level begins to rise and initiates a protection mechanism; p53 binds to many regulatory areas of the DNA and causes the production of protective proteins. If the DNA damage is too great, p53 initiates programmed cell death, apoptosis. If p53 mutates, this control mechanism is lost and so is protection against uncontrolled cell division.

Tumor cells have a series of **common characteristics** which are used to target and fight them:

- They are physically and morphologically different in appearance from normal cells
- They have lost their original function (dedifferentiation)
- They are practically "immortal" in cell culture
- They interact with neighboring cells in a different way
- They do not attach to surfaces
- They possess chemically altered cell membranes
- They can contain chromosomal changes
- They secrete different proteins

An **early diagnosis** (Ch. 10) is the key to successful anticancer treatment.

■ 9.9 New Cancer Treatments

30% of cancers are treated surgically, especially in the early stages. The other new forms of treatment benefit from genetic engineering and genomic research. In **radio-immunotherapy** (RIT) antibodies marked with isotopes (Ch. 5) are used to localize tumors. Yttrium-90 and iodine-131 can also directly kill cancer cells. The antibodies concentrate these isotopes on the tumor.

In **chemotherapy**, "small molecules" are used. Microbial antitumor agents such as mitomycin, daunomycin, daunorubicin, and the plant agent paclitaxel have been used for many years. The first mass-produced drug of the biopharmaceutical industry was **Glivec®** (Gleevec®) by Novartis in Basel (Box 9.4). Glivec® binds to the ABL-BCR fusion product in **chronic myeloid leukemia** (**CML**). The latter is a tyrosine kinase, such as the Src protein, which usually phosphorylates protein with ATP and is overactivated in cancer.

In **biotherapy**, proteins are used against cancer: interferons, interleukins, monoclonal antibodies. Interferons and interleukins stimulate the immune system. **Interleukin 2** (IL-2), for example, stimulates the production of lymphokine-activated killer cells which penetrate tumors and release toxins. In adoptive biotherapy, the patient's T-lymphocytes are stored in cell culture, activated by IL-2, and then transferred back into the patient. **Monoclonal antibodies** recognize tumor antigens and set off the complement cascade and other cytotoxic effectors. **Rituximab** is an excellent example of this (details Ch. 5).

Immunotoxins are antibodies (Ch. 5) which transport a small molecule (such as ricin in chemotherapy-resistant forms of leukemia and lymphoma or the bacteria *Pseudomonas* enterotoxin) to the site where it takes effect (Figs. 9.19 and 9.20). They are particularly effective for forms of blood cancer, because these are more easily accessible than fixed tumors. A single molecule of ricin can kill a whole cell as it "jumps" from one molecule to another (in contrast to poisons such as cyanide which work 1:1).

Recombinant antibodies can also carry cytokines such as IL-2 or else they are constructed as abzymes/catalytic monoclonal antibodies (Ch. 5).

The development of **anticancer drugs** has been accelerated with the help of genetic engineering. Plants possess an astoundingly great potential for anticancer active substances. **Acetylsalicylic acid** (Aspirin) does not just have anti-inflammatory, analgesic, and anticoagulant effects helping to prevent coronary infarction, it may also be effective against cancer. Acetylsalicylic acid attacks the first stage of the synthesis of important messenger materials, the **prostaglandins**. This changes the fatty acid arachidonic acid into prostaglandin using the membrane enzyme cyclooxygenase (Fig. 9.18).

Some drugs such as **paclitaxel** (Taxol®) (Box 9.3) (from yew), **vinblastine** and **vincristine** (from the Madagascar periwinkle, Ch. 7) are synthesized in plants in tiny quantities. They have fantastically complicated structures. The chemical designation for the drug paclitaxel, for example, is

5β,20-Epoxy-1,2-α,4,7-β,10β,13-α-hexahydroxytax-11-en-9-one 4,10-diacetate 2-benzoat 13-ester with (2R, 3 S)-N-benzoyl-3-phenyl-isoserine

Attempts are being made to reprogram plants using metabolic engineering. An active substance from yew (*Taxus*) was the first anticancer drug from plants.

■ 9.10 Paclitaxel against Cancer

Yew is an evergreen conifer which bears brilliant red fruits in autumn. All parts of the conifer are highly poisonous – with one exception: the red fleshy seed coat is edible, although the kernel is poisonous. The trees conceal in their gnarled bark a drug that can save lives. During the life of the tree, the yew (Box 9.3) continuously stores an alkaloid in its bark, **paclitaxel**, also known as **Taxol®**.

Paclitaxel has become an **indispensable cytostatic** in medical treatment. Cells react to inhibition of cell division with programmed cell death – known as apoptosis. After intensive research by the US National Cancer Institute (NCI), paclitaxel has been very successfully used for **breast and ovarian cancer** in humans since the early 1990s, but this drug has one disadvantage: it is rare and therefore very expensive.

The **Pacific yew** (*Taxus brevifolia*) supplied the much sought after paclitaxel. It can be found in the remaining primeval forests on the coast of Northern California. Unfortunately, it is predominantly the

particularly ancient trees that have accumulated the alkaloid in higher concentrations.

Two grams of the drug are required to treat one cancer patient. The demand is enormous, because to provide sufficient treatment for one year for all those women in the USA who have breast cancer, all the remaining yew trees in California would have to be felled. A tree of about 2 feet in diameter takes 200 years to reach this size and supplies about 5½ pounds (2.5 kilograms (5½ lb) of paclitaxel. That corresponds to a yield of only 0.004%.

As resistance from conservationists against deforestation gradually became more intense, the protection of the yew and primeval forest was set against the profit interests of commercial companies. In January 1993, three weeks after the approval of the active substance by the FDA, it was announced that there would be no further felling of yews in the US National Parks.

In the meantime, the active substance was **semisynthetically produced from the regrown needles** of the yew and was initially used as mentioned in women, for the treatment of malignant ovarian and breast tumors.

The treatment of **lung cancer**, which normally leads to a rapid death, has also made some small but important progress over the last five years with the drugs Taxol® or Taxotere® based on **paclitaxel**. The World Health Organization (WHO) says that lung cancer is the most frequently occurring malignant disease in the world. Almost a million men and 333,000 women become ill with the disease each year, which according to expert opinion is caused by cigarette smoke in about 80% of cases. It is estimated that around 160,000 deaths in the United States in 2006 were due to lung cancer. It is also estimated that 90 percent of lung cancer cases are caused by smoking. Other causes include radon, asbestos, and air pollution.

In the 1980s and the early 1990s, more than 100 leading chemists competed to totally synthesize paclitaxel in the laboratory. In the end, **Robert A. Holton**'s team (Box 9.3) at Florida State University won and the scientists were able to report on the complicated process of total synthesis in 1994, which produced a yield of 2%.

What plant cells create at normal temperature, normal pressure and in water, chemists "cook up" in **40 small steps**. The only thing the two have in

Fig. 9.19 The castor oil plant (*Ricinus communis*) and its seeds are suppliers of ricin.

An attempt was made in London in 1978 to assassinate the exiled Bulgarian Georgi Markov. He was waiting for a bus on Waterloo Bridge and was stabbed from behind in his calf with an umbrella. That same evening, he developed a fever and his blood pressure dropped. Four days later, he died due to cardiac arrest. At the autopsy, a 1.52 millimeter diameter ball of platinum and iridium was found which had two holes in it filled with ricin.

This assassination technique was used in at least six more cases in the late 70s and early 80s.

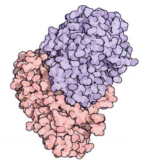

Fig. 9.20 Ricin is a potent phytotoxin that protects the seeds of the castor oil plant from mammals. The two chains have different tasks: the B-chain (shown in pink) binds strongly to sugars on the cell surface and directs the A-chain (blue) into the inside of the cell where it enzymatically attacks the ribosomes.

Box 9.4 Glivec® – the first tailor-made cancer drug

As a twelve-year-old, his heroes were **Louis Pasteur** and **Marie Curie** and he wanted to have as great an effect on the world as they had, perhaps by finding a new drug. **Alex Matter** was born in Basel in 1940 and after medical training in Basel and Geneva and then training in Pathology and Immunology in the USA and Europe, he pounced upon cancer, which was considered an almost hopeless case. Matter thought that we should be able to find out from nature how unwanted cells are killed, and that "We should be able to cure cancer like an infectious disease!"

In 1980, he was working in his laboratory in France on interferon (Box 9.2), which was being celebrated at that time as a wonder drug against cancer. Unfortunately, these hopes were naïve. Even worse, to Matter's disappointment, no-one seemed to be keen on cancer research at that time. The pharmaceutical "giant" Ciba-Geigy in Basel had closed its cancer research department in 1980.

Nevertheless, Matter was enticed back to Basel in 1983 to head the re-opened department, although no-one expected anything useful from cancer research in the short term. Matter was warned that this would be the end of his scientific career. He formed a team only to find out that the project was "blacklisted" – nothing unusual in the industry! However, Matter fell upon molecular biology, which was now booming, believing it had to hold the key to the cancer problem.

He was particularly fascinated by **kinases** (Ch. 3). Some Chinese doctors and some colleagues in Basel believed that kinases might have something to do with cell proliferation. What would happen if these kinases were inhibited? Alex Matter found some Japanese work in the library which showed that the natural substance staurosporin made from fungi inhibits a series of kinases.

The idea of fighting cancer like this seemed to his colleagues rather outlandish and too simple. After two years in Basel, he hired **Nick Lydon** from Boston for kinase research and they concentrated on **tyrosine kinase** (Fig. page 241).

Today, we know that tyrosine kinases are essential in healthy cells. These enzymes are important for **signal transduction** (sending signals). A chain of communication in the inside of the cell is set in motion upon receipt of a suitable external message: signals reaching the cell from the outside are taken up and distributed as far as the cell nucleus and the genes.

Tyrosine kinases are the most important intracellular message transmitters. They transfer an ATP phosphate group to the tyrosine modules of proteins.

Alex Matter

In Boston, Matter was working with the post-doctorate medical oncologist **Brian Druker**, who was testing inhibitors. The only person missing from the team was the chemist **Jürg Zimmermann**, who was working on them in Basel, and a biologist who was still testing them, **Elisabeth Buchdunger**.

Jürg Zimmermann

Brian Druker knew that **chronic myeloid leukemia** (**CML**) was the most promising cancer and was the only one out of more than 100 types of cancer where the genetic cause was known. Druker predicted in 1988 that CML could be the first disease on which a protein kinase inhibitor worked. With chronic myeloid leukemia, the white blood cells (leukocytes) multiply out of control. The obvious cause of excessive growth with this type of leukemia is a genetic defect visible with a microscope. The degenerate white cells can be seen on the changed chromosome, the **Philadelphia chromosome**. It develops as a result of an accident in one of the stem cells in the bone marrow, where white blood cells are produced. As the cells divide, chromosome 9 wrongly receives a piece of chromosome 22 instead of a piece of chromosome 9. This **translocation** of genetic material is characteristic of most cases of chronic myeloid leukemia. The Philadelphia chromosome is so typical of the disease that doctors have long used it as a marker chromosome for the diagnosis.

With the fusion of the chromosome pieces, two genes come together whereas they do not normally have anything to do with each other. These two genes are the *abl* gene from chromosome 9 and the *bcr* gene from chromosome 22. The two genes become immediate neighbors – with fatal consequences. Together, they form an **oncogene, a gene that causes cancer, in this case** *bcr-abl*.

Druker doubted that they could create an inhibitor that selectively inhibited a cancer-causing tyrosine kinase without affecting the other 150 or so tyrosine kinases of the cells. He read in the journal *Science* that a group had specifically inhibited the receptor for Epidermal Growth Factor (EGF, Ch. 9). Druker now thought that if a **tyrosinase kinase inhibitor** could stop the *bcr-abl* oncogene, that would be proof of a molecular biology concept for fighting cancer.

By 1990, there were 100 researchers employed at Alex Matter's Ciba-Geigy laboratory, compared to half a dozen in 1983, in spite of the direct and indirect resistance of the company management. Because of the low number of patients with CML, a pharmaceutical blockbuster was not expected, but for Alex Matter it was a matter of the principle. If the inhibitor worked with this cancer, others could also succeed by following the molecular biology route. Matter and Lydon had hundreds of substances they wanted to test, but all turned out to be poor inhibitors.

Jürg Zimmermann joined Matter's department in 1990. His compounds fitted the bcr-abl enzyme "pocket". This active center is filled with ATP. However, the compounds must not inhibit the "good" or important tyrosine kinases. Every week, Zimmermann synthesized about 10 compounds which were then tested by biologists over the next two weeks. The first results were shattering, as they showed good inhibition but

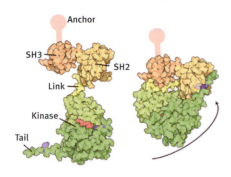

The Src-protein, a tyrosine kinase, has a number of mobile parts. In its inactive form (right) it is ball-shaped. The active form (left) is open so that it can phosphorylate proteins.
It binds ATP (red) which supplies the phosphate group. Tyrosine (blue) is phosphorylated in the first reaction. When this site is phosphorylated, the enzyme is fully active.
Loss of control in tyrosine kinases: if the tail is capped, the enzyme can no longer close up into its inactive form (right); Src is then permanently active, control of cell growth is lost, and cancer can arise.

affected all kinases, so these substances were poisonous!

Zimmermann then became aware of a class of substance already on the market, phenyl-aminopyrimidines. His "gut feeling" as a chemist told him that something could be done with these.

When Zimmermann arrived at the laboratory on 26th August 1992, he could feel success in the air. He knew that there was a series of substances that were active against the *bcr-abl* oncogene and even showed *in vivo* activity in laboratory animals!

A gene encoding tyrosine kinase is disturbed by chromosome translocation coded for tyrosine kinase (see illustration above). Tyrosine kinase is deregulated and remains permanently "switched on". The **"hyperactive" enzyme phosphorylates all proteins in the cascade** of reactions which tell the cell to divide. Uncontrolled phosphorylation is thus the source of a catastrophe that leads to cancer. Phosphorylation must therefore be brought under control by kinases in order to fight CML.

The test by Elisabeth Buchdunger took only a few hours. Cell cultures were incubated with the substance that had been found, **imatinib mesylate**. The level of phosphorylation was measured to see if the substance had had an effect on the *bcr-abl* oncogene. Indeed, phosphorylation had been inhibited! The researchers still did not know that they were writing medical history. In Spring 1993, the substance was classified

as a "drug candidate" and was given the **code name STI571**.

The drug had one disadvantage in that low patient numbers meant low profits, even if they were successful. Four similar compounds, including STI57, were tested by Brian Druker in the USA in 1993; the first results were produced in February 1994 and showed that 90% of the leukemia cells had been inhibited in cell culture. Animal trials started in 1995, all with a high level of success. In March 1996, Ciba-Geigy merged with Sandoz and **Daniel Vasella** became Chief Executive Officer (CEO) and, in 1999, President of Novartis.

Alex Matter's research capacity was doubled by colleagues from the Sandoz oncology department. People remained sceptical about his concept and pressure on him grew endlessly, while the marketing management insisted that the market was much too small.

Alex Matter believed the project had to work. "I was like a dog with a bone," he says. "I could not let it go!"

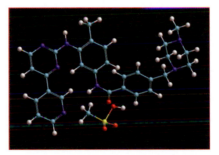

The structure of Glivec® (*Gleevec*®) or imatinib mesylate

A hundred million dollars had been poured into the project. The Glivec concept had already been thought up in 1983, the imatinib mesylate molecule had already been synthesized in 1993, and after some ups and downs tested on the first patients in 1998, then approved in record time by the Food and Drug Administration (FDA) in 2001 because cancer patients had organized themselves so well.

STI or Glivec® is one of the **first drugs tailor-made on a molecular level** and is a representative of a new class of cancer drugs, "signal transduction inhibitors". Compared to traditional cancer drugs, which could only partly distinguish between healthy and altered cells and are frequently associated with severe side effects, the new active substances try to attach themselves precisely to the molecular information pathways typical of tumor cells.

Disturbed communication pathways inside cells are not just significant for chronic myeloid leukemia. The active substance STI571 interferes with overactive tyrosine kinase. It is incorrectly formed because of the Philadelphia chromosome. Communication disturbed in this way is blocked in the cell at one of the first and most important molecular transduction stages. Thanks to the drug, the growth message no longer reaches the cell nucleus with the genes. Uncontrolled cell proliferation is now stopped.

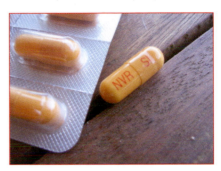

The small life-saving Glivec pills

The results were so promising that the American health authorities approved the new cancer drug in May 2001 for the treatment of patients with chronic myeloid leukemia. The phase-II trials with more than 1100 patients in 16 countries have now been concluded. After five years, the survival rate for CML patients is 95 %. Novartis is proud of the fact that, after 2002, 15,810 CML patients in 81 countries were given the drug free of charge because they could not afford it for themselves.

Glivec® affects other tyrosine kinases such as c-kit. This kinase promotes the growth of gastrointestinal stromal tumors (GIST). Until now there has not been a drug treatment for this abdominal cancer.

In the meantime, the successor to Glivec, AMN107 or Nilotinib, is also well under way and was scheduled to be presented to the health authorities for approval towards the end of 2006.

Cited Literature

Eberhard-Metzger C (2001) *Die neuen Medikamente gegen Krebs. [New drugs for cancer]* Spektrum der Wissenschaft, PD 46 ff
and material from Novartis
www.glivec.com

Fig. 9.21 Aspirin® is at present the most commonly taken drug and 80 billion tablets of it are swallowed annually.

Salicylic acid was isolated from willow bark (*Salix alba*) in 1828 as a yellow, crystalline mass and named salicin. Felix Hoffmann (1868-1946) then patented acetyl salicylic acid (ASS) in 1889 and offered the concept to the Bayer company. Aspirin® blocks the production of prostaglandins in the body, mediators that pass on and strengthen local pain signals. All prostaglandins come from the precursor, arachidonic acid. The enzyme cyclooxygenase (COX) inserts two oxygen molecules into arachidonic acid resulting in the production of prostaglandins in the body. If this enzyme is inhibited, pain signals do not arise as reactions to inflammation and blood clotting is also delayed. Aspirin® inhibits this enzyme and blocks prostaglandin formation.

Evidently, "non-steroidal anti-inflammatories" such as Aspirin® are also a protection factor against some forms of cancer. It is seen as a wonder drug against coronary infarction, but Aspirin® cannot prevent a first infarction, it can only reduce the danger. From a purely financial point of view, Aspirin® is probably the most cost-effective precaution against some treatment-intensive and expensive diseases.

Fig. 9.22 Human Growth Hormone (above). Would Napoleon have been so great if he had been taller?

common in the end is the **low yield** of the material which is so much sought after. As the Californian yews could not all be felled at once, alternative sources had to be found.

Taxol can now be made in fermenters with plant cell cultures from yew cells. The microbiologist **Gary Strobel** tracked down new sources of paclitaxel in symbiotic (endophytic) fungi (Box 9.3) which live in the internal tissues of the tree. However, fungal production of paclitaxel has not been developed as an industrial method.

■ 9.11 Human Growth Hormone

A lack of **growth hormone** (Fig. 9.19), which is formed in the pea-sized pituitary gland (hypophysis), leads in extreme cases to dwarfism/stunted growth or infertility. Until the end of the 1950s, dwarfism could not be remedied by medical treatment. It was not until 1958 that more and more children began to receive human growth hormone, hGH. In contrast to insulin, which can also come from pigs or cattle for the treatment of diabetes, the growth hormone is species-specific; only **human growth hormone** can therefore be used for treatment. It therefore had to be taken from the brains of human corpses. The previously two-year course of treatment for one child required 50 to 100 pituitary glands. Today, we assume that treatment with growth hormone must be continued beyond the 18th year of life, in order to be successful.

After **two patients died**, the sale and use of natural hGH was prohibited at the start of 1985 because, while the hormone was being isolated, infectious particles from the corpses supposedly contaminated the drug and caused **Creutzfeldt-Jakob disease** (similar to BSE).

Fortunately, there is now a new genetically engineered hGH. The Swedish company Kabi Vitrum, for example, formerly the biggest producer of "natural" human growth hormone in the world, now uses a genetic engineering process. Bacteria in a 120 gallon (450 liter) bioreactor now produce the same quantity as was previously produced from 60,000 pituitary glands. Previously there was so little hormone available that only children with severe dwarfism could be treated.

The market for growth hormone is **over 1 billion US dollars per year**. New findings have produced a definite increase. Of economic interest is the fact that growth hormones promote milk production in

cows (Ch. 8), but in mast cells it produces an "anabolic effect", i.e., growth hormone increases protein (muscle) and reduces the formation of fat, which is exciting news to bodybuilders and the world of sport.

In Paris in 2003, the leading German doping specialist **Werner W. Franke** declared at the Athletics World Championships: "This World Championship was a festival of dental brace wearers. This is a sure sign that human growth hormone is being taken, as the jaw bones grow so that the teeth jut out, as happens with malformations in children. Dental braces are a definite indicator of hormone misuse."

■ 9.12 Epidermal Growth Hormone – Wrinkles Disappear and Diabetic Feet Heal

Huge sums of money are spent all over the world on making wrinkles and creases disappear from the face. Women even allow themselves to be given injections of the **botulism toxin** (**Botox**). In principle, this is a biological weapon! The toxin is formed by bacteria that are notorious for food poisoning. It is injected under the skin to paralyze the muscles so that wrinkles disappear, although the effect is temporary and lasts only for as long as the muscles remain paralyzed.

The cosmetic industry also uses other costly anti-wrinkle creams containing fats and vitamins. For the most part, young girls with flawless complexions are used as advertising media (Fig. 9.20). **Epidermal growth factor** (EGF) works differently. It is a peptide that stimulates the skin cells to regenerate.

Wan Keung Wong of the Hong Kong University of Science and Technology (HKUST) had human EGF genetically engineered using *E. coli*. The relatively expensive EGF cream (one little jar costs 450 HK dollars or 50 Euros) is applied daily and after four to six weeks, the wrinkles are replaced by regenerated skin cells (Fig. 9.23).

There has also been a dramatic medical breakthrough with EGF reported, involving patients with severe diabetes, who often suffer from diabetic feet ulcers, open wounds which will not heal. In the US, 600,000 patients are affected by this. In a Hong Kong study with EGF, definite healing could be seen after eight weeks in almost hopeless diabetics in Hong Kong. Their only alternative had been amputation of the foot! EGF also helps the skin to form new cells faster in burns cases and is also successfully used for eye injuries and for healing gastric ulcers.

9.13 Stem Cells, the Ultimate Fountain of Youth?

Stem cells (Fig. 9.24) continuously renew our body cells, although the regeneration of muscle and nerve tissue remains problematic. It has therefore been believed until now that the brain cannot regenerate after a stroke nor can the heart muscle recover after a coronary infarction. However, it now seems that there are stem cells that can do this.

How is it that stem cells are capable of **regeneration**? Not only can they divide many times and keep themselves alive, but they can also form new specialized body cells. Stem cells from the blood system divide in the bone marrow when signaled by the body (e.g., with erythropoietin, EPO), as a result of blood loss, Ch. 4) and form secondary stem cells as well as new blood cells.

Tissue stem cells are therefore "primeval cells" which continuously allow new specialized cells to arise where they are needed. Stem cells can also be found in the early embryo, five to ten days after fertilization. These **embryonic stem cells (ES)** led to bitter discussion from 2000 onwards in the media and various parliaments, because in principle, all cell types can arise from them, as they are **pluripotent**.

How do embryonic stem cells arise? The fertilized egg (**zygote**) forms a compact ball of 12 cells, the morula ("mulberry"), in three to five days after rapid cell division. After five to seven days, an embryo arises in the form of a 100 cell **blastocyst**.

The blastocyst measures only about one seventh of a millimeter. Its outer cells are called the **trophoblast**. The small clump of cells inside is the inner cell mass and the source of the embryonic stem cells. These ESs have the ability to differentiate into about 200 cell types; they are, then, pluripotent.

James Thomson of the University of Wisconsin in Madison was the first to isolate and cultivate human ES. In the same year, **John Gearhart** and his team from Johns Hopkins University managed to produce primitive embryonic gamete cells (sperm and egg cells), which also differentiated.

Human ESs have two main properties:

- They can regenerate themselves indefinitely and form more stem cells.

- They can differentiate themselves under certain conditions to form a number of mature specialized cells.

Human ESs do not age because they express a high level of **telomerase**. Various working groups have kept cell lines alive for more than three years and more than 600 rounds of division without problems. However, stem cells can also be frozen for long periods and do not lose their properties. If they are stimulated by growth factors, they can differentiate into different cell types such as skin cells, brain cells (neurones and glia cells), cartilage (chondrocytes), osteoblasts (bone forming cells), hepatocytes (liver cells), muscle cells (including smooth muscle which covers the walls of blood vessels), cells of the muscles of the skeleton, and heart muscle cells (myocytes).

If stem cells are transplanted into a sick organ, they can form the necessary replacement tissue on site. In rats and mice, dead heart muscle cells have been regenerated with embryonic stem cells after an experimental **coronary infarction** (Fig. 9.23). This has even been achieved in humans with coronary infarction patients. It is therefore no wonder that everyone is interested in these versatile cells.

It is already possible to breed specific cell types with embryonic and tissue stem cells in a nutrient solution. For example, attempts have been made to breed islets of Langerhans in the pancreas from embryonic stem cells. Instead of injecting insulin, diabetics could be treated with new insulin-forming cells resulting from deposits of stem cells under the skin.

Tissue stem cells (**adult stem cells**) have a lower development potential than embryonic stem cells. Their ability to multiply and their life-span are limited. However, adult stem cells have been successfully isolated from the brain and neurons have been grown from them in cell culture. With injuries to the spinal cord in rats and mice, neurons produced from stem cells have been injected to improve nerve function.

Embryonic stem cells also have some disadvantages. There are ethical problems in harvesting them because embryos must be used for this. Recipients may also suffer a rejection reaction and malignant growth can occur. This is not a problem with adult stem cells or tissue stem cells. They can be taken from the patient and safely reintroduced later without fear of rejection. However, they still bear the patient's congenital or inherited defects. There are no ethical objections to these and they can be extracted from a bone marrow puncture.

Fig. 9.23 Above: EGF is already used in lipstick in China.
Middle: production of EGF with *E. coli* in a bioreactor (right WK Wong).
Below: after weeks of expensive EGF treatment, wrinkles disappear, only to reappear after the cream is stopped... (self-experiment).

Fig. 9.24 Lucas Cranach's "fountain of youth"; does biotechnology make it possible?

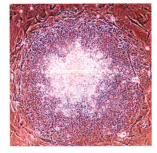

Fig. 9.25 Human stem cells.

Box 9.5 The Expert's View:
James W. Larrick about the search for "Magic Bullets". Is it over? Antibodies versus small molecule chemical pharmaceuticals

The past century has witnessed the development of a large pharmaceutical industry with an appetite for new "pills", so-called "magic bullets" promising to cure our ills.

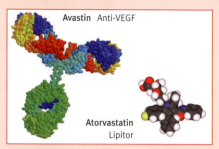

Avastin (Anti-VEGF, monoclonal antibody against Vascular endothelial growth factor) stops solid tumor growth. Atorvastin (Lipitor), a cholesterol-lowering drug made by Pfizer is the largest-selling pharmaceutical in history, surpassing Pfizer's other wonder drug, Viagra.

Recently "biopharmaceutical bullets" such as recombinant proteins and antibodies, RNA and DNA-based therapies and so-called "regenerative medicines" using cellular therapies have shown promise. What is the future of this search for magic bullets and the pot of gold they promise to provide at the end of the pharmaceutical rainbow?

The author with his Waorani friend Kampati. Kampati is preparing the curare mixture (left).

"Polypharmacy"

The history of pharmaceutical use began long ago among our tribal forefathers who lived a life not dissimilar from the **Waorani Indians** native to the headwaters of the Amazon River in the trackless rainforests of Eastern Ecuador. Around 600 individuals call themselves Waorani, living their lives as hunter-horticulturalists in groups of 30 to 50 people, probably like our ancestors over the past 10,000 generations. I spent several months with these people, was adopted by them and was able to study the ways in which they use pharmaceuticals from Nature. Until recently, the group fiercely defended an area of some 20,000 km² by spearing all foreigners who entered their territory, a long way away from **Rousseau**'s cliché of the "noble savage"!

Over the past 200 millennia modern human groups like the Waorani used trial and error to find small molecule "pharmaceuticals" adapted for healing purposes and other beneficial uses. **Drug therapy began as "polypharmacy"**: the Waorani's and better developed pharmacopeias such as those from China and India are usually mixtures of extracts from many sources: minerals, plants, animals, fungi, etc. The active substance is seldom known in detail. We observed in action in the jungle one of the most exciting examples: **blowgun dart poison**. The Waorani prepare a mixture of perhaps 50 plants for application to the tips of their blowgun darts. Blowguns permit the Indians to hunt the rainforest canopy for woolly monkey (and other primates), which form a major part of their diet. Western medical science has identified **curare** (*Strychnos toxifera*) as the active substance in the lethal mixture. Derivatives of curare are widely used in modern hospitals to "paralyze" patients for various surgical procedures. However, among the Waorani, this "pharmaceutical product" may "contribute" to a third or more of the tribe's calories.

Paul Ehrlich's Magic Bullets

In the late 19th century, the German professor **Paul Ehrlich** first argued that certain "chemoreceptors" on parasites, microorganisms, and cancer cells would differ from analogous structures in host tissues. He believed that these differences could be exploited therapeutically to make **"magic bullets"** (see main text 5.14). From this concept grew the idea of small molecule chemical therapeutics, the basis of the modern pharmaceutical industries which accounts for perhaps 2% of global economic activity! In the late 19th century German dye companies such as Bayer and Hoechst pioneered the **first rationally sought small organic drugs**, each directed to a specific pharmacological target and confirmed by biological or animal screening assays.

Paul Ehrlich on the old 200 DM German banknote

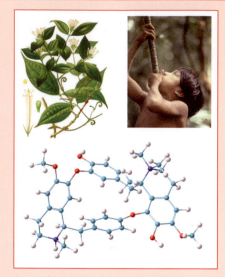

Curare: the plant *Strychnos toxifera* and the structure of turbocurarine

A rival of Ehrlich's, also German, **Emil von Behring** won the first Nobel prize for his demonstration of passive immunotherapy using specific antisera obtained from immunized animals. While quite promising, this alternative approach to obtain "magic bullets" was limited by serum sickness, the immune response of an individual to the administered antibodies, and fell into disuse.

A renaissance in antibodies and other protein therapeutics had to wait for the development of **recombinant DNA technology** and **human monoclonal antibody technology** over the past 25 years. These novel therapies are expensive, often require parenteral administration and presently account for only about 10-15% of pharmaceutical revenue. Although worldwide sales of therapeutic and diagnostic imaging antibodies have experienced logarithmic growth, presently they represent only 2.5% of worldwide pharmaceutical sales. Although most pharmaceuticals continue to be small organic molecules consumed in the form of a pill, sales of **biopharmaceuticals are growing at twice the rate** of conventional pharmaceuticals!

Within the past 50 years **pharmaceuticals have become a most profitable industry**, outperforming the oil and gas industries. This has been driven, at least in part, by the development of so-called "blockbuster" drugs, highly profitable "magic bullets" with annual sales of >$1 billion. Will the pharmaceutical industry be able to identify more "blockbuster magic bullets" in the years ahead or is the ecology of this "resource" limited like oil and gas?

World Pharmaceutical Market – 2005

	US$ Billions	Percentage of Total
North America	268.8	44.5%
EU	180.4	29.9%
Japan	69.3	11.5%
SE Asia and China	28.8	4.7%
Latin America and Caribbean	26.6	4.4%
Middle East	4.9	0.8%
Africa	6.7	1.1%
Indian Subcontinent	5.7	0.9%
Oceania	7.7	1.3%
Russia	5.0	0.8%
Total	603.9	100%

European Alcohol Sales~$600B Source:IMS Health, Inc.

Per capita Pharmaceutical Markets

	Population Millions	Per Capita Consumption US$ 2000 est.
United States	276	342
Japan	126.5	409
Germany	82.8	178
France	59.3	231
Italy	57.6	161
United Kingdom	59.5	153
Spain	40	135
Canada	31.3	172
Brazil	172.9	30
Mexico	100.4	47
CHINA	1300	14
Australia/NZ	23	126

Sources: IMS Health, September 2000, CIA The World Factbook, 2000.

BIOPHARMA growing faster

GLOBAL BioPharmaceutical Sales

Yr.	2000	2001	2002	2003	2004
Sales ($B)	23	27	33	40	48
Growth (%)	16	19	23	21	20

GROWTH BioPharmaceutical vs. Pharmaceutical Industry

Yr.	2000	2001	2002	2003	2004
Global Pharma ($B)	354	392	430	492	550
Growth (%)	6.6	10.7	9.7	14.4	11.8
Biotech share	6.4	6.9	7.6	8.1	8.7

Source: Intercontinental Medical Statistics

10 companies ~40% of Sales

World Pharma Market	$550.0 Billions
1 Pfizer (U.S.)	46.1
2 Sanofi-Aventis (Europe)	31.8
3 GlaxoSmithKline (Europe)	31.4
4 Johnson & Johnson (U.S.)	22.1
5 Merck (U.S.)	21.5
6 AstraZeneca (Europe)	21.4
7 Novartis (Europe)	18,5
8 Roche (Europe)	17.3
9 Bristol-Myers Squibb (U.S.)	15.5
10 Wyeth (U.S.)	14.5

Subtotal Top 10 Companies $229.6B 41.7%

Here are some impressive numbers:

The scale of PHARMA

World Economy	~$44.0 Trillions
US Economy	~$12.0 Trillions
Pharma	~$ 0.6 Trillions

1-2% of Global Economic Activity

The Universe of Drug Targets

A recent survey (Nat. Drug Discovery, Dec. 2006) of the roughly **21,000 US FDA registered drug products** found less than 1200 novel molecular entities (NMEs). These were remarkably few *unique* molecular pharmaceutical drugs: only 324 for approved drugs! Among these targets, only 266 were human molecules with the remaining being found in pathogens. Considering the complexity of human physiology this is remarkably few!

Despite huge investment (>$30B annually), on average only 5 new targets have been identified each year since the 1980s.

Is the pharma industry going to run out of drug targets; how big is the "Universe" of chemical drug targets?

New therapeutic targets from the Human Genome Project?

The completion of the human genome project revealed around 29,000 genes, far fewer than originally predicted given the complexity of human beings. Based on the structures of currently known drug targets, bioinformatics methods have estimated that perhaps 8-10% of these genes could encode "druggable" protein targets, i.e., a protein having a crevice that can bind a small organic molecule. Although perhaps 10-20% of proteins are still functionally unclassified, there are no undiscovered large protein families. Any remaining novel protein targets are likely to be members of very small families. Although post-translational modifications and assembly of functional complexes may increase the target number, the basic protein folds are well known among druggable targets: the majority of targets are G protein coupled receptors (GPCRs) or ion channels.

The metabolome: source of new leads?

Most drugs compete against small molecules for binding sites on proteins or enzymes, i.e., statins inhibit HMG-CoA reductase and bronchodilators bind beta adrenergic receptors. The number of binding sites is proportional to the function of the size of the "metabolome", i.e., the total set of small molecules in an organism. One route to novel target discovery might lie in identifying enzymes and receptors from metabolomic profiling. By contrast, the "druggability" of targets identified by proteomic or transcription profiling is likely to be low because protein folds are conserved.

Unfortunately, "druggable" does not equal drug target!

Protein binding of small molecules with the appropriate chemical properties at the required binding affinity MIGHT make it druggable, but does NOT necessarily make it a potential drug target. That "honor" belongs only to proteins that are also linked to a target disease and can somehow modify that disease!

It is NOT LIKELY that a large number of specific drug-binding sites will be discovered because of the conservation of protein folds and the modular nature of proteins. The McKusick genetic disease database lists 1,620 proteins DIRECTLY linked to the pathogenesis of disease yet only 105 are drug targets!

Thus many "targets" are "associated" with certain diseases: leptin or leptin receptor with obesity, LDL receptor with atherosclerosis, complement receptors with inflammation, interleukin-4 (IL-4) with allergic diseases. But this does not mean that they represent suitable intervention targets for new drugs!

Various bioinformatics analyses suggest that **the "universe" of drug targets** may be 2-3 fold that presently known with a ceiling of 1,500.

Why, then, is the number of drug targets so small? Perhaps biological structures such as proteins have been selected with an innate 'protection' against external chemicals, especially against small molecules? Since Ehrlich's time many magic bullets have been identified, characterized and utilized for pharmacotherapy. The fact that the numbers are limited may suggest that only a limited fraction of biological responses can be manipulated so simply!

Can the low frequency of "druggable sites" be overcome by novel biopharmaceutical technologies such as antibodies?

Perhaps. We expect that the "**Epitope Universe**", i.e., sites available for antibody binding is likely to be bigger than the "Druggable Universe". Novel targets from genomics, proteomics, tran-

See next page

Worldwide Sales of Therapeutic and Diagnostic Imaging Antibodies

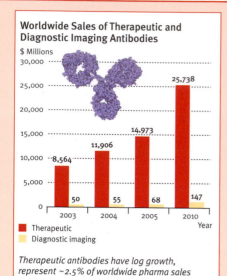

$ Millions

Therapeutic values: 8,564 (2003), 11,906 (2004), 14,973 (2005), 25,738 (2010)
Diagnostic imaging values: 50 (2003), 55 (2004), 68 (2005), 147 (2010)

■ Therapeutic
■ Diagnostic imaging

Therapeutic antibodies have log growth, represent ~2.5% of worldwide pharma sales

Drug targets

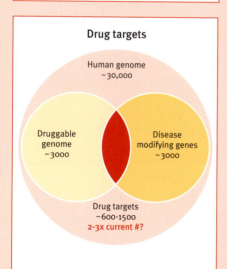

Human genome ~30,000

Druggable genome ~3000

Disease modifying genes ~3000

Drug targets ~600-1500
2-3x current #?

How large is the protein-bindable genome?

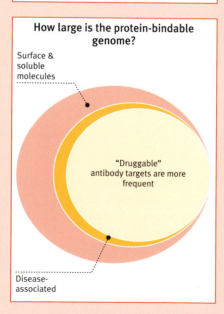

Surface & soluble molecules

"Druggable" antibody targets are more frequent

Disease-associated

scriptomics, etc. will show up that are uniquely approachable by the newer technologies. What about redundancy of antibody targets versus targets of small chemical molecules? Until now, there are ONLY 9 targets modulated by BOTH antibodies and small organic molecules. Take as a successful example Epidermal Growth Factor-Receptor (EGF-R). The recombinant antibodies cetuximab and panitumumab target the extracellular domain of the receptor to modulate tumor cell growth. Small chemical drugs, gefitinib and erlotinib target the adenine portion of the ATP-binding site of the cytosolic catalytic kinase domain.

Antibody drugs to the rescue?

There are quite a large number of questions to be answered: What is the size of the **"antibody-able" genome**, proteome? What is the "Universe of epitopes"? How many of these are associated with disease? What fraction can serve as a lever to control the disease? Can antibodies penetrate the "chemical invasion" barrier?

Five Prime Inc., a San Francisco-based pharmaceutical company created perhaps the world's largest cDNA library database of >280,000 full length cDNA clones from >200 tissues. Bioinformatics methods identified around 2500 genes for secreted and around 5000 genes encoding membrane proteins, giving perhaps 7500 potential antibody targets.

Are Mabs best for inhibiting protein-protein interactions?

Immunoglobulin E (IgE), vascular endothelian growth factor (VEGF), interleukin-2 and IL-5 or their respective receptors, may represent very attractive drug targets in the case of allergies, cancer, autoimmune diseases or asthma. Traditional small-molecule drug discovery has largely failed with these types of targets. However, protein-protein interfaces have "hot spots," small regions that are critical to binding. They are the same size as small molecules. Targeting "hot spots" with small molecules may block protein-protein interactions. One such example is the development of VLA4 (Very late antigen 4, VLA4) inhibitors by Biogen-IDEC and Merck.

The diversity of antibody therapy versus drug therapy is huge. The number of targets approachable with antibodies may challenge the metabolome. One can assume that each protein touches 4-5 others providing many points of attack. Developments of antibody fragments and

intrabody gene therapy make complex cytoplasmic targets approachable.

Unlike small molecules, antibodies can be designed with diverse effector functions. Antibodies can vary in size, affinity, or isotype, e.g., secretory IgA for topical or enteric use. Antibody catalysis, conjugates and enzyme fusions are possible. Thus it is no surprise that we observe a **growth of Mab technology and products** with more than 250 antibody-based products in clinical trials.

Polypharmacy and polypharmacology replace blockbuster monopharmacy?

The traditional pharmacopeias are on to something. In the next generation, **polypharmacy** and **polypharmacology** will give optimal clinical benefits by combined use of different drugs/pharmacophores. Polypharmacy, the combination of more than one drug each with a different mechanism or complementary toxicity, once used primarily in cancer chemotherapy will find use throughout medicine.

Polypharmacology seeks to identify a **single drug to bind and modulate multiple targets**. The clinical effects are mediated by a set of targets. Multikinase inhibitors Sorafenib/Sunitinib are one example, with activity versus RAF-kinase (and other kinases).

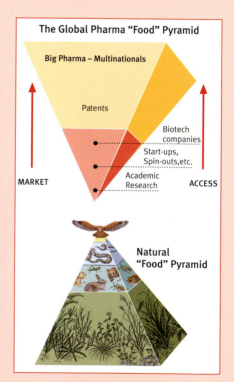

The Global Pharma "Food" Pyramid

Big Pharma – Multinationals

Patents

Biotech companies

Start-ups, Spin-outs, etc.

Academic Research

MARKET

ACCESS

Natural "Food" Pyramid

The changing ECOLOGY of the pharmaceutical industry

The 1990s have been characterized as the "tools decade": there was a REVOLUTION in genomics, proteomics, transcriptomics, etc. Reforms of the early Clinton administration (1993-95) downgraded me-too products. Only novel drugs were considered worth pursuing and the definition of novelty became stricter. The explosion in available novel biotech tools increased research productivity.

However, new discovery tools have NOT increased the successful identification and development of magic bullets by the pharma industry and product pipelines are being challenged by generic versions of blockbusters. Expiration of critical patents has further aggravated the monopolies once enjoyed by these highly profitable companies.

One result of this changing landscape has been a remarkable consolidation of the industry with mega mergers and acquisitions. The largest companies have become multinational marketing organizations that are "branding" drugs, e.g., Lipitor by Pfizer, to protect market share. These companies now often buy "content" from smaller companies and research organizations. There are huge problems with this unhealthy "dinosaur ecology". It is characterized by a loss in diversity and a loss in creativity. Drug discovery is in jeopardy. The indicators for a grave crisis are clearly visible:

- Low productivity
- Management -- top down
- Conformity
- Research "managers" not leaders
- Blockbuster mania
- Merger mania
- Shareholder pressure
- Shift from R&D to marketing

The BIG PHARMA has made numerous misjudgments in the last few years. Drugs once declared as "orphans" are now "mainstream". Some

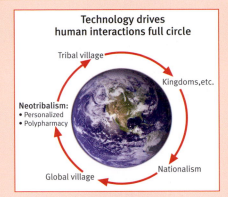

Technology drives human interactions full circle

Tribal village

Kingdoms, etc.

Neotribalism:
- Personalized
- Polypharmacy

Nationalism

Global village

examples of labels put onto certain of these drugs[1] are shown in the Footnote.

After this, should we trust the BIG PHARMA to choose the right way forward?

...and the future? "Tribalized Medicine?"

Another challenge to "blockbuster mania" is so-called personalized (tribalized?) medicine.

New therapeutics are being associated with diagnostics. Think Trastuzumab (Herceptin), Genentech's antibody therapy for breast cancer which is only administered to women who have been shown to have a tumor positive for the HER-2/neu oncoprotein.

In fact, biomarker-diagnostics are growing faster than drug discovery. Increasingly, therapeutics will be targeted for specific conditions and patient populations; does this signify the end of the blockbuster? Biomarker diagnostics will be used for screening and monitoring. Biological/antibody therapeutics will be more common. Individualized regenerative and stem cell medicines will be increasingly applied.

How to capitalize on these trends?

- Develop technologies that facilitate complex disease analysis
- Create biomarker discovery applications
- Develop easy-to-use products for deployment in diagnostic setting
- Market solutions for the production and testing of biologics

The age of the "Internet SHAMAN"?

The 21st century is characterized by complexity, isolation, alienation, and an aging global population. Pharmaceuticals will be increasingly used to address the medical problems associated with aging and alienation. Internet-driven global connectivity will drive the market.

Who will pay for new drugs? The government or the consumer? Pharmaceuticals are being branded and "consumerized": Designer drugs, Nutriceuticals, Cosmeceuticals, and Regenerative medicine. In fact, the Internet already has a "tribal structure": many individuals find their "virtual tribe" on the net with personal treatments or therapies administered "in silico" by one or more bloggin "Shamen" from their tribe!

When reflecting upon my personal experience in the rainforests of South America with the Waorani Indians, I believe that, good or bad, the age of the "Hightech Internet SHAMAN" is coming...
Back to the Future!

Jim Larrick is a Founder and Head of Panorama Research Institute in Silicon Valley, USA. As an active biomedical entrepreneur he has founded >15 biopharmaceutical companies. He has contributed to an improvement of world health having organized biomedical research and service expeditions to Nepal, Papua New Guinea, China, Ecuador, Peru, Ghana, etc.

---

1 "Market too small": Acyclovir – HSV2; AZT – HIV
   "Mechanism too dangerous": Lovastatin – HMG CoA inhibitor; Atorvastatin – HMG CoA inhibitor
   "Current therapy works fine": Propanolol –beta blocker; Cimetidine – H2 blocker, Captopril – ACE inhibitor
   "Tricyclics work fine": Bupropion – antidepressant
   "Epilepsy is Not a big problem": Gabapentin – GABA mimetic; Fosphenytoin – dilantin prodrug; Pregabalin – 2 subunit, voltage dep. Ca channel

Box 9.6 Even bacteria age

Theoretically, bacteria neither age nor die. A single *E. coli* cell can produce two identical copies in 20 minutes, and these copies are able to continue dividing in their turn for an unlimited number of times. However, we no longer need to feel jealous of bacteria. Two years ago, researchers at the Biocenter of the University of Basel found evidence of the first bacteria to age.

Bacteria are prokaryotes, the simplest form of life, and do not have true cell nuclei. They are assumed to be immortal as long as they are sufficiently nourished and are not exposed to damage from environmental influences. All higher cells on the other hand (eukaryotes) appear to have a "use-by" date. They fulfill their task and divide a few times before beginning to age and finally die. Their telomeres, special ends on the chromosomes, become shorter with each division.

A prerequisite for aging in bacteria is asymmetrical cell division, where the two cells resulting from division are not completely identical. This was observed with the bacterium *Caulobacter crescentus*, which occurs in streams where there is a low level of nutrients. There are two versions of this organism: the "swarmer", which is flagellated and motile but not capable of reproduction and the non-motile stalk cell, which is able to reproduce. All swarmer cells eventually change into stalk cells, attach to a suitable site, and begin to form swarmer cells again. Experiments have shown that the ability of the stalk cells to reproduce was significantly reduced over a period of about two weeks. Up to 130 offsprings were produced, but these did not divide uniformly over the two weeks. Some of the stalk cells had stopped forming swarmer cells by the end of the tests while others only divided sporadically. It was assumed that the *Caulobacter* that was examined did not in any way represent a curiosity in the world of bacteria. What happens, though, with symmetrical cell division?

The bowel bacterium *Escherichia coli* has long been known to divide completely symmetrically. Did this continuous division make it immortal? Like most bacteria, it is constricted in the middle to produce two equally sized cell halves. Each new half is then newly synthesized. Each cell consists of an old cell pole inherited from its predecessor and a new pole.

Eric Stewart of the French Medical Research Institute in Paris, INSERM, and his colleagues followed the fate of dividing bacterial cells depending on the age of their cell pole. In order to do this, they marked individual *E. coli* cells with fluorescent colorant and observed the growth of colonies using an automatic time-lapse microscope. The subsequent evaluation of more than 35,000 cells showed that the bacteria with the oldest cell poles (shown in red in the photo) had a lower rate of growth and separation than those with newly synthesized cell halves.

In the opinion of the researchers, signs of aging appear in all organisms. Even bacteria are therefore not immortal.

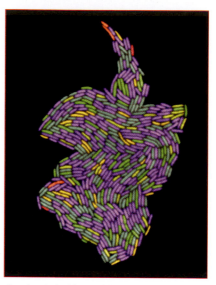

Experimental evidence of the aging of bacteria

Coronary infarction of female mouse induces damage to the left ventricle

Bone marrow cells from adult male mouse

Stem cells isolated from bone marrow cells

Fig. 9.26 Stem cell therapy for coronary infarction (in mice).

The blood from the **placenta or the umbilical cord** is also rich in stem cells. There are companies today that offer reliable deep freezing of the blood from the umbilical cord for a lengthy period of time so that it may be used if the donor should ever require it. These stem cells can only be returned to the donor. Embryonic stem cells on the other hand are universal.

There are **three ways of harvesting embryonic stem cells**:

- From "surplus" embryos as a result of artificial insemination. There are hundreds of thousands of cases of artificial insemination all over the world.

- From aborted or spontaneously aborted embryos or fetuses; these are fetal stem cells.

- Through **"therapeutic cloning"**, transfer of a cell nucleus into an enucleated embryo; this is how Dolly the cloned sheep was made.

Recently, US scientists have reported the harvesting of stem cells from amniotic fluid.

Any cloning of human embryos or the harvesting of stem cells from them is forbidden in Germany. In Great Britain and South Korea, therapeutic cloning is expressly allowed. In 2006, President Bush vetoed the Stem Cell Research Enhancement Act, a bill that would have reversed the Clinton-era law which made it illegal for Federal money to be used for research where stem cells are derived from the destruction of an embryo. The people of the U.S. state of Missouri passed Amendment 2, which allows usage of any stem cell research and therapy allowed under Federal law, but prohibits human reproductive cloning. In February 2007 the California Institute for Regenerative Medicine became the biggest financial backer of human embryonic stem cell research in the United States when they awarded nearly $45 million in research grants.

> The more a scientific field deals with human affairs, the greater the chance that scientific theories will clash with traditions and beliefs.
>
> **François Jacob** (Nobel Prize Physiology or Medicine 1965)

9.14 Gene Therapy

George Brown, chairman of the Scientific Committee in the US House of Representatives, called the history of little **Ashanti De Silva** living proof that miracles can still happen. She appeared before the Committee in September 1995 to much media attention. As a two-year-old, she had received the first gene therapy in history on September 14, 1990. She suffered from extreme immune deficiency and had to grow up in sterile rooms in a hospital, a so-called "bubble baby". A simple influenza infection could have killed her. She was missing one single gene and she was a carrier of **ADA syndrome** (adenosine deaminase). White blood cells need ADA protein to grow and divide. Without this gene product most of a patient's immune system fails. The disease can be treated with the administration of ADA, although there are only about 100 patients worldwide, so the treatment is extremely expensive and amounts to about 40,000 US dollars per month.

In an *ex vivo* (outside the body) experiment, Ashanti was given a **transfusion of her own T-lymphocytes** into which the intact ADA gene had been transferred beforehand (Fig. 9.26). The result was sensational, as the patient was able to leave hospital soon afterwards and finally went to school as normal. About one half of her blood corpuscles contained the new ADA gene, so this was a super start for gene therapy.

However, we are still miles away from a standard treatment for this rare disease. ADA was a comparatively easy case, as **one defective gene** just had to be replaced by its natural variation.

With other diseases such as cancer, **multiple missing and mutated genes** have to be replaced. Cancer occurs in a process of many stages with multiple mutated genes, and all affected genes are altered and silenced in the end, a result of a combination of unlucky "coincidences" or "DNA accidents".

There are also hereditary (genetic) predispositions to some cancers. **Napoleon** died, like his father, his grandfather, and his three siblings of stomach cancer. **Breast cancer** also occasionally occurs more frequently in families. The genes *brca1* and *brca2* are responsible for this. Even if only one gene, the *brca1* gene, is missing, the risk of developing cancer is increased.

There is not as yet any gene therapy for cancer because of the many stages of its development. It is not yet possible to influence so many genes at the same time. The best way forward seems to be the strengthening of the immune system and formation of repair enzymes.

The crux of the matter is the development of effective procedures for smuggling genes into the cells. Until now, harmless retroviruses (viruses with single-stranded RNA) have been used as gene vectors, as retroviruses integrate their own genotype into the infected cells. The idea of a gene vector is attractive, but it is not easily controlled.

After some initial teething problems due to poorly targeted vectors, new **gene vectors** are being tried. It is important to inject the gene vectors directly into the patients, and they must either be introduced into the chromosome at a safe site or they must directly replace the defective gene.

Finally, the new gene must be able to react to physiological changes, such as in diabetes, where the gene must react to rising and falling glucose levels, as the natural genes do. Modern gene therapy (Fig. 9.26 and Box 9.6) is no more than a modern form of applying a drug.

An alternative type of gene therapy is RNA interference (RNAi).

9.15 The Junk Yields its Treasures: RNAi, RNA Interference

Besides functional genomic or proteomic research (Ch. 10), a new technology is taking the bioscience market by storm – **RNA interference**. For the first time, individual genes can be switched off in a highly effective manner and more quickly than with any other procedure. This does not disturb protein formation by neighboring genes. For the first time, it is possible to determine the functions of genes by silencing them in a high throughput manner and to derive economic benefits from the DNA sequences discovered in the genome project (Ch. 10).

According to expert estimates, RNAi technologies are about 80% cheaper than the creation of so-called **knockout mice** (Ch. 8) which have previously been used for discovering gene function. Every gene can be examined with the help of RNAi technology, which means that RNAi opens com-

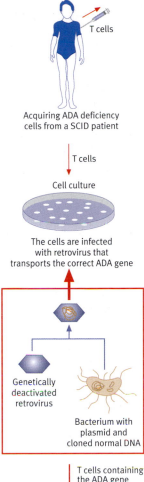

Acquiring ADA deficiency cells from a SCID patient

T cells

Cell culture

The cells are infected with retrovirus that transports the correct ADA gene

Genetically deactivated retrovirus

Bacterium with plasmid and cloned normal DNA

T cells containing the ADA gene

Reinfusion of the T cells containing the ADA gene

Fig. 9.27 How gene therapy is carried out for ADA deficiency.

Fig. 9.28 Thomas Tuschl, RNAi pioneer.

Box 9.7 The Expert's View:
Fetal gene therapy

Most families are usually completely unprepared when they become affected by genetic disease. **Prenatal genetic tests** have opened up the opportunity to find out before birth if a fetus is affected and many families choose to avoid having a sick child by terminating the pregnancy. Many have then been able to have healthy children after further pregnancies.

The **strategy of somatic gene therapy** is persuasively simple: the disease is caused by abnormal gene (DNA) expression, so treatment of the cause should be possible by correcting this abnormal expression using DNA or more generally nucleic acids.

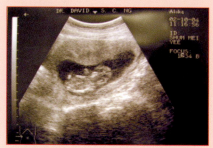

Ultrasound picture of my student's son Theo Alex Kwong (with the permission of Dr. Alex Kwong)

DNA with specific information is administered as a "drug" and takes effect in the treated organism by producing the therapeutic protein encoded by it. The efficacy and safety of the vector system which carries the foreign DNA into the target organism was and is the decisive factor for the speed of progress in gene therapy. The gene treatment of genetically caused diseases must be effective throughout life, either by repeated gene application or by use of a permanently active gene transfer system.

Vectors of the retrovirus group are currently the best example for the latter as they can integrate into the host genome. However, even if the development of effective gene therapy is successful, it will often not remedy any irreversible organ damage that has already occurred. These and other considerations have led to the concept of in utero gene therapy (or fetal or pre-natal gene therapy). This approach could prevent the occurrence of premature organ damage, enable better access to organs and cell systems (including rapidly expanding stem cell populations), and reduce the development of immune reactions. In the ideal case, this should make it possible to correct the relevant tissue stem cells and achieve a lifelong cure for the particular disease.

As mentioned above, it is possible for an increasing number of genetic diseases, to diagnose affected fetuses *in utero* by prenatal tests. Such tests are normally carried out for families that have already had one affected child, but prenatal screening is also carried out in some countries for diseases that occur more frequently in certain populations, such as haemoglobinopathies, Tay-Sachs disease and cystic fibrosis.

Presently, the diagnosis of an affected fetus, confronts the family with the difficult decision to terminate the pregnancy or to care life-long for an affected child. The possibility of **fetal gene therapy would offer these families a third option**.

Fetal gene therapy is of course not an alternative to postnatal treatment but the extension of this novel therapeutic concept to a preventive intervention.

Fetoscopy or ultrasound-supported *in utero* interventions (see illustrations) are an excellent basis for targeted prenatal vector application. In the last few years, it has been shown in different animal models and with various gene transfer vectors that it is possible to reach practically all organs relevant for gene therapy of genetic diseases in utero.

The adeno-associated virus (**adenovirus**) has recently been accepted for clinical trials in gene therapy for hemophilia. This virus is only slightly immunogenic and lingers longer in infected cells.

Adenovirus

Retroviral group vectors that integrate into the DNA of the host organism are currently the most promising candidates; however, due to their random integration into the genome they also carry a certain risk of triggering oncogenesis (development of a cancer).

With the aim to develop the technology for clinical application of in utero gene transfer, our interdisciplinary research team at Imperial College, and University College London were the first in 1999, to use minimally invasive ultrasound-supported injection into the umbilical vein for gene transfer to fetal sheep.

Sheep have a constant gestation period of about 145 days with predominantly singleton pregnancies. They show good tolerance to in utero manipulation and their anatomy allows the performance of procedures that can be used on the human fetus.

The fetal medicine specialists in our team developed novel methods for *in utero* gene delivery at various gestation times of the sheep. Several of the new techniques have not yet been applied to the human fetus. In all these procedures, maternal mortality was extremely low and fetal mortality about 15%, usually as a result of intra-operative infection.

The **first therapeutically successful fetal gene transfer** was carried out in 2003 on rat fetuses by **Jurgen Seppen** (Netherlands). The so-called Gunn rat is a natural model of the very rare Crigler Najar Type I syndrome (CN 1) in humans. This disease is caused by mutations on human chromosome 2. The mutations lead to toxic bilirubin levels in the blood and to severe brain damage by the accumulation of bilirubin in the brain stem. Direct injection of a viral vector (which controls production of the human enzyme bilirubin UDP glucuronyl transferase) into the liver of Gunn rat fetuses, succeeded in reducing the toxic bilirubin levels over a period of one year by about 45%.

Permanent correction of a clotting factor disorder was achieved by the Imperial College group in a mouse model of **hemophilia B** (see also 9.4 in running text) by a single injection of a viral vector into the yolk sack blood vessel (which brings the vector to the liver where clotting factor IX is produced). This mouse strain does not have functioning factor IX protein, but when the animals were treated fetally they had therapeutic levels of the transgenic factor IX in their blood throughout their lives. It is particularly important that no antibodies against the human HX protein were found in the in utero treated animals.

How far away from clinical application in human fetuses are these principles that have been confirmed by animal experiments?

Still a long way to go....

We believe that in spite of the great progress of the last ten years, there is still a long way to go. The differences between species and a range of general and specific safety requirements and ethical considerations related to prenatal gene therapy must be taken into account first.

Fetal gene therapy carries several specific **procedural risks** which are obviously different from

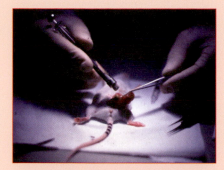

Intra-amniotic injection into mouse fetus

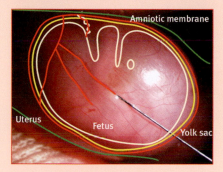

Amniotic membrane

Uterus

Fetus

Yolk sac

Yolk sac vessel injection in fetal mouse

postnatal gene therapy. As with most obstetrical interventions, these affect both the mother and the fetus, and the life and wellbeing of the mother, of course, has precedence.

These risks include infection, premature contractions and miscarriage. Other frequently discussed risks include the possibilities of unintended germ line transmission, developmental disorders and oncogenesis (the development of cancer).

Exclusion of such risk factors requires careful exclusion of birth defects and long-term post natal studies on in utero treated animals, to exclude oncogenesis or other adverse effects before clinical application can be considered.

Healthy twins, a dream for Chinese families under the one-child rule

Designer babies

Germ line modification is one of the most frequently expressed fears related to prenatal gene treatment, and is often confused with the cre-

ation of so-called "designer" babies. The present aim of both prenatal and postnatal gene therapy is **exclusively the treatment of severe diseases** for which no other effective treatment exists.

We are still in the initial stages of implementing this aim. However, once gene therapy shows sufficient safety and efficacy it may well be applied also to less severe diseases and even for postnatal cosmetic use, as long as it is ethically acceptable and the costs can be justified.

The aim of prenatal gene therapy is to **prevent severe suffering** caused by mutations in a precisely defined gene in an individual. The assumption that this technique could be used to create "drawing board people" with "desirable" physical and intellectual characteristics is simply an illusion.

Neither can we predict on a fertilized oocyte such characteristics, which are dependent on the interaction of many different genes, nor do we know which genes would need to be manipulated and in what manner; leave alone the **safety considerations and ethical restrictions which rightly prohibit such attempts.**

For the same reasons it is even more erroneous to assume that such manipulations could be used to create a "super race" by manipulating the germ cells for transmission to future generations. Germ line gene therapy in humans is at present neither medically necessary nor technically reliable and is therefore neither ethically acceptable nor legally admissible. The direct gene manipulation of unfertilized germ cells is currently simply not technically practical or possible. At best, the chance insertion of a therapeutic gene sequence into the germ cell genome might be achieved but the consequences for gene expression of the therapeutic protein would not be predictable.

It would also only be possible to provide evidence of successful insertion by either destroying the germ cell or using it to create a new organism. Increasing knowledge about the function of the human genome and technical perfection of such gene manipulation, ensuring accuracy and safety, as well as real medical needs, could in the future justify germ line modification.

Ethical and legal frame work

It would then be the responsibility of future generations to create the necessary ethical framework, safety regulations and legal definitions.

An approach to increase the safety of *in utero* gene therapy could be based on the use of gene

modified autologous fetal stem cells, for example, from fetal liver biopsy after the third week of pregnancy. This *ex vivo* treatment option would allow lower vector doses to be used and prevent vector spread in the fetal or maternal organism. In the future, *ex vivo* screening and the selection of cells with harmless integration sites might be possible in order to decrease the risk of ontogenesis. Such safe cell clones would then be expanded *ex vivo* before re-infusion.

Under what circumstances may parents who are confronted with the diagnosis of a fetus with a severe genetic disease decide for in utero gene therapy instead of a termination of pregnancy or acceptance of a sick child?

Although there is obviously a need for a therapeutic or preventative solution to the problem, we are also aware that initially **only a few affected families will probably be prepared** to choose this option, even if its safety and efficacy has been proven in animal trials. We must be sure that gene therapy will both prevent the disease and not cause additional damage. Fetal gene therapy will thus be subject to even stricter medical criteria and safety standards than most postnatal gene therapy applications.

The clinical introduction of fetal gene therapy will also **require broad public acceptance**. Our knowledge of the views on this among the population or affected patients and their families currently rests on hearsay at best.

Studies in preparation of human fetal gene therapy should therefore include the questioning of patients/families, health staff and the general population after detailed explanation of the aims, possibilities and limits of this new type of preventative gene therapy. This should help to communicate the necessary information, disclose anxieties and highlight considerations that must be taken into account and thereby create the scientific basis for a rational risk-to-benefit assessment.

Charles Coutelle is a Professor at Imperial College London.

287

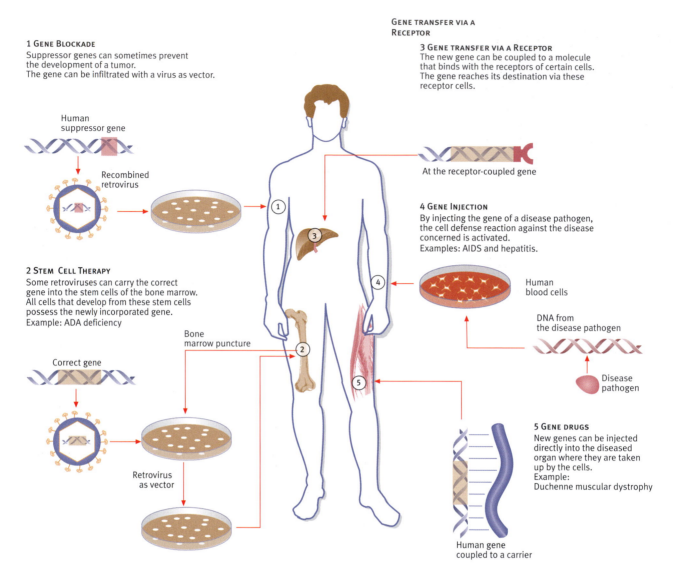

1 GENE BLOCKADE
Suppressor genes can sometimes prevent the development of a tumor.
The gene can be infiltrated with a virus as vector.

Human suppressor gene

Recombined retrovirus

GENE TRANSFER VIA A RECEPTOR

3 GENE TRANSFER VIA A RECEPTOR
The new gene can be coupled to a molecule that binds with the receptors of certain cells. The gene reaches its destination via these receptor cells.

At the receptor-coupled gene

4 GENE INJECTION
By injecting the gene of a disease pathogen, the cell defense reaction against the disease concerned is activated.
Examples: AIDS and hepatitis.

Human blood cells

DNA from the disease pathogen

Disease pathogen

2 STEM CELL THERAPY
Some retroviruses can carry the correct gene into the stem cells of the bone marrow. All cells that develop from these stem cells possess the newly incorporated gene.
Example: ADA deficiency

Correct gene

Bone marrow puncture

Retrovirus as vector

5 GENE DRUGS
New genes can be injected directly into the diseased organ where they are taken up by the cells.
Example:
Duchenne muscular dystrophy

Human gene coupled to a carrier

Fig. 9.29 Possibilities for gene therapy

Fig. 9.30 Craig C. Mello (b. 1960) shared the Nobel Prize in Physiology or Medicine 2006 "for their discovery of RNA interference – gene silencing by double-stranded RNA" with Andrew Z. Fire. Fire (b. 1959) is congratulated by father and son Kornberg. Roger Kornberg's Nobel Prize 2006 came 47 years after his father Arthur received the award for his work.

pletely new perspectives for post-genomic research. However, RNAi also shows all the key properties one would expect from a sequence-specific therapy. RNAi technology is used to examine the genotype of model organisms and humans for new targets for drugs which are then validated and sold to the pharmaceutical industry.

Approval of the very first active substance is not expected until 2008 at the earliest. The foundation stone for this was laid five years ago by **Andrew Fire** and **Craig Mello** with research work on the nematode *Caenorhabditis elegans*. For the first time, they described a new technology that enables the prevention of the formation of certain proteins in living cells with the help of a few double-stranded RNA molecules (**dsRNA**). Fire and Mello were awarded the Nobel Prize in 2006 (Fig. 9.30). One strand of the dsRNAs needs to carry the same sequence as the gene encoding the protein that is to

be silenced. The dsRNAs do not prevent the gene from being read, but switches on the cell's own mechanism that destroys the mRNA read from the gene. It thus prevents the production of the relevant protein (Post-Transcriptional Gene Silencing, PTGS).

Targeted mRNA breakdown is triggered by an **"enzyme dicer"**, which cleaves double-stranded RNA (dsRNA) to short double-stranded fragments of 21–23 nucleotides. One of the two strands of each fragment is then incorporated into the **RNA-induced silencing complex** (**RISC**) and base-pairs with complementary sequences. The most well-studied outcome of this recognition event is a form of post-transcriptional gene silencing. This occurs when the guide strand base pairs with a messenger RNA (mRNA) molecule and induces degradation of the mRNA by **"argonaute"**, the catalytic component of the RISC complex. The short RNA fragments are known as small interfering RNA

Box 9.8. The Expert's View:
Analytical Biotechnology predicts heart risk and diagnoses AMI

Acute myocardial infarction (AMI), commonly known as a heart attack, is a disease state that occurs when the blood supply to a part of the heart is interrupted. It is a medical emergency, and one of the leading causes of death all over the world.

Classical **symptoms of AMI** include chest pain, shortness of breath, nausea, vomiting, palpitations, sweating, and anxiety or a feeling of impending doom. Patients frequently feel suddenly ill. However, approximately one third of all myocardial infarctions are 'silent' i.e., without chest pain or other symptoms.

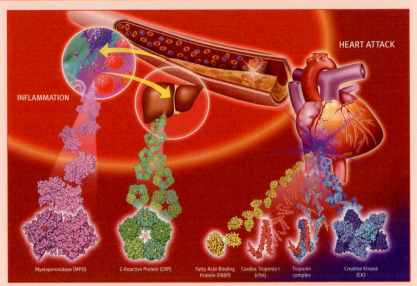

How can one tell that this is an AMI?

Nowadays, a patient receives a number of diagnostic tests, such as an **electrocardiogram** (**ECG**) and **blood tests** to detect elevated creatine kinase or troponin levels. These are biomarkers that normally occur in cells but that are released by damaged tissues, especially the myocardium during periods of ischaemia or necrosis. These two late markers appear in blood only 3-6 hours after the onset of symptoms. More recently a new marker by the name of **heart fatty acid-binding protein** (**H-FABP**) has been discovered. This is a much smaller protein which is abundant in heart muscle, released after severe ischaemia and which may be detected in the blood stream as early as 1 hour after the onset of symptoms. In patients with an AMI, the earlier that the condition is diagnosed and treated the better. The amount of permanently damaged cardiac muscle is reduced, the heart is stronger, and the patient is less likely to die prematurely and less likely to have complications. H-FABP has great potential as a very early, sensitive marker for the early diagnosis of AMI (see Ch.10).

We have developed and tested a **rapid and sensitive immunotest** for the detection of H-FABP in blood samples. The combination of H-FABP with with the later marker troponin makes an excellent combination for the early diagnosis of AMI.

The **Accident & Emergency Department** of a hospital is the place where rapid tests are required to make a fast decision which may save lives and where such a test may be most useful. It can also be used in ambulances or even at home if patients wish to purchase their own personal test.

In the case of AMI, the damage has been done already. But **how can we predict an AMI?**

The key term here is **inflammation** (from Latin, *inflammatio*, to set on fire). Inflammation is the complex biological response of vascular tissues to harmful stimuli (pathogens, damaged cells, or irritants). It is a protective attempt by the organism to remove the injurious stimuli as well as to initiate the healing process for the tissue.

Inflammation is also a key component of heart disease. In the past, researchers believed that chronically inflamed blood vessels set the stage for atherosclerosis (hardening of the arteries). Inflamed patches become "sticky" and start collecting plaque in the heart vessel "pipe". The pipe finally becomes clogged. US-cardiologists **Paul Ridker** and **Peter Libby** reported a radical new approach. Plaques together with inflammation affect the arteries. Investigations begun more than 20 years ago have now demonstrated that arteries bear little resemblance to inanimate pipes. They contain living cells that communicate constantly with one another and their environment. These cells participate in the development and growth of atherosclerotic deposits, which arise in, not on, vessel walls. Further, relatively few of the deposits expand so much that they shrink the bloodstream to a pinpoint. Most heart attacks and strokes stem instead from less obtrusive plaques that rupture suddenly, triggering the emergence of a blood clot or thrombus, which blocks blood flow.

When tissues become inflamed, they release several different proteins, including **C-reactive protein** (**CRP**). CRP has the greatest amount of supportive data and the strongest evidence that its increase can allow us to predict risk at all levels of cholesterol and all levels of the Framingham Risk Score. After adjusting for other variables, such as age, smoking status, and diabetes, women with high levels of CRP were twice as likely to develop future cardiovascular events as women with low CRP. Women with high CRP were more likely to develop a heart attack or stroke than those with high cholesterol. Combined both cholesterol and CRP tests appear to be superior to the use of either alone. Many well-known risk factors for heart disease seem to raise the levels of CRP. According to a recent report, smoking, high blood pressure, extra weight, and lack of exercise are all associated with high CRP. Genes also seem to play a strong role. If your parents had high levels of CRP, there's a good chance you will too...

We are working on a whole range of **new rapid immunotests**, not only for AMI, but also for new inflammation markers like CRP, neopterin, and myeloperoxidase (MPO).

These tests can allow prediction of future cardiovascular events and may push YOU to change your life style!!

Dr Cangel Pui Yee Chan is Research Director of R&C Biogenius Ltd (Hong Kong, China). Prof Tim Rainer is Head of Accident and Emergency Medicine Academic Unit, The Chinese University of Hong Kong, Prince of Wales Hospital, Hong Kong.

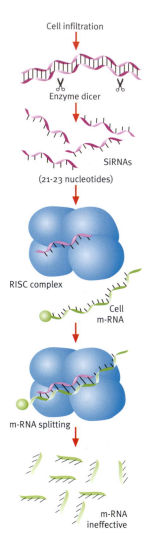

Cell infiltration

Enzyme dicer

SiRNAs
(21-23 nucleotides)

RISC complex

Cell
m-RNA

m-RNA splitting

m-RNA
ineffective

Fig. 9.31 Hypothetical process of RNA interference. It is not clear whether short RNA pieces are single or double stranded.

Fig. 9.32 Although the details of RNA interference are still uncertain, it is clear that molecular biologists have completely underestimated the role of short RNA pieces in the cell, as they appear to have huge potential.

(siRNA) when they derive from exogenous sources and microRNA (miRNA) when they are produced from RNA-coding genes in the cell's own genome (Fig. 9.31).

Barely two years ago, attempts to use longer dsRNAs (>50 RNA components) to silence the most promising genes for drug development were still being foiled by the interferon response. This is an immune reaction to viral infection which causes mammalian cells to stop producing proteins and to degrade mRNA nonspecifically (as described earlier in this Chapter).

Then in 2001 a breakthrough was achieved by the German scientist **Thomas Tuschl** (Fig. 9.28), who at that time was at the Max Planck Institute for Biophysical Chemistry in Göttingen. Tuschl found short RNA molecules consisting of only 21 to 23 components. He discovered "diamonds" in – of all things – that 98.5% of the genotype which was regarded as "junk DNA". After extracting the waste material, he could effectively block mRNA in the cell nucleus. Tuschl immediately recognized the potential of his discovery. With the aid of the small interfering RNA of mammalian cells, genes could be switched off as required – even though this could only be done in cell cultures for the time being.

With the artificially created siRNAs of 21 to 23 nucleotides in length, his team silenced mammalian genes for the first time, without eliciting an **interferon response**. Since then, there has been one research success after another – for example, the silencing of various HIV genes, and the fights against influenza and the hepatitis C virus.

Researchers certainly knew by 1998 at the latest that genes could be silenced with the help of RNA molecules. However, it was not clear how this mechanism, named **RNA interference** (**RNAi**), functioned.

Two years later, siRNA was being used in laboratories all over the world. Tailor-made artificial RNA strands were used in the meantime for testing of gene function in large quantities – within weeks instead of over a period of years. In the process, it was not just individual genes that were examined, but hundreds at the same time.

Tuschl's methods did not just extend the repertoire of basic researchers who were able to use them to unravel the function of approximately 30,000 genes of the human genotype; large pharmaceutical companies were also interested in the technique.

Genes are an important starting point for therapeutic products. Approximately 5,000 human genes are of interest as drug targets, and it is hoped that they will be found with the aid of "short" RNA strands.

Even siRNA can be used as a drug itself. Once a faulty gene has been identified using siRNA, it can be used to suppress the relevant mRNA for therapeutic purposes. That is exactly what researchers had wanted for 30 years with so-called **antisense technology** (Ch. 7 "Anti-mush tomatoes"). Scientists had been trying to use pieces of RNA that were too large, so the cell nuclei produced an allergic reaction to what appeared to them to be destructive viral genetic material. The cell apparatus therefore attacked the intruder and destroyed it. There was just one possibility for evading this "cell nucleus immune system": the RNA had to be tiny, no longer than 21 to 23 components. RNA of this length occurs naturally in the cell nucleus and one of its functions is to control whether and how often a gene is expressed. Tuschl's small molecules reduced the production of proteins 1000 times more efficiently than the old antisense technique.

The list of possible **areas of use** includes rheumatism, Alzheimer's disease, Parkinson's disease, cancer, metabolic and auto-immune disorders, and infectious diseases.

However, the route from laboratory to clinical application is a long one. The greatest difficulty with regard to siRNA treatment is to get the fragile molecules inserted into the cell nuclei where they can be effective. Sensitive material is attacked in the blood stream and quickly degraded by enzymes. Some researchers will package it in liposomes; others will transport the molecules with viruses into the cell nucleus or try to do it with naked siRNA that has been chemically stabilized.

Perhaps they will resolve the puzzle as to why we have only eight times more genes than *E. coli*, but are genetically 1000 times more complex than the bacterium. Are the small RNAs and all the genes that are not directly encoding protein indeed the true rulers in the cell nucleus and responsible for the engine of evolution? Not until this question has been answered can the great flexibility of the metabolism of higher lifeforms be explained.

Answers should come from the annotation of the genomes of important lifeforms and finally of humans.

Cited and recommended Literature

- **Thieman WJ, Palladino MA** (2004) *Introduction to Biotechnology*, Benjamin Cummings, San Francisco. Good introducing textbook for undergraduate students, clear illustrations, easy to read.

- **Crawford MH** (2004) *Essentials of Diagnosis & Treatment in Cardiology.* McGraw-Hill Medical, New York
Solid knowledge for beginners

- **Kayser O, Müller RH** (2004) *Pharmaceutical Biotechnology.* Wiley-VCH, Weinheim
Good overview about Pharma Biotech

- **Walsh G** (2003) *Biopharmaceuticals. Biochemistry and Biotechnology* (2nd edn.) Wiley, Chichester
Biopharma products and manufacturing for advanced readers

- **Scott CT** (2005) *Stem Cell Now: From the experiment that shook the world to the new politics of life.* Pi Press, London

- **Lanza R** and **Rosenthal N** (2004) *The Stem Cell Challenge. What hurdles stand between the promise of human stem cell therapies and real treatments in the clinic*? Scientific American, June issue

- **Kleinsmith LJ** (2006) *Principles of Cancer Biology.* Pearson-Benjamin Cummings San Francisco. Comprehensive and well-illustrated introduction.

- **Berg, JM, Tymoczko JL, Stryer L** (2007) *Biochemistry*, 6th edn., WH Freeman, NY. Basic dynamic understanding about DNA and genes

- **Sohail M** (ed.) *(2004) Gene Silencing by RNA Interference: Technology and Application* CRC Press, Boca Raton

- **Fire A, Xu SQ, Montgomery MK, Kostas SA, Driver SE, Mello CC** (1998) *Potent and specific genetic interference by double-stranded RNA in Caenorhabditis elegans.* Nature. 391, 806-811
The original paper in *Nature* of the Nobel Prize winners Fire and Mello

Useful weblinks

- For a starting point, the Wikipedia, with many useful weblinks like, for example for AMI, stem cells, or taxol:
http://en.wikipedia.org/wiki/Myocardial_infarction or/stemcells or/Paclitaxel

- Risk assessment tool for estimating your 10-year risk of having a heart attack
http://hp2010.nhlbihin.net/

- Heart attack warning signals from the Heart and Stroke Foundation
http://ww2.heartandstroke.ca/

- The National Institute of Health (NIH) about stem cells:
http://stemcells.nih.gov/

- *Understanding Stem Cells*, was released by the National Academies in October 2006
http://dels.nas.edu/bls/stemcells/booklet.shtml

- American Cancer Society Homepage:
http://www.cancer.org/

- Glivec website from Novartis:
http://www.gleevec.com/

- Hope Clinic Australia. A huge database of various cancers:
http://smile.org.au/

Eight Self-Test Questions

1. What should an emergency doctor do for a myocardial infarction, and which biochemical marker is faster in the blood: FABP or Troponin?

2. How could Count Dracula theoretically have helped with stroke?

3. Can you really get rid of wrinkles with EGF? Are there more sensible applications?

4. Should stem cell research be restricted? Can it be misused?

5. How do some sportspeople use the achievements of biotechnology to enhance their performance?

6. How can an utterly safe drug for hemophiliacs be provided?

7. How do Pacific yew trees save the lives of cancer sufferers?

8. What is interfering RNA and how can it be used to shut down genes?

ANALYTICAL BIOTECHNOLOGY AND THE HUMAN GENOME

CHAPTER **10**

Fig. 10.1 Bad nutrition habits and lack of exercise are the main causes of obesity and diabetes.

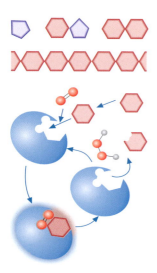

Fig. 10.2 GOD recognizes glucose from a sugar mixture, binds to it and converts it (described in more detail in Ch. 2).

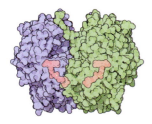

Fig. 10.3 GOD is a glycoprotein dimer (sugar chains not shown here). The flavin adenine dinucleotide (FAD, red here) in the active site takes up two electrons from glucose and transfers them to oxygen. The reduced oxygen (O_2^-) binds two protons (H^+) from water to form hydrogen peroxide (H_2O_2). Glucose is reduced to gluconolactone, which reacts with water to form gluconic acid.

10.1 Enzyme Tests for Millions of Diabetics

With the discovery of enzymes, the ancient dream of being able to give an accurate diagnosis from sampling body fluids came true. From a mixture of hundreds of substances present in blood or urine, individual compounds can be specifically targeted – e.g., β-D-glucose in **diabetes** – using **glucose dehydrogenase** (GDH). Together with cofactor nicotinamide adenine dinucleotide (NAD^+), GDH converts glucose into gluconolactone, reducing NAD^+ to NADH ($+H^+$). NADH can be easily quantified in a photometer, using Warburg's optical test (Ch. 2) at a wavelength of 340 nm. NAD^+, by contrast, does not absorb light of this wavelength.

Glucose oxidase (**GOD**, Figs. 10.2 and 10.3) is another enzyme, also an oxidoreductase, which reduces oxygen to hydrogen peroxide (H_2O_2) with the help of electrons from glucose. From a mixture of sugars, GOD specifically acts on α-D-glucose. Therefore, if H_2O_2 can be detected in a body fluid (blood, serum or urine) after the addition of GOD, it is an indicator of the presence of glucose. The higher the amount of H_2O_2 – the amount of enzyme added remaining constant – the higher is the glucose content. Solutions with a known glucose concentration provide a calibration curve that makes it possible to work out the glucose content of the mixture at hand. Portable **biosensors** have been developed which can be used by diabetic patients (Fig. 10.12).

Approximately seven to eight percent of the populations of developed nations are diabetic. Diabetes mellitus is a metabolic disorder with constantly raised blood sugar levels. An excess of glucose cannot be filtered out by the kidneys and is excreted in the urine.

Diabetes is caused by a lack of the hormone **insulin** (Ch. 3) or by some disruption of its action. Insulin is produced in the pancreas and lowers the blood sugar level by transporting glucose into cells. It is also involved in fat and protein metabolism. Thus, a lack of insulin has more repercussions than just a disrupted sugar metabolism.

Normal **blood sugar levels** lie between 60 and 110 milligrams/deciliter and should not rise above 140 milligrams/deciliter even after a meal. In diabetics, however, even the fasting blood sugar level lies above 126 milligrams/deciliter and rises to 200 milligrams/deciliter and more after meals.

In **type 1 diabetes**, there is a lack of insulin due to the destruction of insulin-producing cells in the pancreas. The cause is probably an underlying autoimmune reaction which makes the body attack its own cells. What causes the immune reaction is as yet unclear. One of the possible causes that are being discussed is a viral infection during childhood. Additional genetic factors come into play as well, as 20 percent of patients have a close relative suffering from type 1 diabetes.

In **type 2 diabetes**, the patients initially produce sufficient amounts of insulin, but its action on the metabolism is reduced (insulin resistance). The insulin receptors become increasingly insensitive. Thus, although the amount of glucose made available to the cells decreases, blood glucose concentration increases, stimulating the pancreas to produce even larger amounts of insulin in order to lower the blood sugar level. This leads to a further reduction in the number of receptors on the cells, lowering insulin receptivity levels even further. The constant overproduction of insulin exhausts the β-cells, leading eventually to a lack of insulin.

type 2 diabetes is a disease that mostly goes unnoticed for a while before it is finally detected, as there are no very specific clues. General symptoms include tiredness, lack of initiative, eating bouts, weight loss, perspiration and headaches, which, on their own, will not point to the right diagnosis. This is where **analytical tests** come in. More than half of all diabetics are ignorant of their disease. Untreated, diabetes can have disastrous consequences. In Germany alone, it is responsible for 30,000 strokes, 3,000 cases of blindness, and 35,000 heart attacks every year. Kidneys are also at risk – 3,000 patients must undergo dialysis every year as a consequence of diabetes. Furthermore, diabetes is to blame for 70 percent of all foot and leg amputations.

10.2 Biosensors

How can glucose levels be measured quickly and accurately? With biosensors, of course!

Six hundred years after the first mention of a glucose biosensor in literature by **Giovanni Boccaccio** (Fig. 10.6), **Leland Clark Jr.** and **George Wilson** from the US, **Isao Karube** from Japan, **Anthony P. F. Turner** from the U.K. and **Frieder W. Scheller** from Germany (Fig. 10.9) became pioneers in the field.

In biosensors, the basic idea is to couple biomolecules (enzymes, antibodies) or cells directly with sensors (electrodes or optical sensors), amounting to their quasi-**immobilization** (Ch. 2). **Direct coupling** enables the analyte (e.g., glucose) to be converted by a biocomponent (e.g., GOD) into a biochemical signal (e.g., H_2O_2) (Fig. 10.8) that is immediately passed on to the sensor. It is then transformed into an electronic signal (electricity) and amplified.

The immobilized **biocomponent is regenerated after use**. In glucose sensors, for example, the same GOD can be used to measure glucose in blood samples approximately 10,000 times. Reusability was an important aspect for hospital labs which have to deal with hundreds of patient samples quickly and economically (Fig. 10.10).

Single-use sensors, however, are the ideal option for patient selftests, as they protect from infection (HIV, hepatitis). The biochip is only used once and then discarded. Patients are often given an electronic monitor as a free gift and are thus tied to a particular company for buying biochips.

Most glucose biosensors use a specific enzyme obtained from a fungus (*Aspergillus*), **glucose oxidase** (**GOD**), which only binds to β-D-glucose and oxygen, converting them into gluconic acid and hydrogen peroxide (H_2O_2) within fractions of a second (Fig. 10.2). All glucose measurements are based on the principle that more glucose in the sample means more of the product, and that more oxygen will be reduced. The amount of product is monitored with color tests, test strips, and modern glucose sensors.

Glucose sensors represent the **first generation of biochips**. Two cutting-edge technologies – microelectronics and biotechnology – were combined for the first time.

In order to protect the patient from infection (HIV, hepatitis), light-weight and low-price glucose monitors with single-use chips (Fig. 10.12) are widely used. In a kind of immobilization (Ch. 2), glucose oxidase has been screen-printed on the chips. Permanent immobilization, as required for reusable biosensors, is not needed, as the chips are only used once.

The patient puts a new biochip into the monitor (Fig. 10.12), then pricks the tip of a finger (or, in other models, the forearm) with an aseptically packed automatic lancet. A minute drop of blood is adsorbed into the biochip through capillary action. The blood, being liquid, activates GOD the before dry, and glucose is converted in seconds. When binding to glucose, the enzyme passes electrons on to a **mediator** substance, e.g., ferrocene (Fig. 10.7).

The use of mediators makes the sensor independent of oxygen. The oxygen concentration in a drop of blood is difficult to calibrate. The enzyme binds to the mediator rather than to the natural co-substrate oxygen. Via the active site of GOD (flavin adenine dinucleotide, FAD), two ferrocene molecules each take up an electron of a glucose molecule (Fig. 10.7). The reduced ferricinium diffuses from GOD, reaching the electrode (Fig. 10.7). The electrons produce an electric signal in the chip sensor, and the display shows the glucose concentration, calculated from the strength of the current.

Other biosensors followed, such as a **lactate** sensor to measure the fitness of athletes and race horses. **Glucose** and **lactate** levels were measured in **race horses** in Hong Kong before and after three to five-minute races, using enzyme sensors. Lactate rises sharply if the muscles are not sufficiently supplied with oxygen and glucose must be broken down anaerobically (Ch.1). The fitter the horse, the lower the increase in lactate will be. This is not to say that the fittest horse will always win – after all, the jockey must be taken into account as well.

Anyhow, glucose testing remains the most widespread application of biosensors, having reached a market volume of $300 million, and the market is set to grow rapidly, due to emerging economies such as China and India.

Fig. 10.4 Claudius Galenus von Pergamon (129 – 199 A.D.), alongside Hippocrates, the most famous medic in the ancient world, originated the doctrine of the four humors.

Fig. 10.5 Tropical butterflies find the urine that tastes sweeter than the control in the experiment.

Fig. 10.6 Giovanni Boccaccio (1313 – 1375) wrote the *Decamerone* between 1348 and 1353 (it was only published long after his death, in 1470) where he gave the first literary description of a biosensor – the tongue of a doctor detects glucose in the urine of his beautiful lady patient.

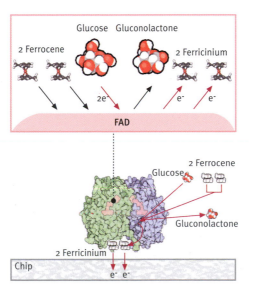

Fig. 10.7 Single-use chip for glucose. In the active site of GOD absorbed by the chip, glucose binds to the flavin adenine dinucleotide (FAD). Two electrons are transferred via FAD on to two mediator molecules (ferrocene) which leave the active site in a reduced state. They diffuse to the chip surface and pass on the electrons. The strength of the current is proportional to the number of transmitted electrons, which, in turn, is proportional to the mediator and thus to the glucose concentration.

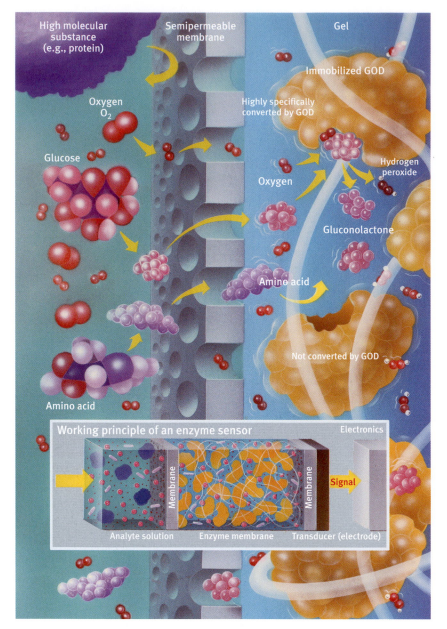

Working principle of an enzyme sensor — Electronics — Membrane — Signal — Membrane — Analyte solution — Enzyme membrane — Transducer (electrode)

High molecular substance (e.g., protein) — Semipermeable membrane — Gel — Immobilized GOD — Highly specifically converted by GOD — Oxygen O_2 — Glucose — Oxygen — Hydrogen peroxide — Gluconolactone — Amino acid — Not converted by GOD — Amino acid

10.3 Microbial Sensors – Yeasts Measuring Water Pollution in Five Minutes

Biosensors using living immobilized microbes play an important part in environmental testing. Yeasts such as *Trichosporon cutaneum* and *Arxula adenivorans* are the most frequently used species to measure the **organic pollution** of wastewater. *Arxula* is a yeast that was originally used in the Soviet Union for a *Single Cell Protein* project (Ch. 7). The idea was to break down Siberian cellulose with acid and turn it into carbohydrates. Then the acid would be neutralized with a base, resulting in high concentrations of salt. Researchers were looking for a microbe able to digest large amounts of carbohydrate, while also being salt-tolerant (halophilic).

The cellulose – *Single-Cell Protein* project failed eventually (Ch. 7), as did the petroleum SCP projects, but the yeast turned out to be very suitable for wastewater monitoring. In tropical coastal countries with a lack of freshwater, seawater is often used to flush toilets. Thus, wastewater is high in salt, which inactivates many microbes – but not *Arxula*!

The idea of using microbial sensors (Fig.10.11) developed from the traditional measuring of the **Biochemical Oxygen Demand** in five days, **BOD5**. In commercial BOD tests (Box 6.2, Ch. 6), wastewater samples are incubated with a mixture of microbes for five days at 20-25°C (68-77°F), while the oxygen content is measured at the beginning and at the end of the incubation period, using a Clark electrode. In clean water, the microbes have got nothing to process, going into standby mode and not using up oxygen. The oxygen content remains unchanged over the five days.

If, however, the sample contains many aerobically degradable compounds, the microbes reproduce in proportion to the available food supply, using up the oxygen. It is thus possible to calculate how much oxygen would be needed to completely degrade the biodegradable compounds in a wastewater sample, i.e., to produce clean water. In certain developed countries, treated wastewater must not exceed a BOD5 of 20 milligrams of O_2 per liter.

While the **BOD test takes five days**, microbial biosensors only take **five minutes** for their measurements. The living yeast cells are immobilized in a polymer gel (as in the glucose sensor, Fig. 10.8) and mounted on an oxygen electrode (Fig. 10.13).

Fig. 10.8 Reusable enzymatic sensor for glucose. Clinical sensors that produce thousands of measurements with the same enzyme (Figs. 10.10 and 10.11) consist of a sensor (electrode) covered by a thin membrane containing immobilized glucose oxidase (GOD) which is sandwiched between polyurethane gel (right-hand side of the diagram).

Only low-molecular-weight compounds and oxygen (red) from the sample can diffuse through the dialysis membrane into the gel (light blue). Large-molecule analytes cannot penetrate the membrane. GOD converts only β-D-glucose, reducing oxygen and producing gluconolactone and H_2O_2. The immobilized GOD cannot be washed out of the membrane. The resulting H_2O_2 (hydrogen peroxide) is electrode-active, i.e., its concentration can be established using an electrode. The inserted schematic diagram gives an overview of the signaling process.

Glucose concentration is proportional to H_2O_2 concentration, which, in turn, is proportional to the strength of the current.
To measure the glucose content, the biosensor is immersed into the solution to be analyzed. The amount of hydrogen peroxide produced is the parameter that determines the amount of glucose.

After the measurement, the enzyme membrane is rinsed with clear solutions that do not contain compounds that react with GOD. Thus, diffused substances from the GOD reaction are washed out, and the biosensor is regenerated, ready to take new measurements.

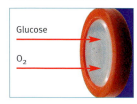

Glucose — O_2

The same enzyme membrane can be used for 10,000–20,000 rapid, cheap, and accurate measurements.

The sensor measures how much oxygen the "starving" cells are taking up. When clean water, containing no degradable substances, is added, the yeast does not take up additional oxygen. When a sample with "food" is added, containing carbohydrates, amino acids, and/or fatty acids, the cells become active, ingesting and degrading the compounds through respiration. The amount of oxygen used up is proportional to the amount of food.

Such microbial sensors are ideal for the **monitoring** of sewage treatment plants. They indicate how polluted the incoming water is and help regulate the aeration pumps for the activated sludge basins. Thus, energy can be saved. A biosensor at the outlet of the treatment plant shows if the water has been treated effectively.

10.4 Immunological Pregnancy Tests

Around 1600 A.D., the average woman in Central Europe became pregnant about 20 times during her lifetime. Nowadays, birth control methods (about the pill, see Ch. 4) as well as rapid pregnancy tests are available in industrialized countries.

The injection of the urine of supposedly pregnant women into frogs ("the frog has reacted positively") is now medical history. The **frog** or **toad test**, also known as Galli Mainini test was a biological pregnancy test. A male frog (or toad) was injected subcutaneously or into his dorsal lymph pouch with the urine or blood serum of a female test person. If, after 24 hours, the microscope showed semen in the urine of the tested frog, the tested woman was pregnant (not the frog). The frog or toad test remained the method of choice into the 1960s for establishing pregnancy in its early stages. The animal, usually an African clawed frog (see Ch. 3), was given a break to recover and then used again for more tests. Apart from injections, celibacy and imprisonment, it did not suffer any torture. Nowadays, the test has been replaced by **immunological pregnancy tests**.

Alongside enzymatic glucose tests, immunological pregnancy tests are the most widely sold biotests. Antibodies infallibly recognize substances like **human chorionic gonadotropin (hcg)**, which is responsible for the secretion of the pregnancy hormone progesterone.

What happens biochemically during pregnancy? The fertilized egg finds its place in the mucosa of the uterus six days after fertilization. This is called nidation and elicits a dramatic release of hormones in mother and embryo. The placenta rapidly produces the hormone humane chorionic gonadotropin. Hcg overrides the normal hormone cycle that would culminate in menstruation. The hcg is produced in such vast amounts that its concentration in the blood is extremely high and it is also excreted through kidneys and urine.

By definition, **urine** is a body fluid that is excreted voluntarily in large amounts. It is thus freely and – in contrast to the use of blood in tests – painlessly available and suitable for selftests. Many such tests are available in pharmacies (Figs. 10.14 and 10.15). They are easy to carry out and have a reliability of 90 to 98 percent.

The earlier a test with urine is carried out, the more uncertain its result will be. In a medical context, a pregnancy is usually established by a blood test.

A **blood test** is more sensitive by far than a urine test and can therefore provide reliable results at an early stage of a pregnancy. As early as ten days after fertilization, hcg can be detected in the blood as a certain indicator of an existing pregnancy.

A window in the plastic casing of the **urine test kit** shows a colored control line (Fig. 10.14, test kit is working), and a second window displays the crucial colored line – baby on its way. If this line does not appear, no hcg is detected in the urine, and thus, there is no pregnancy.

This test, which takes no longer than a minute, will not work from day one after fertilization. The embryo must have time to embed itself in the uterus. The test will only work after menstruation should have occurred.

The pregnancy test uses **monoclonal antibodies** (Ch. 5) for the detection of hormones. They specifically recognize the hcg secreted by the placenta in any mix of thousands of components. Similar to enzymes, the antibodies fish for substances that bind accurately. Antibodies bind to the relevant antigens without degrading their substrate.

Suppose an antibody has found an antigen – **how is this visualized**?

In an enzyme-linked immunosorbent assay (**ELISA**, described in more detail in Ch. 5), enzymes are used to detect the formation of an antibody-antigen complex. This is **wet chemistry**, which requires

Fig. 10.9 The author (second from right) had the privilege and the pleasure of working as a doctoral student in Frieder W. Scheller's research laboratory. Scheller (left; at present prorector of the University of Potsdam) began his research into biosensors in Berlin-Buch (then East Germany). Although material and technical conditions were not the best, Scheller was ahead of his time and managed to produce the most rapid glucose sensor in the world – GKM-01.

Fig. 10.10 Top: A 10,000 times regenerable thick layer biosensor has been developed by EKF diagnostics, Magdeburg, Germany. Bottom: A handheld glucose meter by the Berlin company BioSensor Technologie, BST. Both are based on Frieder Scheller's glucometer GKM-01.

Fig. 10.11 A reusable wastewater sensor (top), using immobilized yeast cells (bottom) to determine BOD5 in five minutes instead of five days.

Fig. 10.12 Glucose selftest. How to measure blood glucose yourself. A glucose level of 104 mg/dL (or 5.6 mmol/dL) is normal.

Fig. 10.13 Sensors that are used in biosensor production. Printed thick layer sensors (BioSensor Technologie, Berlin).

equipment such as readers and provides an accurate assay of antigen concentration.

Where just a clear and fast **yes/no response** is called for, **test strips** come into their own, and no machinery is involved. Antibody material is bonded to small colored **latex globules or colloidal gold** and put at the end of a filter paper strip and left to dry. They are called **detector antibodies** because their color makes them detectable.

When blotting paper is dipped into a liquid with one end, the liquid rises into the pores of capillaries of the paper – a phenomenon which is utilized in **immunochromatography**. When the end of the strip is dipped into urine, the liquid rises, wetting the whole strip in the process and transporting hcg to the "waiting" detector antibody.

hcg... detector antibody with colored globule

The antibody binds to hcg, and the resulting construct wriggles its way through the pores of the paper. In the middle of the strip, catcher antibody has been firmly bonded to the paper as a line. The catcher fishes the hcg-detector-pigment construct out of the liquid and holds on to it.

Both the detector as well as the catcher antibody bind to hcg – one on each side. The whole complex is therefore also called a sandwich construct. Here it is:

Paper... catcher antibody... hcg... detector with colored globule

A colored line is clearly visible. If the urine contains no hcg, it cannot bind to the detector color complex. When the detector reaches the catcher, it will not be fished out, as the hcg is missing, so no sandwich forms. The control line is there to show if the test works at all. A second catcher antibody has been bonded to the paper which recognizes the detector antibody in the absence of hcg and makes it visible at the control line:

Paper... catcher antibody for detector antibody... detector antibody with colored globule

If the control does not work, the test result is invalid.

10.5 AIDS Tests

Would you believe it – approximately 15,000 people become infected with a deadly virus every day! Why is there no worldwide panic, and what happened to aid initiatives? We have learned to live with the virus and the threat it poses. Approxi-

mately 40,000,000 people worldwide suffer from AIDS (Ch. 5). In South Africa alone, 5.3 million people are HIV-positive – about 12 percent of the total population. More than half a million people with AIDS have died in the USA. In total, an estimated 988,376 people have been diagnosed with AIDS in the USA until 2006.

What makes AIDS such a treacherous disease is that the immune system itself, which should give protection, is affected. Upon an HIV infection, the body activates its immune system, as it would in any other infection. In contrast to other infections, however, it takes the immune system at least four to six weeks to produce **antibodies against** the virus.

This is the signal to attack the intruders. The antibodies can be detected in the blood of an infected person. As the first symptoms of AIDS are often hardly noticeable or nonexistent, tests have been developed. Existing antibodies can be identified by an immune test, while the viral RNA itself can be detected through PCR. The **major drawback of the immune test** is that it cannot detect the infection before antibodies have been produced. Recent infections can only be detected through **PCR**. Other tests for viruses work in a way similar to the AIDS test, e.g., the **hepatitis B test**. But there has also been an increase in easily manageable yes/no test strips for various viral infections. They are structurally similar to pregnancy tests.

■ 10.6 Myocardial Infarction Tests

The onset of a heart attack (acute myocardial infarction, AMI, see Ch. 9) is characterized by a feeling of being unwell and strong pressure on the breastbone. The pain radiates into the left arm, a fear of death arises, and the patient breaks out in a cold sweat. Well, if only the symptoms were always as clear-cut as this! Sometimes people feel just a little unwell, and in some women, it seems to be no more than a stomach ache. In approximately 40 percent of all infarctions, the **electrocardiogram** (**ECG**) does not show unambiguous results, i.e., an acute infarction. This is where the rapid immune test comes in (see end of Ch. 9).

In an infarction, the blood flow towards the heart is reduced or even disrupted by blood clots (thrombi). Starved of food and oxygen, the heart cells begin to die. The **dying heart cells** release proteins into the blood. Heart muscle enzymes such as **creatine** kinase (**CK**) and proteins such as various **troponins** have been measured over the past years.

These comparatively large proteins are called **late markers**. Once they are found in the blood, disaster has already struck, and the heart attack has been ongoing for at least an hour. Reliable as they are as indicators, they come a bit late.

"**Time is muscle**" is a cardiologist's adage. The earlier the thrombus can be dissolved, the smaller the damage to heart tissue. Genetically engineered enzyme (tissue plasminogen activator tPA, Ch. 9) can be injected to dissolve the blood clot, but there is a risk of brain hemorrhage. However small the risk may be, the injection can only be justified if there is an actual heart attack. Hence, the search for more rapid markers.

Since the 1990s, I have been working on developing such a marker with my Hong Kong research team, in collaboration with **Jan F.D. Glatz** in Maastricht (The Netherlands) and two companies in Berlin. The **fatty acid-binding protein** (**FABP**) is a very small protein (MW 15,000) that, due to its small size, appears in the blood almost immediately after the infarction, one to two hours before late infarction markers such as creatine kinase and the troponins T and I appear. It is therefore classified as an **early marker**.

The fastest heart attack test in the world works on the same lines as a pregnancy test, using monoclonal antibodies binding to the fatty acid-binding protein, which are able to recognize heart-specific FABP (h-FABP) out of thousands of compounds in the blood (Fig. 10.17).

How can h-FABP recognition be visualized?

As in a pregnancy test, the detector antibody is put on the end of a narrow strip of filter paper and left to dry. The antibody has not been bonded to colored latex globules, but to **colloidal gold particles**. In solution, these particles convey a lovely red color. Thus, the detector antibody carries a **red label**. The test strip is put in a flat plastic casing in credit card format (Fig. 10.16). The card has a funnel-like opening that takes up the blood and a window that shows the result. Just as in the diabetes test, a finger is pricked with a sterile lancet. Three drops of blood go into the funnel opening. The filter paper absorbs the blood by capillary action, transporting h-FABP to the "waiting" gold-labeled detector antibodies, which bind to FABP, and move on as a construct through the pores of the paper:

FABP... detector antibody with red gold globule

In the middle of the strip, a catcher antibody is firmly bonded as a line on the paper:

Paper... catcher antibody

The catcher "fishes" the FABP-detector antibody construct from the liquid and holds on to it by binding to it to form a sandwich (Fig. 10.19):

Paper... catcher antibody... FABP... detector antibody with red gold globule

A clearly visible **red line** appears, indicating an **infarction**.

In the **absence of FABP** in the blood, nothing will bind to the gold-labeled detector. The detector moves to the catcher on its own, but cannot be caught because there is no FABP to form a sandwich. Thus, **no red line** appears, indicating that there has not been an infarction. **A control line** shows whether the test has been working. The whole test takes 10 to 15 minutes. The test has already helped doctors make accurate diagnoses within the critical time window (Fig. 10.17).

■ 10.7 Point of Care (POC) Tests

New immune tests have appeared, which assess, for example, the **risk of a heart attack or stroke**. So far, this has been done on a blood lipid level basis. In some pharmacies, it is already possible to determine the lipid profile within five minutes. Gone are the days when the total cholesterol level was the main parameter. Nowadays, it is common knowledge that a high cholesterol level on its own does not say anything about the risk of atherosclerosis. Thus, the levels of various lipids in the blood need to be measured.

Triglycerides, "good" HDL (high-density lipoprotein) cholesterol must be distinguished from "bad" LDL (low-density lipoprotein) cholesterol and very low density lipoprotein. The critical parameter is the total cholesterol/HDL ratio, which should be below 4.0. Higher levels indicate a risk of atherosclerosis.

Huge automated analyzers in central laboratories, measuring hundreds of components simultaneously, are not always needed. There is a clear trend towards decentralization and rapid testing on location, also called **point-of-care** (**POC**) testing.

Fig. 10.14 Pregnancy selftest from a bioanalytics lecture. The urine test shows only a single line in the window, the control line. The second line for hcg does not appear. A sigh of relief: "No, our professor is not pregnant!"

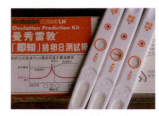

Fig. 10.15 An ovulation immune test indicates the presence of luteinizing hormone and thus imminent ovulation. The urine test begins 17 days before the expected menstruation date. A positive test means that an egg will be available for fertilization within the next 36 hours. These are not the author's test results!

Fig. 10.16 Top: Jan F. C. Glatz (University of Maastricht). The Netherlands) found the fastest myocardial infarction marker so far, FABP. Bottom: the "life-saving credit card", a rapid test produced by Rennesens GmbH, Berlin-Buch.

Fig. 10.17 The strip lies in a credit card-shaped casing with a funnel-like opening to take up the blood and a window to show the result. As in a diabetes test, the finger is pricked with a sterile lancet. Three drops of blood set off the test. If the result is ambiguous, it should be repeated half an hour later. The life-saving credit card fits into any wallet or handbag and can be used in the ambulance or the Accident & Emergency department as well.

Fig. 10.18 Working principle of gel electrophoresis.

How do I know before it is too late that I am at risk of a heart attack or stroke? Warning signals come from inflammation markers. C-reactive protein CRP has been recognized in the US as a further risk indicator in addition to the blood lipid levels. Even US president **George W. Bush** has been CRP-tested and found to be not at risk.

Further rapid tests in the pipeline look very promising. They will tell a patient with a runny nose and a temperature whether it is a **viral or a bacterial infection**. The markers are **procalcitonin** (**PCT**) for bacterial and **neopterin** for viral infections (Ch. 5).

As is generally known, **antibiotics** are effective in bacterial infections only. Over-prescription of antibiotics, as well as patients deciding to stop taking them as soon as the symptoms have disappeared, have resulted in resistant bacterial strains. New antibiotics or higher doses are needed – a vicious cycle. New biotests help to stop the inappropriate use of antibiotics.

Rapid immune tests are essential for the detection of bacteria and viruses and the targeted use of antibiotics. DNA analysis has become another new tool for disease testing.

■ 10.8 How DNA is Analyzed – Gel Electrophoresis Separates DNA Fragments According to Size

How can a DNA sequence be identified? In order to analyze genes, genomic DNA, or other DNA, e.g., bacterial plasmids, the sequence is cut with one or several **restriction enzymes** (Ch. 3). The resulting fragments are separated on an **agarose gel** according to size.

Agarose is a polysaccharide obtained from red algae (Ch. 7) and liquidized by boiling it in buffer. When it cools down, it turns into a solid gel with large pores. When an electric current is run through the gel to create an **electric field**, nucleic acids, carrying negatively charged phosphate groups, move to the positive pole (anode). The smaller DNA fragments move more quickly through the pores during this **gel electrophoresis** (Fig. 10.18 and 10.20).

The resulting DNA fragments are separated in the gel according to size. A **DNA standard** is run in parallel under the same conditions, containing DNA fragments of known size and a ladder containing, for example, fragments between 1,000 and 5,000 base pairs (Fig. 10.18), which gives a rough idea of fragment sizes.

DNA is visualized by the DNA-binding dye **ethidium bromide**. It intercalates with the bases of nucleic acids and **fluoresces under UV light**. The DNA is either dyed after electrophoresis, or ethidium bromide can be added to the agarose gel beforehand (Fig. 10.21). It is also possible to use radioactively labeled DNA and visualize it on an X-ray film (**autoradiography**).

When a DNA molecule is cleaved by one or several restriction enzymes, a restriction map can be created, i.e., the binding locations and distances between restriction sites within the DNA molecule are identified. Gel electrophoresis has been an essential method in **genetic fingerprinting**.

■ 10.9 Life And Death – Genetic Fingerprinting in Establishing Paternity and Investigating Murders

Since 1892, fingerprints have been used to identify people (**dactyloscopy**).

The case of the US football star **O.J. Simpson** gave molecular **DNA fingerprinting** a high media profile. Simpson was not proven guilty in the end, thought to be mainly due to negligence of the police when collecting evidence and the skill of Simpson's lawyers who pointed out contradictions in the prosecutors' arguments. Thus, the case could not be made in spite of overwhelming evidence.

When the Swedish foreign secretary, **Anna Lindh**, was assassinated in 2003, police secured the weapon – in contrast to the still unsolved murder of prime minister **Olof Palme** – and although no fin-

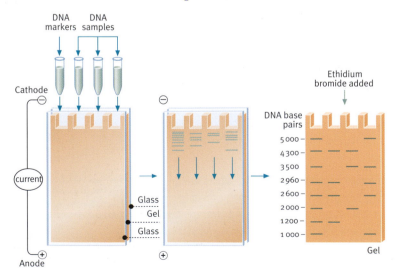

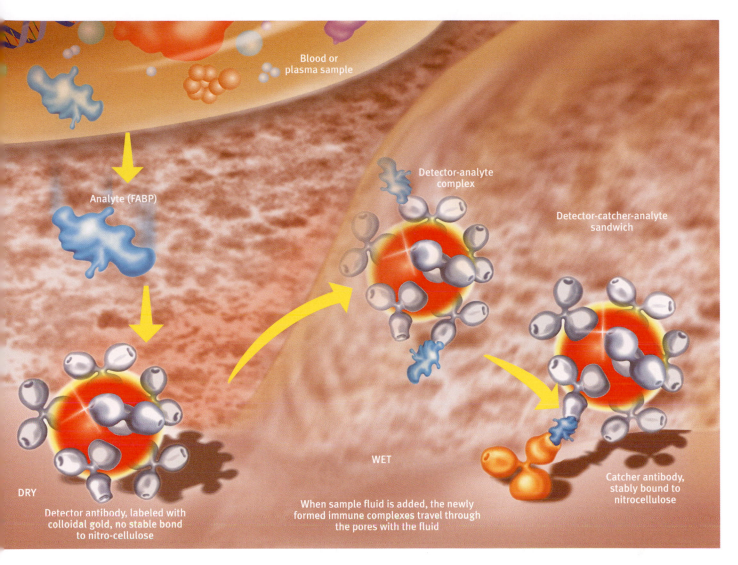

Blood or plasma sample

Analyte (FABP)

Detector-analyte complex

Detector-catcher-analyte sandwich

WET

DRY

Detector antibody, labeled with colloidal gold, no stable bond to nitro-cellulose

When sample fluid is added, the newly formed immune complexes travel through the pores with the fluid

Catcher antibody, stably bound to nitrocellulose

gerprints were found on it, only skin particles, the assassin was convicted on the basis of his DNA analysis.

The **DNA of two individuals** differs by just **0.1 percent**. In people who are not related to each other, there will be only a difference of **one letter in a thousand base pairs**. This small difference, however, is sufficient to provide an unmistakable genetic fingerprint. It is interesting that the genetic difference between people from different continents is far smaller than has been assumed. An individual from Africa may well be genetically closer to a European or Asian individual than to another African.

DNA fingerprinting was described for the first time by **Alec J. Jeffreys** at the Unversity of Leicester, England, in the 1970s (more details in Box 10.1). The test is based on the observation that DNA, once cut up into fragments, differs in number and size between individuals, and relies on **restriction fragment length polymorphisms analysis** (**RFLP analysis**, see Box 10.2, pronounced "*riflip*"). What RFLP describes is a **method of identifying DNA variants** – the use of highly specific DNA-cutting enzymes, restriction endonucleases (Ch. 3). For example, if two DNAs differ in their recognition sequence for a restriction enzyme, one will be cut by the restriction enzyme, but perhaps not the other. Thus, the fragments of cut variants differ in length (more detail in Box 10.2).

Most DNA in higher organisms (97 percent) does not carry information on protein formation. Genes are like oases in a desert. The surrounding **non-coding DNA** (**introns**) carries more mutations than coding DNA. These **mutations**, of course, usually do not have a life-threatening effect on the cell and can be passed on from one generation to the next without any change in the phenotype of the organism.

Fig. 10.19 When sample fluid is added, the newly formed immune complexes travel through the pores with the fluid.

Fig. 10.20 Machinery used in agarose gel electrophoresis.

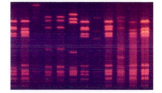

Fig. 10.21 DNA bands visualized by ethidium bromide in UV light after agarose gel electrophoresis.

Box 10.1 Biotech History:
Alec Jeffreys, DNA Profiling and the Colin Pitchfork Case

In 1983, fifteen-year-old **Lynda Mann** was found raped and strangled in the village of Narborough, England. Three years later, Dawn Ashwoth, also 15 years old, was found in nearby Enderby. Since both girls were found near a footpath called the "Black Pad", the newspapers called the killer "*The Black Pad Killer*".

Alec J. Jeffreys with the first DNA fingerprint in history.

The police could not find any traces. It was around this time that **Alec Jeffreys**, genetic scientist at the University of Leicester, England, publicized his **restriction fragment length polymorphisms, RFLP** method.

It is said that the idea struck him in the darkroom at 9 a.m. on Monday, September 15th, 1984. Jeffreys was investigating the evolution of a gene for the protein responsible for oxygen transport in muscles, myoglobin. After gel electrophoresis, he had photographed the relevant DNA fragments. When he looked at the just developed image, it showed DNA in various bands, resembling a bar code on packaged goods. Gosh, he thought, what have we got here? These are very varied, unique patterns. It should be possible to identify every single individual with these! Only a few hours later did Jeffreys and his team coin the name "genetic fingerprinting" for their accidental finding.

DNA from the sperm traces was isolated and the DNA profile of the killer established. 5,000 men between 16 and 34 who did not have alibis were asked to give a blood sample. Of course, the police did not expect the killer to come forward to give a blood sample of his own accord, but a coincidence set them on the right track. A woman who worked in a local bakery told police that in a pub, one of her co-workers said he had taken the blood test for another man. When police questioned the man, **Ian Kelly**, he did not deny it. His mate, 27-year-old **Colin Pitchfork**, had convinced Kelly he couldn't take the test because he'd already helped out someone else who was in trouble.

The real reason was that Pitchfork was the killer. In January 1988, he pleaded guilty and received a life sentence. He was the first killer in history who was convicted on the basis of DNA evidence. Earlier on, a man who was mentally ill, **Rodney Buckley,** had already pleaded guilty for the murder of Dawn Ashworth. He was released because his evidence did not match at all the sperm DNA found at the scene of the crime.

Thus, Buckley became the first suspect in human history who was acquitted on the basis of DNA analysis.

The killer's genetic fingerprint matches that of Colin Pitchfork.

In 1987, DNA tests became officially approved as evidence in the US and the UK. The British database now contains over 3.5 million entries. It has been used in 75,000 investigations, about 500 per week. In 10,000 rape cases between 1989 and 1996, 25% of the suspects initially arrested could be excluded on the basis of DNA analysis.

It also became apparent that eye witnesses and even the US Court of Justice are not always reliable. In his *Innocence Project*, the New York defense lawyer **Barry Scheck** calls DNA "the gold standard of innocence". Since 1992, he has been able to achieve the release of 201 innocent prisoners. Scheck says that at least one in seven people sentenced to death is innocent.

Alec J. Jeffreys was knighted in 1994 by the Queen for his merits for humanity.

Fig. 10.22 DNA fingerprints reveal how loyal birds are. Here are some lovebirds (*Agapornis*) in the author's aviary where it does not matter who is the father of this little chap.

Thus, the differences in the **non-coding RFLPs** are greater between individuals than those of coding DNA sequences. This makes them ideal diagnostic tools.

■ 10.10 DNA Markers – Short Tandem Repeats and SNPs

RFLPs have their **limitations** where comprehensive analyses are required. After all, each restriction site offers no more than two possibilities. Microsatellite DNA is dispersed throughout the chromosome, whereas mini- and macrosatellite DNA is found in the centromeres and telomeres of chromosomes only.

Microsatellite DNA plays an important part, representing five percent of sequence polymorphisms. These are short DNA sequences, repeated in tandem-like fashion, e.g., CACACACACA – hence the name **short tandem repeats**, **STRs**. The repeats comprise two to ten one-, two-, three-, or four-nucleotide repeats which are repeated between five and twenty times. The number of repeats per satellite varies. The human genome contains at least 650,000 STRs. Microsatellite DNA is rarely found in genes, but if so, it has far-reaching implications, e.g., in the genetically transmitted Huntington's disease. Establishing the number of STRs is a useful tool to achieve an acceptable resolution when mapping the human genome.

RFLPs as well as satellite markers are good **physical mapping tools** – i.e., they can be identified on a genome map and thus found on a chromosome. When a chromosome is mapped, it must be completely covered by markers, e.g., in order to make it easier to identify the locus of disease genes. Thus, it can be determined how often the markers (RFLPs, STRs) are passed on with the disease in certain families.

The most important markers, however, appear to be **single nucleotide polymorphisms** or **SNPs**, pronounced '*snips*'.

Polymorphism means that the copies of a gene within a population are not exactly identical. Single-nucleotide polymorphism means that the difference affects no more than **one base pair**.

SNPs are the markers of choice in large studies of humans or other populations because they can easily be identified using **gene chips** (see below).

SNPs represent 95 percent of polymorphous sequence variations. 1.5 million SNPs have already been identified. Their only drawback is that any SNP will only affect one of the base pairs A-T or G-C, which makes it difficult to find the genetic difference between two people.

SNP blocks must be identified in a way similar to a bar code. These SNP combinations are also called haplotypes. **Haplotype mapping** is now part of all gene mapping programs. In the 1990s, RFLP fingerprinting was complemented by **SNP fingerprinting**.

DNA is becoming increasingly important in **forensics**. 20–50 nanograms of DNA are sufficient for a DNA fingerprint. A DNA sample, e.g., of a rape victim, containing sperm, and a DNA sample of a suspect are compared using fingerprinting. In order to catch a criminal, **swabs** of thousands of people have been taken – an approach that has often led to results (Figs. 10.23 and 10.25, Box 10.1)

Police forces in most countries now have a **DNA database** of convicted criminals. How extensive they are and for how long the data will remain in the database varies from country to country. While the U.K. database had nearly 4 million entries in early 2007, Germany's database had under 500,000. In the US, there are federal DNA databases as well as databases on state level.

DNA fingerprinting is becoming more and more sensitive. A droplet of saliva on a mobile phone or a single hair root provides enough material. The **hair root method** was developed by the German Federal Criminal Police Office BKA.

DNA analysis was used to identify victims of the 9/11 attack in New York in 2001. It also helped "*Las Abuelas*" (the grandmothers), a group of

activist women in Argentina to reunite some of the 2000 children, displaced during the military regime, with their families. In this case, it was **mitochondrial DNA** of the children and of the grandmothers that was analyzed. Mitochondrial DNA is only passed on through the maternal line. Other high-profile DNA analysis cases included that of the German concentration camp doctor **Mengele** and of the family of the Russian czar, the **Romanovs**, which led to the exposure of imposter **Anne Anderson**, who had claimed to be the last remaining daughter of the czar.

Other applications include, for example, **fiber analytics**. DNA analysis can establish if a Cashmere sweater really contains wool from a Cashmere goat rather than ordinary sheep. DNA fingerprinting can also be applied to artificial DNA in order to produce specific tags that help **prevent counterfeiting**. A 20-base DNA sequence offers ten quadrillion possible variants for coding. The tags are used to protect valuable paintings and cars from theft.

If DNA is available only in small amounts, it must first be replicated – and the magic formula to do this is the polymerase chain reaction (PCR).

■ 10.11 Polymerase Chain Reaction – Copying DNA on a Mega Scale

PCR has become the DNA-replicating method par excellence. What happens during a **polymerase chain reaction** is the same as what happens during the division of a cell into two daughter cells. As the genetic information provided to each of the daughter cells must be identical, the entire set must be copied from the mother cell. The two strands of the double helix are separated, and both single-stranded DNA molecules serve as templates for the formation of new single strands. An enzyme called **DNA polymerase** (Fig. 10.25) helps synthesize the complementary strands within the cell. The polymerase adds the nucleotide that matches the template. Thus, the DNA in both daughter cells is identical with that of the mother cell.

Thousands of biochemists and biotechnologists had been trying for years to mimic the process *in vitro*, but it took **Kary Mullis** (b. 1944) to hit on the crucial idea – PCR (Box 10.3).

Here is a brief run-through of the main stages of PCR:

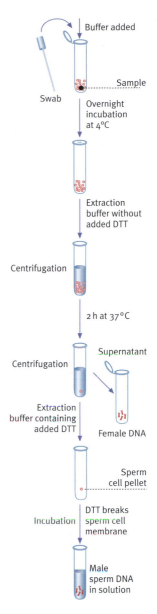

Fig. 10.23 Forensic DNA test, separating female DNA from sperm DNA. DTT (dithiothreitol) frees sperm DNA by breaking up disulfide bridges.

Fig. 10.24 A buccal swab is taken from Fortune, my male kitten.

Fig. 10.25 Action of DNA polymerase in PCR.

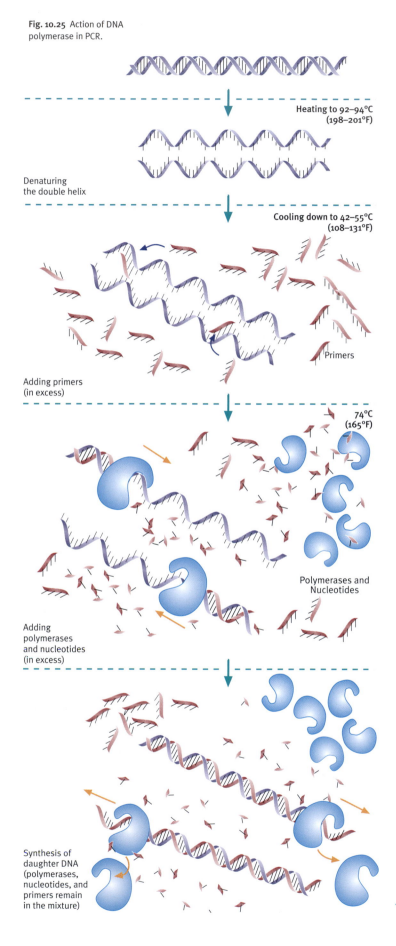

Heating to 92–94°C
(198–201°F)

Denaturing
the double helix

Cooling down to 42–55°C
(108–131°F)

Adding primers
(in excess)

Primers

74°C
(165°F)

Polymerases and
Nucleotides

Adding
polymerases
and nucleotides
(in excess)

Synthesis of
daughter DNA
(polymerases,
nucleotides, and
primers remain
in the mixture)

- Separating the DNA strands through heating (denaturation)
- Cooling and binding of primers
- Reheating and synthesizing new double-stranded DNA through polymerase
- Denaturing of the synthesized DNA
- Repetition of the process

This is an automatic cycle that only takes a few minutes.

Initially, fresh DNA polymerase had to be added for each new cycle because the enzyme – which had been obtained from *E. coli* bacteria – lost its activity at 94°C (201°F) during the denaturation process. Then, in boiling hot springs such as the Yellowstone National Park geysers (Fig. 10.36), microorganisms were found that also rely on a polymerase for their replication.

The enzyme isolated from *Thermus aquaticus* (**Taq polymerase**) was genetically modified and produced in large quantities. It works most effectively at 72°C (162°F) and can survive 94°C (201°F) undamaged. Its ability to **remain active in the test tube during the whole cycle** gives the method its decisive edge. Box 10.4 and Fig. 10.25 describe the PCR method in detail.

■ 10.12 A New Lease of Life For Dinosaurs and Mammoths?

The analysis of extinct animals with PCR is a fascinating story. Wherever genetic material is found, it can be amplified.

The DNA analysis of a **mammoth** that was found in Siberian ice 20,000 years after its death confirmed the expected close relationship between mammoths and elephants. The Russian-French-American team that is cutting out a Woolly Mammoth in huge blocks from the permafrost soil is hoping to find intact mammoth DNA which could be transferred into elephant egg cells for cloning (Ch. 8).

Thus, one of the first species made extinct by man could be recreated, as an act of gratitude. After all, we owe it mostly to the mammoth that humans survived the Ice Age. Mammoths were virtually the only source of meat.

Insects enclosed in amber (Fig. 10.32) might actually have ingested dinosaur blood and thus their DNA – *Jurassic Park* stuff. However, the few fragments of DNA will be far too scarce to give us back *Tyrannosaurus rex*. Prehistoric mummies provide DNA that can be compared with living humans. Interestingly, there is a DNA database for *Homo neandertalensis* being built up alongside the database for *Homo sapiens*.

■ 10.13 The Sequencing of Genes

Knowledge of the nucleotide sequence, i.e., the sequence of A, G, T, and C, can be crucial, for the following reasons:

- To derive the amino acid sequence of a protein encoded by the gene in question
- To determine the exact sequence of a gene
- To identify regulatory elements such as promoter genes
- To identify differences within genes
- To identify mutations such as polymorphisms.

Nowadays, several methods of DNA-sequencing are available. The most widely used method was developed by **Frederick Sanger** in 1977 (Fig. 10.27) and is also called the **chain termination method**. A radioactively labeled DNA primer is hybridized with denatured template DNA (Fig. 10.26) in a tube containing the four deoxyribonucleotides (dNTPs) and DNA polymerase. The polymerase copies the strands, starting at the 3′termini of the primers. A **modified nucleotide** (**dideoxyribonucleotide, ddNTP**) is added. In a ddNTP, the 3′terminus carbon of the sugar binds to only a hydrogen atom (-H) instead of a hydroxyl group (-OH). The insertion of a ddNTP into DNA leads to the **termination** of chain elongation because phosphodiester bonds with new nucleotides can no longer be formed.

Four different tubes are used, each containing DNA, primers and all four dNTPs and a small amount of only one of the four ddNTPs. Thus, the **polymerase inserts randomly ddNTPs**, and a number of fragments of varying length arises, all of them terminated by ddNTPs.

In a subsequent electrophoresis, a polyacrylamide gel separates the fragments according to their size. They are radioactively labeled and visualized on an X-ray film blackened by radiation (**autoradiogram**, Fig. 10.26).

The Sanger chain termination method can only cope with sequences between 200 and 400 base

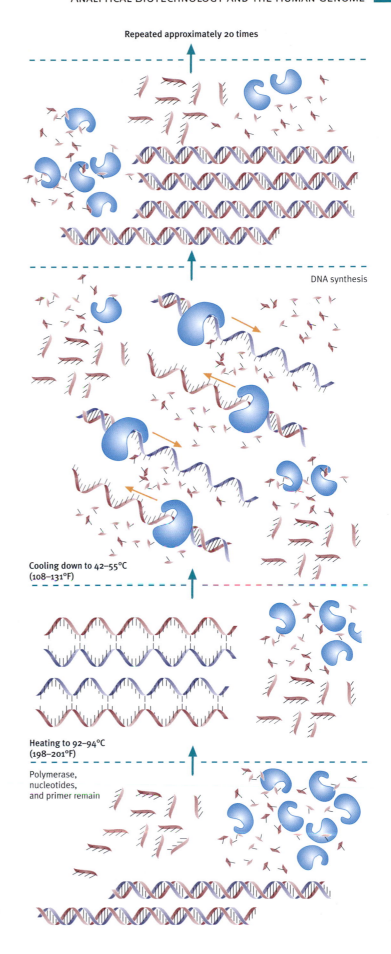

Repeated approximately 20 times

DNA synthesis

Cooling down to 42–55°C (108–131°F)

Heating to 92–94°C (198–201°F)

Polymerase, nucleotides, and primer remain

Box 10.2 RFLP And Paternity Testing

How is it possible that homologous DNA (i.e., DNA from the same section of a chromosome) from various individuals results in fragments of varying length, when cut with the same restriction enzymes? The restriction endonucleases always cut DNA at the same places, e.g., after guanine (G) in sequence GAATTC (Ch. 3). If a person called "**Renneberg**" carries a DNA fragment with an intron carrying the following sequence

…TTTTGAATTCTTTTGAATTC…

then two places where the enzyme can cleave are easily identified:

…TTTTG/AATTCTTTTG/AATTC…

Thus, three fragments arise:

…TTTTG and AATTCTTTTG and AATTC…

Another person called "**Darja Süßbier**", by contrast, carries a simple mutation or SNP, see below. One of the Gs has been replaced by an A:

…TTTTGAATTCTTTTAAATTC…

Here, the enzyme can only cut once:

…TTTTG/AATTCTTTTAAATTC…

Resulting in only two DNA fragments:

…TTTTG and AATTGTTTTAAATTC…

The outcome is a different pattern. The Süßbier DNA sample contains only two fragments, the second of which will move more slowly in gel electrophoretic separation. The three small Renneberg fragments will move faster through the gel.

These differences in DNA between individuals are visualized by blotting, i.e., by pipetting a solution that contains the cut DNA fragments into wells on a gel and running an electric current through it (gel electrophoresis). The negatively charged DNA fragments migrate through the pores of the gel to the anode (positive pole) and are separated on their way according to size.

For the blotting process, nitrocellulose paper is laid over the gel with the sorted DNA fragments. These are absorbed by the paper in the same distribution and firmly bind to the nitrocellulose. A radioactive label is added which makes the DNA fragments visible on a superimposed X-ray film. The result looks like a ladder with steps of irregular thickness which are unevenly distributed. It is reminiscent of the bar codes on supermarket packaging that are run through readers at the cash points. The steps of the ladder are referred to as DNA bands.

The blotting results of Süßbier and Renneberg thus show clearly distinctive band patterns. The band "TTTTG" is present in both, but the two other smaller Renneberg bands are missing in Süßbier who has a larger step on the ladder elsewhere. This would explain a small genetic difference between the graphics designer and the author of this book. One is an ingenious painter and hates writing, the other loves writing, but is only able to draw formulae and cartoons for his Chinese students.

From a scientific as well as a personal view, it would be interesting to make a further RFLP comparison of DNA bands between Süßbier and Renneberg (from former East Germany), our US-Editor **Arny Demain** from New York, **Merlet Behncke-Braunbeck** from former West Germany, and our two translators **Renate FitzRoy** from Sylt, the northernmost German island, and **Jackie Jones** from Hampshire, UK. However, it is not as simple as that, as not all DNA sequences are coding sequences.

To the great joy of US historians, it is now possible to use DNA testing to find out more about President Thomas Jefferson's moral conduct a hundred years later. Left: 1802 caricature featuring Jefferson and Sally Hemings.

History, rewritten by DNA?

In a paternity test, DNA fingerprints of mother, child, and putative father are put side by side and compared. The bands of mother and child at the same height are identified and the remaining bands in the child's fingerprint with those of the supposed father, as the bands that do not match those of the mother must come from the father. If the bands do not match, it is very likely that the father is a different person (see also Box about Sir Alec Jeffreys).

Apparently, there are virtually unlimited possibilities for the use of genetic fingerprinting.

A bizarre example comes from the United States where in 1802, president **Thomas Jefferson**, father of the Declaration of Independence, was accused by a newspaper of having an affair with his slave **Sally Hemings** who was of mixed black and white origin. It was therefore possible that Jefferson was the father of one of Sally

Heming's seven children. 19 DNA samples of Jefferson descendant were collected in 1997, including blood samples of male descendants of Jefferson's paternal uncle, Field Jefferson. Each sample was searched for polymorphic markers on the Y-chromosome, which are only passed on via the male line. The analysis showed that Jefferson appeared to have been the father of the youngest child, Eston Hemings.

Another interesting question had arisen from this investigation: In the DNA from Jefferson's male relatives, taken in 1997 by Professor **Mark Jobling** from the University of Leicester, it was found the president had a rare genetic signature. The special haplotype (see also Boxes 10.3 and 10.7) is found mainly in the Middle East and Africa. This raised questions about Jefferson's claim of Welsh ancestry.

Now this DNA type has been found in two Britons with the Jefferson surname. It was discovered the two British Jeffersons carried the same rare male chromosome haplotype K2 as the third US president.

DNA-analysis shows clearly that the British men shared a common ancestor with Thomas Jefferson about 11 generations ago. But neither knew of any family links to the US.

Thomas Jefferson – who was president from 1801 to 1809 – must have had recent paternal ancestors from the Middle East. K2 makes up about 7% of the Y chromosome types found in Somalia, Oman, Egypt, and Iraq. It has now been found at low frequencies in France, Spain, Portugal, and Britain. Of the K2s looked at by the study, Jefferson's Y chromosome was most similar to that of a man from Egypt. Could he have had a relatively recent origin in the Middle East?

But genetic relationships between different K2s are poorly understood, and this may have little significance. Instead, say the researchers, their study makes Jefferson's claim to be of Welsh extraction much more plausible.

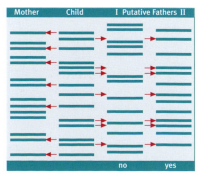

Example of a paternity test.

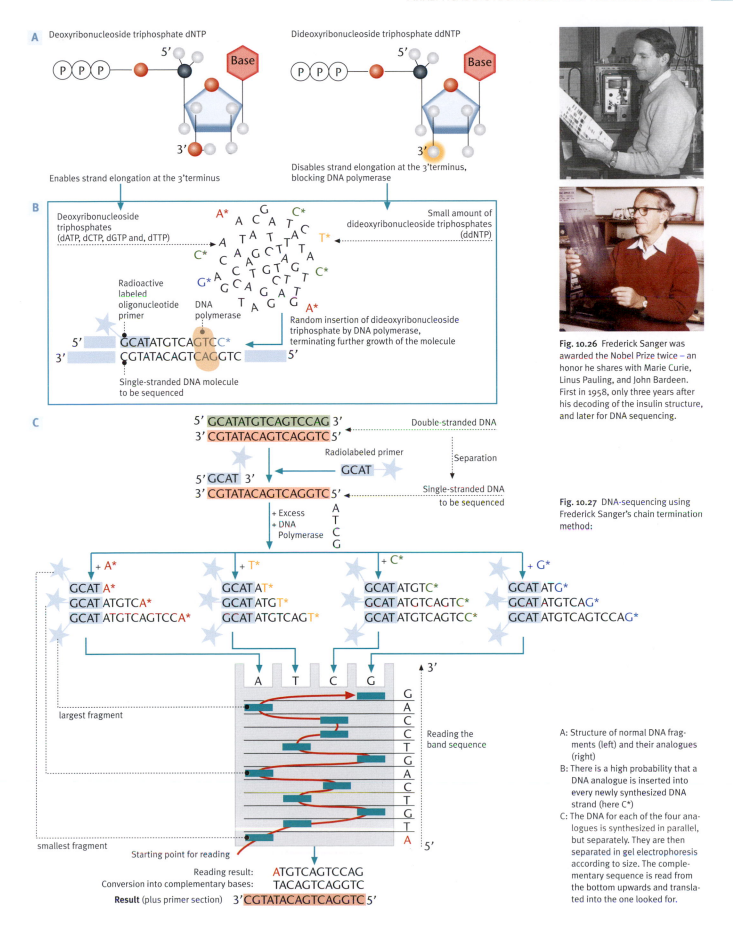

A Deoxyribonucleoside triphosphate dNTP

Dideoxyribonucleoside triphosphate ddNTP

Enables strand elongation at the 3'terminus

Disables strand elongation at the 3'terminus, blocking DNA polymerase

B Deoxyribonucleoside triphosphates (dATP, dCTP, dGTP and, dTTP)

Small amount of dideoxyribonucleoside triphosphates (ddNTP)

Radioactive labeled oligonucleotide primer

DNA polymerase

Random insertion of dideoxyribonucleoside triphosphate by DNA polymerase, terminating further growth of the molecule

5' GCATATGTCAGTCC*
3' CGTATACAGTCAGGTC 5'

Single-stranded DNA molecule to be sequenced

C
5' GCATATGTCAGTCCAG 3' ← Double-stranded DNA
3' CGTATACAGTCAGGTC 5'

Radiolabeled primer

GCAT Separation

5'GCAT 3'
3' CGTATACAGTCAGGTC 5' ← Single-stranded DNA to be sequenced

+ Excess
+ DNA
Polymerase

A
T
C
G

+ A*
GCAT A*
GCAT ATGTCA*
GCAT ATGTCAGTCCA*

+ T*
GCAT AT*
GCAT ATGT*
GCAT ATGTCAGT*

+ C*
GCAT ATGTC*
GCAT ATGTCAGTC*
GCAT ATGTCAGTCC*

+ G*
GCAT ATG*
GCAT ATGTCAG*
GCAT ATGTCAGTCCAG*

A T C G 3'

largest fragment

G
A
C
C
T
G
A
C
T
G
T
A 5'

Reading the band sequence

smallest fragment

Starting point for reading

Reading result: ATGTCAGTCCAG
Conversion into complementary bases: TACAGTCAGGTC
Result (plus primer section) 3' CGTATACAGTCAGGTC 5'

Fig. 10.26 Frederick Sanger was awarded the Nobel Prize twice – an honor he shares with Marie Curie, Linus Pauling, and John Bardeen. First in 1958, only three years after his decoding of the insulin structure, and later for DNA sequencing.

Fig. 10.27 DNA-sequencing using Frederick Sanger's chain termination method:

A: Structure of normal DNA fragments (left) and their analogues (right)

B: There is a high probability that a DNA analogue is inserted into every newly synthesized DNA strand (here C*)

C: The DNA for each of the four analogues is synthesized in parallel, but separately. They are then separated in gel electrophoresis according to size. The complementary sequence is read from the bottom upwards and translated into the one looked for.

Fig. 10.28 DNA analysis through gel electrophoresis and Southern blotting.

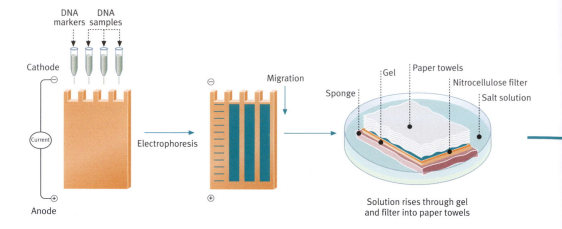

Fig. 10.29 Sir Edwin Southern (University of Oxford) who developed Southern blotting.

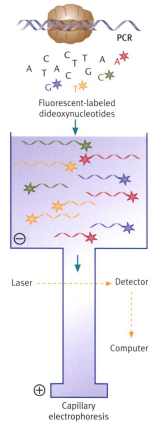

Fig. 10.30 Automatic high-throughput sequencing: The sequencing reaction products are separated from each other by capillary electrophoresis. DNA molecules are negatively charged. In an electric field, all the DNA moves toward the positive electrode. Shorter DNA fragments move more quickly through a capillary than do larger DNA fragments. As each DNA fragment reaches the end of the capillary, a laser excites its fluorescent dye. A camera detects the color of the emitted light and passes that information to a computer. One by one, the machine records the colors of the DNA fragments that pass through.

pairs in a single reaction process. If, for example, 1000 base pairs are to be analyzed, several cycles are required, and overlapping sequences are put together as in a jigsaw puzzle.

■ 10.14 Southern Blotting

In 1975, **Edwin Southern** (b. 1938, Fig. 10.29) in Oxford developed a dramatic improvement in DNA gel electrophoresis. The method has been named **Southern Blotting** after him and begins with the cleavage of DNA by restriction enzymes and the separation of the fragments by agarose gel electrophoresis. If, for example, chromosomal DNA is cleaved, the number of fragments is so large that they cannot be resolved in a gel. The bands become blurred or smeared. Southern blotting, however, makes it possible to locate specific fragments.

Double-stranded DNA is denatured in sodium hydroxide (NaOH) solution. Single-stranded DNA is formed for later hybridization (Fig. 10.28). After electrophoresis, the gel is placed on a **nitrocellulose filter** or a nylon membrane. Pressure is applied evenly to the gel (either using suction, or by placing a stack of paper towels and a weight on top of the membrane and gel). Capillary action transfers the single-stranded DNA onto the filter or membrane, to which it binds. Thus, an image of the DNA in the gel appears firmly bonded to the filter.

The main advantage of blotting techniques is that it makes the DNA analytes more accessible. In gel electrophoresis, the reaction partners (e.g., DNA probes) can only slowly diffuse into the gel, while the analytes are already wandering off. On the nitrocellulose filter, the analytes will not move away and are ready to be analyzed.

The filter is taken out, and a radioactively labeled **DNA probe** is added. Hybridization begins. Free probe molecules are removed by **washing**. The filter is then put on an X-ray film. The fragments carrying hybridized DNA show as black marks in the autoradiogram.

To honor Ed Southern, similar blotting techniques were named after the other directions. While Southern blotting is the transfer of DNA onto nitrocellulose, **Northern blotting** is the transfer of **RNA**, while in **Western blotting**, **a protein rather** than a nucleic acid is transferred from an SDS polyacrylamide gel onto nitrocellulose (see end of this chapter). The wanted protein is then visualized by labeled antibodies or other protein detection methods.

■ 10.15 Automatic DNA Sequencing

For the amount of sequencing required for the Human Genome Project, methods like gel electrophoresis and Southern blotting were far too slow. What was needed were computer-aided sequencing machines that could handle sequences larger than 500 base pairs in a single reaction.

In the above-mentioned Sanger method, the synthesized molecules with terminated chains are radioactively labeled. The DNA sequence is then read from the autoradiogram. Over the last years, radiolabeling has been increasingly **replaced by fluorescent labeling**. The labels can be attached either to ddNTPs instead of radioactive labels or to the 5'terminus of a sequencing primer. Where different ddNTP labels are used, the reaction can take place in a single test tube. The sample is then separated in a capillary gel (extra-thin hollow fiber used in **capillary electrophoresis**) and scanned with a laser beam. The laser beam excites the fluorescent dyes which then emit their specific color patterns for each nucleotide. A fluorescence detector records

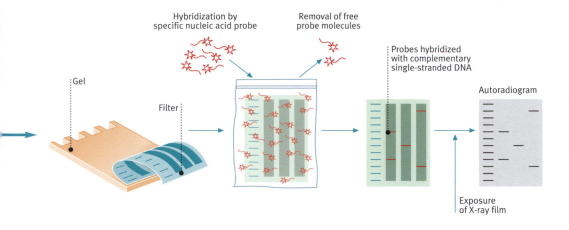

Hybridization by specific nucleic acid probe

Removal of free probe molecules

Probes hybridized with complementary single-stranded DNA

Gel

Filter

Autoradiogram

Exposure of X-ray film

Fig. 10.31 Successful "mammoth revival" in the author's office (model courtesy of the British Museum).

Will you turn me into a mammoth? I have already got my proboscis full of DNA!

Fig. 10.32 Insects found in amber inclusions may have ingested dinosaur blood and may contain their DNA.

The Insect shown here, found on www.amberworld.com, however, is only a fungus gnat (Mycetophilidae family), a harmless vegetarian not containing any dinosaur DNA. Baltic amber is roughly between 28 and 54 million years old. The Cretaceous amber found in New Jersey, in Lebanon, or Burma might well contain bloodsucking insects, but such inclusions are very rare. This is why the model for the film *Jurassic Park* came from a Baltic amber inclusion.

the signals emitted from the individual bands, and a computer converts the four different color signals into a DNA sequence (Figs. 10.30, 10.38 and 10.39).

By replacing gel electrophoresis with capillary electrophoresis, it became possible to analyze 96 capillaries containing approximately 40,000 nucleotides within four hours. Machines containing 384 capillaries were developed. Sometimes, over a hundred of these machines were used in the genome project, and even allowing for multiple sequencing due to reading errors, entire genomes could be read within a very short timespan.

■ 10.16 FISH – Identifying the Location on a Chromosome and the Number of Gene Copies

How can one find out on which chromosome a gene is located or whether a gene is stored in the genome as a single copy or if it occurs in multiple copies? **FISH (fluorescence in situ hybridization)** can be used to determine which chromosome carries a specified gene. The chromosomes are spread out on a slide. The complementary DNA of the gene concerned is enzymatically labeled with fluorescent nucleotides and incubated with the chromosomes.

The probe hybridizes with the relevant genes on the chromosome, and the labeled genes light up under a fluorescent microscope. Thus, the 23 human chromosome pairs can be sorted according to sizes and banding of chromatides – a **karyotype** (Fig. 10.35).

In prenatal diagnosis, it is now possible to analyze cells directly, without having to first grow them in a culture. The cells can be amnion cells from the **amniotic fluid**. The chromosome analysis of amnion cells could, for example, easily detect trisomy 21 (Down's syndrome). However, aberrations in single cells cannot reliably predict a chromosomal disorder in a human being; at least 50 cells in an amniotic fluid sample must be analyzed in order to secure a reasonable result.

FISH tests can detect numeric chromosomal aberrations of autosomal chromsomes 13, 18, and 21 as well as sex chromosomes X and Y in non-cultured cells from amniotic fluid. A method that scrutinizes the structure and composition of chromosomes has been developed by scientists of the University of Jena, Germany, and the company Metasystems (Fig. 10.35).

The technique called **multicolor banding** (Fig. 10.35) stands to gain particular significance in cancer diagnostics and treatment. Cancer cells carry characteristic aberrations in their genetic information that distinguish them from normal cells. These are passed on in an uncontrolled manner with each cancer cell division. In tumor cells, the natural order in the 23 chromosome pairs is upset. Parts of chromosomes are missing, having been reversed or attached to other chromosomes. These are the changes that are picked up by multicolor banding. The labeled chromosomes of a tumor cell fluoresce in all colors. Cytogenetic defects thus become apparent, for example, if in both chromosomes number five a blue ring is missing, which is usually there in a normal cell. The color sequence could also be altered in the tumor cell, or several such defects could come together. The diagnosis requires extensive computing programs – and some tricks from the genetic engineering toolbox. In order to make the chromosome fragments glow in all colors,

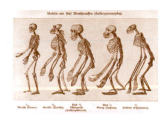

Fig. 10.33 Five million years ago, primates and humans had common ancestors. Today, the difference in DNA between humans and primates is one to two percent.

BOX 10.3 *DNA Travel Map 1*: What the Genographic Project Found out About my DNA's Journey out of Africa

Dear Professor Reinhard Renneberg, here are your results!

Type: Y-Chromosome
Haplogroup: *E3b (M35)*
Your STRs

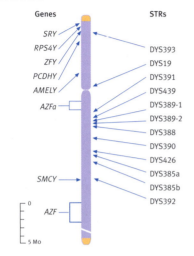

How to Interpret Your Results

DYS393:13	DYS439:12	DYS388:12	DYS385a:17
DYS19:13	DYS389-1:13	DYS390:25	DYS385b:18
DYS391:10	DYS389-2:18	DYS426:11	DYS392:11

Above are results from the laboratory analysis of your **Y-chromosome**. Your DNA was analyzed for **Short Tandem Repeats (STRs)**, which are repeating segments of your genome that have a high mutation rate. The location on the Y chromosome of each of these markers is depicted in the image, with the number of repeats for each of your STRs presented to the right of the marker. For example, DYS19 is a repeat of TAGA, so if your DNA repeated that sequence 12 times at that location, it would appear: DYS19 12. Studying the combination of these STR lengths in your Y Chromosome allows researchers to place you in a haplogroup, which reveals the complex migratory journeys of your ancestors. Y-SNP: In the event that the analysis of your STRs was inconclusive, your Y chromosome was also tested for the presence of an **informative Single Nucleotide Polymorphism (SNP)**. These are mutational changes in a single nucleotide base, and allow researchers to definitively place you in a genetic haplogroup.

Your Y-chromosome results identify you as a member of haplogroup *E3b*.

The genetic markers that define your ancestral history reach **back roughly 60,000 years** to

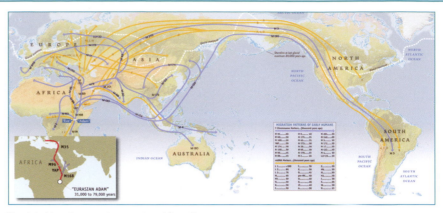

The global travel map and my own travel (insert).

the first common marker of all non-African men, *M168*, and follow your lineage to present day, ending with *M35*, the defining marker of haplogroup *E3b*.

If you look at the map highlighting your ancestors' route, you will see that members of haplogroup *E3b* carry the following Y-chromosome markers:

M168 > YAP > M96 > M35

Today, the *E3b* line of descent is most heavily represented in **Mediterranean populations**. Approximately 10 percent of the men in Spain belong to this haplogroup, as do 12 percent of the men in northern Italy, and 13 percent of the men in central and southern Italy. Roughly 20 percent of the men in Sicily belong to this group. In the Balkans and Greece, between 20 to 30 percent of the men belong to *E3b*, as do nearly 75 percent of the men in North Africa. The haplogroup is rarely found in India or East Asia. Around 10 percent of all European men trace their descent to this line. For example, in Ireland, 3 to 4 percent of the men belong; in England, 4 to 5 percent; Hungary, 7 percent; and Poland, 8 to 9 percent. Nearly 25 percent of Jewish men belong to this haplogroup.

Back to our African roots!

What's a haplogroup, and why do geneticists concentrate on the Y-chromosome in their search for markers? For that matter, what's a marker?

Each of us carries DNA that is a combination of genes passed from both our mother and father, giving us traits that range from eye color and height to athleticism and disease susceptibility. One exception is the **Y-chromosome, which is passed directly from father to son, unchanged, from generation to generation.**

The Director of the program, Dr Spencer Wells, speaking onstage with representatives of indigenous communities, who are participating in the Genographic field research, during the April 13, 2005, launch of the Genographic Project at National Geographic's headquarters in Washington, D.C..

Unchanged, that is unless a mutation – a random, naturally occurring, usually harmless change – occurs. The mutation, known as a marker, acts as a beacon; it can be mapped through generations because it will be passed down from the man in whom it occurred to his sons, their sons, and every male in his family for thousands of years.

In some instances there may be more than one mutational event that defines a particular branch on the tree. What this means is that any of these markers can be used to determine your particular haplogroup, since every individual who has one of these markers also has the others. When geneticists identify such a marker, they try to figure out when it first occurred, and in which geographic region of the world. Each marker is essentially the beginning of a new lineage on the family tree of the human race. Tracking the lineages provides a picture of how small tribes of modern humans in Africa tens of thousands of years ago diversified and spread to populate the world.

A haplogroup is defined by a series of markers that are shared by other men who carry the same random mutations. The markers trace the path your ancestors took as they moved.

One of the goals of the five-year Genographic Project is to build a large enough database of anthropological genetic data to answer some of these questions.

Your Ancestral Journey: What We Know Now

M168: Your Earliest Ancestor

Skeletal and archaeological evidence suggest that anatomically modern humans evolved in Africa **around 200,000 years ago**, and began moving out of Africa to colonize the rest of the world around **60,000 years ago**.

The man who gave rise to the **first genetic marker in** your lineage probably lived in northeast Africa in the region of the **Rift Valley**, perhaps in present-day Ethiopia, Kenya, or Tanzania, some 31,000 to 79,000 years ago. Scientists put the most likely date for when he lived at **around 50,000 years ago**. His descendants became the only lineage to survive outside of Africa, making him the common ancestor of every non-African man living today.

But why would man have first ventured out of the familiar African hunting grounds and into unexplored lands? It is likely that a **fluctuation in climate** may have provided the impetus for your ancestors' exodus out of Africa.

The **African ice age** was characterized by drought rather than by cold. It was around **50,000 years ago** that the ice sheets of northern Europe began to melt, introducing a period of warmer temperatures and moister climate in Africa. Parts of the inhospitable Sahara briefly became habitable. As the drought-ridden desert changed to a savanna, the animals hunted by your ancestors expanded their range and began moving through the **newly emerging green corridor of grasslands**. Your nomadic ancestors followed the good weather and the animals they hunted, although the exact route they followed remains to be determined.

In addition to a favorable change in climate, around this same time there was a **great leap forward in modern humans' intellectual capacity**.

YAP: An Ancient Mutation

Sub-Saharan populations living today are characterized by one of three distinct Y-chromosome

My family history from the male side: My grandparents Anna and Alfred Renneberg.

branches on the human tree. Your paternal lineage *E3a* falls under one of these ancient branches and is referred to by geneticists as *YAP*.

YAP occurred around northeast Africa and is the most common of the three ancient genetic branches found in sub-Saharan Africa. It is characterized by a mutational event known as an Alu insertion, a 300-nucleotide fragment of DNA which, on rare occasion, gets inserted into different parts of the human genome during cell replication.

A man living around 50,000 years ago, your distant ancestor, acquired this fragment on his Y-chromosome and passed it on to his descendants. Over time this lineage split into two distinct groups. One is found primarily in Africa and the Mediterranean, is defined by marker *M96* and is called haplogroup E. The other group, haplogroup D, is found in Asia and defined by the *M174* mutation. Your genetic lineage lies within the group that remained close to home, and was carried by men who likely played an integral role in recent cultural and migratory events within Africa.

M96: Moving Out of Africa

The next man in your ancestral lineage was born **around 30,000 to 40,000 years ago** in northeast Africa and gave rise to marker *M96*. The origins of *M96* are unclear; further data may shed light on the precise origin of this lineage.

You are descended from an ancient African lineage that chose to move north into the Middle East. Your kinsmen may have accompanied the Middle Eastern Clan (with marker *M89*) as they followed the great herds of large mammals north through the grassy plains and savannas of the Sahara gateway. Alternatively, a group of your ancestors may have undertaken their own migration at a later date, following the same route previously traveled by the Middle Eastern Clan peoples.

Beginning **about 40,000 years ago**, the climate shifted once again and became colder and more

arid. Drought hit Africa and the grasslands reverted to desert; for the **next 20,000 years, the Saharan Gateway was effectively closed**. With the desert impassable, your ancestors had two options: remain in the Middle East, or move on. Retreat back to the home continent was not an option.

M35: Neolithic Farmers

The final common ancestor in your haplogroup, the man who gave rise to marker *M35*, was born around 20,000 years ago in the Middle East. His descendants were among the first farmers and helped spread agriculture from the Middle East into the Mediterranean region. At the end of the last ice age around 10,000 years ago, the climate changed once again and became more conducive to plant production. This probably helped spur the Neolithic Revolution, the point at which the human way of living changed **from nomadic hunter-gatherers to settled agriculturists**.

The early farming successes in the Fertile Crescent of the Middle East beginning **around 8,000 years ago** spawned population booms and encouraged migration throughout much of the Mediterranean world. Control over their food supply marks a major turning point for the human species. Rather than small clans of 30 to 50 people who were highly mobile and informally organized, agriculture brought the first trappings of civilization. Occupying a single territory required more complex social organization, moving from the kinship ties of a small tribe to the more elaborate relations of a larger community. It spurred trade, writing, calendars, and pioneered the rise of modern sedentary communities and cities.

These ancient farmers, your ancestors, helped bring the Neolithic Revolution into the Mediterranean.

Further reading:
www. national geographic/genographic.com

Box 10.4 Biotech History: At Night on the Californian Highway

Karry Mullis who was working for the biotech company Cetus was on his way home for the weekend, zipping along the moon-lit Californian Highway.

The bright lights of a highway sparked Mullis's bright idea.

He spent three hours thinking about how one could produce millions or even billions of copies of a specific DNA fragment. This would allow scientists to find the proverbial needle in a haystack in molecular terms. One possibility involved inserting the DNA into ring-shaped plasmids which could be introduced into bacteria, which, in turn, would replicate millions of times. The plasmids could then be reisolated and the cloned DNA cut out – an extremely laborious process.

Mullis watched the lights of the cars going in both directions meet, pass each other and drive off at the various junctions. This symphony of interweaving lights sparked off the crucial idea that earned him the Nobel Prize only eight years later. He stopped his car and began to draw lines: DNA duplicating in the test tube, the product of each cycle providing material for the next. Just 20 cycles would be enough to generate 1,000,000 identical DNA molecules. Mullis woke his sleeping passenger: "You won't believe this. It is just incredible." She mumbled something not terribly friendly and went back to sleep – something impossible for Mullis to achieve that night! "It was difficult for me to sleep with deoxyribonuclear bombs exploding in my brain," he later wrote.

Coming back to the Cetus lab on Monday, Mullis began to work feverishly on his new idea – and lo and behold, it worked! However, his colleagues were not terribly impressed. It was such a simple idea – surely somebody somewhere must have tried it out before. Soon afterwards, Nobel Prize winner and Cetus consultant Joshua Lederberg (Ch. 3) took a thorough look at the poster Mullis presented at a conference, and asked, more or less in passing "Does it work?" When Mullis said yes, he finally got the reaction he had hoped for. The iconic figure of molecular genetics, Joshua Lederberg, pulled his sparse hair and exclaimed "Oh my god, why didn't I think of that?"

But what had really set the ball rolling, leading to the discovery of the polymerase chain reaction (PCR), was the first Russian satellite SPUTNIK! In his autobiography, Nobel Prize winner Mullis describes how he and his class mates practiced hiding under their desks as fast as possible in case the Russians dropped a nuclear bomb on Columbia, South Carolina. The Sputnik shock caused the US government to fund the education of their citizens. 13-year-old Kary Mullis asked Santa to bring him a chemistry set for Christmas, and that's how the history of PCR began.

Kary Mullis (b. 1944).

Fig. 10.34 Automatic sequencer in the Human Genome Project.

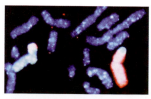

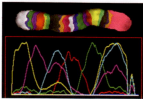

Fig. 10.35 FISH (top) and multicolor banding of a chromosome, performed by Ilse Chudoba (*Metasystems* Jena, Germany).

they are hybridized with fluorescent-labeled DNA probes. A cytogenetic multicolor banding analysis of cancer cells provides the basis of a more accurate prognosis for the individual patient, and thus for a treatment appropriate to the risk.

■ 10.17 The Ultimate Biotechnological Achievement – The Human Genome Project

In October 1990, the most comprehensive biotechnological project of all times was launched. It is said that approximately $3,000,000,000 went into the funding of the project. Biotechnologists and molecular biologists worked around the clock to map the approximately **3.4 billion base pairs**, distributed on 23 human chromosome pairs (Box 10.5). The 23 pairs are composed of two sets of chromosomes from the parents – 22 autosomal chromosome pairs and two sex chromosomes (female XX or male XY). The 3,400 million base pairs contain an incredible amount of information – or 750 megabytes of digital information. It could be stored on a single DVD.

The smallest chromosome, the **Y-chromosome**, responsible for the minute difference between man and woman, carries 50 million base pairs, the largest, chromosome 1, contains 250 million base pairs. Lab pet *Drosophila*, the fruit fly, would take up a storage space of only 10 phone books, yeast just one book, and *Escherichia coli* would even fit into 300 pages in the New York phone book.

About 6,000 diseases are caused by **aberrations of single genes**. If a single word in the genetic text is misspelled, the cell will not produce the correct protein or will produce the wrong quantities of a protein.

Monogenetic diseases include phenylketonuria (PKU), Huntington's disease, and cystic fibrosis.

In more complex diseases, several genes, often dozens of them, are involved. These diseases include heart attack, atherosclerosis, asthma, and cancer and are further complicated by interactions with nutrients and environmental toxins.

Researchers are convinced that precise knowledge of the human genome will open up completely new

Box 10.5 PCR – the DNA-Copying Mechanism *par excellence*

In a polymerase chain reaction (PCR), a selected DNA fragment is augmented at an exponential rate. It could be any fragment of any DNA, as long as its base sequence at both ends is known. These ends are the starting points for DNA polymerase to start the copying process. The DNA polymerase used in the DNA replication process is the heat-stable polymerase from *Thermus aquaticus*, known as Taq polymerase (Fig. 10.36). The heat stability of the enzyme is a major boon when it comes to the automation of the process. As the polymerase is not destroyed by heat and can be reused, it need not be replaced at each new cycle.

Thermal cyclers: Top: MJ Research Model *Tetriad*
Bottom: Roche *Light Cycler.*

If the ends of the DNA fragment to be replicated are known, short fragments of single-stranded DNA, known as primers or starting sequences, can be synthesized. They are tailor-made to complement the starting loci near both ends of the DNA fragment.

To begin the PCR process, all reagents, the DNA template, both primers, the polymerase and the DNA constituents (the four nucleotides A, G, C, and T, also referred to as dNTPs) are dissolved in a tube in a suitable buffer. In a thermal cycler, the following automated reaction takes place:

- The double helix is heated to 94°C (201°F) and separates into two single-stranded DNAs.

- After cooling down to 40 to 60°C (104–140°F), both primers bind to their appropriate starting loci near the ends of the single-stranded DNAs (hybridization). The short double-stranded fragments that arise also serve as docking points for the polymerase to start the copying process.

- Two polymerase molecules work their way towards each other from both ends of the fragment, building matching nucleotides and copying the requested DNA fragment. Thus, two identical double-stranded daughter DNAs have been created.

- Again, the temperature is raised to 94°C (201°F). Both daughter DNAs split into a total of four single-stranded DNAs.

- After cooling, four primers bind to the now four existing ends. Polymerase produces four identical double-helix DNAs.

- Repetition of the cycle, and an exponential row develops: 8, 16, 32, 64, 128, 256 copies and so on.

Suppose a cycle lasts no longer than 3 minutes, then it must be possible to run 20 cycles within an hour and produce a million copies. In theory, a single molecule would be enough to start the replication process, but in practice, at least three to five molecules of the DNA to be replicated are required in order to set off the PCR process.

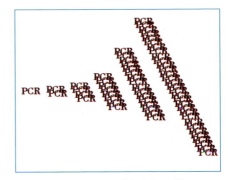

Schematic representation of the PCR working principle.

PCR is invaluable not only in research, but also in diagnostics. Bacteria and viruses can be identified directly; no artificial replication process is required. PCR is increasingly used in the diagnosis of hereditary diseases and cancer.

Real-time polymerase chain reaction, also called quantitative real time polymerase chain reaction (QRT-PCR), is a modification of the polymerase chain reaction.

The procedure follows the general pattern of polymerase chain reactions, but the DNA is quantified after each amplification round. This can be done either by using fluorescent dyes that intercalate with double-stranded DNA, or modified DNA oligonucleotide probes that fluoresce when hybridized with complementary DNA.

---

pathways for treatment and prevention of widespread diseases such as cancer and asthma.

In order to decipher a genome, three types of analysis must be carried out:

- **Genomics**: DNA sequence data are collected. The data are obtained as individual sequences of 500 to 800 bp which are put together to produce a contiguous genomic sequence. This requires an organizational strategy.

- **Postgenomics** or functional genomics: genes, promoters, and other relevant factors are located, and the functions of genes discovered in the process are investigated.

- **Bioinformatics**: high-performance computers are used for genomics and postgenomics. Contiguous DNA sequences are put together and checked for possible genes. The function of genes is predicted. The whole process involves huge amounts of data.

■ 10.18 Genetic Genome Maps

The genome of an organism consists of billions of base pairs. As the sequencing techniques described above only yield sequences up to 750 bp, the genome must be put together like a jigsaw puzzle. The publicly financed genome project relied on what is known as **genomic maps** to identify the

Fig. 10.36 *Thermus aquaticus* (bottom) from the hot springs in the Yellowstone National Park.

Box 10.6 Biotech History:
The Human Genome Project

As late as the 1980s, it was still inconceivable that it would be possible to map the entire human genetic material. At that time, it was just possible to map the genome of a very small virus. The complete genome analysis of phage 174, consisting of 5,375 bases, had been carried out by Fred Sanger in 1977, but it took another twenty years until the genome of the first free-living organism (the bacterium *Haemophilus influenzae*) was completely sequenced.

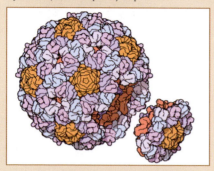

The genome of bacteriophage ΦX 174 – of which the protein envelope, not the DNA, is shown here – was the first to be sequenced in human history.

In 1985, **Robert Sinsheimer**, Chancellor of the University of California in Santa Cruz, was looking for a new big biological project and assembled the cream of genome researchers on his campus. "A bold and exciting idea, but not feasible" was the general verdict, but Walter Gilbert, who was to be awarded the Nobel Prize for his alternative DNA sequencing method (he shared the Nobel Prize with Sanger), did not give up and found a powerful ally in **James D. Watson** (Ch. 3).

After the Cold War had died down, the US Department of Energy (DOE) was looking for new projects and made new funding available. However, many researchers were still doubtful about the point and the use of the whole project. It looked like "incredibly boring industrial-scale research". Nobel Prize winner **Sydney Brenner** once jokingly suggested outsourcing DNA-sequencing to prison inmates. The worse the crime, the larger the chromosome to be sequenced!

Another reservation was that it would swallow up so much money that nothing would be left for other biological research. What was one supposed to do with a sequence of three billion let-ters of which 97 percent probably had no (direct) function at all? Did such a project fall into the purview of the Department of Energy, anyway? The National Institute of Health (NIH), umbrella organization of all biomedical research in the US, was not forthcoming with support. A special committee of the American Academy of Sciences (AAS), composed of supporters as well as opponents of the project finally recommended a gradual approach. First, a rough map of the genome was to be put together. Other genomes of various simple organisms, such as *Escherichia coli*, the yeast *Saccharomyces cerevisiae* and the nematode *Caenorhabditis elegans* were to be sequenced in parallel, and new techniques were to be explored.

The US Congress agreed to the funding, and the NIH came on board. In 1988, it established an Office for Genome Research under Watson's leadership. Watson seized the unique opportunity in his scientific career to be involved in the research from double helix to the three billion steps on the ladder of the human genome. Thus, the NIH took again the lead in genome research. Watson pursued a dual strategy of developing new mapping techniques, while also focusing on the rapid identification of pathogenic genes.

In 1989, the Human Genome Organization (HUGO) was founded, a worldwide organization with members from 30 nations. Its task was to coordinate all efforts in order to avoid competition and double work. Everything seemed to go smoothly until **Craig Venter** appeared on the scene. Venter, then still employed by NIH, did not want to waste time on sequencing junk, but go straight for the genes that might be patentable and thus profitable. Just before the publication of Venter's EST (Expressed Sequence Tags) article, NIH's patent department filed an application for the first 347 ESTs.

The other scientists were outraged. Watson publicly doubted the value of Venter's sequencing method, calling it "the work of a brainless robot". The EST program "might as well be carried out by a monkey." As a response, Venter's lab staff posed for photographs, wearing gorilla masks. Nobel laureate **Paul Berg**, Vice President **Al Gore,** and everybody else agreed that the human genome should not be patentable.

Watson fell out with NIH director **Bernadine Healy** who insisted on the patent application. In 1992, he resigned from his position in the genome project. The Patent Office rejected all patent applications in the first round in August 1992. The new director of NIH, Nobel Prize laureate **Harald Varmus** took a final decision not to file for patents in 1994.

An event that took center stage – similar to the first human landings on the moon – was the joint announcement by Craig Venter, President Bill Clinton, and Francis Collins that the sequencing of the human genome had been completed. Bill Clinton: "Today we are learning the language in which God created life."

Craig Venter had left NIH in July 1992 in order to follow a different route. A venture capital fund offered him $70 million (later increased to $85 million) to help him put his plans into practice. Venter was delighted: "This is the dream of every scientist – having a benefactor who invests in their ideas, dreams, and abilities."

Competition is the soul of business – state-funded versus industrial research into the human genome became a real thriller, captured by **Kevin Davies** in *The Sequence*. In this box, we have only room for the major milestones.

In 1995, Venter announced that in collaboration with Johns Hopkins University, he had been able to map for the first time the genome of a free-living organism, *Haemophilus influenzae*. It had taken them just a year, using the shotgun technique. NIH had rejected the shotgun approach as unreliable and therefore not fundable.

In 1996, NIH-funded researchers published the complete genome of bakers' yeast (*Saccharomyces cerevisiae*). Venter responded in May 1998 by claiming that with his new company (Celera Genomics), he would be able to sequence the entire human genome at a cost of only $300 million, using the contentious shotgun technique, and he would be able to do this within three years. Watson's successor, **Francis Collins**, a pioneer in the search for pathogenicity genes, set the stakes higher. By spring 2001, an outline of the human genome (90 percent of the whole sequence) would be mapped, and the entire genome by 2003, two years earlier than originally anticipated. A year after this announcement, Venter produced the complete sequence of the *Drosophila* (fruit fly) genome. In 2000, Celera claimed to have mapped 90 percent of the humane genome. Two months later, the publicly funded project announced they had deciphered two billion base pairs.

As pressure on the two rival genome research organizations kept increasing, the DOE intervened, suggesting that both parties should agree to publicize their results jointly – which was done on June 26th, 2000. Craig Venter, Bill Clinton, and Frank Collins gave a joint press conference. In the short term, Clinton (and Tony Blair on the British side) had been able to bring the opponents together. However, only five months later, Venter refused to make his version accessible to a public database. In February 2001, the US journal *Science* published Venter's map of the human genome, and *Nature* that of the public funded project.

The two projects differed not only in their handling of finances and public relations, but also in their basic approach. Researchers of the publicly funded project first divided the genome into more manageable portions which they reorganized before they began the sequencing process. Venter took the cheaper approach by first mechanically dissecting the genome, then sequenced the fragments and only put them together at the end.

The publicly funded project analyzed the DNA of twelve anonymous donors, whereas it was rumored that Celera mainly relied on the genome of one single human.

It also turned out that the oh-so-cost-effective private researchers would not have got anywhere with their shotgun approach without additional information from the publicly funded project. The Celera supercomputer used partial sequences made available by their competitors.

As so often in life, the rivals were in fact complementing each other.

Cited Literature:

Davies K (2001) *Cracking The Genome: Inside The Race To Unlock Human DNA.* The Free Press, New York, London.

location of short partial sequences. Genomic maps must have many unambiguous and evenly distributed **markers**, so that the pieces of the jigsaw puzzle that have been sequenced can be related to a marker. There are two kinds of genome maps – genetic and physical maps.

Gene loci can be identified by genetic methods (crossbreeding of lab organisms, family tree analyses in humans). In 1994, the Human Genome Project had achieved the goal of producing a genetic map with a marker every 1,000,000 base pairs. Two years later, the map had become more detailed with one marker per 600,000 bp. It was based on 5,264 microsatellites or **STRs, short tandem repeats** of the sequence CACACA. This tandem DNA is found on all chromosomes, and their number in a gene locus can be established using primers that hybridize on both sides of the STR – for example:

Primer-CACACA-Primer

and

Primer-CACACACACACACACA-Primer

PCR is carried out and the size of the products established by agarose gel electrophoresis. The first PCR product is significantly shorter than the second and will pass more rapidly through the gel. The human genome contains at least 650,000 STRs.

RFPLs and **SNPs** are also used as markers. RFPLs can be detected with PCR primers, PCR, restriction cleavage and electrophoresis. This shows the absence or presence of a restriction site. SNPs are loci in the genome where at least two different nucleotides can be expected, i.e., point mutations. They are detected with short oligonucleotide probes that hybridize only one SNP variant.

All these markers are variable, i.e., they occur in at least two versions or allele forms. How these are passed on can be shown through **family tree analysis** or **crossbreeding experiments.**

■ 10.19 Physical Genome Mapping

A physical genome map shows the position of specific DNA sequences on the DNA molecule of a chromosome. The DNA markers are **expressed sequence tags** (**ESTs**) (Box 10.6) – short sequences obtained from complementary DNA (cDNA), i.e., partial gene sequences. The method is based on the idea that only expressed genes are

Fig. 10.37 Craig Venter and Bill Clinton (top) and Francis Collins (bottom).

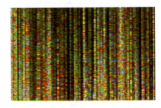

Fig. 10.38 Sequencing gel with fluorescent markers.

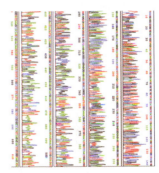

Fig. 10.39 Typical DNA sequence printout.

Fig. 10.40 Craig Venter's autograph on the 48-capillary 3730 DNA Analyzer in a Hong Kong laboratory. This analyzer is the gold standard in medium- to-high throughput genetic analysis. It is used for DNA fragment analysis applications such as microsatellites, SNP analysis, mutation detection, and traditional DNA sequencing.

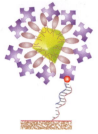

Fig. 10.41 The SuperNova principle: Can PCR be replaced by simpler methods? Dieter Trau and 8sens. biognostic. AG, Berlin-Buch, Germany, came up with an idea in 1997

Atoms are tightly packed in crystals (e.g., a diamond, pictured above). Fluophore crystals are used to label DNA probes. A 100 nm crystal holds 10,000,000 molecules of the fluophore.

A fluorophore crystal used as label is made water-insoluble. It binds to a single-stranded DNA. After several washing steps, the DNA molecule is bound to – in theory – 100 million detector molecules.

When an organic solvent is added, the crystal dissolves, releasing the fluorescent molecules. The solution lights up in the brightest fluorescent colors… almost like a supernova in space!

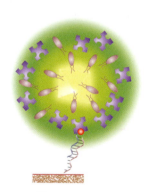

transcribed into mRNA in the cell. Using the mRNA as a starting point and converting it into DNA would produce a library of expressed genes. The cDNA is obtained from mRNA using reverse transcriptase (Ch. 3).

Although the EST sequence does not identify the gene itself, it helps to find its **location**.

Two methods are in use:

- *Fluorescence in situ hybridization* (**FISH**, see above): Two DNA probes labeled with different fluorochromes are used in parallel. When the chromosomal DNA is stretched and spread, it is possible to locate the labels very accurately.

- A **mapping agent** is used which consists of a collection of overlapping DNA fragments covering the relevant gene or chromosome.

10.20 Which Method – Contig Versus Shot Gun?

In 1998, Craig Venter, who had left the human genome project six years earlier, founded his own company Celera Genomics (more see Box 10.5). His plan was to sequence human DNA faster than it was done in the publicly funded genome project.

After two years of fierce competition, an agreement was reached between NIH and Celera Genomics in 2000 that both sides would publicize their data at the same time. Most of the differences between the two projects can be put down to the difference in the methods used, i.e., clone-contig versus shotgun.

The approach of the publicly funded Human Genome Project is characterized by catch phrases like "mapping before sequencing" and "clone for clone".

In the clone **contig technique**, the actual sequencing is preceded by a stage during which a number of overlapping clones are identified. Many copies of the genome are cut by restriction endonucleases, resulting in fragments of approximately 150,000 base pairs. These fragments are cloned into **bacterial artificial chromosomes, BACs,** and introduced into bacteria.

The BACs are newfangled vectors on the basis of the F-plasmid and can carry DNA fragments up to 300 kb.

With the division of the bacterial cells, the human DNA fragments are also replicated. Each bacterium

contains a DNA copy. Thus, if you compare the human genome to a 23-volume encyclopedia, you would find, for example

Clone A contains pages 26 to 49 of volume 12
Clone B contains pages 35 to 52 of volume 12
Clone C contains pages 18 to 36 of volume 12
Clone D contains pages 27 to 50 of volume 12

However, we do not know which pages of which volume are found in which clone. We only know that the clone library covers the entire genome.

Thus, each of the clones must be put in a spatial context with the other clones. The clones are again cut up with several restriction endonucleases to produce a clone fingerprint. PCR is used to detect the absence or presence of restriction loci (this used to be done with Southern blotting).

The individual cloned DNA segments are then sequenced and the sequence is stored in the right location on the **contig map**.

If, for example, clone A shows markers (e.g., STRs) on pages 42 and 46, which are also found in clone B, pages 26 to 52 can be reconstructed. If clone A carries markers on pages 27 and 30, which are also found on clone C, reconstruction can be extended to page 18. The information on clone D, however, is redundant. What is required is the **smallest possible number of partially overlapping clones that cover the entire genome**. This is called a **contig**. In the human genome, it amounts to 30,000 BAC clones.

The fragments in the BACs are too large for direct sequencing. They are therefore split up into subclones which are replicated and sequenced using a modern version of the Sanger chain termination method. It is obvious from this description that the clone contig method is laborious and expensive. Is there any faster and cheaper way?

Craig Venter and Celera Genomics preferred a method known as **shotgun sequencing**, which had already been successfully used in the sequencing of *Haemophilus influenzae* in 1995. Without any previous mapping, the genome (1830 kb) was **randomly fragmented using ultrasound** and analyzed. The sequences were searched for **overlaps**. This only works if overlaps are found between all sequences. The ultrasound-treated DNA was separated in gel electrophoresis. The purified DNA fragments were then cloned into *E. coli.* The resulting **clone library** contained 19,687 clones. In total, 28,643 sequencing experiments were carried out. Sequences that were too small were discarded. The remaining 24,304 sequences were fed into a computer, which analyzed the data for 36 hours and came up with 140 contiguous sequences (contigs). There were still gaps that could have been closed by sequencing still more ultrasound-treated fragments, but the decision was taken to close those gaps by using oligonucleotide probes and a phage library (Fig. 5).

Bacterial genomes were very suitable for the shotgun approach, as they are small and contain no or very few repetitive sequences. Such **repetitive sequences**, however, can wreak havoc in a eukaryotic genome if the shotgun approach is used because it gives rise to false overlaps.

Putting the sequence data together without any reference system is very difficult in such a large genome. Venter developed complex computing schemes to sort out the sequences, but in the end, he had to resort to the publicly accessible physical maps of the Human Genome Project.

The shotgun method was a treat for bacterial genomes which, apart from being small, contain only very few repetitive sequences. When repetitive sequences in eukaryotic genes are put together, they may be wrongly allocated. Such errors can be avoided if a genome map is used to put together the sequences obtained by the shotgun approach. A **targeted shotgun method** would be a reasonable compromise for the sequencing of large genomes, in which the shotgun-obtained sequences are constantly checked against a gene map.

■ 10.21 The Human Genome Project – Where Do We Go from Here?

In spite of its great significance in life, the male Y chromosome consists mainly of junk DNA, to **James Watson** jokingly remarked. It is true that on the whole, human genes are thinly spread like oases in a desert of non-coding DNA and make up just 3 percent of the total DNA.

3.4 billion letters of the human genome must be searched for coding sequences. This is where **bioinformatics** comes in. As we cannot go into too much detail here, suffice it to say that disappointingly, the human, the "pinnacle of creation", has no

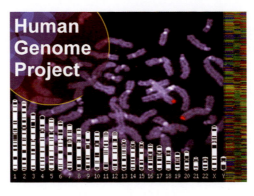

more than 20,000 to 30,000 genes (as of the beginning of 2005). This is still considerably more than bakers' yeast (6,000 genes) or the nematode *Caenorhabditis elegans* (18,000).

In intestinal bacteria, 78 percent of DNA is coding, compared to just 3 percent in humans. In terms of the Manhattan phone book, 78 percent of the 300 bacteria pages, i.e., 234 pages, contain readable information, whereas in humans, the 6,000 pages of genetic information are distributed over two million pages. What does this imply? It is definitely not information to be read by aliens, as some SciFi authors have been speculating, nor is it where the soul or the *Chinese CHI* can be found (see also Ch. 9, RNA).

An international consortium has launched the $100 million *HapMap* **project** – the map of haplotypes (Fig. 10.42). Its mission is to deliver the next generation Human Genome Project, the discovery of those genes that are involved in widespread diseases, such as asthma, cancer, and diabetes. Participants in the consortium are the US, Japan, China, Canada, Nigeria, and the U.K., according to the National Human Genome Research Institute (NHGRI).

HapMap is expected to be accomplished soon. By then, it should have recorded genetic variations within the human genome. By comparing genetic differences between individuals, it is hoped, researchers will obtain a tool that allows them to assess the impact of genes on many diseases. While the Human Genome Project provided the basis of genetic discoveries, *HapMap* is expected to be a first step on the way to practical applications of genome research results.

The DNA used for *HapMap* had come from blood samples. Initially, researchers are provided with samples of 200 to 400 people. The results of the analyzed genes will be freely available on the Inter-

---

The Top Ten Surprises in Genome Research

With apologies to well-known Late-Night-Show presenter David Letterman, Francis Collins presented a Top Ten list at the Annual Conference of the American Association of the Advancement of Science (AAAS) in February 2001. Here is a slightly abridged version:

10. The genome appears to be lumpy with some chromosomes having a higher gene density than others.

9. The total number of human genes is far lower than expected.

8. Due to their capacity for alternative splicing, human genes are able to produce a larger number of proteins.

7. Due to an increase in larger areas of conserved proteins, human genes are more sophisticated than those of other organisms.

6. More than 200 human genes are not related to genes found in their close relatives, but to those in bacteria. This suggests that they may be a result of horizontal transfer from bacteria.

5. Repetitive DNA sequences are a kind of fossil record reaching back 800 million years into the past.

4. Junk DNA is eminently functional.

3. The mutation rate in men is twice as high as in women. Thus, the majority of gene mutations in the gene pool come from men who are the movers and shakers in evolutionary terms.

2. Humans share 99.9 percent of their genome. Most of our genetic differences appear to be racially and ethnically linked. There is, however, no scientific definition of race.

And now to the **NUMBER ONE**: These results emphasize the importance of free, unhindered access to human genomic sequences.

Francis Collins was given a standing ovation.

Box 10. 7 *DNA Travel Maps 2*: The Journey of Female Mitochondrial DNA

After analyzing my own DNA by the Geographic Project (see Box 3), I convinced my friend Claire (who is a dolphin trainer in Hong Kong's Ocean Park) to analyze her DNA in order to compare male and female, European and Asian samples. Here is an extract from her report (with her kind permission)

Dear Claire Ma, here is the analysis of your mitochondrial DNA

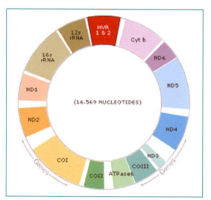

Type: mtDNA , **Haplogroup**: B (subclade B)

Your Mitochondrial HVR I Sequence
Key: C Substitution (transition), C Substitution (transversion) C Insertion / Deletion
Markers: 16140C, 16183C, 16189C, 16243C, 16311C, 16355T, 16519C

How to Interpret Your Results?

Above is displayed the sequence of your mitochondrial genome that was analyzed in the laboratory. Your sequence is compared against the **Cambridge Reference Sequence (CRS)**, which is the standard mitochondrial sequence initially determined by researchers at Cambridge, UK. The differences between your DNA and the CRS are highlighted, and these data allow researchers to reconstruct the migratory paths of your genetic lineage. Substitution (transition): a nucleotide base mutation in which a pyrimidine base (C or T) is exchanged for another pyrimidine, or a purine base (A or G) is replaced with another purine. This is the most common type of single point mutation. Substitu-

```
ATTCTAATTTAAACTATTCTCTGTTCT
TTCATGGGGAAGCAGATTTGGGTAC
CACCCAAGTATTGACTCACCCATCAA
CAACCGCTATGTATTTCGTACATTAC
TGCCAGCCACCATGAATATTGTACG
GTAC CATAAACACTTGACCACCTGT
AGTACATAAAAACCCAATCCACATC
AAACCCCCCCCCCATGCTTACAAG
CAAGTACAGCAATCAACCCTCAACTA
TCACACATCAACCGCAACTCCAAAG
CCACCCCTCACCCACTAGGATACCA
ACAAACCTACCCACCCTTAACAGTAC
ATAGCACATAAAGCCATTTACCGTAC
ATAGCACATTACAGTCAAATCCTTTC
TCGTCCCCATGGATGACCCCCCTCA
GATAGGGGTCCCTTGACCACCATCC
TCCGTGAAATCAATATCCCGCACAA
GAGTGCTACTCTCCTCGCTCCGGGC
CCATAACACTTGGGGGTAGCTAAAG
TGAACTGTATCCGACATCTGGTTCCT
ACTTCAGGGCCATAAAGCCTAAATA
GCCCACACGTTCCCCTTAAATAAGAC
ATCACGATG
```

tion (transversion): a base substitution in which a pyrimidine base (C or T) is exchanged for a purine base (A or G), or vice versa. Insertion: a mutation caused by the insertion of at least one extra nucleotide base in the DNA sequence. Deletion: a mutation caused by the deletion of at least one extra nucleotide base from the DNA sequence.

Your Branch on the Human Family Tree

Your DNA results identify you as belonging to a specific branch of the human family tree called **haplogroup B**. The map above shows the direction that your maternal ancestors took as they set out from their original homeland in East Africa. While humans did travel many different paths during a journey that took tens of thousands of years, the lines above represent the dominant trends in this migration. Over time, the descendants of your ancestors spread throughout East Asia, with some groups making it as far as Polynesia and the Americas. But before we can take you back in time and tell their stories, we must first understand how modern science makes this analysis possible.

The string of 569 letters shown above is **your mitochondrial sequence**, with the letters A, C, T, and G representing the four nucleotides – the chemical building blocks of life – that make up your DNA. The numbers at the top of the page refer to the positions in your sequence where **informative mutations have occurred in your ancestors**, and tell us a great deal about the history of your genetic lineage.

Here's how it works. Every once in a while a mutation – a random, natural (and usually harmless) change – occurs in the sequence of your

Adam and Eve by the German painter Lucas Cranach (1472-1553). This celebrated work brilliantly combines devotional meaning with pictorial elegance and invention. The scene is set in a forest clearing where Eve stands in front of the Tree of Knowledge, caught in the act of handing an apple to a bewildered Adam. Entwined in the tree's branches above, the serpent looks on as Adam succumbs to temptation. A rich menagerie of birds and animals are grouped around the tree completing this beautifully seductive vision of Paradise in the moments before Man's Fall.

mitochondrial DNA. Think of it as a spelling mistake: one of the "letters" in your sequence may change from a C to a T, or from an A to a G. After one of these mutations occurs in a particular woman, she then passes it on to her daughters, and her daughters' daughters, and so on. (Mothers also pass on their mitochondrial DNA to their sons, but the sons in turn do not pass it on.) Geneticists use these markers from people all over the world to **construct one giant mitochondrial family tree**. As you can imagine, the tree is very complex, but scientists can now determine both the age and geographic spread of each branch to reconstruct the prehistoric movements of our ancestors. By looking at the mutations that you carry, we can trace your lineage, ancestor by ancestor, to reveal the path they traveled as they moved out of Africa. Our story begins with your earliest ancestor. Who was she, where did she live, and what is her story?

Mitochondrial Eve: The Mother of Us All

Our story begins in **Africa between 150,000 and 170,000 years ago**, with a woman whom anthropologists have nicknamed "Mitochondrial Eve." She was awarded this mythic epithet in 1987 when population geneticists discovered that all people alive on the planet today can trace their maternal lineage back to her. But Mitochondrial Eve was **not the first female human**. *Homo sapiens* evolved in Africa around

200,000 years ago, and the first hominids – characterized by their unique bipedal stature – appeared nearly two million years before that. Yet despite humans having been around for almost 30,000 years, Eve is exceptional: because hers is the only lineage from that distant time to survive to the present day.

Native Americans of the Archaic Period (8000-1000 BC) were hunters and gatherers, and their settlements reflect an adaptation to the abundant natural resources of the Tennessee region. Sites varied in function from base settlements to transient hunting or collecting camps. A camp is shown here: Women are processing hickory nuts, whose meats are high in proteins and fats. Plant foods were supplemented by such animals as the white tail deer, turkeys, bears, and smaller game like rabbits. This scene is based upon research by University of Tennessee, Knoxville, archaeologists on Archaic Period sites in Benton County, Tennessee, in 1940-41 and in Monroe and Loudon counties in the 1970s.

"So why Eve?" Simply put, Eve was a survivor. A maternal line can become extinct for a number of reasons. A woman may not have children, or she may bear only sons (who do not pass her mtDNA to the next generation). She may fall victim to a catastrophic event such as a volcanic eruption, flood, or famine, all of which have plagued humans since the dawn of our species. None of these extinction events happened to Eve's line.

It may have been simple luck, or it may have been something much more. It was around this same time that modern humans' intellectual capacity underwent what author **Jared Diamond** (see Chapter 1) coined the **Great Leap Forward**. Many anthropologists believe that the emergence of **language** gave us a huge advantage over other early human species. Improved tools and weapons, the ability to plan ahead and cooperate with one another, and an increased capacity to exploit resources in ways we hadn't been able to earlier, all allowed modern humans to rapidly migrate to new territories, exploit new

resources, and outcompete and replace other hominids, such as the Neandertals.

It is difficult to pinpoint the chain of events that led to Eve's unique success, but we can say with certainty that all of us trace our maternal lineage back to this one woman.

The Deepest Branches: Ancestral line "Eve" > L1/L0

Mitochondrial Eve represents the root of the human family tree. Her descendents, moving around within Africa, eventually split into two distinct groups, characterized by a different set of mutations their members carry. These groups are referred to as **haplogroups L0 and L1**, and these individuals have the most divergent genetic sequences of anybody alive today, meaning they represent the deepest branches of the mitochondrial tree. Importantly, current genetic data indicates that **indigenous people belonging to these groups are found exclusively in Africa**. This means that, because all humans have a common female ancestor, "Eve," and because the genetic data shows that Africans are the oldest groups on the planet, we know our species originated there.

Haplogroups *L1* and *L0* likely originated in East Africa and then spread throughout the rest of the continent. Today, these lineages are found at highest frequencies in Africa's indigenous populations, the hunter-gatherer groups who have maintained their ancestors' culture, language, and customs for thousands of years.

At some point, after these two groups had coexisted in Africa for a few thousand years, something important happened. The mitochondrial sequence of a woman in one of these groups, *L1*, mutated. A letter in her DNA changed, and because many of her descendants have survived to the present, this change has become a window into the past.

The descendants of this woman, characterized by this signpost mutation, went on to form their own group, called **haplogroup L2**. Because the ancestor of *L2* was herself a member of *L1*, we can say something about the emergence of these important groups: Eve begat *L1*, and L1 begat *L2*. Now we're starting to move down your ancestral line.

Haplogroup *L2:* West Africa

L2 individuals are found in sub-Saharan Africa, and like their *L1* predecessors, they also live in Central Africa and as far south as South Africa. But whereas *L1/L0* individuals remained predominantly in eastern and southern Africa, your

ancestors broke off into a different direction, which you can follow on the map above.

L2 individuals are most predominant in West Africa, where they constitute the majority of female lineages. And because *L2* individuals are found at high frequencies and widely distributed along western Africa, **L2 haplotypes represent one of the predominant lineages in African-Americans.** Unfortunately, it is difficult to pinpoint where a specific *L2* lineage might have arisen. For an African-American who is *L2* – the likely result of West Africans being brought to America during the **slave trade** – it is difficult to say with certainty exactly where in Africa that lineage arose.

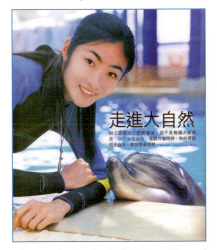

走進大自然

A long way for mitochondrial DNA from Africa to the Ocean Park of Hong Kong... Claire Ma is a most successful dolphin trainer here.

Haplogroup *L3:* moving North, out of Africa

Your next signpost ancestor is the woman whose birth **around 80,000 years ago** began haplogroup *L3*. It is a similar story: an individual in *L2* underwent a mutation to her mitochondrial DNA, which was passed onto her children. The children were successful, and their descendants ultimately broke away from the *L2* clan, eventually separating into a new group called *L3*. You can see above that this has revealed another step in your ancestral line.

While *L3* individuals are found all over Africa, including the southern reaches of sub-Sahara, *L3* is important for its movements north. You can follow this movement of the map above, seeing first the expansions of *L1/L0*, then *L2*, and followed by the northward migration of *L3*. **Your L3 ancestors are the first modern humans to have left Africa**, representing the deepest branches of the tree found outside of that continent.

See next page

Today, *L3* individuals are found at **high frequencies in populations across North Africa**. From there, members of this group went in a few different directions. Some lineages within *L3* testify to a distinct expansion event in the mid-Holocene that headed south, and are predominant in **many Bantu** groups found all over Africa. One group of individuals headed west and is primarily restricted to Atlantic western Africa, including the islands of Cabo Verde. Other *L3* individuals, your ancestors, kept moving northward, eventually **leaving the African continent completely**. These people currently make up around 10% of the Middle Eastern population, and gave rise to two important haplogroups that went on to populate the rest of the world.

Haplogroup *N:* The Incubation Period

Your next signpost ancestor is the woman whose descendants formed haplogroup *N*. Haplogroup *N* comprises one of two groups that were created by the descendants of *L3*.

Haplogroup *M* was the result of the first great wave of migration of modern humans to leave Africa. These people likely **left the continent across the Horn of Africa near Ethiopia**, and their descendants followed a **coastal route eastward, eventually making it all the way to Australia and Polynesia**.

Haplogroup *B:* Your Branch on the Tree

One group of these early *N* individuals broke away in the Central Asian steppes and set out on their own journey following herds of game across vast expanses. Around **50,000 years ago**, the first members of your **haplogroup *B* began moving into East Asia**, the beginnings of a journey that would not stop until finally reaching **both continents of the Americas and much of Polynesia**. Your haplogroup likely arose on the high plains of Central Asia between the Caspian Sea and Lake Baikal. It is one of the founding East Asian lineages and, along with haplogroups *F* and *M*, comprises around three quarters of all mitochondrial lineages found there today. **Radiating out from the Central Asian homeland**, haplogroup *B*-bearing individuals, your own distant ancestors, began migrating into the surrounding areas and quickly headed south, making their way throughout East Asia.

Today **haplogroup *B* makes up around 17% of people from Southeast Asia, and around 20% of the entire Chinese gene pool**. It exhibits a very wide distribution along the

Representatives of indigenous communities who are participating in the Genographic field research and are attending the April 13, 2005, launch of the Genographic Project at National Geographic's headquarters in Washington, D.C., to speak on behalf of their communities.
Left to right: Battur Tumur, descendant of Genghis Khan, Mongolia/San Francisco, Calif., USA; Julius Indaaya Hunume, Hadza Chieftain, Tanzania; Phil Bluehouse Jr., Navajo Indian, Arizona, USA.

Pacific coast, from Vietnam to Japan, as well as at lower frequencies (about 3%) among native Siberians. Because of its old age and high frequency throughout east Eurasia, it is widely accepted that this lineage was carried by the first humans to settle the region.

In northern Eurasia and Siberia, haplogroup *B* individuals with experience surviving the harsh Central and East Asian winters would have been ideally suited for the arduous **crossing of the recently formed Behring land bridge**. During the last glacial maximum, 15,000 to 20,000 years ago, colder temperatures and a drier global climate locked much of the world's freshwater at the polar ice caps, making living conditions near impossible for much of the northern hemisphere. But an important result of this glaciation was that **eastern Siberia and northwestern Alaska became temporarily connected** by a vast ice sheet. Haplogroup *B*-bearing individuals, fishing along the coastline, followed it.

Today, **haplogroup *B* is one of 5 mitochondrial lineages found in aboriginal Americans**, and is found in both North and South America. While haplogroup *B* is very old (around 50,000 years), the reduced genetic diversity found in the Americas indicates that those lineages arrived only within the last 15,000 to 20,000 years and quickly spread once there. Better understanding exactly how many waves of humans crossed into the Americas, and where they migrated to once there, remains the focus of much interest and is central to Genographic's ongoing research in the Americas.

Recent population expansions appear to have brought a subgroup of haplogroup *B* lineages

from Southeast Asia into Polynesia. This lineage is referred to as *B4* (meaning the fourth subgroup within *B*) and is characterized by a set of mutations that took a significant amount of time to accumulate on the Eurasian continent. This closely related subset of lineages likely spread **from Southeast Asia into Polynesia within the last 5,000 years**, and is seen extensively throughout the islands at high frequency. Intermediate lineages – those containing some, but not all, of the B4 mutations – are found in **Vietnamese, Malaysian, and Bornean populations**, further supporting the likelihood that the Polynesian lineages originated in these parts of Southeast Asia.

Looking Forward (Into the Past): Where Do We Go From Here?

Although the arrow of your haplogroup currently ends across sub-Saharan Africa, this isn't the end of the journey for haplogroup *B*. This is where the genetic clues get murky and your DNA trail goes cold. Your initial results shown here are based upon the best information available today – but this is just the beginning. A fundamental goal of the Genographic Project is to extend these arrows further toward the present day.

DNA Shotgun Approach Charting New Waters.

Personal conclusions from RR

After reading this detailed DNA-report, Claire and I started to compare our DNA-trails (see also Box 10.3). It was quite educational to see the place and the time when possibly her great-great-great(etc)-grandma told my great-great-great(etc) – grandpa in the unknown language of that time:

"Okay, my stubborn dearest, here we must part! You go West and I go East – but don't worry, our great-great-greatgrandchildren will meet again a few thousand years from now in a big city! Good luck!!"

She was (as females in most cases are) right! RR

net. Once the *HapMap* project is completed, the data will be available to scientists worldwide who do research into genetic risk factors for a wide range of diseases. Another project is the **National Geographic Genographic Project** (Box 10. and 6) to whom anyone can send in their DNA and can find out about the migration routes of their ancestors, for $130.

The scope of the Human Genome Project far exceeds the investigation of genetic disease. New action pathways for medication are expected to be found, as well as cures for cancer.

10.22 … And How Can the Sequence of the Genome Be Understood?

All right, the DNA sequence has been stored in our computers now, but how are the genes, the three percent that make up the oases in the desert, to be found? As long as the amino acid sequence of the protein product is known, it is not hard to predict the DNA sequence. However, for most genes, that information is not available.

Where does the reading begin? The DNA sequence of a gene is an **open reading frame**, **ORF**. It is a sequence of triplets, often starting with a **start codon,** often TAC (AUG in mRNA) and ending with a **stop codon** ATT (UAA in mRNA), ATC (UAG) or ACT (UGA).

Six reading frames need to be checked as genes can be aligned in both directions on the double helix. Let us take an idealized DNA sequence as a starting point.

5'-TACATAGGAGTTGCCGTTAAATCCCATCTTACCACGACT-3'
3'-ATGTATCCTCAACGGCAATTTAGGGTAGAATGGTGCTGA-5'

This can be read in six frames as follows:

TAC/A ———▶ 1st frame
ACA/T ———▶ 2nd frame
CAT/A ———▶ 3rd frame
5'-TACATAGGAGTTGCCGTTAAATCCCATCTTACCACGACT-3'

Starting from the other end:

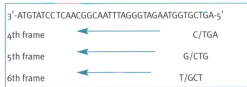

3'-ATGTATCCTCAACGGCAATTTAGGGTAGAATGGTGCTGA-5'
4th frame ◀——— C/TGA
5th frame ◀——— G/CTG
6th frame ◀——— T/GCT

Let us now translate the third frame into mRNA (green) and then into amino acids (blue):

5'-TA/CATAGGAGTTGCCGTTAAATCCCATCTTACCACGACT-3'
AU/GUA/UCC/UCA/ACG/GCA/AUU/UAG/GGU/AGA/AUG/GUG/CUG/A
Val Ser Ser Lys Ala Ile STOP!

This is not an open frame. The sequence UAG on the mRNA acts as a stop signal. The sequence is only partially translated. We could go through all the reading frames in this way. To cut a long story short – only the first of the six reading frames is open.

5'-TACATAGGAGTTGCCGTTAAATCCCATCTTACCACGACT-3'
AUG/UAU/CCU/CAA/CGG/CAA/UUU/AGG/GUA/GAA/UGG/UGC/UGA
Start Tyr Pro Gln Arg Gln Phe Arg Val Glu Trp Cys STOP!

Here it is – the first complete peptide, tiny though it may exist.

This is how it might work in bacteria, but in **eukaryotes**, there are huge gaps between genes, and often there are **random ORFs** that turn out not to be genes at all. Many genes are divided into exons and introns.

Once a gene has been provisionally identified, the database is searched for homologies, similar sequences in other organisms. When a homology is found, it often hints at the function of the gene. Even the nematode *Caenorhabditis elegans* and the fruit fly *Drosophila* contain genes that exist in humans.

10.23 Pharmacogenomics

Less than half of the patients who are prescribed extremely expensive drugs benefit from them, said **Allen Roses**, vice president of the Genetics Section at GlaxoSmithKline (GSK).

This is an open secret in the pharmaceutical industry (Fig. 10.46). Medication against Alzheimer's disease is effective in less than one in three patients, medication against cancer only in one in four. Drugs against migraine, osteoporosis, and arthritis work in only half the patients. "This is because the patients have genes that interfere with the action of the drugs. The overwhelming majority of all pharmaceutical products – more than 90 percent – are effective in only 30 to 50 percent of patients" says Roses. He is an expert on pharmacogenomics.

It is a well-known fact that the same medication can act differently in different people. The basis on

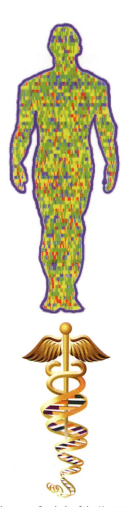

Fig. 10.42 Symbols of the Human Genome Project.

Fig. 10.43 The HapMap project is to include genetic variation in the human genome.

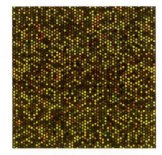

Fig. 10.44 Gene chips (DNA arrays) facilitate the genomic analysis of all organisms.

Box 10.8 Biotech History:
Craig Venter's Impatience or The Tagging of the Sequence

Enlightenment came to Craig Venter 10 kilometers (33,000 feet) above the Pacific when returning to the US from Japan. Venter's job at the National Institute of Health (NIH) was to identify a receptor for the stress hormone adrenalin on the surface of heart cells. Researchers had been trying for years to purify the proteins and find the gene – a slow and expensive business! In 1986, Venter flew to a company called Applied Biosystems (ABI) in California, which had developed a new sequencing machine. The new system was based on **Fred Sanger**'s method, but used fluorescent dyes instead of radioactive labels. ABI 373 A was able to analyze 24 samples simultaneously, deciphering approximately 12,000 letters of DNA per day. The machine, however, had a price tag of $100,000.

Craig Venter today

Early in 1987, Venter's NIH lab tested the new "toy" by sequencing rat genes that were related to the adrenalin receptor. The machine proved to be quicker and cheaper than the manual method used hitherto. However, Venter's ambition went further. His aim was to detect the most interesting genes, e.g., a segment on the long arm of the X-chromosome (Xq28) that seemed to be responsible for dozens of genetically transmitted diseases. Venter applied for further funding from the NIH Genome Center under the direction of the great **James Watson**. The response to his application was delayed further and further.

In the meantime, Venter started on two smaller projects that were to become of eminent importance for human genetics – one on the hereditary Huntington's disease on chromosome 4 and the other on muscular dystrophy on chromosome 19. In both cases, the genes he found were not the main culprits, but the sequencing machines did an excellent job on 60,000 and 106,000 bases respectively. One of Venter's collaborators at the time was **Francis Collins** who

was to become his rival in the Human Genome Project. In spite of this success, the sequencing procedure was still not fast enough for Craig Venter.

And what was the enlightenment we mentioned earlier? The trick was to enrich genes in order to avoid sequencing long stretches of non-coding or junk DNA, the way nature does it when transcribing DNA into RNA (Ch.3). Double-stranded DNA is transcribed into single-stranded messenger RNA, and the non-coding junk sequences or introns are eliminated. Only mature mRNA that makes sense, i.e., that codes for proteins, is transferred from the nucleus to the ribosomes where it is read (translated).

If the relatively instable mRNA molecules could be isolated from a cell and purified, they could be transcribed into complementary DNA, cDNA. Thus, a whole cDNA library could be established. As we have seen in Chapter 3, this could be done with reverse transcriptase. Eukaryotic mRNA (only one to three percent of the total RNA in a cell) could be easily separated through affinity chromatography, as it carries an oligo-A sequence at its 3' terminus. (….AAAAA-3'). The column contains an oligo-T matrix with TTTTT where the AAAAAs hybridize. They elute from the column with a delay.

Venter's idea was shockingly simple. With his colleague **Mark Adams**, he produced the cDNA library of a brain cell, containing tens of thousands of copies of genes that are active in the brain. This was done by transcribing mRNA into cDNA using reverse transcriptase. The cDNA was then inserted into bacterial plasmids and introduced into bacteria. The bacteria could now be grown in vitro into colonies, and individual colonies could be selected. Each colony contained the cDNA of an, as yet, unknown gene that is expressed in the brain. This DNA could now be sequenced. The 200 to 300 base long sequence was then compared to genes in other organisms that had already been identified by the publicly funded gene database. Every single cDNA fragment was registered as an **expressed sequence tag or** (**EST**). The idea was as simple as it was elegant.

Venter had dealt with the problem highly efficiently. He withdrew his funding application for the sequencing of the long arm of the X-chromosome, which still had not been processed after two years, complaining to Watson that in those two years, he could have sequenced two million letters. Then, in June 1991, he publicized his new EST strategy in *Science*. He had discovered

330 new genes that are active in the human brain. The nervous system is affected by approximately a quarter of the over 5,000 genetically transmitted diseases. At that time, the public NIH gene database contained the sequences of less than 3,000 human genes. Within a few months, going at alone, Venter sequenced more than ten percent of all known genes.

Shotgun Approach Charting New Waters

Once the human genome had been deciphered, Venter, a passionate sailor, bought himself a 90-foot yacht, Sorcerer II. In the summer of 2002, Venter and his crew took it for a test run into the Sargasso sea near the Bermuda islands.

Venter's interest lay in the ocean microbes. His expedition yielded astonishing results. The first six samples alone revealed more than 1.2 million new genes – almost ten times more than what had been known to date. The genes included 782 photoreceptor genes which code for enzymes that enable the microscopic organisms to turn sunlight into energy.

Instead of following the usual procedure of growing individual microbe cultures, the researchers fed the gene mix into their DNA sequencing machines at home – the entire genetic material, filtered out of 1500 liters of water. Venter was using his shotgun approach again that stood him in such good stead in the Human Genome Project.

It was indeed possible to put together complete gene sequences of entire organisms, and according to his analysis, the samples contained at least 1800 species, among which 148 totally unknown bacteria species. Venter has started off a new discipline in research – ecological metagenomics. Increasingly, biologists, and geneticists in particular, are homing in on the gene pool of entire ecosystems.

Box 10.9 DNA Chips

DNA chips and DNA μ arrays are nothing but an organized collection of DNA molecules, the sequences of which are known. They are usually arranged in a rectangle or square and may contain between a few hundred and tens of thousands of units (e.g., 60x40, 100x100 or 300x500. The location of each unit is precisely defined as a dot on the glass surface with a diameter of less than 200 μm. Each unit contains millions of copies of a clearly defined short DNA fragment. The information about where to find which DNA in the array is retrievable from a computer.

There are two distinct types of microarrays, using different technology:

The oligonucleotide probe is 'grown' onto silicon wafers one nucleotide at a time. Affymetrix uses this technology (shown in the figures).

In the second type, the oligonucleotide probes are pre-synthesized and are spotted onto microscope slides (glass or plastic) with a special coating.

The result in both cases are surfaces containing many thousand single-stranded short DNA pieces bearing different oligonucleotide sequences. The sample (RNA or DNA) is amplified (by polymerase chain reaction) and labeled with fluorescent labels (or tags). The labeled sample is then applied to the microarray. Each piece of sample DNA (or RNA) can only hybridize to its complementary oligonucleotide probe. In our simplified example, a single-stranded DNA-oligonucleotide GTACTA is bound to the chip and "fishes" the complimentary sample RNA fragment CAUGAU* carrying at the end a fluorescent label (*). It hybridizes well. However, an RNA fragment like GAGACA* would not hybridize here.

As each of the spots contains millions of oligonucleotide probes, the amount of labeled sample that binds within the spot is comparable to the amount contained within the original sample.

When the microarray is scanned and the fluorescence of each feature is recorded, the intensity of the spots is proportional to the hybridized sample.

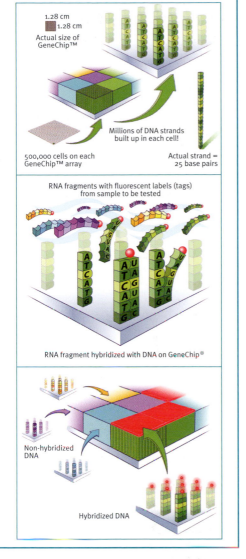

1.28 cm
1.28 cm
Actual size of GeneChip™

500,000 cells on each GeneChip™ array

Millions of DNA strands built up in each cell!

Actual strand = 25 base pairs

RNA fragments with fluorescent labels (tags) from sample to be tested

RNA fragment hybridized with DNA on GeneChip®

Non-hybridized DNA

Hybridized DNA

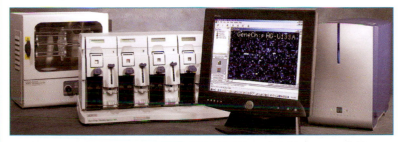

The Affymetrix GeneChip system consists of a gene or probe array, hybridization oven, fluidics station, scanner, and a computer workstation.

---

which the medication is prescribed is the disease. If the doctor could take into account the genetic disposition of the patient as well, it would revolutionize medical practice.

Side effects of drugs could also be reduced. If, for example, genome researchers could identify a group of genes involved in the development of lung cancer, these genes could be compared in healthy people and cancer patients. The differences (polymorphisms) between the two gene sequences may influence the probability of an individual of developing cancer. Often, it is just a matter of mutated individual base pairs, e.g., from A to G or from T to C (single nucleotide polymorphism, SNP).

The information can then be used for **diagnostic screening**. Patients with a higher risk of cancer can be warned, and doctors can find out which drugs

work best in which patients (Fig. 10.48). The -2AR gene, for example, determines how well asthma patients respond to albuterol. Albuterol opens the respiratory tract by relaxing the lung muscles. The relevant gene exists in four to five different variants (alleles), which explains why, in approximately 25 percent of asthma patients, albuterol does not work.

A test to find out the variant of the enzyme cytochrome P-450 (Ch. 4) in a patient could predict whether the patient will respond to certain antidepressants. Other tests could predict the reaction to frequently prescribed drugs for hypertension or migraine (Fig. 10.48).

Human Genome pioneer **Francis Collins** (Fig. 10.37) predicts that by 2010, tests for 25 common diseases will be available worldwide which will

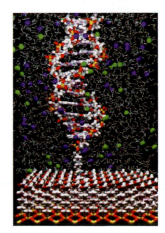

Fig. 10.45 A single-stranded DNA hybridizes with a DNA probe on a DNA chip.

Box 10.10 The Expert's view:
Alan Guttmacher about the Dawn of the Genomic Era

On April 14, 2003 the Human Genome Project officially ended, with the achievement of its ultimate goal, completion of the final sequence of the human genome. Completing the human genome sequence was a truly historic technological and scientific achievement and has already changed the face of biomedical research. As important as was the sequence itself, the Human Genome Project also made other similarly important, even if less obvious, contributions to biomedical research.

Genome **The End of the Beginning**

The Human Genome Project demonstrated that large, centrally managed projects could be beneficial not just in other areas of science such as physics, but also in biomedical research. It also demonstrated the related principle that while hypothesis-driven, investigator-initiated science will continue to be the major avenue for productive biomedical research, there is also a key role for research designed not to answer a specific hypothesis but, instead, to create a community resource that can then enable many lines of investigator-initiated research that will answer a broad array of questions – often questions that cannot even be imagined at the time that creating the resource is originally envisioned.

And the "community" nature of the resource created by the Human Genome Project was both a key to the success of the Project and a major contribution to biomedical research. A hallmark of the Project was that all of its data was made available every 24 hours to the entire global research community. This was a major move away from the still prevalent model that research data "belong" to the investigator. While needing to find appropriate ways to recognize, respect, and reward the effort, time, and intellectual cre-

ativity that the investigator invests in research, science moves more quickly – and society profits more rapidly – if research data are treated less as a personal trove only to be mined by the principal investigator and more as a community resource. This is particularly important in the current era of science, for, as the Human Genome Project exemplifies, we have entered an era when mining scientific data is perhaps even more of a challenge than gathering it.

The Human Genome Project also demonstrated the benefit of acknowledging and addressing the societal context and implications of scientific research. Through its Ethical, Legal, and Social Implications (ELSI) research, the Project ensured that a broad array of individuals with many different life experiences, types of training, and areas of expertise actively considered its potentially far-reaching societal impact. Both the Project itself and society benefited from this rich area of research and thinking that, even after the completion of the Human Genome Project, has continued to be a valuable part of genomics.

One more lesson of the Human Genome Project is that its completion marks not an ending, but only a beginning. As soon as the Human Genome Project was completed in 2003, many voices began to talk of us having now entered the "Post-Genome Era." While we do now have the human genome sequence in hand, it is more accurate to speak of us having just entered the "Genome Era." And the distinction is important.

We are just now poised on the brink of the era in which we can apply genomics - our knowledge of the human genome sequence and the many other scientific and technological advances wrought by the Human Genome Project - to better understanding biology and human health and disease and, perhaps even more importantly, to improving human health around the world.

Genome **Medical Futures**

Alan E. Guttmacher, M.D., is the Deputy Director of the National Human Genome Research Institute (NHGRI) and helps oversee the institute's efforts in advancing genome research, integrating the benefits of genome research into health care, and exploring the ethical, legal, and social implications of human genomics.

In 2003, Dr Guttmacher and the NHGRI's director, Dr Francis S. Collins, co-edited a series about the application of advances in genomics to medical care titled: Genomic Medicine for The New England Journal of Medicine.

Alan Guttmacher also oversees the NIH's involvement in the U.S. Surgeon General's Family History Initiative, an effort to encourage all Americans to learn about and use their families' health histories to promote personal health and prevent disease.

Dr Guttmacher received his M.D. from Harvard Medical School, completed an internship and residency in pediatrics and a fellowship in medical genetics at Children's Hospital Boston and Harvard Medical School. He is a member of the Institute of Medicine.

Recommended Literature:

Guttmacher AE, Collins FS (2002) *Genomic Medicine – A Primer*. New England Journal of Medicine, 19:1512-1520 can be downloaded from the web
http://content.nejm.org/cgi/content/full/347/19/1512
http://content.nejm.org/cgi/content/full/347/19/1512

Collins FC, Green ED, Guttmacher AE, Guyer MS (2005) *A Vision for the Future of Genomics Research*. Nature, 422:835-847, download from the web
http://www.nature.com/nature/journal/v422/n6934/full/nature01626.html
http://www.nature.com/nature/journal/v422/n6934/full/nature01626.html
The article contains 44 of the most important scientific papers with weblinks

help people to change their lifestyle. By 2020, genetically designed drugs will be available for diabetes, hypertension, and many other widespread diseases. By 2030, the genes that control the aging process will have been identified, and individual DNA sequencing on demand will cost less than $1,000. By 2040, medical treatment will generally be genetically based, and many diseases will be recognized even before symptoms appear, and individual drug or gene treatment plans will be worked out.

Personalized medical care is on the horizon. Apart from patients, however, there are other groups who would welcome such detailed genetic information – insurance companies, Human Resources Departments of corporations, governments and, secret services.

Collins argues that since each of us carries four or five totally messed-up genes and another half dozen of not very good genes that may involve certain risks, this information should not be used against us. After all, nobody can choose their genes.

Due to the high cost for clinical drug testing (hundreds of millions of dollars), the pharmaceutical industry has an interest in developing highly effective SNP diagnostic tools. This is particularly urgent because many drug trials fail in the later evaluation stages. At the heart of the medical revolution are **DNA chips** or **DNA microarrays**, no larger than a postage stamp (Fig. 10.47).

■ 10.24 DNA Chips

With the development of DNA chips, another revolution has begun. Using chip technology and laser scanning or CCD image analysis, the DNA patterns in a sample can easily be identified (Box 10.7).

DNA chips, also known as gene chips or DNA microarrays, were developed in the early 1990s. **Stephen Fodor** of Affymetrix Inc., a company in Santa Clara, California, came up with the idea of putting thousands of DNA probes on glass microchips in a similar way transistors are put on silicon (Fig. 10.47). Computer chip production methods were especially adapted for this purpose. Since then, DNA chip technology has developed in leaps and bounds. In ten years' time at the latest, our daily life would not be imaginable without DNA chips.

Gene chips are broadly used in genomic DNA analysis and the analysis of gene activities. Many DNA tests harness dozens or even hundreds of hybridization reactions to provide all the information required – all from a single minute sample. Thanks to sophisticated technology, the chip itself is extremely small and everything fits on it. The scope of its applications is colossal:

- In expression profiles, the entire RNA in a cell is assessed. "Which genes produce how much of what RNA, leading to which proteins at a certain time?"
- Mutation tests for DNA sequences, e.g., for AIDS and cancer research or pharmacogenomics
- SNP analysis

If a biochip carries enough DNA probes, DNA can be sequenced. This is known as **sequencing by hybridization** (**SBH**). The most straightforward practical application, however, is SNP analysis for tailor-made drugs.

■ 10.25 Identifying the Causes of Disease – Gene Expression Profiles

Expression profiles (see the yeast example, Fig. 10.43) are used to investigate the changes in gene activity underlying certain human diseases.

In order to find out which genes are active in healthy and diseased tissue, the **mRNAs** are isolated from both tissue samples. Using **reverse transcriptase**, cDNA copies are made and fluorescent labels added. These are transferred onto **gene chips** containing thousands of gene fragments. If the labeled cDNA fragments bind to their complementary fragments on the chip, the active genes become visible. A comparison of **expression profiles** can show the differences in activity in several hundreds of genes, and the question is which of these genes cause the disease.

Several methods are used in functional genome research to assess the candidate genes, i.e., those genes that are suspected to be the cause of the disease. In a first step, **databases** are consulted about the function of the genes that have been identified. Thus, genes that are known not to upset cell metabolism can be excluded right away. At the end of serial trials, only a few candidates remain which are then tested in **animal models**. The idea is to find out if the genes elicit the same symptoms in animals

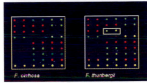

Fig. 10.46 A practical example of genotyping. The gene chip of Michael Yang (Hong Kong City University) can be used for the identification of Chinese medicinal herbs (top). *Fritillaria cirrhosa* (Chin. Chuan bei mu, center) is used for the treatment of chronic cough and tuberculosis.

The worryingly low efficacy of drugs, according to Allen Roses:

- Alzheimer's disease 30%
- Analgesics 80%
- Asthma 60%
- Cardiac arrhythmia 60%
- Diabetes 57%
- Hepatitis C (HCV) 47%
- Incontinence 40%
- Acute migraine 52%
- Oncology 25%
- Arthritis 50%
- Schizophrenia 60%

Fig. 10.47 Less than half of the patients who are prescribed extremely expensive drugs benefit from them.

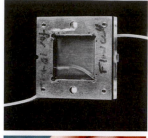

Fig. 10.48 Top: The first Affymetryx gene chip, developed by Stephen Fodor's team.
Bottom: a modern gene chip.

Fig. 10.49 Gene-based medication.

as in humans, and only if this is the case can the molecular involvement of the relevant gene in the disease be confirmed.

Since 1995, there has been a movement towards global gene expression analysis. The expression profiles of thousands of genes are monitored simultaneously. It will be possible to monitor the entire transcriptome, i.e., the entire mRNA of a cell. Thus, all genes involved in pathogenic processes could be identified.

■ 10.26 Proteomics

Transcriptome (mRNA) analysis, as described above, is a useful way of studying and characterizing disease and predicting reactions to medication or environmental changes. However, the actually functional molecules in a cell are not mRNAs, but proteins. A transcript does not always correspond to the relevant protein because protein synthesis is often regulated independently.

The terms **proteomics** and proteome analysis were suggested in 1995 and they represent the **assessment of the entire protein of a cell, tissue, or whole organism**. It is assumed that in a higher organism, each gene gives rise to the production of

an average of ten proteins. How the proteins are expressed, how they function and interact is the subject of proteomics or functional genomics.

Generally speaking, proteome analysis can be divided into two interlinked areas – isolation of samples and separation of single proteins and then the identification and detailed structural analysis of the proteins (looking for modifications).

Sample separation is usually carried out in two-dimensional (2D) **polyacrylamide gel electrophoresis** (**PAGE**). As in the agarose gel electrophoresis of nucleic acids, the proteins are separated according to hydrodynamic mobility (small proteins move faster), but also according to electric charge.

In contrast to nucleic acids, the charges of proteins are not uniform. Proteins with a negative overall charge migrate to the positive pole (anode), whereas the positively charged proteins migrate to the cathode during electrophoresis.

A particular variant of this method, **sodium dodecyl sulfate** (**SDS**) **PAGE** denatures proteins in the presence of mercaptoethanol (which reduces possibly existing disulfide bonds into SH groups). SDS is an anionic detergent (surfactant) that is completely dissociated. It disrupts nearly all non-covalent interaction within proteins (particularly of hydrogen bonds), thus "unfolding" all protein structures. **Micelles** form which mask the different charges in a protein. The negative charge of the micelle is thus – just as in nucleic acids – proportional to the size of the molecule. SDS-charged denatured proteins can therefore be separated in gels in the same way as nucleic acids.

Then the proteins are visualized in gel by dying them with Coomassie Blue. Blue bands appear, and even as little as 0.1 micrograms of a protein will be visible as a band.

Two-dimensional gel electrophoresis (**2D-GE**) (Fig. 10.51) is a more recent development which makes use of isoelectric focusing (IEF). A gel with a pH gradient (from low to high pH) is put in a vertical position, and the protein sample is applied. When a current is run through the gel, the proteins migrate to their isoelectric pH value, i.e., a position where their net charge is zero. A band pattern arises. Then the single-lane gel is placed horizontally on top of a SDS-polyacrylamide slab, and electrophoresis is carried out in the second dimension, perpendicular to the original separation. Proteins

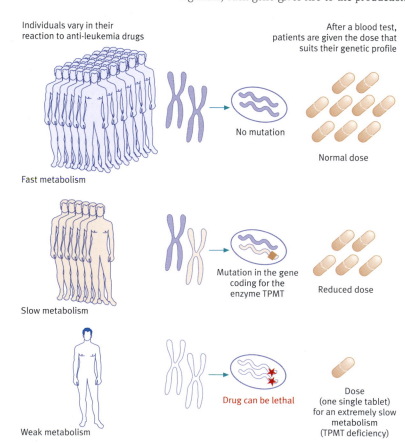

Individuals vary in their reaction to anti-leukemia drugs

After a blood test, patients are given the dose that suits their genetic profile

Fast metabolism

No mutation

Normal dose

Slow metabolism

Mutation in the gene coding for the enzyme TPMT

Reduced dose

Weak metabolism

Drug can be lethal

Dose (one single tablet) for an extremely slow metabolism (TPMT deficiency)

with the same charge are now additionally separated on the basis of mass. Thus, a two-dimensional dot pattern develops. The proteins have first been sorted in one direction, according to their charges and then, perpendicular to that analysis, they are separated according to their mass.

However, most researchers are not satisfied with 2D-GE, as it leaves much to be desired regarding accuracy and reproducibility. Another option is to add mass spectrometry. After separation through 2D gel electrophoresis, the resulting protein spots are cut out of the gel and digested by a protease (such as trypsin) into peptides. The protein can be identified by mass spectrometry via these peptides.

■ 10.27 MALDI TOF – A Gas from Protein Ions

Originally, mass spectrometers had been developed as a particularly sensitive method to identify ionized atoms. They have also been used for a long time in the analysis of small inorganic and organic molecules. However, the analysis of large biomolecules with mass spectrometry is only a recent achievement, as it was not easy to detach these molecules individually from their watery environment, transfer them into the vacuum of a mass spectrometer, and load them with an electric charge, as well as prevent them from disintegrating.

This feat could be compared to sending an astronaut into space without a protective suit. Such large charged molecules do not become volatile just like that! They need some coaxing. One of the two techniques currently in use – **matrix-assisted laser desorption ionization** (**MALDI**) was developed by **Franz Hillenkamp** (Fig. 10.53) and his collaborators at the University of Münster, Germany.

The proteins are inserted into UV-absorbing molecule crystals. This is done by mixing the matrix solution with another solution containing the protein molecules, on a metal carrier and waiting until the solvents have evaporated. The carrier with the protein-doped crystals is put into a high-vacuum device and exposed to short, intensive UV laser radiation. In a quasi-explosive reaction, the UV-absorbing matrix molecules, and thus the proteins are released into the vacuum, and a positive or negative charge is conveyed to the proteins in the process.

The ions produced by MALDI are analyzed in a **time-of-flight** (TOF) **spectrometer**. The ions fly

through an evacuated tube of roughly a meter (1.09 yard) in length. Their time of flight – typically around one millionth of a second – is measured. The protein ions have been accelerated beforehand by an electric field. Proteins with equal charge, but different mass, vary in flight speed. A protein ion with smaller mass will fly faster than one with larger mass, and a protein with two charges is twice as fast as the same protein with a single charge. Thus, the time of flight correlates with the mass/charge ratio (m/z) of the proteins. The time of flight is measured by the time-of-flight analyzer. A good MALDI TOF mass spectrometer measures the mass of a protein with an accuracy of up to 0.001 percent.

A time-of-flight mass spectrometer does the same job as a very fast and accurate SDS gel electrophoresis. Both determine the flight times of charged molecules. Mass spectrometry is the ideal complement to isoelectric focusing. The machines, however, are expensive.

There is a second technique for creating ions of large molecules. It is known as **ESI – electrospray ionization**.

■ 10.28 Aptamers and Protein Chips

Protein microarrays are produced in similar fashion as DNA arrays (Fig. 10.49). The arrays contain specific catcher agents such as antibodies that can be read in mass spectrometers such as MALDI TOF or recognized via fluorescent labels. However, protein chips have not reached the standard of DNA chips by a long shot. DNA chips can carry 10,000 or more spots that indicate the presence and the proportion of RNA in a cell extract.

Protein biochemist **Hubert Rehm**, editor of a review called *Laborjournal*, comments: "Protein biochemists would love to have such a toy – ideally a chip you just chuck into a cell extract, and half an hour later, it would tell you the number, variety, modification, and concentration of every single pro-

Fig. 10.50 Visionary idea of a protein chip.

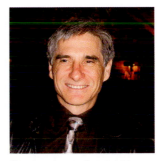

Fig. 10.51 The first 2-D gel electrophoresis.

Fig. 10.52 As a student, Patrick O'Farrell invented 2D-PAGE and succeeded in separating thousands of proteins from *Escherichia coli*. This is an important method, but unfortunately not easily reproducible.

Fig. 10.53 Franz Hillenkamp, University of Münster, built the first MALDI machine. Bottom: a high performance mass spectrometer.

Box 10.11 The Expert's View: David Goodsell About the Future of Nanobiotechnology

Richard Feynman: *"What would happen if we could arrange atoms one by one the way we want them?"*

Nanotechnology is the science of manipulating materials and creating tools on the atomic scale. The prefix "nano" means one-billionth. Atoms are about 0.10 nanometers, so the term nanotechnology seems fitting. Nanotechnology deals with materials which are less than one-thousandth the diameter of a human hair. In 1959, scientist Richard Feynman first highlighted the idea of nanoscience, arguing that studies must begin about equipment which can work at the atomic level.

Since then, the field has grown enormously. Worldwide spending on research and development of nanotechnology surpasses now $3 billion. The U.S. government has invested about $2 billion since 2000, when President Clinton announced the National Nanotech Initiative. And other countries, such as Japan and the European Union have likewise dramatically increased their funding. Within the decade, nanotechnology is expected to grow into a $1 trillion industry.

What will the nanotechnology of the twenty-first century look like?

Many people envision a technology of **macroscopic machines shrunken to nanometer size**: nanorobots with nanometer-scale gears, pulleys, gates, and latches; assemblers with rigid rectilinear struts and circular bearings; storage tanks surrounded by rigid walls of diamond. These machines emulate in every detail the machines that we use today in the macroscopic world. It is a compelling vision, filled with exciting prospects. **But, is it the only course?**

Instead, will nanotechnology be a forest of trees, capturing light to create plastic building materials, ceramic components, or entire dwellings? Will nanotechnology look like a stagnant pool, where cell-like nanomachines feverishly create tailored medicinal compounds, packaging them in custom delivery vessels? Nanotechnology may be a computer that runs not on electricity but instead on sugar and oxygen. Or nanotechnology may look exactly like a virus, but a virus customized to seek out and destroy cancerous tissues in each patient.

Bionanotechnology is a reality today.

Through a confluence of experimental knowledge from biology, chemistry, physics, and computer science, we understand the processes of like in sufficient detail to harness biomolecules for our own use. An entirely new realm of wetware, nanoscale machines that operate under physiological conditions, is open for the taking. Many of the goals of bionanotechnology and nanomachines may be described as augmented biology: We are looking for nanomachines to perform functions normally done by biomolecules or by entire cells, but to do these jobs even better. Wet-ware is perfect for these applications, because these jobs will be performed in environments that are consistent with biological molecules – warm, wet, and salty.

Looking forward, the possibilities, some speculative science and some still the realm of science fiction, are tremendous. We have barely scratched the surface.

A timetable for bionanotechnology

What might we expect in the future?

Of course, it is always dangerous to speculate, because unforeseen developments are at the hearts of most cultural and scientific shifts. Automobiles, trains, and airplanes are steadily shrinking and linking the world. The discovery of microscopic life and the subsequent international effort in antisepsis have doubled the length of our lives. Computers have made entire worlds of inquiry possible and raised important questions about our won minds. The world wide web has expanded our ideas of information and communication. In each case, a scientific or engineering advance opened a previously unimagined world. That said, what might we expect of bionanotechnology given foreseeable advances in our current understanding?

First of all, we can expect a solution to the **protein folding** problem. This will allow prediction of structure and function for a protein of arbitrary sequence, allowing the design of novel bionanomachines. This is a key step in bionanotechnology – extrapolating from existing machinery can only take us so far. By current expectations, effective computational methods for protein structure prediction should be expected in the next decade or so. And once natural proteins are understood, we can move on to the larger problem of improving and expanding the natural building materials of increasingly nonbiological applications.

Cellular engineering is another probability. Given the rapidly growing number of genomes and proteomes, we will have a full parts list of living cells in the near future. The coming decades will yield an understanding of how these parts are arranged and how they interact to perform the processes of life. With this understanding will come the ability to modify, tailoring new cells for custom applications. Already, bacterial cells have been engineered to create specific products, such as growth hormone and insulin. In the future, we will see cells that clean pollution, that make plastic and other raw materials, that fight disease, and that are used for countless other applications.

Organismic engineering will open many doors, but so far only natural methods are available. We have a long history of artificially accelerating evolution. People have bred organisms for centuries, creating organisms better fit to human welfare. With understanding of the molecular processes of development, the exciting possibility of engineering organisms from scratch will become possible. By directly modifying the genome of organisms, all manner of changes might be made. Already, genetic engineering is improving the properties of agricultural plants and animals, although raising important questions about safety. We may also move on to engineering our own bodies, for improved health and welfare.

What about unforeseeable advances?

For this, we might look to the gray areas of biology and speculate on areas that might lead to advances as further secrets of nature are revealed.

Thus far, biology appears to act at a level larger than the quantum scale, using deterministic processes. The indeterminacy of quantum mechanics is intimately involved in covalent bonding, reaction kinetics, and electron transfer, but in very predictable ways. We can express these in terms of bond lengths and rates of reaction or transfer, so that none of the miracles of quantum mechanics – such as information traveling faster than the speed of light when quantum mechanical states collapse – need to be taken into account. This does not mean, however, that they cannot be harnessed. Given that bionanomachines operate so close to the quantum scale, they are the perfect candidates for creation of a new quantum technology. Exciting concepts in quantum computer and quantum communication are being studied in theory and in physics laboratories. Bionanomachines may provide the pathway to translate these ideas into practical applications.

Consciousness is still a mystery that may hold unforeseeable surprises in the future. Consciousness is thought by some to be a consciousness as an irreducible property, perhaps a consequence of quantum indeterminacy or perhaps relying on something more metaphysical. As the study of neurobiology expands, thought and memory appear to be solidly rooted in cellular and molecular structure. If consciousness also turns out to be reducible to physical principles, creation of consciousness in artificial objects (beings?) will creative diverse opportunities. The subtlety and range of response in biological systems may be the most successful way to create this mysterious property.

Lessons for Molecular Nanotechnology

The speculative nanotechnology proposed by **Drexler** and other molecular nanotechnologists is based on a method of mechanically adding atoms one at a time to a growing structure, through the use of an assembler. **Richard E. Smalley** has presented two problems with this approach, which he terms the "thick fingers problem" and the "sticky fingers problem."

The "**thick fingers problem**" is based on the atomicity of all nanomachinery. Because the machine is made of atoms, it cannot have structural details that are finer than the size of the component atoms. This reduces the options for the small constellation of atoms in an assembler that directly interact with the atoms being added to a growing product. Smalley points out that to generate a general-purpose assembler, a variety of chemical environments will be needed and it will prove impossible to design stable robot fingers that can be placed together to create an atom-sized space.

The "**sticky fingers problem**" is based on the interaction of atoms, which is far different from the interactions of familiar objects.

Atoms are sticky. When they get close to one another, they form stable interactions through dispersion forces, hydrogen bonds, and electrostatics. Smalley points out that the interaction with the robot may be just as strong as the interaction with the product, so atoms may remain stuck to the robot arm. Imagine an analogy in our real world. Take a bag of marbles and coat each with a thick layer of rubber cement. Now, with fingers also coated in glue, try to build a pyramid on the table. Challenging.

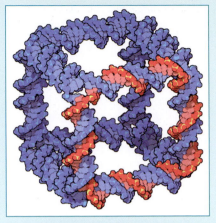

DNA can be used as an attractive nanoscale building material for creating defined multidimensional structures: DNA is programmable and predictable and DNA oligonucleotides of any desired sequence are readily synthesized by automated solid support techniques. DNA double helices are stiff polymers, at least in the range of a few turns of the double helix, or about 10 nm. Under physiological conditions DNA is relatively stable and the many natural enzymes are available for its manipulation.

Seeman has pioneered the use of branched DNA structures to create multidimensional objects taking a modular approach. The building blocks are designed from several strands that assemble to form a cross-shaped complex. The sequences of each leg of the cross must be unique to ensure that only one final structure is formed.

Sticky ends are left on each branch and are used to link blocks together into larger structures. By matching the sequences of these sticky ends, assemblies of any shape may be designed. Large structures of DNA are built with modular subunits with sticky ends.

A cube is created of eight three-armed modules as shown at the bottom. When DNA ligase is used to link the modules together, the structure is composed of eight topologically linked DNA circles.

Origami (from Japanese *ori*, folded, and *kami* paper) as a traditional folk art pervaded the Japanese culture more strongly than any other, but had its origins in China two thousand years ago.

This modern origami shape is an octahedron with a cube-shaped opening in the center. It was originally created by Lewis Simon, a member of the West Coast Origami Guild, the replica was made by Louiza Law, Hong Kong.

Of course, the fact that you are sitting here reading these words is proof that both of these problems have solutions. These problems may prove insurmountable for the diamondoid-based mechanosynthesis as envisioned by molecular nanotechnology, but real, working solutions for creating objects from the bottom up by direct manipulation of individual atoms have been developed by biological systems.

The solution discovered by nature is hierarchical, using atom-level nanotechnology when possible and self-assembly when it is not.

Enzymes must solve both of these problems. They must create a molecule-sized active site using only atoms, and they must capture their substrates and release their products. Active sites are created with a large overhead of protein infrastructure around the active site, using hundreds of times as many atoms as are needed to interact directly with the substrate. This large infrastructure allows for the precise placement of atoms to form the active site, to tolerances much smaller than the radius of the atom.

Products are released by carefully tailoring the binding strength: The active site is designed to bind tightly to the unstable transition state but not to the products. The shape of the active site favors reaction and then release. In many cases, however, the release of products remains the slowest step of the reaction, showing that nature is still plagued by sticky fingers.

Enzymatic assembly is useful for a certain class of reactions. In general, the enzyme must be able to surround the molecules being modified, creating an enclosed chemical environment. This is excellent for creation of custom organic molecules and for creation of linear polymers.

But for the creation of large, three-dimensional objects, the approach fails. Enzymes are not generally effective when faced with a flat wall and asked to make changes. So self-assembly is used to create larger structures.

Design of protein and nucleic acid polymers that spontaneously fold into stable, globular structure allows the design of modular units, which then self-assemble into objects of any desired size. These 10- to 100-nm modules are far easier to manipulate than individual atoms, and a variety of modification machinery can lay bricks, modify surfaces, activate modules on site, cross-link modules once assembled, and countless other variations.

See next page

As we work to design synthetic methods for nanotechnology, there are **two key lessons to be learned from biology,** lessons that are also learned from chemistry.

First, **combination of specific atoms into molecules** is a difficult and challenging task. In both biology and chemistry, each new molecule, each new molecule, each new bond, requires the design of a custom technique. By all expectations, if we desire to build objects atom by atom, we will have to employ a large set of construction tools tailored specifically for each new assembly task. If, however, we are willing to step up one level coarser and use polymers to build our nanoscale objects, the construction task becomes immeasurably easier.

Then a single synthetic reaction may be used in all cases, but a wide variety of monomer units can be used to create a variety of final products. With polymers, we cannot choose any arbitrary combination of atoms. We are limited to the polymeric linkage scheme that we choose. But we gain incredible ease of synthesis and flexibility of design specification. Looking to nature and chemistry, we see a combination of the two tech-niques: designing specific molecules atom by atom through laborious design of appropriate enzymes and use of proteins, plastics, and other polymers when larger structures are needed.

Final Thoughts

The potential of bionanotechnology for feeding the world, for improving our health, for providing rapid and cheap manufacturing with environmental mindfulness, is immense. But we must temper this excitement with careful thought.

The philosophy of manifest destiny, that all things are provided for our use, is an integral part of Western culture, and is often followed without introspection and though as to the potential outcomes. It would serve us well to look to Nature - to her world-spanning interconnectedness, to her unassuming creativity, to the sheer wonder of her accomplishment-for guidance as we proceed, tempering the strong cultural forces of novelty and capital gain.

Cited Literature:

Goodsell D S (2004) *"Bionanotechnology-Lessons from Nature"*, Wiley-Liss, Hoboken

David S. Goodsell is a molecular biologist and Associate Professor at Scripps Research Institute in La Jolla, California. His lab researches drug resistance in HIV, which involves studying both the structure and function of biomolecules in the disease. With funding from the National Science Foundation, Goodsell also writes and creates the illustrations for a column called " Molecule of the Month" for the Protein Data Bank, an archive of the 3-D protein structures of 20,000 different molecules. Professor Goodsell has a lifelong passion for art. His paintings, drawings, and computer-generated illustrations of molecules and cells have been displayed in galleries and featured on the covers of magazines and science journals.

We are very fortunate that he contributed his phantastic art and accurate science and his creative passion to this book. Thank you, David! RR

Legend:
- CHAPTER 1
- CHAPTER 2
- CHAPTER 3
- CHAPTER 4
- CHAPTER 5
- CHAPTER 6
- CHAPTER 7
- CHAPTER 8
- CHAPTER 9
- CHAPTER 10

Biotech Tree:
The Roots and Fruits of Biotechnology

Reinhard Renneberg and Manfred Bofinger, from „Liebling, Du hast die Katze geklont", (Honey, you have cloned the cat) with kind permission from Wiley-VCH Weinheim 2006

tein in the mixture. Some more demanding colleagues even ask for chips that would also give the concentration of metabolites such as glucose, lactate, etc as well as that of the oligosaccharides and polysaccharides […] However, this would require more than a hundred thousand different monoclonal antibodies…" So the conclusion is – it is simply unfeasible!

Never mind – there are other possibilities, such as **aptamers**. Aptamers are DNA or RNA oligonucleotides (15–60 nucleotides) that bind specifically to proteins. Copious amounts of them are produced very easily and cheaply in DNA synthesizers. Aptamers have come to replace antibodies in many areas. Aptamers that bind to Zn^{2+} can be just as easily produced as aptamers binding to ATP, peptides, proteins, and glycoproteins. Although they are less specific than antibodies, they can be used as binding indicators (Fig. 10.54). At the 5' terminus of a nucleotide, the aptamer is labeled with a fluorescent dye, and a quencher is added to the 3' terminus.

A quencher is a molecule that "extinguishes" the fluorescent dye. In the absence of the analyte, the aptamer forms a loop, and the fluorescence is quenched. By contrast, if the analyte binds to the molecule, the quencher is separated from the fluorophore, and the aptamer glows brightly.

10.29 Is Total Control Over the Human Genome Possible?

Pessimists like **Stephen Hawking** think that there haven't been any significant changes in human DNA in the past 10,000 years. However, he feels that we soon will be able to increase the complexity of human DNA, without having to wait for biological evolution which is slow.

He thinks we will be able to completely redesign it in 1000 years, possibly by increasing brain size, among other things. He considers the possibility that human genetic engineering will be banned but doubts that it can be stopped. For economic considerations, he feels that plant and animal genetic engineering will be permitted and that it will then be tried on humans and humans will thus be improved.

Optimists like **James D. Watson**, on the other hand, think that genetics *per se* cannot be bad and that the moral consideration only applies when we misuse genetics.

10.30 *Quo vadis*, Biotech?

"In the future, computers may only weigh one and a half tons" was a prognosis given in 1949.

With this in mind, we will try to tread carefully when predicting what's in store for us in the future. I am basing this forecast on an analysis by **Oliver Kayser** at the University of Groningen, The Netherlands.

I leave it to the reader to come to their own verdict. Here is an overview of **prognoses** coming from a large variety of sources.

- **In beer, wine, bread, and cheese** production (Ch. 1), the use of genetically engineered microbes will increase. Good, reliable strains will be modified, and the taste of the products will, if anything, be enhanced!

- In **agriculture** (Ch. 7), significantly fewer pesticides and fertilizers will be used and farmers will no longer be exposed to so many harmful substances. On the downside, biodiversity will continue to decrease dramatically, and the dependency on all-in-one seed-fertilizer-pesticide-biotechnology companies will increase. The latter, however, is an economic trend that cannot be blamed on new technologies.

- **Biodegradable plastic** and microbes that clean up the environment will reduce the burden on the environment (Ch. 6).

- Biotech will have an enormous impact on the **pharmaceutical industry** (Ch. 4 and 9). Genetic engineering, vaccine production (Ch. 5), and diagnostics will revolutionize the treatment of heretofore untreatable diseases. Entirely new biotech drugs are being developed.

- Biotech pharmaceutics make up currently five percent of the enormous pharmaceutical market worldwide. This share will probably rise to 15 percent by 2050.

- Since 1995, the number of **biotech patents** has been increasing by 25 percent per year.

- Effective **vaccinations against AIDS and malaria** will become possible (Ch. 5).

- With the incredibly fast growth of **gene diagnostics** (Ch. 10), it will be possible to produce an individual genetic profile within the hour for no more than 100 Dollars.

Fig. 10.54 ENIAC (Electronic Numerical Integrator And Computer), was the first large-scale electronic digital computer that could be reprogrammed to solve a full range of computing problems. ENIAC was designed and built to calculate artillery firing tables for the U.S. Army. The machine marked a turning point in computing. ENIAC weighed 30 tons, drawing enough energy to dim the lights of Philadelphia when it was run (operated at 5000 instructions per second).

Fig. 10.55 Stephen Hawking (b. 1942) communicates with eye movements via his computer: "…someone will improve humans somewhere…"

Fig. 10.56 Cutting-edge research by Peter Fromherz: A neuron from a freshwater snail (*Lymnaea stagnalis*) is part of a neuronal network, immobilized in a cage on a chip which measures the stimulation of the nerve cells as well as the resulting signals. The neurons grow together, forming electric synapses.

Fig. 10.57 Bioelectronics is at the interface between chip technology, biotechnology, and informatics and is set to produce major biotechnological breakthroughs in the near future.

Fig. 10.58 Nerve cell from a rat brain on a silicon chip. In the center, field effect transistors are visible, which measure the neuronal signals.

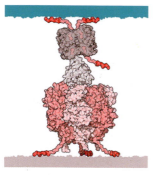

Fig. 10.59 Researchers at Cornell University built a nanopropeller out of inorganic materials fuelled by ATP synthase. The tricky bit was finding a way to connect the biological to the inorganic components. First, a support for the machine was created by electron beam lithography. It was composed of an array of nanoscale posts, 50-120 nm in diameter and 200 nm tall.

These were coated with nickel to provide a hold to attach the motor. Then an engineered version of ATP synthase was created, with histidine-rich tails attaching the motor to the posts. Finally, the nanopropeller was attached to the axle of the motor.

The propeller was a rod 150 nm in diameter and 740-1400 nm in length, coated with nickel. A molecular linker was created, composed of a histidine-rich peptide and biotin. The peptide bound to the propeller, while streptavidin was used to link the biotin on the nanopropeller. Of the 400 posts created on the chip, 5 sites were seen to rotate when ATP was added. The working propellers turned for several hours before the propellers broke loose.

- Hundreds of new genetic and immune tests will make blood products safer. Cheap **biotest strips and biochips** will be able to test several parameters simultaneously within minutes (e.g., the case of heart attacks). Medically untrained people will be able to carry out these tests.

- **Personalized medicine and diagnostics** on biochips will be the main focus of the industry over the coming decades.

- **Proteomics and pharmacogenomics** will detect heretofore unknown markers that will enable the treatment of illnesses before the first symptoms show.

- There will be an end to the chronic shortage of donor organs, due to the availability of organ **xenotransplants** from transgenic animals.

- **Stem cells** (Ch. 9) will be crucial for the treatment of Parkinson's and Alzheimer's disease, leukemia and genetic defects such as adenosine deaminase deficiency (ADA) and cystic fibrosis (CF). However, a decisive breakthrough is not expected to happen within the next ten years. The discussion of the ethical and social implications of stem cell treatment will continue. The public needs to be won over, and the risks involved need to be carefully assessed. The obvious benefit for patients will certainly work in favor of the new technology. However, serious reservations that do exist in various quarters must be respected.

- The benefit of biotech for the individual as well as society will become apparent. More and more people will benefit from life-saving biotech, which will raise the acceptance level of the industry.

- The development costs for a pharmaceutical product are currently $880 million dollars, and it takes 15 years from the beginnings to a market-ready product. 75% of the costs are caused by drug failures. Genome technology could help reduce the cost to $500 million and cut development time by 15%.

- The total global market for protein drugs was worth $47.4 billion in 2006. Sales will reach $55.7 billion by the end of 2011 – an average annual growth rate (AAGR) of 3.3%.

- By 2015, 30% of the current low-molecular drugs will be replaced by genetically engineered medications.

- From 2010, essential human proteins will be produced by **gene-pharmed plants and animals** (Ch. 7 and 8).

- **Gene therapy** and **nanorobots** are expected to become available between 2010 and 2018.

The most intriguing arena in modern science seems to be **nanobiotechnology**. Biomolecules are used to build molecular machines, e.g., nanobio-engines that are run on ATP just like bacterial flagella (Fig. 10.59 and Box 10.11).

Biosensors for glucose are the first bioelectronic products on the market. The latest development are neurons growing on chips that transfer signals to the actual chip (Figs. 10.57 and 10.58).

Epilogue

As prophesized Victor Hugo –
There is nothing so powerful as an idea whose time has come.
The time for the biotechnology idea has come, and we will all witness amazing developments!

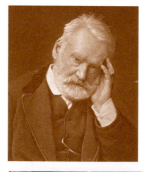

Victor Hugo (1802 - 1885) was a French poet, novelist, playwright, essayist, visual artist, statesman, and human rights campaigner.

Hugo is sometimes identified as the greatest French poet. His work touches upon most of the political and social issues and artistic trends of his time.

Cited and Recommended Literature

- **Watson JD, Berry A** (2004) *DNA. The secret of life*. Arrow Books, London. The whole story of DNA with an outlook. Perfect reading, addictive!

- **Campbell AM, Heyer LJ** (2006) *Discovering Genomics, Proteomics, and Bioinformatics* (2nd edition) Benjamin Cummings, San Francisco. Combines integrated Web exercises with a problem-solving approach to train readers in basic hands-on genomic analysis. The authors present global problems, then provide the tools of genomic analysis.

- **Boyer RF** (2000) *Modern Experimental Biochemistry* (3rd edn.) Prentice Hall, Upper Saddle River. Best practical introduction to important lab techniques in biochemistry.

- **Primrose SB, Twyman RM** (2004) *Genomics. Application in Human Biology*. Blackwell, Maden. The whole story of DNA with an outlook. Perfect reading, addictive. Best textbook in genomics, maybe a bit difficult for beginners.

- **Goodsell DS** (2004) *Bionanotechnology. Lessons from Nature*. Wiley-Liss, Hoboken. By the ingenious creator of the biomolecule graphics in this book, Scripps Professor David Goodsell.

- **Watson JD, Baker TA, Bell SP, Gann A, Levine M, Losick R** (2004) *Molecular Biology of the Gene*. (5th edition) Pearson/Benjamin Cummings, San Francisco. The updated classical DNA book with interactive animations, structural tutorials and exercises in critical thinking.

- **Davies K** (2001) *Cracking The Genome: Inside The Race To Unlock Human DNA*. The Free Press, New York, London. Kevin Davies, founding editor of *Nature Genetics*, gives an insight into the personalities and technologies involved in the great DNA sequence race, which makes exciting reading.

- **Cummings MR** (2003) *Human Heredity. Principles and Issues* (6th edn.). Thomson-Brooks, Belmont. The fascinating world of human genetics, well explained.

- **Jeffreys AJ** (2005) *Genetic Fingerprinting*, Nature Medicine 11, 1035-1039. Can be downloaded as pdf from: *http://www.nature.com/nm/journal/v11/n10/full/nm1005-1035.html* The best overview, written by the inventor of DNA profiling himself.

- **Schena M, Shalon D, Davis RW, Brown PO** (1995) *Quantitative Monitoring of Gene Expression Patterns with a Complementary DNA Microarray*. Science. 270 (5235): 467-70. Essential article about DNA chips.

- **Guttmacher AE, Collins FS** (2002) *Genomic Medicine – A Primer*. New England Journal of Medicine 19:1512-1520; can be downloaded from web *http://content.nejm.org/cgi/content/full/347/19/1512*

- **Collins FC, Green ED, Guttmacher AE, Guyer MS** (2005) A Vision for the Future of Genomics Research. Nature 422:835-847; download from web *http://www.nature.com/nature/journal/v422/n6934/full/nature01626.html* The article contains 44 of most important scientific papers with weblinks.

- **Rehm H** (2006) *Protein, Biochemistry and Proteomics* (The Experimenter Series) Academic Press, New York. By the legendary publisher of Laborjournal (now available in English *http://www.labtimes.org*). It provides the reader with tips and tricks for more successful lab experiments using a relaxed yet professional tone. A must-have for students and researchers interested in protein biochemistry and proteomics!

- **Müller HJ, Roeder T** (2005) *Microarrays* (The Experimenter Series). Academic Press, New York. Protein and DNA microarrays in the lab.

Useful weblinks

- For a start, the Wikipedia, many useful weblinks: *http://en.wikipedia.org/wiki/DNA*

- Perfect online teaching about DNA: The Cold Spring Harbor Laboratory Dolan Learning Center: *http://www.dnalc.org/home.html*

- Phantastic website of National Human Genome Research Institute with great animations for beginners: *www.genome.gov/*

- Sir Alec Jeffreys' website: *http://www.le.ac.uk/genetics/ajj/index.html*

- Pharmacogenomics explained, with weblinks: *http://www.ornl.gov/sci/techresources/Human_Genome/medicine/pharma.shtml*

Eight Self-Test Questions

1. Which disease is on its way to becoming the most common disease of all, and how can biosensors help in its identification and control?

2. Which chemical substance is detected in a pregnancy test? Does the test work immediately after the fertilization of an egg cell?

3. Which property of DNA and its fragments permits separation in a gel through which an electric current runs?

4. Which are the two basic methods of multiplying DNA?

5. How many DNA copies can be produced within an hour, using one of the methods in the previous question, supposing once cycle runs for three minutes?

6. What happens with protein molecules in MALDI?

7. How does the shotgun method differ from the contig method in the Human Genome Project, and which of them yielded quicker reliable results?

8. What do genomics and proteomics stand for?

CREDITS

Ian Billings, Norvic Philatelics, design by John Billings; Direcção dos Serviços de Correios Macao; Australia Post

Chapter 4

4.2 DG; 4.3 © The Nobel Foundation; 4.4 Kyoto Encyclopedia of Genomes and Genes (KEGG); 4.6 Biochemical Pathways: with permission of Dr Friedhelm Huebner, Roche Diagnostics GmbH, *www.roche-applied-science.com*; 4.7 DG and RR; 4.8 DG; 4.9 DG and RR; Box 4.1: KS; 4.10 DG and The Oncologist; 4.11 DG, RR and FB; Box 4.2: Woodruff B; FB; Ausmees N, Uppsala University (photo Streptomyces); 4.12 DG; 4.13 FB; 4.14 Ellis D,Univ Adelaide; CM; 4.16 FB; 4.17 Kyoto Encyclopedia of Genomes and Genes (KEGG); Box 4.3: FB, RR und WACKER Fine Chemicals, CM; 4.18 und 4.19 FB; 4.20 Kyowa Hakko Kogyo; 4.21 modified after Hopwood DA (1984); Corynebacterium by Sahm H, Forschungszentrum Jülich; 4.22 Paul Struck (graphics „Walpurgisnacht"); Pühler A (genome), British Mail; Box 4.4: Kyowa Hakko; Kinoshita S; 4.24 Wan C, Hong Kong; 4.27 Gerhard Gottschalk; 4.28 David Malin; 4.29 Walther A and Wendland J; 4.31 FB und © The Nobel Foundation ; 4.33 FZ Jülich; Box 4.5: Roche Basel, RR und MB; 4.34 and 4.35 Coca Cola Inc.; 4.36 FB and RR; 4.37 Fraunhofer Institut für Grenzflächen und Bioverfahrenstechnik Stuttgart und Semartec Ltd.; Box 4.6: DG, The Oncologist, and RR; 4.38 akademie spectrum, Berlin; Box 4.7: modified after Hopwood DA (1984), photos by Gesellschaft für Biotechnologische Forschung (GBF); 4.39 GBF Braunschweig; Box 4.8: © The Nobel Foundation; The New York Botanical Garden; University of Pennsylvania; 4.40 © The Nobel Foundation; MB; 4.42 Gist-Brocades; 4.43 DG Box 4.10: Fa.WECK and AKD; 4.45 Dept. Agriculture and Agrifood, Government of Canada; 4.46 Goodsell D (2004) Bionanotechnology, with permission from Wiley-Liss, Hoboken; 4.48 FB; Box 4.11: Goodsell D (2004) Wiley-Liss and RR; photo: GBF; Box 4.12: modified after Gaden EL jr. (1984) in: Gruss P et al; photos: Roche Penzberg, and GBF; 4.49 GBF; 4.50 © The Nobel Foundation; 4.52 Ellis D, Univ Adelaide; 4.54 Novartis AG, Basel; 4.55 after Aharonowitz Y and Cohen G (1984) in: Gruss P et al; FB (structure); Mayo Clinic, Rochester (photo); Box 4.13: Carl Djerassi; FB;

Portrait of Carl Djerassi © David Loveall, *www.loveallphoto.com.*; 4.57 FB; 4.58 KS; Ellis D, Univ Adelaide; cartoon page 135: CM and RR; map on page 136: Roche Biochemical Pathways , with permission of Dr Friedhelm Hübner, Roche Diagnostics GmbH, *www.roche-applied-science.com.*

Chapter 5

5.3 South China Morning Post, Hong Kong; 5.4 CM; Box 5.1: RR and from Goodsell (2004) Wiley-Liss; Box 5.2: Preiser W, and Korsman S; , DG, Dennis Kunkel (photo HIV), Mark Newman (maps); 5.8 to5.10 Goodsell D (2004) and Wiley-Liss; Box 5.3: RR and DG; modified after Breitling F and Dübel S (1997) Rekombinante Antikörper, SAV, Heidelberg; Box 5.4: From Guns, Germs And Steel, The Fates Of Human Societies by Jared Diamond. Copyright © 1997 by Jared Diamond. Used by permission of W. W. Norton & Company, Inc. Diamond J; AKD; Internet, Diamond J (portrait); 5.12 and 5.13 DG; Box 5.5: © The Nobel Foundation and UV; 5.14 modified after Brown TA (2002); 5.15 UV; 5.16 and 5.17 World Health Organization; 5.19 Bayer AG; Box 5.6: modified after Breitling F und Dübel S (1997); 5.21 GBF; Box 5.7: © The Nobel Foundation; 5.22 RR und FB; 5.26 DG und UV; 5.27 MB; Box 5.9: Ligler FS; BioHawk: Courtesy of Research International of Redmonton, WA USA;5.28 and 5.29 modified after Watson JD et al(1993); GBF; 5.30 Smith GP; 5.32 DG; GBF; 5.32 DG; 5.33 UV, Proceedings of the Royal Society London

Kapitel 6

6.1 © The Nobel Foundation; 6.2 AKD und MB; 6.3 KS; 6.4 AKD, Meinicke I und Bernitz H-M (1996) Der Gemüsegarten Berlins. Ausstellungskatalog. Rangsdorf; Box 6.2; Strobel G, FB; CM; 6.9 Bayer Leverkusen, Werk Bürrig; 6.12 United Nations Environmental Program (UNEP); 6.17 Internet; 6.18 Chakrabarty A; 6.19 modified after Hopwood DA (1984) in: Gruss P et al; 6.20 Gundlach E (*www.oil-spill-info.com*); Box 6.4: Reprinted from: Pespectives on properties of the Human Genome Project (Kieff, FS, Olin, JM, eds.) (2003) Elsevier Academic Press; title of the article: Chakrabarty A M: Patenting life forms: yesterday, today, and tomorrow; excerpts from pages 1-11; with permission from Elsevier; Chakrabarty A (portrait); Kunkel D

(Pseudomonas); British Mail; 6.24 FB; Box 6.5: KS, VW do Brazil, CM; 6.25 FB; 6.26 KS; Box 6.6: AKD; 6.28 to 6.30 Kennecott Utah Copper/Minerals Corp.; Box 6.7: UV, RR, Saab and Ghisalba O; 6.37 DG und RR; 6.39 RR, MB, FB; page 194 maps by Mark Newman, University Michigan

Chapter 7

7.1 Goldscheider S, Biothemen, Karlsruhe; 7.3 Grassmeier D, Spirulife; Box 7.1: DG and RR; 7.7 Polle J and Hutt Farm (Australia); 7.10 Imperial Chemical Industries (ICI); Box 7.2; ICI, and Petrolchemisches Kombinat (PCK) Schwedt; 7.12 ICI; 7.13 Marlow Foods; 7.14 modified after Bourgaize D, Jewell, T R and Buiser RG (2000) Biotechnology. Demystifying the concepts. Addison Wesley Longman, San Francisco; 7.16 Gratschow W; 7.17 KS; Wan MKC; 7.18 Wan MKC; 7.20 KS; Box 7.3: excerpts from Biotechnology, An Introduction (with InfoTrac), 2nd edition by Barnum SR 2005. Reprinted with permission of Brooks/Cole, a division of Thomson Learning: *www.thomsonrights.com.*; 7.22 Stanley J; 7.25 Max-Planck-Institut für Züchtungsforschung, Köln (MPIZ); Box 7.4: RR; Lewen-Doerr I,GreenTec, Heide L; MB; 7.29 KS; 7.31 RR, and Wellmann E, Universität Freiburg; 7.32 van Montagu M; 7.33 MPIZ Köln; 7.35 Wang Z-Y, reproduced from Critical Reviews in Plant Sciences, 2001, 20 (6) by permission of Taylor & Francis Group, LLC, *www.taylorandfrancis.com.*, Inc.; Box 7.5: Ingo Potrykus, Vatican; Box 7.6: Matthias Lehmann, 8sens biognostic GmbH; 7.37 Renneberg I, Berlin; 7.39 FB and DG; 7.40 MB; Box 7.7: from Greenfield S (2004) Tomorrow's people. Penguin Books , London , pages 23-25, 27-28, © Susan Greenfield. Used with permission by Baroness Susan Greenfield; artwork by Coneyl Jay, Science Photo Library, Motiv T395/126 "Micro-syringe" (c) SPL by permission of Photo-und-Presseagentur GmbH Focus; 7.42 BioSicherheit / Peter Ruth (upper photo), Timo Wolf (below); 7.43 BioSicherheit / Norbert Lehmann; 7.44 bioSicherheit / Stephan Kühne, BBA Kleinmachnow; 7.45 DG; 7.48 Calgene, Inc.; 7.49 attac Germany; 7.50 RR, and BioSicherheit; Box 7.8: Knäblein J; modified after Trends in Biotechnology (2002) KS; 7.54 modified after Biosicherheit; 7.55 Potrykus, I.; 7.56 The Samuel Robert Noble Foundation, Inc.; Box 7.9: Süßbier, D;

YorkSnow, Inc.; 7.57 Lindow S; Box 7.10: Satkula S; KS; TransGen/biosicherheit, Internet; 7.58 TransGen; 7.59 MB

Chapter 8
8.3 Friedrich R, Giessen; 8.5 Ma, KYC; 8.6 and 8.7 Cincinnati Zoo; 8.8 Anderson G, University of Califonia at Davis; 8.12 KS; 8.13 Konrad M; 8.14 Roslin Institute, Edinburgh 8.15 Wadsworth L; 8.17 Hwang W-S; 8.19 Wadsworth L, and RR; 8.23 Andrew Miller;
Box 8.1: Brinster R; Box 8.2: DG; RR; GloFish, Yorktown Technologies; 8.26 and 8.27 Norman McLean N, University Southampton; 8.28 and 8.29 Choy L Hew, National University of Singapore; 8.31 Blüthmann H, Roche Center for Medical Genomics; 8.32 Roslin Institute, Edinburgh; 8.33 Kruszelnicki KS, Great moments in science, full text at *www. abc.net.au/science/k2/moments/default.htm*; Box 8.3: Wolf E, Pfeifer A, Universität München; Köppl, T.; 8.34 biodidac, Ottawa, and RR; Box 8.4: © The Nobel Foundation (Paul Berg), Internet (1), Larry Wadsworth (2); 8.35 Internet; 8 36 KS; 8.38 RR, and Lindholm SB (bottom); Box 8.5: after Dolly, The Uses And Misuses Of Human Cloning by Ian Wilmut and Roger Highfield. Copyright © 2006 by Ian Wilmut and Roger Highfield. Used by permission of W. W. Norton & Company, Inc.; photos Roslin Institute, and National Museum of Scotland; 8.40 Roslin Institute; 8.41 CM, and Wadsworth L; 8.42 Wadsworth L; 8.43 biodidac, Ottawa; Box 8.3 CM: (cartoon), Wikipedia, with permission by author of picture Jon Sullivan; RR; Amdam GV (portrait); 8.44 and 8.45 Wadsworth L; 8.47 CM ; Box 8.7: modified after Thieman WJ and Palladino MA (2004) Introduction to Biotechnology. Pearson Benjamin Cummings, San Francisco; 8.48 KS; 8.50 Kwong A; Box 8.8: Richardson MK, University Leiden (modern embryos), KS (Haeckel); Bank of England; Museum of Natural History, Manhattan (Darwin's sketch); 8.51 Wolf E, Universität München; 8.52 Behncke-Braunbeck M

Chapter 9
9.3 and 9.4 DG; 9.5 modified after Dingermann T (1999); DG; 9.6 KS; 9.8 Paion AG; Box 9.1: Perlitz U ; 9.9 DG; 9.10 AKD; 9.12 DG; Box 9.2: DG; 9.11 DG; 9.13 DG and The Oncologist ; 9.14 DG; Box 9.3: Powell DC (plaque), DG , Stanyard R and The Florida State University Research in: Review Magazin (Holton); FB; Strobel G. Zocher R; 9.16 DG; 9.17 DG and The Oncologist ; 9.18 La Poste Française; 9.19 KS; 9.20 DG; Box 9.4: Matter A, and Zimmermann J (Novartis) ; DG; FB; RR ; 9.21 MB; 9.22 CM; 9.25 Hwang W-S; Box 9.5: Larrick JW; KS, Deutsche Bundesbank; FB; Box 9.6: Stewart E, and Timmermann S, INSERM; 9.27 modified after Thiemann WJ and Palladino MA (2004); 9.28 Tuschl T; Box 9.7: Kwong A (Embryo); Coutelle C; DG (adenovirus); 9.29 modified after Schellekens H et al (1994) Ingenieure des Lebens. Spektrum Akademischer Verlag Heidelberg; 9.30 Carlin R, University of Massachusetts Medical School (photo Mello), Stanford University (photo Fire); 9.31 CM; Box 9.8: R&C Biogenius Ltd., Hong Kong (prmission for poster); RR

Chapter 10
10.1 MB; 10.3 DG; 10.9 archive RR; 9. 10: EKF Magdeburg; BioSensor Technologie GmbH, Berlin; 10.16 Glatz J; rennesens GmbH, Berlin; 10.17 RR und MB; 10.18 modified after Campbell NA (2003); Box 10.1: Sir Jeffreys AJ; 10.23 modified after Thieman WJ and Palladino MA (2004); Box 10.2: modified Cellmark Diagnostics; AKD; 10.26 Wellcome Trust Sanger Institute Cambridge (photos); modified after Alberts B, Bray D, Hopkin K, Johnson A, Lewis J, Raff M, Roberts K, Walter P (2005); 10.27 Wellcome Trust Sanger Institute, Cambridge; 10.28 modified after Watson et al (2003); Sir Southern E; 10.32 MB; von Holt J, *www.amberworld.com*; 10.33 KS; Box 10.3: National Genographic Project; Photo (Spencer Wells) Mark Thiessen, National Geographic, by permission of "National

Geographic Image Collection"; RR (elephant); CM and RR (cartoon); Box 10.4: RR und © The Nobel Foundation; 10.34 and upper photo 10.35 U.S. Department of Energy Human Genome Program; 10.35 (lower photo) Chudoba I, MetaSystems, Jena; Box 10.5: RR (PCR-cyclers), and Rebers J (PCR principle); 10.36 Ministry of Agriculture and Agri-Food, Canada; Box 10.7: National Genographic Project; photo by Mark Thiessen, by permission of "National Geographic Image Collection"; Native Americans: painting by Greg Harlin, © McClung Museum of the University of Tennessee; Ma KYC (portrait), CM and RR (cartoon);
Box 10.5: DG; Pearl TV Hong Kong; 10.37 Craig Venter, and U.S. Department of Energy Human Genome Program (Collins); page 317 (top) Pettet R, Wellcome Trust Sanger Institute, Cambridge; 10.38 und 10.39 U.S. Department of Energy Human Genome Program, www.ornl.gov/hgmis; 10.42 top: Powell D, Wellcome Trust Sanger Institute, Cambridge; U.S. Department of Energy Human Genome Program; 10.43 U.S. Department of Energy Human Genome Program; Box 1.8: Venter C (portrait); CM and RR (cartoon); Box 10.9: Affymetrix; 10.45 Monte Pettitt; Box 10.9: Guttmacher A; British Mail; 10.46 RR und Yang M; 10.47 MB; 10.48 Affymetrix; 10.49 modified after Thieman WJ and Palladino MA (2004); 10.50 Klenz U, Jena; 10.52 O'Farrell PH, reproduced by permission from: O'Farrell, P.H. (1975) High resolution two-dimensional electrophoresis of proteins. J. Biol.Chem. 250: 4007; 10.53 Hillenkamp F; U.S. Department of Energy Human Genome Program; Box 10.11: University Miami, Dept Physics (Feynman); DG; Law MS, Hong Kong (origami); page 330, Biotech Tree, by permission from Wiley-VCH 2006; 10.54 Internet; 10.56 and 10.58 Fromherz P; 10.59 DG and Wiley-Liss

INDEX

disulfide bridges   33, 85, 106
Dixon, B.   151
Djerassi, C.   132f, 135
DNA   8, 58, 64
  arrays   321
  chips   323, 325
  database   303
  fingerprint   302
  hybridization   82, 84
  ligase   34, 75, 77
  microarrays   325
  polymerase   46, 58, 303
  probe   82, 84, 308, 323
  repair enzyme   118
  sequence   63
  sequencing   307
  synthesizer   78, 82, 87
  vaccines   156
  virus   140, 155
dogfish (*Squalus acanthias*)   207
Dolabella auriculata (Sea Hare)   207
dolastatin-10   207
Dolly   254
dolphins   237
domestic animals   20
domestic plants   23
Donahue, J.   70
dormant viral DNA   140
double helix   58
double-stranded RNA (dsRNA)   141, 256, 288
downstream processing   126
doxorubicin (adriamycin)   127
Drake, R.   226
Drexler, E.   329
drones   250
Drosophila   112, 116, 315
druggability   281
Druggable Universe   281
drug targets   281f
Druker, B.   276f
drying and pickling   20
DsRED (Red Fluorescent protein)   225
Dübel, S.   147,   166
Dubock, A.   215
Duclaux, E.   34
Duke William IV of Bavaria   26, 38
Dunaliella   200
Durand, P.   124
Dürer, A.   148
dynamite   188

E
early marker   299
Ebola virus   160
echinoderm   207
*Eco*RI   74f, 77
edible vaccines   157
Edwards, B.   250
efficiency of biofuel programs   187
EGF (Epidermal Growth Factor)   279
Egypt   2, 5, 18, 21
Ehrlich, P.   151, 158, 166, 280
Einstein, A.   188
elastase   41
electrocardiogram (ECG)   290, 298
electrofusion   211
electroporation   91
ELISA   154f, 158, 160, 297

embryo   183, 259
  transfer   238
embryonic stem cells (ES)   238f, 279, 246, 284
emersed culture   36
Emmental   24
Endomycopsis   7
Endomycota   7
endophytes   174
endoplasmic reticulum   98
energy awareness   41
energy balances   192
ENIAC (Electronic Numerical Integrator And Computer)   331
enolase   13
enolpyruvylshikimate phosphate synthase (EPSP synthase)   217
enterotoxin   225, 274
entrapment   45
envelope proteins   140, 157
enzymatic stain remover   41
enzyme   7, 25, 30, 32f, 48, 83, 96, 192, 329
  catalyzed reaction   39
  dicer   288
  from slaughtered animals   35
  inhibitors   123
  membrane   296
  membrane reactors (EMR)   49, 52
Enzyme-linked Immunosorbent Assay (ELISA)   162
epidemic diseases   22
Epidermal Growth Factor (EGF)   276, 278
  Receptor (EGF-R)   282
epitope   146f
Epitope Universe   281
epoxide   100
Erbitux   166
Eremothecium ashbyii   114
Ergül, A.   24
Erwinia   112f, 115, 226
erythromycin   127
erythropoietin (EPO)   246, 268, 279
Escherichia coli   6, 50, 63, 74, 83, 85f, 90, 97, 106, 108, 117, 119f, 126, 143, 154, 156, 163, 177, 194, 265, 278f, 284, 316, 327
essential amino acids   45
Etchells, J. L.   XVII
ethanol   9_11, 13, 17, 51, 101, 178, 185, 188, 192
ethical reservations   246
ethidium bromide   237, 300
ethylene   185
Eubacteria   176
eukaryotes   66, 71, 97
eukaryotic   97
  cell   96_98
  yeast   90
EURO   21
European corn borer   220, 228f
eutrophication   43
exons   60, 65, 71
exponential (log) phase   130
expressed sequence tags (ESTs)   315
expression   66
  profiles   325
extracellular enzymes   30, 122
extracellular hydrolases   42
Exxon Valdez   180, 183

F
Fab fragment   147, 163
Fabian, F. W.   XVI
FABP   158
factor VII   264
factor VIII   246, 267
factor X   264
facultative anaerobes   26
familiy tree analysis   315
Farley, P.   83,   XIX
farmers   204
Father of the pill (Carl Djerassi)   133
fat-soluble vitamins A, D, E. and K   114
fatty acid-binding protein (FABP)   299
Fc (fragment constant)   147
Felis caracal (Caracal)   237
fermentation   2, 5, 9, 11, 17, 20, 25
fermented food   20
ferments   25
Fermosin   202
ferrocene   295
fertilizers   43, 204, 331
fetal gene therapy   286
fetal mouse   287
Feynman, R.   62, 328
fiber analytics   303
fibrin   264
fibrinogen   264
fibrinolytics   264
ficin   36
Fiji   149
Fire, A. Z.   288
First World War   188, 200
Fischer, E.   30f, 37
FISH   309,   312
fish farms   243
fitness drinks   48
FitzRoy, R.   306, XII, XV
fixed bed reactor   129
flavin adenine dinucleotide (FAD)   31, 35, 295
flavor 219
Flavr Savr®   221
  tomatoes   216, 222
Fleming, A.   31, 37, 102, 119f
flex-fuel cars   186f
Florey, H.   102, 120, 122
fluorescence in situ hybridization (FISH)   309, 316
foal   254
Fodor, S.   325f
Food and Drug Administration (FDA)   277
food preserves industry   124
food production   21
Ford, B. F.   3
forensics   303
L-forms   108
formate dehydrogenase (FDH)   49, 54
formic acid (formate)   10, 54
formivirsen   222
D-forms   108
fosphenytoin   283
foxglove (*Digitalis*)   211f
Fraley, R.   213
Francisella tularensis (tularaemia, Rabbit fever bacterium)   160
Franke, W. W.   278
Franklin, F.   70
Franklin, J. Sir   124
Franklin, R.   63, 68

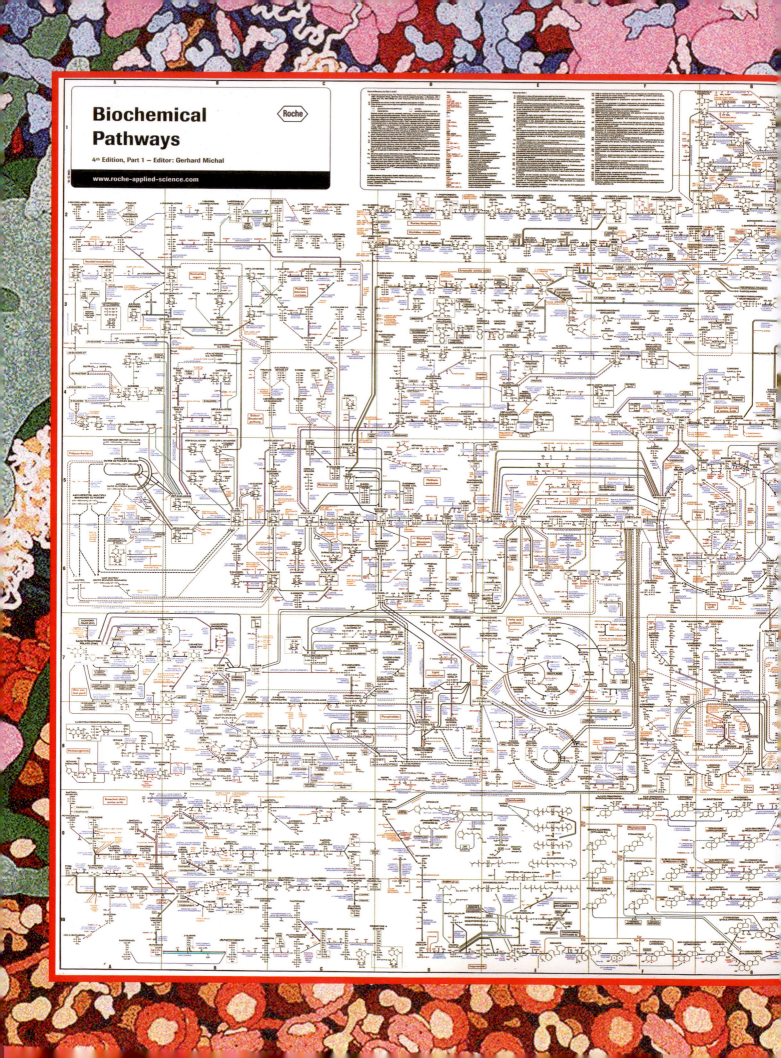

Biochemical Pathways

Roche

4th Edition, Part 1 — Editor: Gerhard Michal

www.roche-applied-science.com

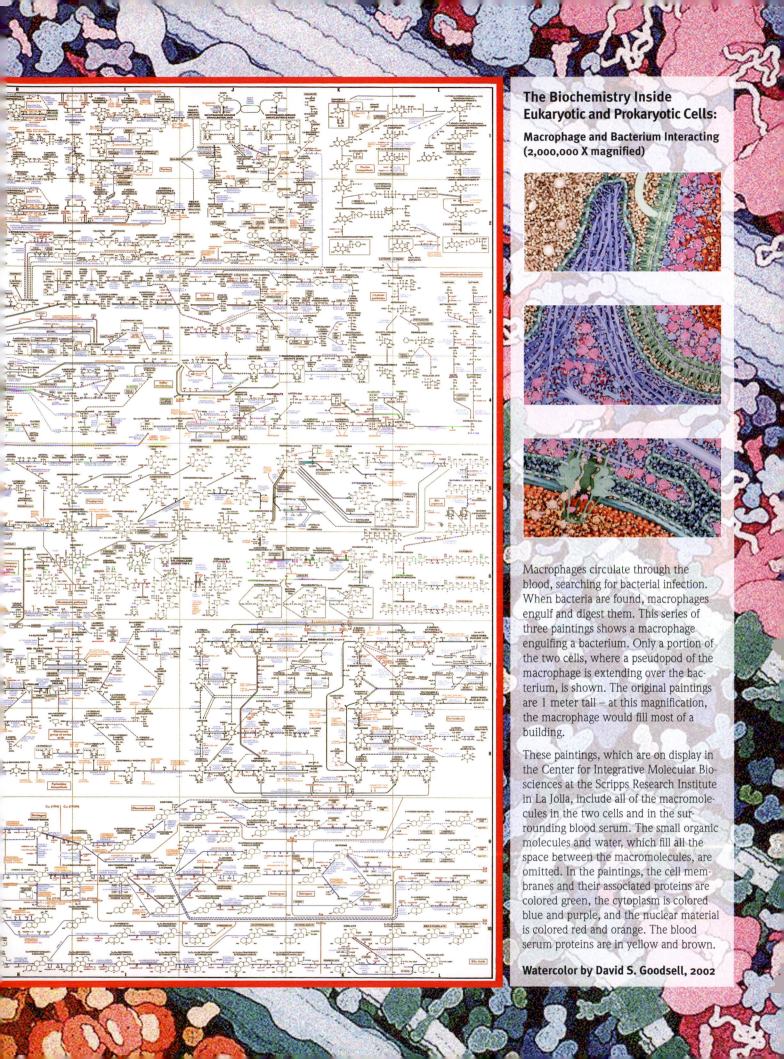

The Biochemistry Inside Eukaryotic and Prokaryotic Cells:

Macrophage and Bacterium Interacting (2,000,000 X magnified)

Macrophages circulate through the blood, searching for bacterial infection. When bacteria are found, macrophages engulf and digest them. This series of three paintings shows a macrophage engulfing a bacterium. Only a portion of the two cells, where a pseudopod of the macrophage is extending over the bacterium, is shown. The original paintings are 1 meter tall – at this magnification, the macrophage would fill most of a building.

These paintings, which are on display in the Center for Integrative Molecular Biosciences at the Scripps Research Institute in La Jolla, include all of the macromolecules in the two cells and in the surrounding blood serum. The small organic molecules and water, which fill all the space between the macromolecules, are omitted. In the paintings, the cell membranes and their associated proteins are colored green, the cytoplasm is colored blue and purple, and the nuclear material is colored red and orange. The blood serum proteins are in yellow and brown.

Watercolor by David S. Goodsell, 2002

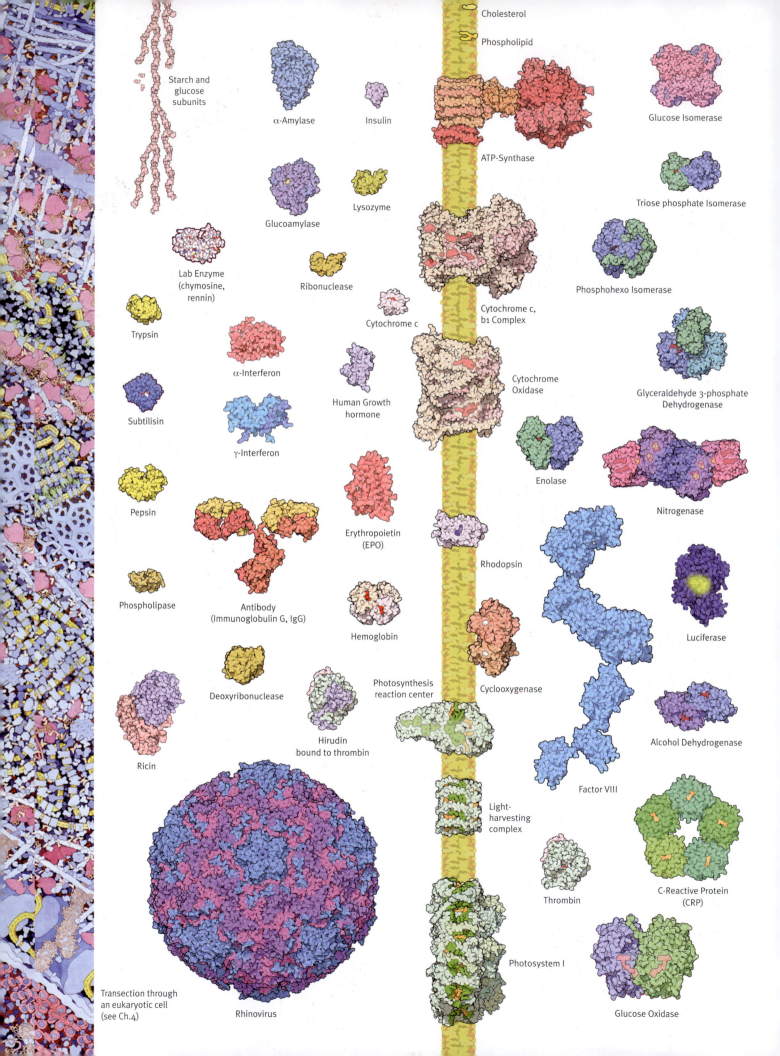

Starch and glucose subunits

α-Amylase

Insulin

Cholesterol

Phospholipid

ATP-Synthase

Glucose Isomerase

Glucoamylase

Lysozyme

Triose phosphate Isomerase

Lab Enzyme (chymosine, rennin)

Ribonuclease

Cytochrome c

Cytochrome c, b1 Complex

Phosphohexo Isomerase

Trypsin

α-Interferon

Human Growth hormone

Cytochrome Oxidase

Glyceraldehyde 3-phosphate Dehydrogenase

Subtilisin

γ-Interferon

Pepsin

Erythropoietin (EPO)

Enolase

Nitrogenase

Rhodopsin

Phospholipase

Antibody (Immunoglobulin G, IgG)

Hemoglobin

Cyclooxygenase

Luciferase

Deoxyribonuclease

Hirudin bound to thrombin

Photosynthesis reaction center

Alcohol Dehydrogenase

Ricin

Light-harvesting complex

Factor VIII

Thrombin

C-Reactive Protein (CRP)

Transection through an eukaryotic cell (see Ch.4)

Rhinovirus

Photosystem I

Glucose Oxidase